Progress in Pesticide Biochemistry and Toxicology

Volume 3

Progress in Pesticide Biochemistry and Toxicology

Volume 3

Edited by
D. H. Hutson
and
T. R. Roberts
Shell Research Limited, Sittingbourne

A Wiley–Interscience Publication

JOHN WILEY & SONS
CHICHESTER · NEW YORK · BRISBANE · TORONTO · SINGAPORE

British Library Cataloguing in Publication Data

Progress in pesticide biochemistry and toxicology.
—Vol. 3
1. Pesticides—Periodicals
668′.65 SB951

ISBN 0 471 90053 2

Set in Monophoto Times by
COMPOSITION HOUSE LTD., SALISBURY
Printed by
PAGE BROS. (NORWICH) LTD.

Contributors to Volume 3

D. COLE — *Department of Agricultural and Forest Sciences, University of Oxford, Parks Road, Oxford, OX1 3PF, UK.*

G. H. DAVIDONIS — *Pesticide Research Laboratory and Graduate Study Center, Department of Entomology, Pennsylvania State University, University Park, Pennsylvania, 16802, USA.*

A. D. DODGE — *School of Biological Sciences, University of Bath, Claverton Down, Bath, BA2 7AY, UK.*

D. E. GAMES — *Department of Chemistry, University College, P.O. Box 78, Cardiff, CF1 1XL.*

P. J. KUHN — *Shell Research Ltd., Sittingbourne Research Centre, Sittingbourne, Kent, ME9 8AG, UK.*

P. LANGCAKE — *Shell Research Ltd., Sittingbourne Research Centre, Sittingbourne, Kent, ME9 8AG, UK.*

P. W. LEE — *Shell Development Company, Biological Sciences Research Center, P.O. Box 4248, Modesto, California, 95352, USA.*

W. MUECKE — *CIBA-GEIGY AG, Agricultural Division, CH-4002 Basle, Switzerland.*

R. O. MUMMA — *Pesticide Research Laboratory and Graduate Study Center, Department of Entomology, Pennsylvania State University, University Park, Pennsylvania, 16802, USA.*

J. R. SANBORN — *Shell Development Company, Biological Sciences Research Center, P.O. Box 4248, Modesto, California, 95352, USA.*

D. M. SODERLUND — *Department of Entomology, New York State Agricultural Experimental Station, Cornell University, Geneva, New York, 14456, USA.*

J. W. VONK — *Institute for Applied Chemistry TNO, Netherlands Organisation of Applied Scientific Research, P.O. Box 5009, 3502 JA Utrecht, The Netherlands.*

M. WADE — *Shell Research Ltd., Sittingbourne Research Centre, Sittingbourne, Kent, ME9 8AG, UK.*

Contents

Preface

The biochemistry of pesticides, for the purposes of this series, is considered to cover the following areas: (1) mode of action (i.e. biocidal action in target species); (2) biotransformation in target species (which may, of course, be related to mode of action); (3) biotransformation in non-target species (which may include soils, bacteria, insects, plants, fish, birds and mammals, including man); (4) environmental effects (e.g. effects on the ecology of treated areas); (5) environmental chemistry (distribution and fate in the environment); and (6) biochemical toxicology in mammals, including man. In selecting areas for review, we take account of new or changing aspects of the science, including techniques, and of changes in the importance or use patterns of the chemical classes of pesticides.

Studies of the modes of action of the various pesticide classes are of fundamental importance to the identification of their targets and to the optimization of biological activity. Often inextricably linked with mode of action is the biotransformation of pesticides in the target and host organisms. The biochemical processes to which pesticides are subjected may limit or enhance their biological action. Effective structure–activity relationships require that the structure of the ultimate toxicant be known and, therefore, that metabolism be taken into account. These two interlinked themes: mode of action and metabolism, together with the techniques used for their study have formed the basis of much of the content of Volumes 1 and 2 of *Progress in Pesticide Biochemistry* (listed at the end of this Volume).

The major part of Volume 3 is dedicated to the mode of action and metabolism of fungicides and herbicides. The utilization of the plants own systems is a very effective means for the distribution of fungicides throughout the organism. Thus, systemicity is a much sought-after property in these agrochemicals. Chapter 1 reviews this aspect together with the mode of action of the systemic fungicides and the metabolic fate of fungicides forms the subject of Chapter 2. The modes of action of the important classes of herbicides are reviewed in Chapter 3. Detoxification and bioactivation of herbicides are effected by a variety of enzyme-catalysed reactions. Oxidation is very important in this context and, while oxidative processes occurring in mammals have received an enormous amount of research and written coverage, that in plants, particularly at the enzyme level, is less well understood. These aspects are covered in Chapter 4. Plant tissue culture (Chapter 5) is finding increasing use in the study of metabolism and, to some extent, of mode of action.

Techniques used for the identification of pesticide metabolites are constantly developing. Identification involves two main areas of activity: isolation of metabolites in a pure state and the elucidation of their structures. In Chapter 6 are described the currently available range of procedures used for the isolation and purification of small amounts of pesticide metabolites from biological sources. Our increasing capability for the handling of the more polar conjugated metabolites becomes clearly evident in this chapter. The improvement in techniques (particularly column chromatography) is paralleled by the development of mass spectral systems which can cope with polar compounds. The original soft ionization method (field desorption) has been complemented with a variety of other processes (Chapter 7).

Finally, continuing the review of various aspects of the behaviour of the pyrethroid insecticides, Chapter 8 covers the available information on the metabolism and kinetics of the pyrethroids in target organisms, the insecta.

Several of the aspects of pesticide science encompassed by this series may be classified as Toxicology in the broadest sense. This mode of toxic action in target species (plants, fungi, insects, etc.,) and in beneficial species (aquatic organisms, plants, insects, land mammals, birds and man) are reviewed. In view of this and of the proposed contents of future volumes we feel it appropriate to expand the title of the series to **Progress in Pesticide Biochemistry and Toxicology**.

Sittingbourne, 1981

D. H. Hutson
T. R. Roberts

Progress in Pesticide Biochemistry and Toxicology, Volume 3
Edited by D. H. Hutson and T. R. Roberts

CHAPTER 1

The mode of action of systemic fungicides

P. Langcake, P. J. Kuhn and M. Wade

INTRODUCTION

Among the three main groups of pesticides (insecticides, fungicides and herbicides) fungicides have probably the longest history. (For a review of the history of systemic fungicides see Wain and Carter, 1977.) They date back to the accidental discovery in 1882 of Bordeaux mixture and the value of copper-based preparations for the control of vine downy mildew disease. Even as late as 1974, it was estimated that approximately 16% of all fungicides used were based on copper preparations (Anon, 1975). In the early 1930s more systematic approaches to the discovery of fungicides were taken up and since the 1940s a range of highly useful compounds, in particular dithiocarbamates and phthalimides, were introduced. They continue to be used in large quantities today. These fungicides have a protectant action being applied to the surface of the plant where they provide a protective antifungal barrier to any potential pathogens. Limited redistribution of the compounds can occur but only by the action of external forces, such as rain splash. The disadvantages of these fungicides therefore lay firstly in their inability to cure an established infection due to their failure to penetrate plant tissue. Secondly, they must be carefully applied to provide maximum cover of the susceptible surfaces of the plant. Needless to say, unsprayed parts are not protected and this necessitates frequent spray applications to protect new growth and to make good any deficiencies in the previous coverage due to weathering.

The discovery and commercialization of the organophosphate insecticides with systemic properties raised the expectation that systemic fungicides would soon be discovered. Systemic compounds are those that are taken up into a plant from the surfaces to which they are applied, are redistributed through the normal transport mechanisms (principally the apoplast but also the symplast) and exert their activity at places remote from their point of application. Several experimental fungicides with systemic properties had been described in the 1940s and 1950s, but it should be no surprise that the first systemic fungicide to be commercialized, triamiphos, in 1960, was an organophosphorus compound that also possessed insecticidal properties. Many more of this 'new generation' of fungicides based on a wide variety of chemical structures have been introduced subsequently (Pearson, 1977; Bent, 1979).

In relation to fungicides, the term 'systemic' has often been misused since it ignores one of the most important features of the 'new' fungicides, namely that they can eradicate established infections, i.e. they are therapeutic. Systemic and therapeutic properties do not necessarily always occur together, but both features require that the fungicide penetrates the plant and that it is selectively toxic to the fungus, as opposed to the plant cells. Studies of their mode of action have shown that therapeutic fungicides are always specific-site directed. In other words their action can be ascribed primarily to the inhibition of a single metabolic process, usually an enzyme, although other less specific secondary effects may also be evident. In contrast, it is normally the case, although there are some exceptions, that protectant fungicides are multi-site inhibitors, affecting a wide range of enzymes.

This review attempts to provide a comprehensive account of the current state of knowledge on the biochemical modes of action of systemic/therapeutic fungicides. For the most part it is restricted to those compounds that have found significant commercial application in agriculture although mention is also made of some experimental compounds, where appropriate.

Fungicides can be broadly classified according to their mode of action as follows:

(a) Fungicides that act against the growth of fungal pathogens. Such compounds may or may not require transformation to the actual fungitoxicant after application to the plant or fungus. They affect metabolic processes that are common to pathogenic and non-pathogenic fungi alike and their activity can therefore be demonstrated against the growth of certain fungi in culture.

(b) Fungicides that act against the host–parasite complex. There are two subdivisions of this category. Firstly there are compounds that affect the host plant in such a way that it is no longer a suitable environment for the growth of a pathogen, e.g. by affecting the expression of host resistance mechanisms. Secondly there are fungicides that may act specifically against the metabolic activities of pathogens that are essential for those fungi to be pathogenic. In theory at least, such fungicides may act after metabolic conversion (activation) of the parent species, or the parent species itself may be active.

Recent reviews of fungicide mode of action include those by Kaars Sijpesteijn (1977), Siegel and Sisler (1977) and Leroux (1981).

FUNGICIDES THAT ACT AGAINST THE GROWTH OF PATHOGENS

Interference with sterol biosynthesis

The greater proportion of the fungicides that have been commercialized during the 1970s, have been identified as inhibitors of sterol biosynthesis and this underlines the importance of this site of action for modern fungicides. However, there are two main categories of fungicide interfering with sterol biosynthesis, namely those that inhibit sterol demethylation and those that inhibit other processes in sterol biosynthesis.

Inhibition of sterol demethylation

The many fungicides that inhibit sterol demethylation represent a wide range of apparently unrelated chemical classes. These are listed in Table 1.† There is no doubt that initially these various chemical classes were discovered quite independently. More recently, the proliferation of compounds known, or presumed, to have this mode of action has resulted from extensive analogue synthesis programmes within the agrochemicals industry. Although most of them have fungicidal properties and are of potential or current use in agriculture, there are several others that also inhibit sterol demethylation that are principally of value in medicine as antimycotics. The conventional chemical representation of their structures suggests few relationships between the classes. Nevertheless, since it is likely that most of them have identical modes of action, from the point of view of their target site, they must have features in common. As yet these features have not been adequately described.

Biologically, the sterol demethylation inhibitors have several similarities. They are all very good agents for the control of powdery mildew diseases of various crops. They have therapeutic activity against these diseases and many of them are, in addition, systemic. They differ somewhat in the spectrum of diseases that are controlled in addition to mildew, but few of them show particularly good activity against diseases caused by Oomycete fungi (*in vivo* or *in vitro*). They are highly active against a range of fungi *in vitro*, especially Ascomycetes and Fungi Imperfecti, and in some cases activity is observed *in vivo* as well.

Studies on their mode of action were pioneered by Ragsdale with triarimol (**14**) and by Kato with buthiobate (**18**) in the early 1970s. These classic studies have been taken as models by subsequent workers and since those of Kato, although slightly later than those of Ragsdale, are in some respects the more

† Structures given Roman numerals are natural compounds. Structures with Arabic numerals are foreign compounds.

comprehensive, they will be described in detail as representative of this type of action. Other compounds, together with any differences in detail noted, will be described later.

Buthiobate The mode of action of buthiobate (**18**, Denmert, S-1358) has been described in a comprehensive series of papers by Kato and co-workers. *In vitro*, buthiobate was only slightly inhibitory to three Basidiomycete fungi tested at 500 μM, while it was inactive against four Phycomycetes (represented by three Oomycetes and one Zygomycete). Several Ascomycetes and Fungi Imperfecti showed varying degrees of sensitivity with *Monilinia fructigena* the most sensitive of those tested, having an ED_{50} value on agar medium of 0.29 μM (Kato *et al.*, 1975b). This organism was used for most of the further studies. Buthiobate had very little effect on the germination of conidia of *M. fructigena* but strongly inhibited the development of germ tubes. The hyphae were distorted and excessively branched and some had swollen hyphal tips (Kato *et al.*, 1974). Similar effects were observed on treated cucumber plants inoculated with powdery mildew, *Sphaerotheca fuliginea*, and here hyphal tips were abnormally stained with vital dyes (methylene blue and neutral red), suggesting an effect on membrane permeability and function (Kato *et al.*, 1975b).

As with other sterol demethylation inhibitors, inhibition of dry weight increase of actively growing fungal cultures is not apparent until after a time lag representing some 1 to 2 generation times, regardless of the fungal species used. Thus, addition of buthiobate at 10 μM to 14 h cultures of *M. fructigena* in liquid medium did not result in total inhibition of growth until approximately 8 h later. During this time the dry weight of the fungus had increased by slightly more than twofold. Nevertheless morphological effects as described above were evident by 2 h after treatment. In a similar experimental system, buthiobate had only very slight or no inhibitory effects over the first 2 h of treatment on the incorporation of (a) radiolabelled amino acids into proteins; (b) uracil and thymine into nucleic acids; (c) glucose, glucosamine and acetate into cell walls; and (d) acetate into lipids. Over more prolonged periods of incubation the effects were more pronounced, corresponding with the onset of morphological effects and ultimately with the inhibition of growth. Buthiobate did not inhibit endogenous or exogenous respiration of glucose or sodium succinate substrates (Kato *et al.*, 1974).

Although the changes in incorporation of acetate into total lipids were small, at least initially, further analysis of the lipid fraction revealed very pronounced and immediate effects on the quantitative distribution of radiolabel among the lipids. The most striking effects were seen on the sterol fractions. The major sterol fractions in *M. fructigena* (as in most Ascomycetes) are: the steryl esters (principally esters of ergosterol (**IV**) (see Figure 2) [a desmethylsterol, i.e. one lacking methyl groups at the C4 and V14 positions] with various fatty acids);

Table 1 Antifungal compounds that inhibit sterol demethylation

Chemical class	Common name	Structure	Other name	Primary use	References
Triazole	Triadimefon	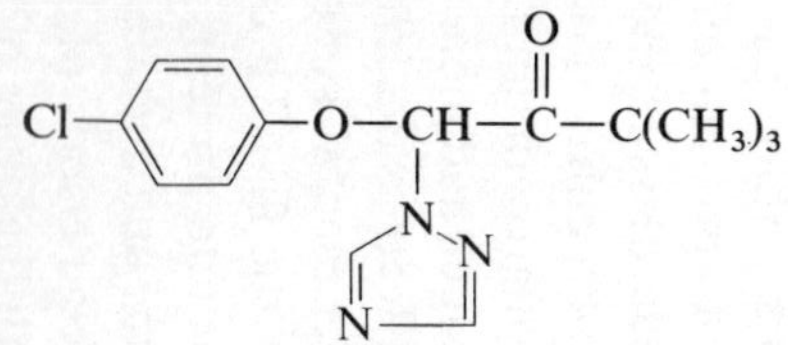 (1)	Bayleton	Agricultural fungicide	Buchenauer, 1976; 1977a; b; 1979; 1980; Buchenauer and Grossmann, 1977; Clark *et al*., 1978; Forster *et al*., 1980a; b; c; Gasztonyi, 1981; Gasztonyi and Josepovits, 1979; Hippe *et al*., 1980; Langcake (unpublished); Leroux *et al*., 1976; Leroux and Gredt, 1976; 1978a; b
Triazole	Triadimenol	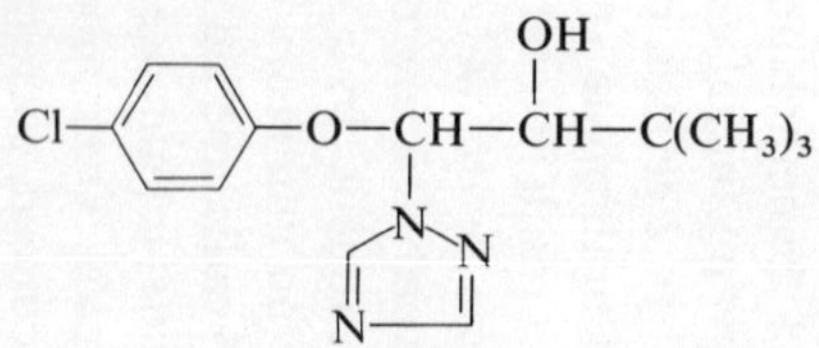 (2)	Baytan	Agricultural fungicide	Buchenauer, 1978a; 1979; 1980; Clark *et al*., 1978; Forster *et al*., 1980a; b; c; Gasztonyi, 1981; Gasztonyi and Josepovits, 1979; Langcake (unpublished); Leroux and Gredt, 1976
Triazole	Biloxazol	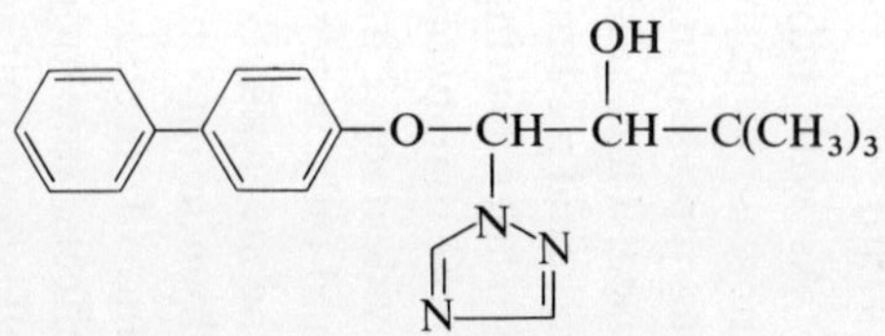 (3)	Baycor	Agricultural fungicide	Kraus, 1979

Triazole	Fluotrimazole	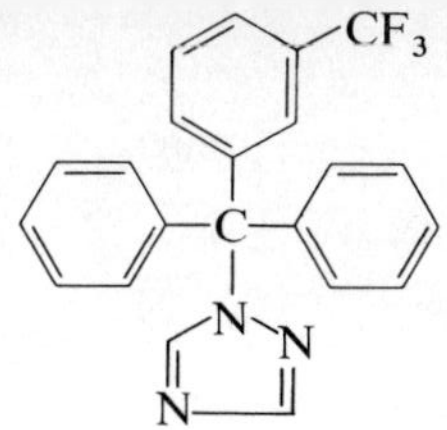 (4)	Persulon	Agricultural fungicide	Buchenauer, 1976; 1978b; 1979
Triazole	Dichlobutrazol	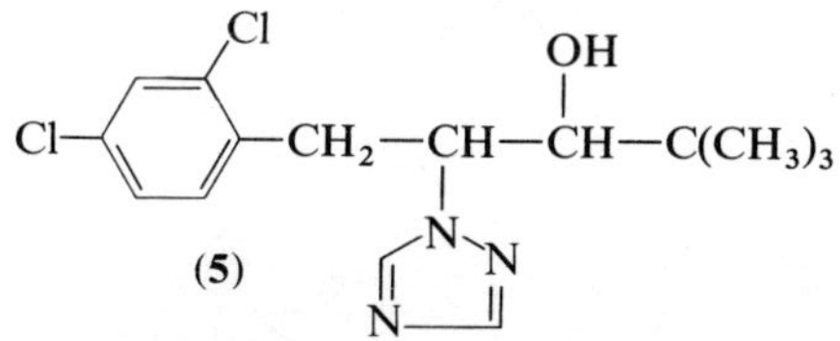 (5)	Vigil	Agricultural fungicide	Bent and Skidmore, 1979
Triazole	Propiconazole	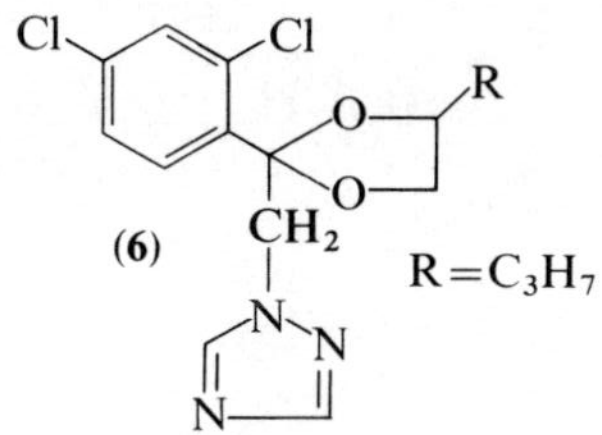 (6)	Tilt (CGA 64250)	Agricultural fungicide	Buchenauer and Kemper, 1981; Langcake (unpublished)
Triazole	Etaconazole	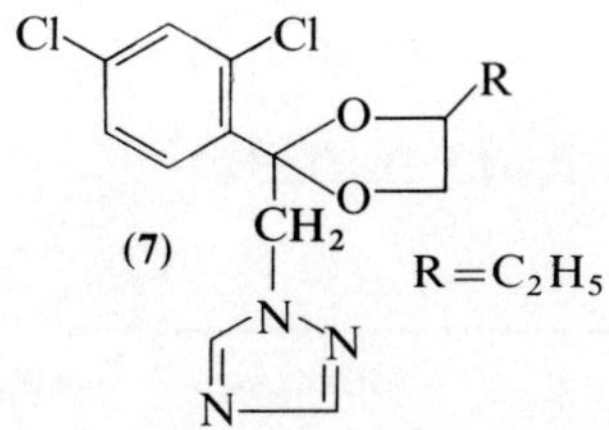 (7)	Vanguard (CGA 64251)	Agricultural fungicide	Henry and Sisler, 1981

(continued)

Table 1 (*continued*)

Chemical class	Common name	Structure	Other name	Primary use	References
Imidazole	Imazalil	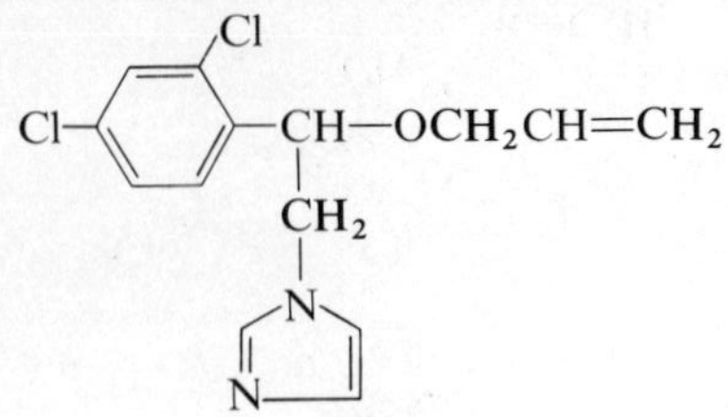 **(8)**	Fungaflor	Agricultural fungicide	Barug and Kerkenaar, 1979; Buchenauer, 1977c; d; 1979; 1980; Leroux *et al.*, 1976; Leroux and Gredt, 1978a; b; c; Siegel *et al.*, 1977; Siegel and Ragsdale, 1978; Siegel and Solel, 1981; Vonk and Dekhuijzen, 1979
Imidazole	Miconazole	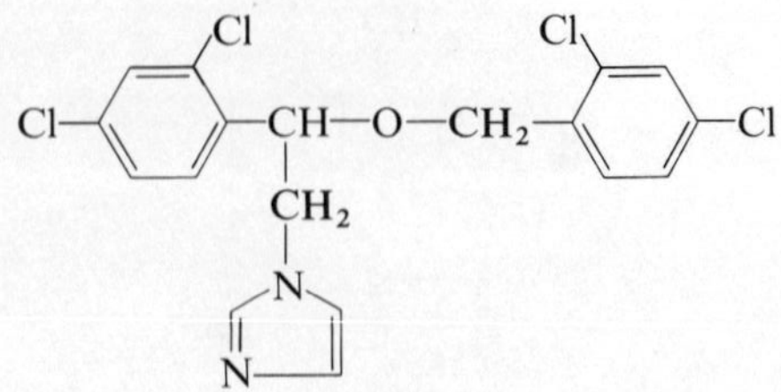 **(9)**		Antimycotic	Barug and Kerkenaar, 1979; Cope, 1980; Henry and Sisler, 1979; Odds, 1980; Sud and Feingold, 1981a; b; Swamy, *et al.*, 1974; van den Bossche *et al.*, 1978; Marriott, 1980
Imidazole	Clotrimazole	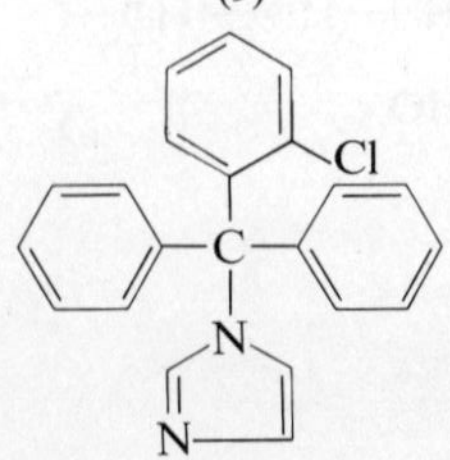 **(10)**	Canestan	Antimycotic	Barug and Kerkenaar, 1979; Buchenauer, 1978b; Odds, 1980; Sud and Feingold, 1981a; b

Imidazole	Prochloraz	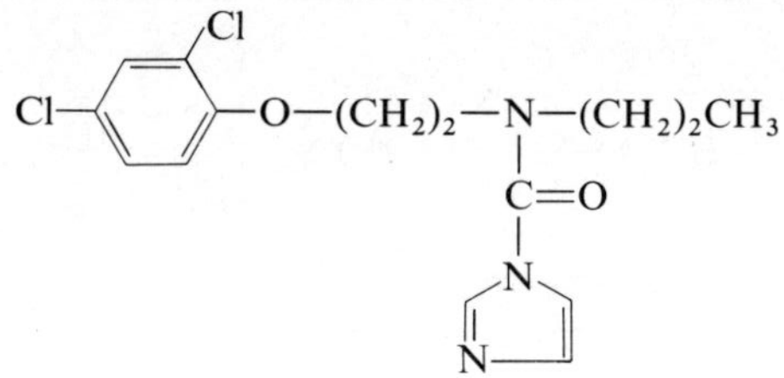 (**11**)	Sportak	Agricultural fungicide	Pappas and Fisher, 1979
Imidazole	Fenapronil	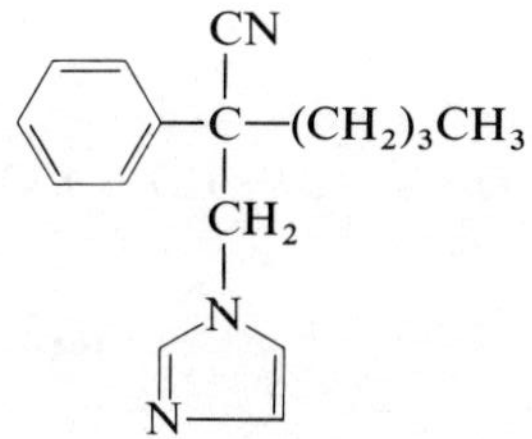 (**12**)	Sisthane RH 2161	Agricultural fungicide	Girardet and Buchenauer, 1980
Imidazole	Dodecyl imidazole	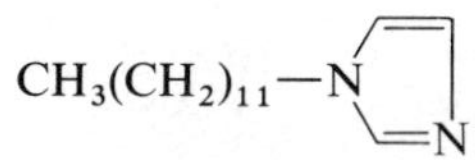 (**13**)		Antimycotic	Henry and Sisler, 1979
Pyrimidine	Triarimol	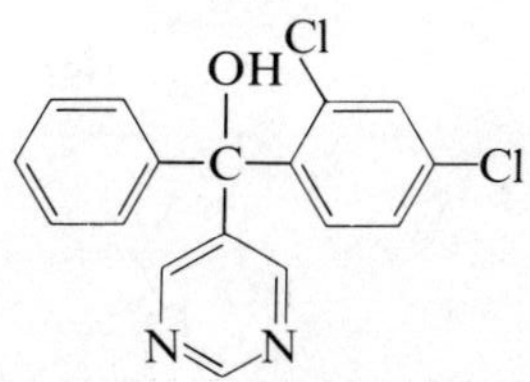 (**14**)	EL-273	Agricultural fungicide (discontinued)	Frasinel *et al.*, 1978; Kato and Kawase, 1976; Leroux and Gredt, 1976, 1978a; b; Leroux *et al.*, 1976; Mitropoulos and Gibbons, 1976; Ragsdale, 1975; Ragsdale and Sisler, 1972; 1973; 1975; Sherald *et al.*, 1973; Shive and Sisler, 1976; Sisler and Ragsdale, 1977

(*continued*)

Table 1 (*continued*)

Chemical class	Common name	Structure	Other name	Primary use	References
Pyrimidine	Fenarimol	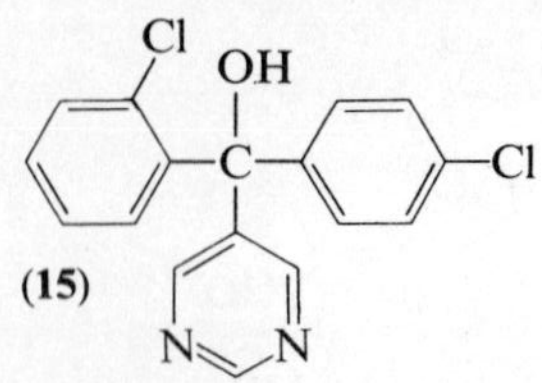 **(15)**	Rubigan	Agricultural fungicide	Barug and Kerkenaar, 1979; Buchenauer, 1977b; c; 1979; de Waard and van Nistelrooy, 1979a; 1980; Leroux and Gredt, 1978b; Ragsdale and de Waard, 1977; Schmitt and Benveniste, 1979
Pyrimidine	Nuarimol	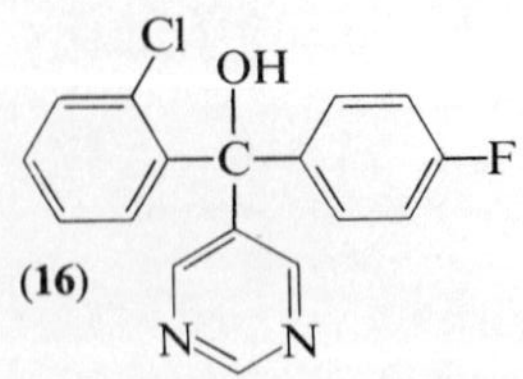 **(16)**	Trimidal (EL-228)	Agricultural fungicide	Barug and Kerkenaar, 1979; Buchenauer, 1977b; c, 1979, 1980; Leroux and Gredt, 1978b
Piperazine	Triforine	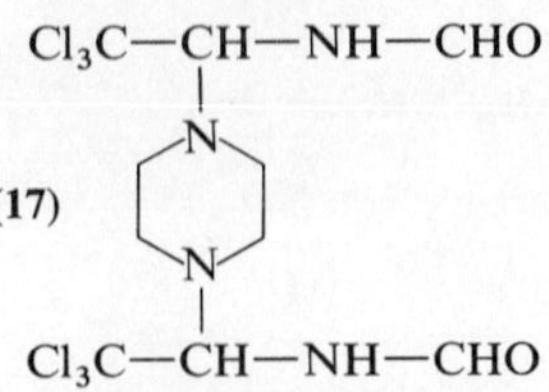 **(17)**	Saprol	Agricultural fungicide	Fuchs *et al.*, 1977; Fuchs and Drandarevski, 1973; Leroux and Gredt, 1976, 1978b; Leroux *et al.*, 1976; Sherald *et al.*, 1973; Sherald and Sisler, 1975
Pyridine	Buthiobate	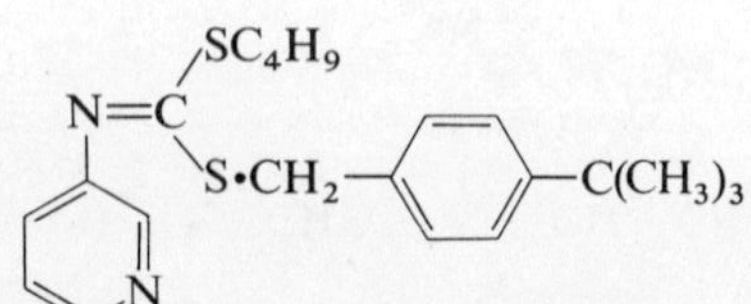 **(18)**	Denmert (S-1358)	Agricultural fungicide	Kato, 1977; Kato and Kawase, 1976; Kato *et al.*, 1974, 1975a; b; 1980; Tanaka *et al.*, 1977

the desmethyl sterols (principally ergosterol); and the 4,4-dimethylsterols (principally 24-methylene dihydrolanosterol (**VI**)). Treatment of fungal cells with low concentrations of buthiobate (10 μM) caused an almost total inhibition of the incorporation of acetate into desmethyl sterols and an accumulation of label in their methylated sterol precursors (both 4-methyl and 4,4-dimethyl sterols). This occurred very rapidly (within minutes) and before other effects on fungal growth and metabolism were evident. There was also an inhibition of incorporation of label into the steryl ester fraction. Effects on other lipid fractions were much less pronounced except after prolonged incubation (Kato *et al.*, 1974; 1975a).

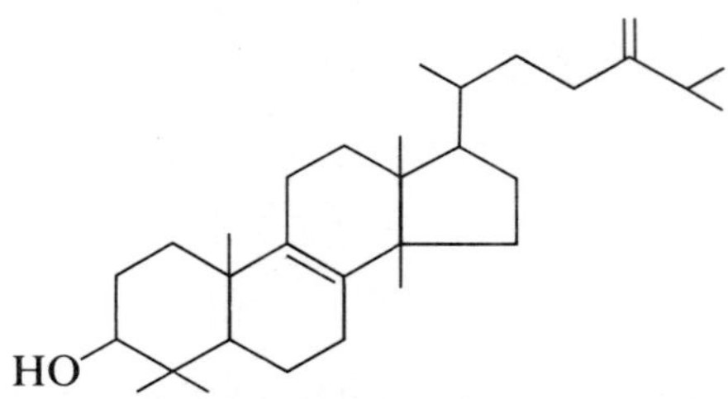

24-Methylene dihydrolanosterol (**VI**)

More critical analyses of the effects on sterol biosynthesis were performed using gas liquid chromatography (GLC) combined with mass spectroscopy (MS). This technique, in combination with information from thin layer chromatography (TLC) mobilities, allows structural assignments of sterols with a reasonable degree of confidence, especially with regard to the number and position of methyl groups on the steroid nucleus.

In control cells, the major sterol was ergosterol (**IV**). The synthesis of ergosterol was almost totally inhibited by buthiobate and the sterols accumulating were identified as obtusifoliol (**V**) (a minor component) and 24-methylene dihydro-lanosterol (**VI**) (the major component) whose identity was confirmed by nuclear magnetic resonance spectroscopy (NMR) and by comparisons with an authentic sample of the semisynthetic compound (Kato *et al.*, 1975a).

To appreciate the significance of these results, it is necessary to consider the biosynthetic pathways to the sterols. The general principles of these pathways are common to all organisms that synthesize sterols, but there are differences of detail. In simplified form they involve (a) the formation of mevalonate from acetate, (b) the formation of squalene (**VII**) from mevalonate, (c) the cyclization of squalene, via 2,3-oxidosqualene (**VIII**) to lanosterol (**I**) and (d) the conversion of lanosterol into ergosterol (**IV**, the most common fungal sterol) by a series of nuclear demethylations, side chain alkylation, hydride shifts and reductions/dehydrogenations (Figures 1 and 2). A large number of possible intermediates exist between lanosterol and ergosterol and many of these have been identified

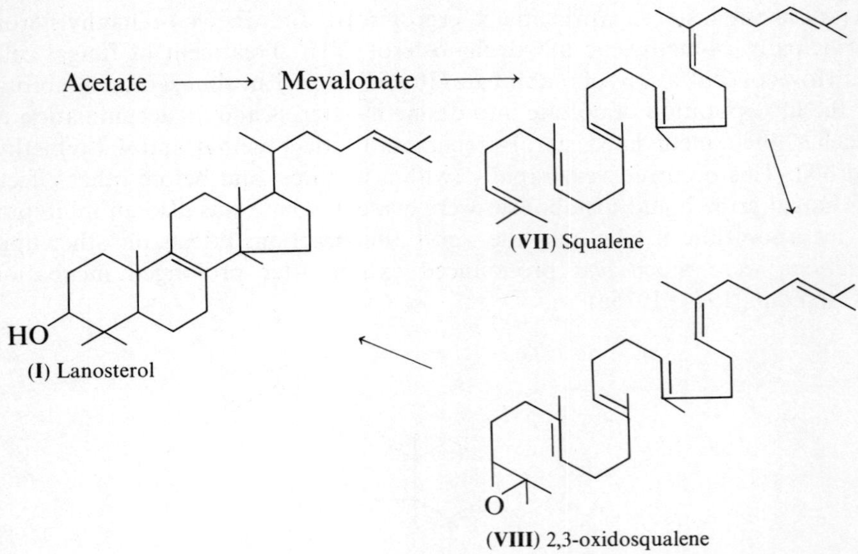

Figure 1 A simplified scheme showing the biosynthesis of lanosterol from acetate

at various times in various fungi (Weete, 1973; Barton *et al.*, 1973). Interconversions between these can occur so that there is a matrix of pathways rather than a single linear sequence. However, some sequences seem to be preferred over others. In most fungi, such as *M. fractigens* and *U. maydis*, C-24 alkylation takes place at the level of lanosterol with 24-methylene-dihydro-lanosterol being the first major product isolatable. In *S. cerevisiae*, lanosterol can be isolated. C-24 Alkylation probably takes place at a later point in the sequence (in green plants, cycloartenol **(XIV)** (see Figure 5) rather than its isomer lanosterol, is the first product of cyclization). It is generally thought that C-14 demethylation precedes that of C-4.

Based on the fact that the sterols in buthiobate-treated cells all retain the methyl group at C-14, Kato *et al.* (1975a) concluded that buthiobate specifically inhibits C-14 demethylation (Figure 2). To confirm this, Kato and Kawase (1976) studied ergosterol biosynthesis in a cell-free multi-enzyme system from *S. cerevisiae*. This organism was used in preference to *M. fructigena* because preparations from *M. fructigena* failed to convert a radiolabelled mevalonate substrate beyond 4,4-dimethyl sterols into ergosterol. With homogenates of *S. cerevisiae*, [^{14}C]mevalonate was converted into squalene **(VII)**, oxidosqualene **(VIII)**, lanosterol **(I)**, 14-desmethyl lanosterol **(II)**, 4-α-methyl zymosterol **(III)** and the demethylated sterol episterol **(IX)**. Apparently ergosterol, the normal sterol component of yeast, was not formed, the preparation being unable to convert episterol into ergosterol (Figure 2).

In the presence of buthiobate at 10^{-4} M, the formation of C-4 and C-14 desmethyl sterols was strongly inhibited and label accumulated in lanosterol.

This parallels the accumulation of 24-methylene dihydrolanosterol in whole cells of *M. fructigena.* Even at 10^{-6} M buthiobate, a slight effect was still detectable. When radiolabelled lanosterol was used as substrate, similar effects were observed, but with radiolabelled 14-desmethyl lanosterol (**II**) there was no inhibition of the incorporation of label into desmethyl sterols or 4-α-methyl sterols. This is compelling evidence for the specific action of buthiobate against sterol C-14 demethylation, with no effects against sterol C-4 demethylation.

Triarimol Ragsdale and Sisler (1972) were the first to describe a fungicide, triarimol (**14**) (see Table 1), as having a specific mode of action against fungal sterol biosynthesis. Using the yeast-like form of growth (sporidia) of *Ustilago maydis* they confirmed the lack of effects against major metabolic processes other than lipid biosynthesis (Ragsdale and Sisler, 1973). The action at the C-14 demethylation step was deduced from GC–MS analysis of the sterols accumulating in treated cells (Ragsdale, 1975). However, it was also concluded from this that the fungicide inhibits the introduction of the C-22(23) double bond, and reduction of the C-24(28) double bond. It is not yet clear whether these are consequences of the block at C-14 demethylation or whether they are primary effects of the fungicide, since its effects on a cell-free enzyme capable of carrying out these conversions was not tested. Kato and Kawase (1976) tested triarimol against the cell-free enzyme system of *S. cerevisiae* and Mitropoulos and Gibbons (1976) tried it on a cholesterol-synthesizing preparation from rat liver. Both confirmed the inhibition of C-14 demethylation. According to the data on *S. cerevisiae*, triarimol was somewhat more active than buthiobate, but on the rat liver demethylase, it was not very active, 50% inhibition requiring a concentration of 70–80 μM.

An important feature of the effects of triarimol on fungal lipid synthesis is the accumulation of up to 100 times the quantity of free fatty acids found in control cells (Ragsdale, 1975). This has been explained on the basis that although fatty acid synthesis is diminished only slightly, the rate of synthesis exceeds the rate at which they are incorporated into glycerides and phospholipids. In addition, breakdown of phospholipids provides an additional source of free fatty acids. The accumulation tends to occur after longer incubation with the toxicant and would appear to be a secondary effect of the inhibition of sterol biosynthesis (Ragsdale and Sisler, 1975). Its significance for the lethal effects of triarimol was suggested by Sisler and Ragsdale (1977). Other fungicidal inhibitors of sterol demethylation reported to cause an accumulation of free fatty acids include triadimefon (**1**) (Buchenauer, 1977a), fluotrimazole (**4**), clotrimazole (**10**) (Buchenauer, 1978b), triforine (**17**) (Sherald and Sisler, 1975), propiconazole (**6**), (Buchenauer and Kemper, 1981) and etaconazole (**7**) (Henry and Sisler, 1981). However, in studies of several other compounds, e.g. those of Kato *et al.*, with buthiobate described above, Buchenauer with fenarimol (**15**) and nuarimol (**16**) and Siegel and Ragsdale (1978) with imazalil (**8**), the effect was not noted (see Table 1 for structures).

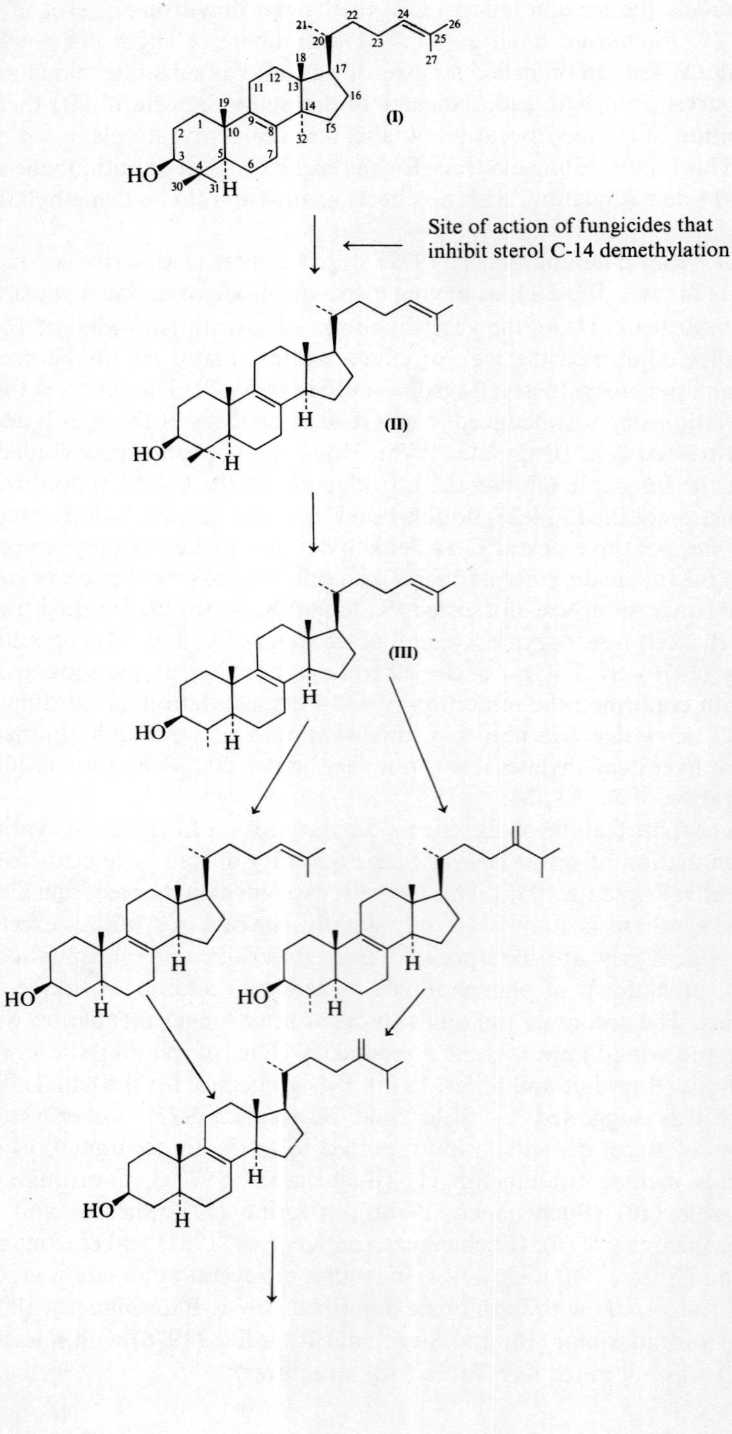
(I)
Site of action of fungicides that inhibit sterol C-14 demethylation
(II)
(III)

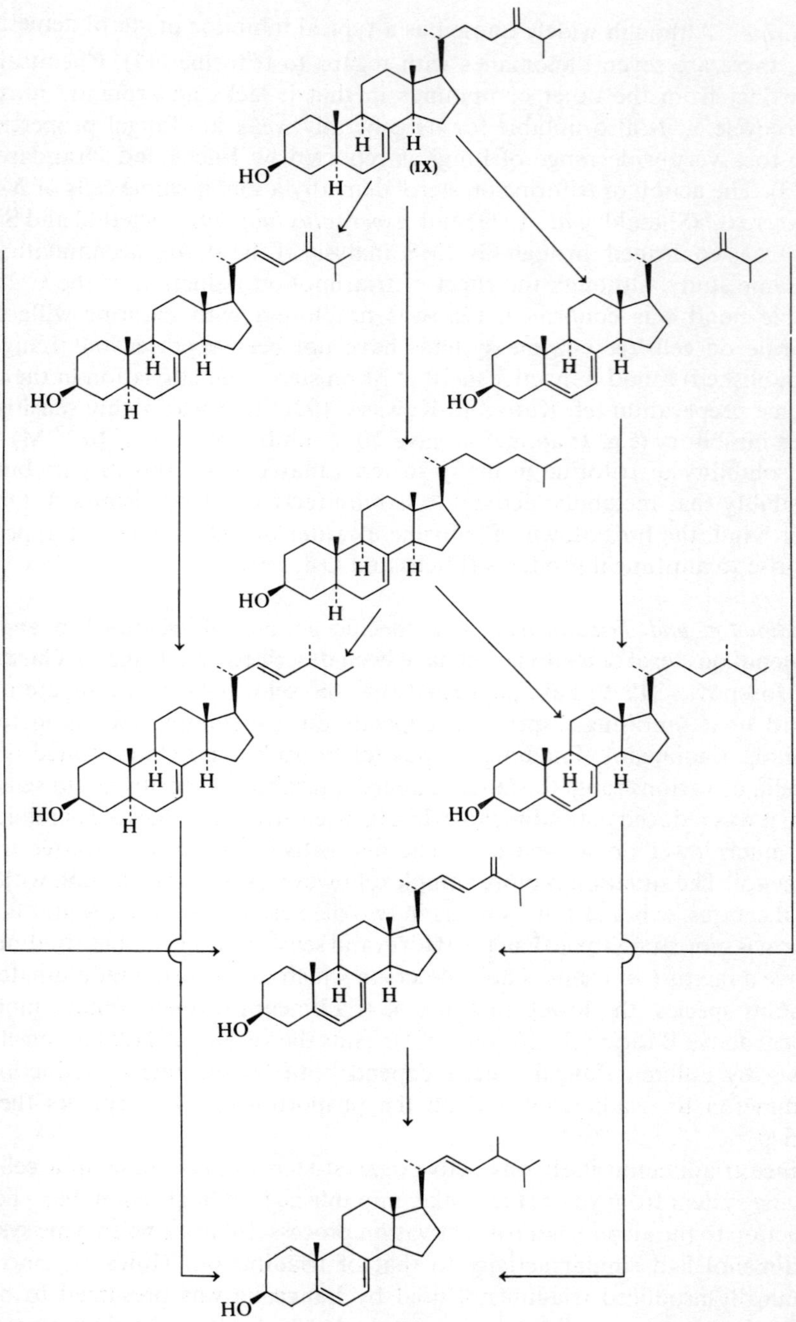

Figure 2 Biosynthesis of ergosterol from lanosterol in yeast (after Barton *et al.*, 1973)

Triforine Although widely quoted as a typical inhibitor of sterol demethylation, there are several anomalies with regard to triforine (**17**). Chemically it is distinct from the other compounds in that it lacks an aromatic nitrogen heterocycle. It is also notable for its generally weak antifungal properties *in vitro* to a very wide range of fungi as reported by Fuchs and Drandarevski (1973). The action of triforine on sterol demethylation in whole cells of *Neurospora crassa* (Sherald *et al.*, 1973) and *Aspergillus fumigatus* (Sherald and Sisler, 1975) was confirmed through GC–MS analysis of the sterols accumulating. In the same study, although the effect of triarimol on reduction of the C-22(23) double bond was confirmed, this was not found with triforine. Effects of triforine on cell-free enzyme systems have not been reported but Langcake (unpublished) found it inactive at 10^{-4} M on sterol demethylation in the yeast enzyme preparation (cf. Kato and Kawase, 1976) that was highly sensitive to other inhibitors (e.g. triadimefon gave 50% inhibition at 7×10^{-7} M). The low solubility of triforine in many solvents may explain this in part but the possibility that metabolic activation is required cannot be dismissed. On the other hand, the breakdown of triforine in water or plants does not appear to give rise to antifungal products (Fuchs and Ost, 1976).

Triadimefon and Triadimenol The specific actions of triadimefon and triadimenol on sterol demethylation have been described (see Table 1). Gasztonyi and Josepovits (1979) have suggested that the selectivity of triadimefon with regard to its antifungal spectrum depends on a metabolic reduction to the alcohol, triadimenol. Triadimefon was taken up by and concentrated within mycelia of various fungi to similar extents regardless of sensitivity. In sensitive fungi it was reduced to triadimenol while in insensitive fungi the rate of reduction was much lower or non-existent. The insensitive fungi were sensitive to triadimenol. The situation is more complex, however, since triadimenol, with two chiral centres, exists as four isomers in two diastereomeric pairs, A and B. The A form is more fungitoxic than the B form and sensitive fungi reduce triadimefon to give a mixture of triadimenol isomers in which the A form predominates. In resistant species, the lower amounts of triadimenol formed contain more of the less active B form (Gasztonyi, 1981). Thus the varying degrees of sensitivity shown by different fungal species depend both on the rate of reduction of triadimefon to triadimenol and on the proportions of the isomers thereby formed.

Since triadimefon itself was active against sterol biosynthesis in a cell-free enzyme system from yeast (Langcake, unpublished), it is questionable whether reduction to the alcohol is a true activation process. In the same enzyme system, triadimenol had similar activity to that of triadimefon. However, since the chemically-produced triadimenol used by Langcake was presumed to be an equimolar mixture of all four isomers, a selective biological reduction to give the more active isomers would indeed represent a degree of activation.

Triadimefon is also reduced to the alcohol in plants (Gasztonyi and Josepovits, 1979; Buchenauer, 1975).

Other compounds There are several studies on the effects of various powdery mildew fungicides that indicate sterol C-14 demethylation as the primary site of action. These are listed in Table 1 together with the appropriate references.

General considerations The work cited above raises several questions with regard to the action of all of these fungicides.

Is sterol demethylation the primary site of action? There are probably no compounds that are truly specific in their action, that is compounds that affect only a single biochemical process and have no effects on other processes at any concentration. It is necessary therefore to distinguish between primary effects (specific site-directed effects) and secondary effects, i.e. those that are consequences of the primary action as well as those that are direct, but subsidiary, effects on other biochemical targets. In relation to the fungicides described above there is good evidence that their primary action is on sterol demethylation since:

(1) Inhibition of sterol demethylation is the most sensitive biochemical process identified so far. In many cases effects on sterols are evident at concentrations that cause only small effects on growth. For example, with the antimycotic miconazole (**9**), the minimum inhibitory concentration for *S. cerevisiae* was 2.0×10^{-7} M, while effects on sterol ratios were found in whole cells at concentrations as low as 1×10^{-9} M (Sud and Feingold, 1981a).
(2) Inhibition of sterol biosynthesis is the first effect of the fungicides detectable, it occurs immediately after addition of the toxicant to fungal cultures, well in advance of effects on growth.
(3) In general, the later effects on other metabolic processes parallel the effects on growth and are therefore secondary consequences.
(4) The inhibition of growth can be alleviated with sterols. The evidence on this point is, however, rather more equivocal. Kato *et al.* (1974) for example, found that ergosterol incorporated into paper discs would alleviate toxicity of buthiobate to *M. fructigena* growing on agar medium. However, toxicity was also annulled by Tween 80 and unsaturated fatty acids and these were more effective than ergosterol. Moreover, in liquid culture tests, ergosterol was inactive while oleic acid was fully effective. Leroux and Gredt (1978a) distinguished between non-specific antagonists (e.g. fatty acids, β-carotene, Tween 20) that alleviated toxicity of several fungicides having a wide range of modes of action (including inhibition of sterol biosynthesis) and specific antagonists, e.g. ergosterol, that only affect sterol biosynthesis inhibitors (imazalil, triadimefon and triarimol).

Again, antagonism by ergosterol was only partial as was shown by Buchenauer (1980) who found that various compounds, including plant sterols such as β-sitosterol and even gibberellin A_3, were more active than ergosterol.

It is possible that these results can be explained by a low permeability to ergosterol. Aerobically grown cells of *S. cerevisiae*, but not anaerobically grown cells, are reported to be impermeable to ergosterol (Trocha and Sprinson, 1976). In other examples, however, permeability to sterols does not appear to be a problem. Trocha *et al.* (1974) isolated various ergosterol-requiring mutants of yeast that were presumably permeable to sterols and Langcake (1974b) found that *Fusarium oxysporum* would incorporate stigmasterol from an agar medium where it replaced some of the normal ergosterol content of the fungus. There is as yet no adequate explanation for the rather weak antagonism by ergosterol.

Can the inhibition of sterol biosynthesis explain the fungitoxicity? Sterols have at least two major roles in fungi (Hendrix, 1970). Firstly, they are major structural components of cell membranes where they may contribute to their selective permeability properties. The pythiaceous fungi, in common with insects and some bacteria, are unable to synthesize sterols. When grown in the absence of sterols, vegetative growth is much reduced and in some cases non-existent, and the fungal cells tend to be leaky (Child *et al.*, 1969). Exogenous sterols are an absolute requirement for reproductive growth (whether sexual or asexual) which reflects on the second major function of sterols in most organisms, and probably fungi too (Nelson *et al.*, 1967), as precursors for steroid hormone-type substances, particularly those governing reproductive processes (cf. antheridiol (Barksdale, 1969)). It seems plausible therefore that inhibition of the biosynthesis of sterols would be detrimental to fungal growth and development.

On the other hand, fungicide-treated cells are not deficient in sterols and may in fact contain more free sterols than untreated cells (Sherald and Sisler, 1975). These sterols contain 'extra' methyl groups and there is considerable evidence that methylated sterols cannot adequately substitute for the demethylated sterols. Many yeast mutants have been isolated with various deficiencies in their abilities to synthesize sterols (Trocha *et al.*, 1974; Barton *et al.*, 1974) and some of these are thereby resistant to the polyene antibiotic nystatin (Johnson and Subden, 1977; Fryberg *et al.*, 1974). Those that are unable to demethylate lanosterol all require an exogenous supply of demethylated sterols for normal growth, whereas those that can demethylate sterols, even if they are unable to synthesize ergosterol, the normal sterol component, can grow without added sterols. Also when *S. cerevisiae* is grown anaerobically, it is unable to synthesize sterols since the cyclization of squalene to lanosterol and lanosterol demethylation require oxygen. Again, for anaerobic growth, a demethylated sterol supplement is required (Trocha and Sprinson, 1976 and references therein).

Other work has demonstrated the adverse effects on membranes of substituting C-14 methylated sterols for demethylated ones (Chang *et al.*, 1977; Nes, 1974; Lala *et al.*, 1978). Langcake (1974) found that lanosterol would not substitute for demethylated sterols in stimulating the growth of *Phytophthora infestans*, a pythiaceous fungus. Lanosterol was somewhat toxic to *P. infestans* and direct toxicity of methylated sterols in addition to that of fatty acids which may accumulate (Sisler and Ragsdale, 1977) is another possibility that has received little consideration and which could explain the poor antagonism by ergosterol.

The above evidence would indicate that inhibition of fungal growth can be attributed to the block at C-14 demethylation. The fact that fungi continue to grow for one or more generations in the presence of the fungicides probably represents the utilization of reserves of demethylated sterols (primarily ergosterol) stored as steryl esters within fatty vacuoles (Rose, 1976).

What is the site of action at the molecular level? Although it is clear that sterol demethylation is the primary action, the process is a complex one and raises the question: with what fungal component do the fungicides interact? There are two contenders for this role, Cytochrome P-450 and the Sterol Carrier Protein (SCP).

The specific inhibition of C-14 demethylation implies that the mechanism of C-14 demethylation is distinct from that of C-4 demethylation and there is evidence that this is the case. C-4 Demethylation proceeds via hydroxylation of the methyl group followed by oxidation and decarboxylation involving a 3-keto steroid intermediate. (Kato and Kawase (1976) had also identified this intermediate in their studies with the cell-free enzyme from yeast.) By contrast, C-14 demethylation involves hydroxylation and elimination of the carbon atom as formic acid rather than carbon dioxide (McCorkindale, 1976) (Figure 3). The initial hydroxylation involves a mixed function oxidase and is mediated by cytochrome P-450. Since no C-32 oxygenated sterols were found in Denmert-treated cell-free homogenates of *S. cerevisiae*, Kato and Kawase (1976) have suggested that buthiobate, and presumably the other fungicides, block the initial hydroxylation of the C-14 methyl group. In this respect, the mode of action of these fungicides is identical with that of carbon monoxide, a compound that also inhibits C-14 demethylation but not C-4 demethylation in sterol biosynthesis (Gibbons and Mitropoulos, 1972; 1975; Gaylor and Mason, 1968). Carbon monoxide binds with cytochrome P-450 and it is possible that the fungicides do this also. The nitrogen atom of the aromatic heterocycle (present in all the compounds except triforine) could well interact directly with the haem of the cytochrome. In support of this, Langcake (unpublished) obtained cytochrome P-450 containing microsomes from *S. cerevisiae* grown anaerobically (cf. Yoshida *et al.*, 1977) and showed that on addition of triadimefon at 10^{-6} M, a typical Type II (Schenkman *et al.*, 1967) spectral change occurred indicative of a direct interaction with the cytochrome (Figure 4).

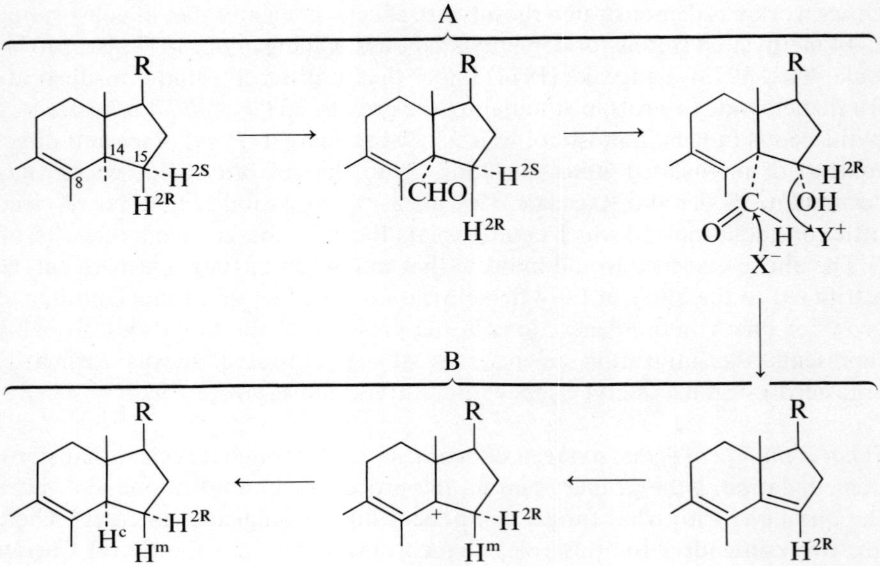

Figure 3 14α-Demethylation of sterols. H^m: hydrogen from the medium; H^c: hydrogen from NADPH. A—proposed site of action of sterol C-14 demethylation inhibitors. B—proposed site of action of sterol Δ^{14} reductase inhibitors. (After McCorkindale, 1976)

An alternative hypothesis concerns the existence of sterol carrier proteins (SCP) (Scallen *et al.*, 1974). These proteins are thought to be involved in all steps of sterol biosynthesis from squalene to cholesterol. They bind the hydrophobic sterols, interact with microsomes and make the sterols accessible for conversion by microsomal enzymes. They may also be involved in sterol transport. At least two carrier proteins have been identified from mammalian liver, one acting on squalene and the other on 7-dehydrocholesterol. It has been suggested that triarimol (Ragsdale, 1975), triforine (Sherald and Sisler, 1975) and the antimycotic 1-dodecylimidazole (**13**) (Henry and Sisler, 1979) may bind with SCP and thereby inhibit enzymatic conversion of the normal substrate lanosterol. Since the fungicides are specific in the enzyme step which they inhibit, this would imply that there are specific carriers for each of the intermediates in the sterol biosynthesis pathway from squalene to cholesterol or ergosterol. The involvement of SCP in fungal sterol biosynthesis has yet to be established. Other sterol solubilizing factors, such as a polysaccharide from yeast (Adams and Parks, 1968) are known to exist.

Are there other sites of action? As discussed already, some effects of the fungicides on metabolism are consequences of the action on sterol demethylation and the subsequent cessation of fungal growth. However, there is evidence

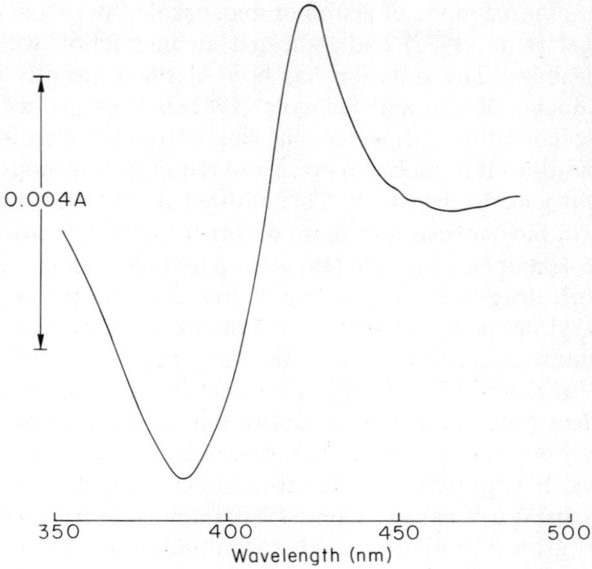

Figure 4 The binding of triadimefon to microsomes from yeast. Samples (3 ml) of microsomal suspension (5.07 mg protein/ml) in the oxidized state were added to sample and reference cuvettes of the spectrophotometer (Pye SP8–100). Then 10 μl acetone solution of triadimefon (final concentration 10^{-5} M) added to sample cuvette. The difference spectrum was recorded. To obtain the baseline of equal absorbance, 10 μl acetone was then added to sample cuvette and 10 μl triadimefon solution in acetone to reference cuvette and difference spectrum recorded. The results show the difference spectrum obtained after averaging 4 scans. (Langcake, unpublished)

for alternative modes of action, especially for the imidazole group of fungicides. This was suggested by the following observations:

(1) Gram-positive bacteria, which do not synthesize sterols, are sensitive to several of the fungicides, e.g. miconazole (**9**) (van Cutsem and Thienpont, 1972);
(2) Some phthiaceous fungi, again organisms that do not synthesize sterols, are susceptible to several of the fungicides, e.g. various species of *Pythium* and *Phytophthora* to triadimefon, triadimenol, fenarimol, nuarimol and imazalil (Buchenauer, 1979).
(3) The rather poor reversal of inhibition frequently observed with ergosterol and other sterols.

Early studies on the mode of action of miconazole (Swamy *et al.*, 1974) and imazalil (Siegel *et al.*, 1977) had indicated an interaction with membranes causing cell leakage. The situation has been clarified recently in simple but elegant experiments of Sud and Feingold (1981a). They utilized the fact that under aerobic conditions, *S. cerevisiae* can synthesize sterols while under anaerobic conditions it is unable to do so and requires exogenous sterol and an unsaturated fatty acid for growth. This allowed them to distinguish between effects on sterol biosynthesis and those on other metabolic processes. Results with miconazole and clotrimazole (**10**) were essentially similar. Under aerobic conditions both drugs were fungistatic at low concentrations (at which the ratios of methyl:desmethyl sterols were altered) but they were fungicidal at higher concentrations. Anaerobically the fungistatic effect was not evident although the fungicidal effect of higher concentrations was still seen. Thus the fungistatic effect occurred at low concentrations and can be attributed to the effects on sterol biosynthesis while the fungicidal effect occurred only at higher concentrations. It is probably related to a direct interaction with membrane components causing cell leakage. Cope (1980) found a close correlation between sensitivity of growth inhibition and susceptibility to miconazole-induced leakage of K^+ from *Candida albicans*. Others have shown that miconazole causes disruption of liposome model membranes, especially those containing free unsaturated fatty acids (Yamaguchi and Iwata, 1979; Sud *et al.*, 1979) and microscopic studies of the effects of miconazole also indicate membrane breakdown in *C. albicans* (de Nollin and Borgers, 1976). Similar conclusions can be inferred for imazalil (Siegel and Solel, 1981) in which toxicity to *Aspergillus nidulans* showed the normal lag period at lower concentrations (up to 1 μg/ml for the wild type strain) but was instantaneous at higher concentrations (20 μg/ml) and associated with an immediate reduction in dry weight.

It would appear, however, that not all sterol biosynthesis inhibitors cause direct membrane damage. In comparative studies of miconazole, clotrimazole and ketoconazole (**23**) (Sud and Feingold, 1981b), ketoconazole alone was found not to be fungicidal nor to cause membrane damage.

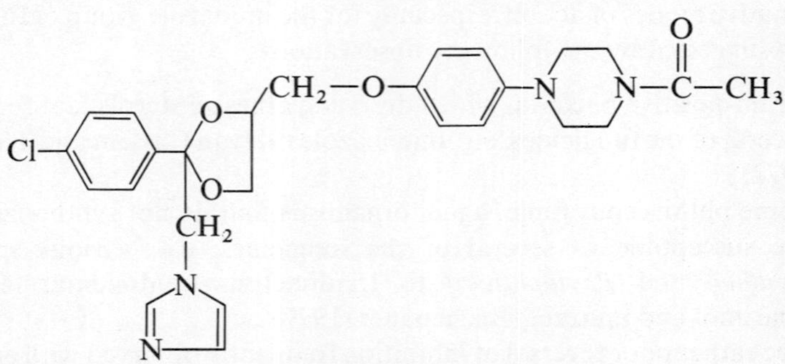

(**23**) Ketoconazole

Other possible modes of action of the fungicides are suggested by the fact that sterol C-14 demethylation is representative of a type of biochemical oxidation (cytochrome P-450 mediated) involved in other metabolic processes that may be equally susceptible to inhibition. Thus, triarimol and ancymidol (**19**) (a closely related plant growth regulator) (Shive and Sisler, 1976) have anti-gibberellin activity in plants and in the case of ancymidol, this has been attributed to inhibition of the cytochrome P-450 mediated oxidation of kaurene to kaurenol, intermediates in gibberellin biosynthesis (Coolbaugh and Hamilton, 1976).

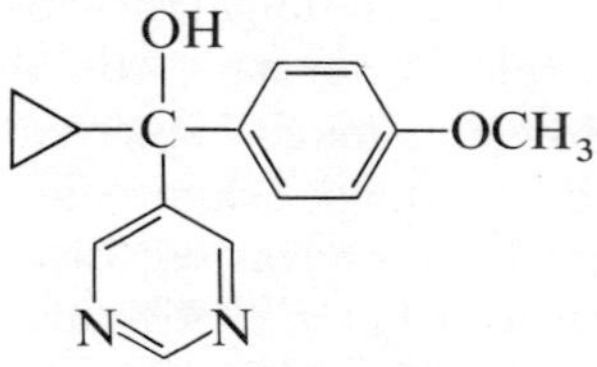

(**19**) Ancymidol

Effects on powdery mildew diseases The sterol demethylation inhibitors are used in agriculture primarily for the control of powdery mildew diseases. A legitimate criticism of the mode of action studies reported above is that powdery mildew fungi have rarely been used as test organism. Triforine (Schlüter and Weltzien, 1971; Clifford and Hislop, 1971), triarimol (Brown, 1970) fenapronil (**12**) (Girardet and Buchenauer, 1980) triadimefon, triadimenol, fluotrimazole, fenarimol, nuarimol, imazalil (Buchenauer, 1977c) and propiconazole (Buchenauer and Kemper, 1981) all interfered with haustorial development of powdery mildews. Appressoria and infection pegs were unaffected but the haustorial sheath was swollen and development of secondary hyphae was inhibited. Biochemical studies on powdery mildew fungi (technically difficult) have not been reported. Henry and Sisler (1981) have suggested that the practical success of the fungicides may not be due to inhibition of fungal growth alone. It may involve modification of secondary metabolism necessary for pathogenesis (cf. tricyclazole, p. 79), and viable but slow-growing fungal hyphae may elicit host defence mechanisms that contribute to the suppression of the disease. In this context, it is interesting that Wolf (1980) has reported the formation of encasements, presumably of host origin, around haustoria of powdery mildew fungi growing on plants treated with triadimefon.

Effects on plants Sterol biosynthesis occurs in many eukaryotic organisms and the ergosterol synthesis inhibitors are potentially toxic to a wide range of organisms. Most of the compounds used in agriculture are reported to reduce the growth of plants (Shive and Sisler, 1976; Buchenauer and Grossmann, 1977; Buchenauer, 1977c; Buchenauer and Rohner, 1981) and the effects were partly, or in some cases fully, reversible with gibberellins. They were only partly reversible with kinetin and not reversible with IAA. Inhibition of gibberellin biosynthesis by a process similar to that for ancymidol (above) may be responsible for this. Extractable gibberellin-like activity in triadimefon-treated plants was reduced, while only very weak effects on GA_3-induced α-amylase in barley endosperm (a test for gibberellin action rather than synthesis) were found (Buchenauer and Rohner, 1981).

Nuarimol, triadimefon, triadimenol and imazalil all inhibited the incorporation of [^{14}C]acetate into lipids of barley plants and the effects on the sterol fractions were analogous to those on fungal systems (Buchenauer and Rohner, 1981; Buchenauer, 1977c) (plants normally contain Δ5 demethylated sterols rather than ergosterol). However, much higher concentrations were required to achieve this as compared with fungi. Schmitt and Benveniste (1979) similarly found that sterol C-14 demethylation was sensitive to fenarimol in suspension cultures of bramble cells. Frasinel *et al.* (1978) also showed that inhibition of *Chlorella* species by triarimol was associated with C-14 demethylation. Plant growth inhibition by triarimol and ancymidol (Shive and Sisler, 1976) occurred in the absence of net sterol biosynthesis and it was concluded that seeds contain enough preformed sterols to satisfy the requirements of the initial stages of germination and growth.

Forster *et al.* (1980a; b; c) examined the side effects of triadimefon and triadimenol on barley. The stunting of growth of seedlings decreased with age and there was no influence on yield in the absence of the pathogen. Other effects noted were weak.

Resistance to sterol demethylation inhibitors Although fungicide resistance is beyond the scope of this review, the occurrence of cross-resistance (i.e. the resistance of a fungal strain to two or more fungicides) has frequently been cited as evidence that compounds have similar modes of action. Caution should be exercised in drawing such conclusions since cross-resistance can occur between fungicides with widely differing modes of action, as well as other pleiotropic effects.

Resistance to sterol demethylation inhibitors has not been a problem in practical use as yet, as predicted by Fuchs and Drandarevski (1976), but resistant strains of culturable fungi are readily obtained in the laboratory. Various forms of cross-resistance have been reported (e.g. van Tuyl, 1977; de Waard and Gieskes, 1977; Barug and Kerkenaar, 1979; Sherald *et al.*, 1973; Sherald and Sisler, 1975; Woloshuk *et al.*, 1979; Girardet and Buchenauer, 1980; Buchenauer and Kemper, 1981; Leroux and Gredt, 1976; Leroux *et al.*, 1976).

Van Tuyl (1977) found that resistance to imazalil is multigenic. This indicates that there are several modes of action or that resistance depends on more than one mechanism and that these are additive in their effects. Studies on the mechanisms of resistance (de Waard and van Nistelrooy, 1980a; Ragsdale and de Waard, 1977; Siegel and Solel, 1981) tend to suggest that they depend on exclusion of the toxicant from the primary target site, rather than on variations in the sterol-synthesizing capabilities of the resistant strains.

Compounds affecting other aspects of sterol biosynthesis

Understandably, there are strong similarities between this group of anti-fungal compounds and those in the section above, but differences in their specific sites of action suggest that they should be treated separately. The group encompasses a wide range of compounds that are of potential use in agriculture (again as fungicides, primarily for the control of powdery mildews), and in medicine as antimycotics (e.g. azasterol (**20**)) and hypocholesteremics (e.g. triparanol (**21**), AY9944 (**22**), Diazacholesterol (**24**)) that have antifungal activity (Aaronson, 1971; Elliott, 1969; Matolcsy *et al.*, 1973; Sobus *et al.*, 1977). The hypocholesteremics tend to function by inhibition of the Δ8 to Δ7 isomerase, but there may be species differences in their action. Attention will be directed here to the agricultural fungicides, principally tridemorph (**25**).

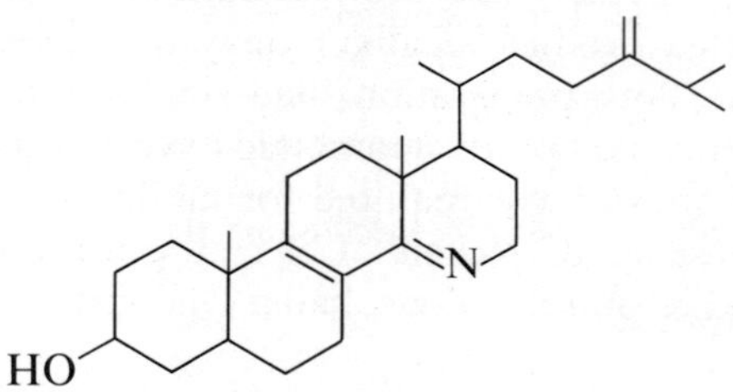

(**20**) Azasterol

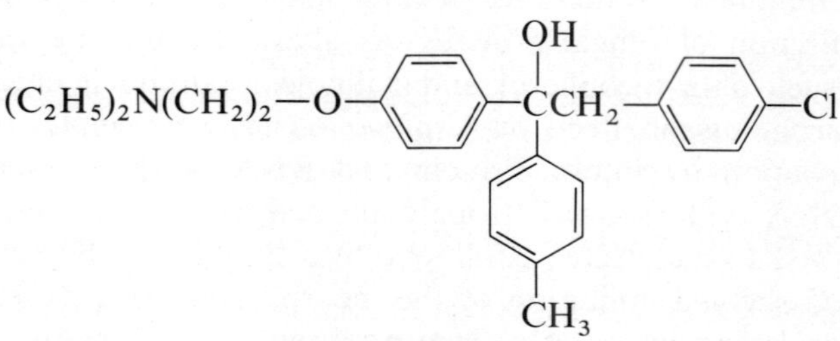

(**21**) Triparanol

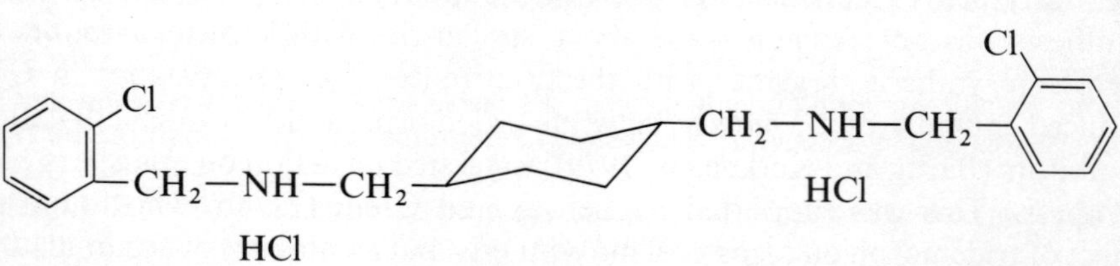

(**22**) AY9944

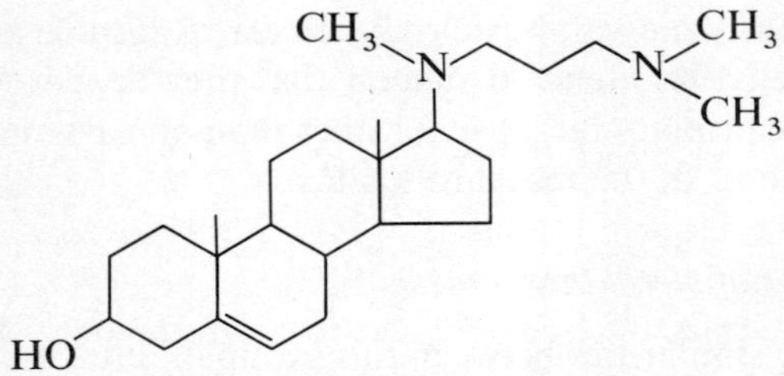

(**24**) Diazacholesterol

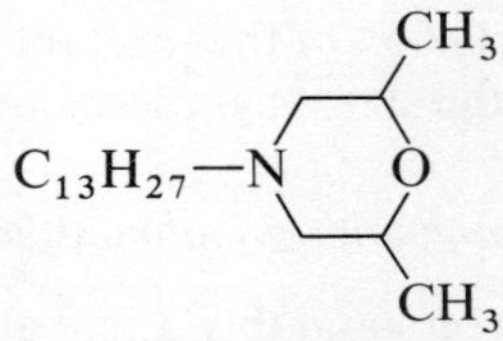

(**25**) Tridemorph

Tridemorph Earlier work on the mode of action of tridemorph (**25**) (quoted in Kerkenaar and Kaars Sijpesteijn, 1979) indicated a primary effect on the respiratory electron transport chain. There was some doubt about this (Garg and Mehrotra, 1975) and actions on cell membrane function (Kaars Sijpesteijn, 1977), lipid biosynthesis (Polter and Casperson, 1979) and protein synthesis (Fisher, 1974) were proposed.

Kerkenaar and Kaars Sijpesteijn (1979) attempted to resolve these conflicting reports. In particular they drew attention to the fact that in some studies, the formulated product 'Calixin' may have been used rather than technical tridemorph. Both preparations had similar antifungal properties *in vitro*, but only the formulated product inhibited oxygen consumption at concentrations just slightly above those required for inhibition of growth. This was attributed to the presence of Nekanil LN in the formulation which, while inhibiting growth and respiration alone, had a synergistic effect in combination with tridemorph.

Tridemorph has several features in common with the sterol demethylation inhibitors. It inhibited a wide range of fungi *in vitro* as well as Gram-positive bacteria. Inhibition of fungal growth was alleviated only by various lipid compounds such as α-tocopherol and trilinolein (although ergosterol was inactive). Morphological effects on *Ustilago maydis* were similar to those for C-14 demethylation inhibitors. Protein and RNA syntheses were virtually unaffected. DNA synthesis was strongly affected within 2 h while total lipid synthesis showed a small decrease followed by a stimulation. Separation of the lipids on TLC showed inhibition of the incorporation of [^{14}C]acetate into desmethyl sterols and increases in incorporation into C4-methyl sterols and free fatty acids (Kerkenaar *et al.*, 1979). Although the inhibition of DNA synthesis was not explained, the above similarities with known inhibitors of sterol biosynthesis, together with the occurrence of cross-resistance in UV-induced mutants of *U. maydis* between tridemorph and C-14 demethylation inhibitors (Barug and Kerkenaar, 1979), suggested an action on ergosterol biosynthesis. This was supported by Leroux and Gredt (1978b) who found an effect of tridemorph on ergosterol biosynthesis, but an absence of accumulation of 4,4-dimethyl sterols. Leroux and Gredt obtained similar results with cycloheximide (see p. 70), a known inhibitor of protein synthesis, and concluded that the effects of tridemorph were secondary consequences of an action on protein synthesis (cf. Fisher, 1974).

Kato *et al.* (1980) had also carried out a comparative study of tridemorph and buthiobate on *Botrytis cinerea*, that largely confirmed the results of Kerkenaar *et al.* (1979). An effect of tridemorph on cell permeability was also noted at high concentrations. Incorporation of [^{14}C]acetate into the various lipid fractions was hardly affected but on further analysis of the 4-desmethyl sterol fraction, large differences between control and treated samples were seen. The amounts of ergosterol (**IV**) (the normal sterol component) and episterol (**IX**) (a normal, but minor sterol component) were reduced and there was an accumulation of fecosterol (**X**), ergosta-8,22,24(28)-trien-3β-ol (**XI**) and ergosta-8,22-dien-3β-ol (**XII**). The identifications were based on mass spectral

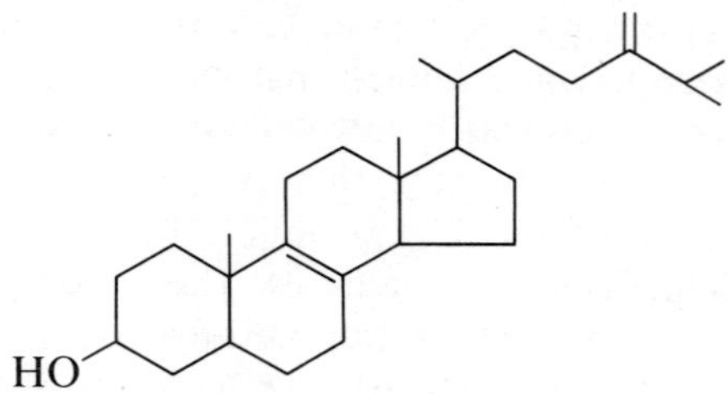

Fecosterol (**X**)

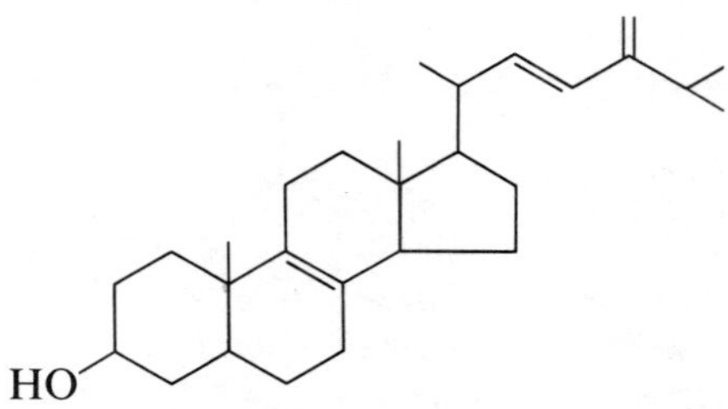

Ergosta-8,22,24(28)-trien-3β-ol. (**XI**)

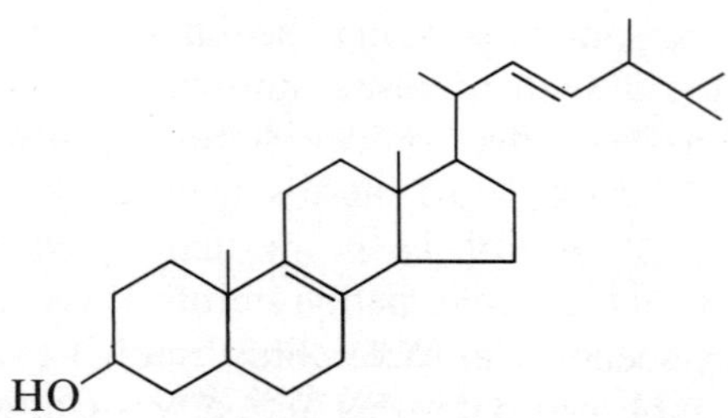

Ergosta-8,22-dien-3β-ol. (**XII**)

fragmentation patterns. Since these sterols retain the C 8(9) double bond, Kato *et al.* (1980) suggested a specific inhibition of the Δ8 to Δ7 isomerization in ergosterol biosynthesis. The same site of action is proposed for the hypocholesteremic drug trifluperidol (**26**) in *S. cerevisiae* (Sobus *et al.*, 1977).

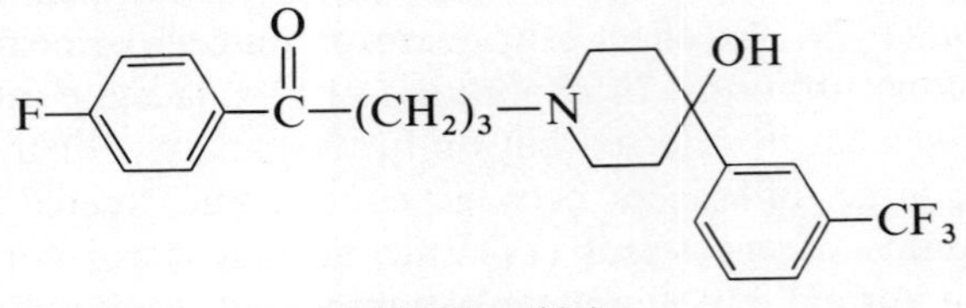

(**26**) Trifluperidol

Near-identical results were obtained in the latest work of Kerkenaar *et al.* (1981) although their interpretation was different. They suggested that the major sterol accumulating in tridemorph-treated cultures was misidentified by Kato *et al.* (1980) as fecosterol. They claimed that its true identity is ergosta-8,14-dien-3β-ol (**XIII**), a compound with very similar mass spectral characteristics to those of fecosterol. Consequently, they proposed that tridemorph specifically inhibits reduction of the 14(15) double bond. This normally occurs directly after C-14 demethylation (Figure 3). This mode of action is identical with that suggested for azasterol (**20**), an antimycotic natural product (Hays *et al.*, 1977; Woloshuk *et al.*, 1979; Schmitt *et al.*, 1980). Kerkenaar *et al.* (1981) gave evidence that inhibition of the Δ14 reductase is more likely to be harmful to fungal growth than inhibition of the Δ8 to Δ7 isomerase.

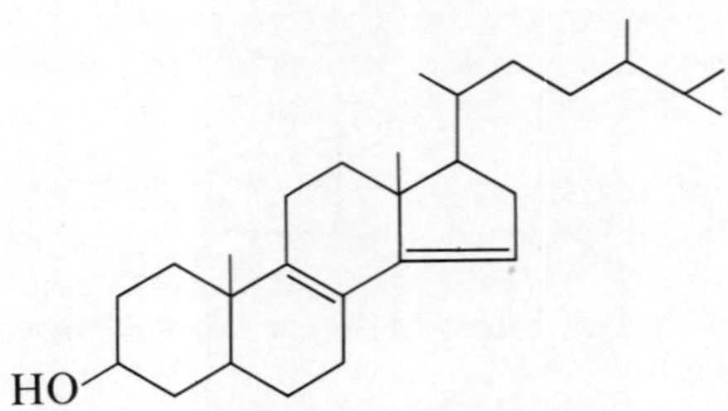

Ergosta-8,14-dien-3β-ol (**XIII**)

The action of tridemorph in plants may be different again. Interestingly, the fungicide impaired the growth of suspension cultures of bramble cells only slightly, yet had a marked effect on the sterol composition. It caused the accumulation of 9β,19-cyclopropyl sterols (cycloartenol (**XIV**), the 9β,19-cyclopropyl isomer of lanosterol, is the normal equivalent of lanosterol in plants) and a depletion of the Δ5-desmethyl sterols. It was concluded that tridemorph inhibits the cycloeucalenol (**XV**)-obtusifoliol (**V**) isomerase (Figure 5). In a reference to unpublished data, this action was confirmed against a cell-free enzyme conducting this conversion (Schmitt *et al.*, 1981).

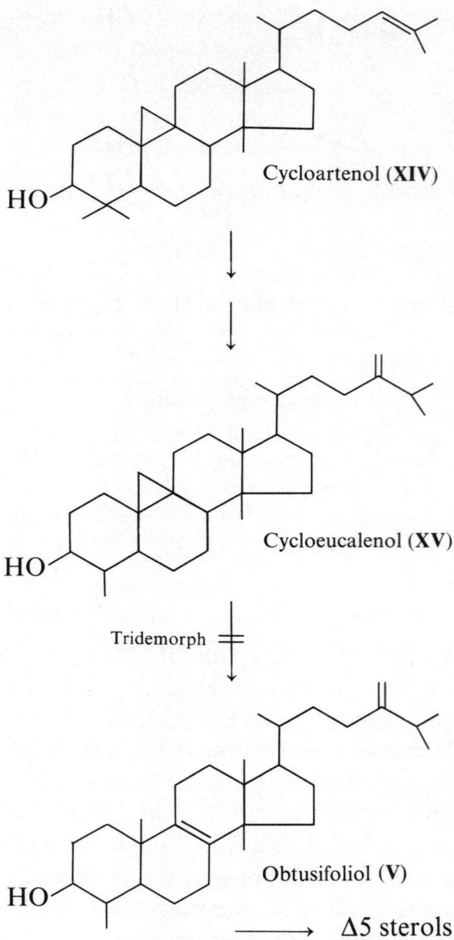

Figure 5 Proposed site of action of tridemorph in suspension cultures of bramble cells (Schmitt *et al.*, 1981)

Dodemorph and Fenpropemorph Whatever the precise mode of action of tridemorph, it seems likely that the closely related compounds dodemorph (**27**) and fenpropemorph (**28**) have similar modes of action.

Interference with respiration

Cis-crotonanilides

The oxathiin derivatives, carboxin (**29**, 'Vitavax') and its sulphone analogue, oxycarboxin (**30**, 'Plantvax') were the first *cis*-crotonanilide compounds shown

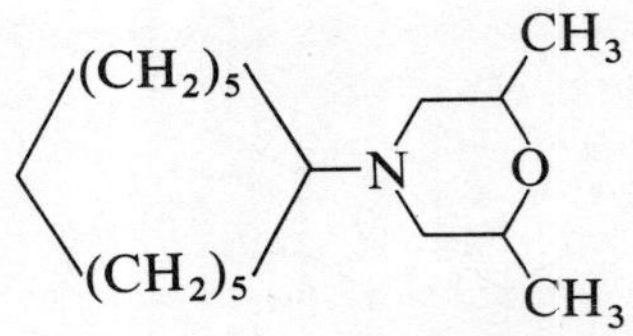

(**27**) Dodemorph

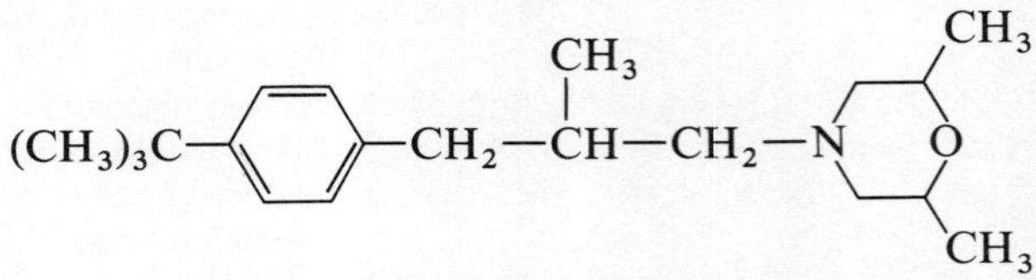

(**28**) Fenpropemorph

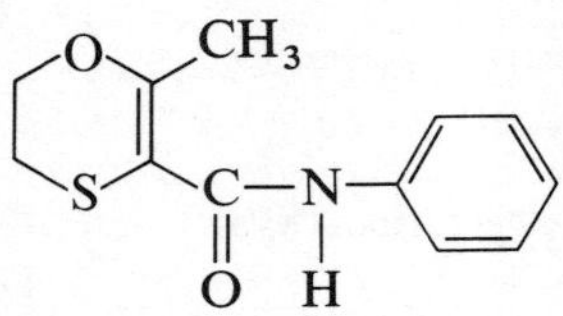

(**29**) Carboxin

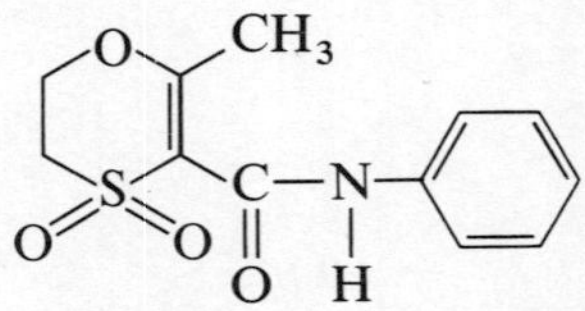

(**30**) Oxycarboxin

to exhibit systemic antifungal activity (von Schmeling and Kulka, 1966). Since the discovery of these two materials a range of other carboxamide fungicides have been synthesized including mebenil (**31**) (Pommer and Kradel, 1969), pyracarbolid (**32**) (Jank and Grossmann, 1971) and fenfuram (**33**) (ten Haken and Dunn, 1971). They are all effective against Basidiomycetes (see, for example, Edgington and Barron, 1967; Edgington *et al.*, 1966), and are considered to possess the same mode of action. Thus, although the majority of information in the literature pertains to carboxin and oxycarboxin, the conclusions can probably be extended to encompass the fungicides mentioned previously and other compounds with closely related structures.

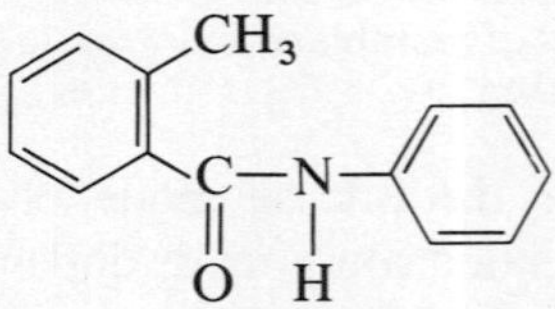

(**31**) Mebenil

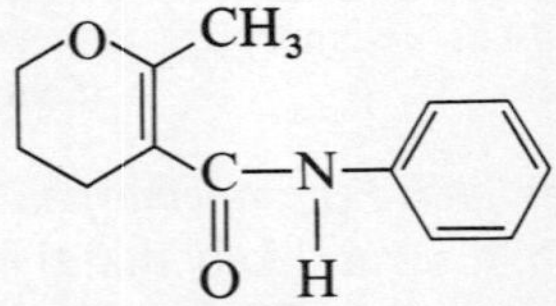

(**32**) Pyracarbolid

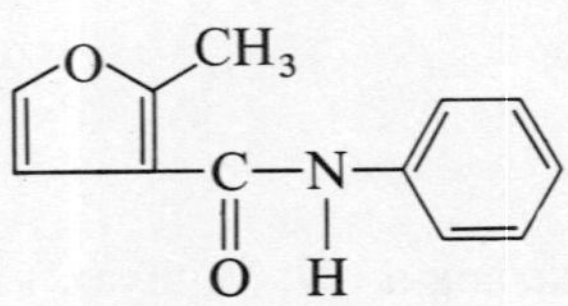

(**33**) Fenfuram

Much of the definitive work on the mechanism of action of carboxin has been carried out using the corn smut pathogen, *Ustilago maydis*. *In vitro* ED_{50} values for this fungus, and for *Rhizoctonia solani* were 1.0 and 0.4 μM respectively (Snel *et al.*, 1970), which provide a striking contrast with the insensitive yeast, *Saccharomyces cerevisiae*, where less than 25% inhibition was observed with fungicide concentrations up to 50 ppm (212.5 μM) (Mathre, 1968). When *R. solani* and *U. nuda* were grown in the presence of carboxin and D-[U-^{14}C]-glucose, there was a decreased incorporation of label into citrate, malate and fumarate, and a concomitant and pronounced accumulation of activity in succinate (Mathre, 1970). This, together with independent experiments by Ragsdale and Sisler (1970), suggested that carboxin probably exerted its primary effect via inhibition of succinate metabolism, while other induced changes such as reduced protein and nucleic acid synthesis were secondary effects. Solute leakage from mycelial discs of *R. solani* exposed to 100 μM carboxin (Shively and Mathre, 1971) must similarly be considered as a follow on effect. In fact, at the more relevant level of 4.25 μM, carboxin induced no detectable leakage from cells of *U. hordei* (Ben-Yephet *et al.*, 1975a).

The view that carboxin probably acted by inhibiting respiration was reinforced by studies employing mitochondrial fractions, which revealed that oxidation of succinate was inhibited by the fungicide, while NADH oxidation was unaffected (Mathre, 1971a; Ulrich and Mathre, 1972). Using *U. maydis*, White (1971) found a reasonably good correlation between inhibition of succinate oxidation by carboxin and related compounds, and inhibition of growth. Collectively, these results suggested that the site of action of carboxin was in the Complex II (succinate–ubiquinone reductase) region of the electron transport system (Figure 6). Strong support for this conclusion has been provided by genetic and biochemical studies with carboxin-resistant mutants of *U. maydis* (Georgopoulos *et al.*, 1972). Resistance to the fungicide was governed by a single-gene mutation which modified the succinate dehydrogenase complex (SDC); the latter term and the variant 'succinate dehydrogenase system' appear quite often in the literature on the mode of action of carboxins, where they are used as synonyms for Complex II. Being unnecessary additions to an already complicated nomenclature, and in an effort to avoid confusion, we have used here the term Complex II throughout.

In addition to studies using fungal material, carboxin has also been shown to inhibit succinate oxidation in bacterial (Tucker and Lillich, 1974), higher plant (Mathre, 1971a) and mammalian (Day *et al.*, 1978) systems. Direct comparison of carboxin activities between different studies is made difficult, however, by the varied nature of the subcellular preparations and electron acceptors employed in assays. Taking such factors into account, available data certainly indicate that succinate oxidizing activities from Basidiomycete fungi, mammalian sources and the bacterium *Micrococcus denitrificans*, are considerably more sensitive than those from higher plants and yeast cells (Table 2).

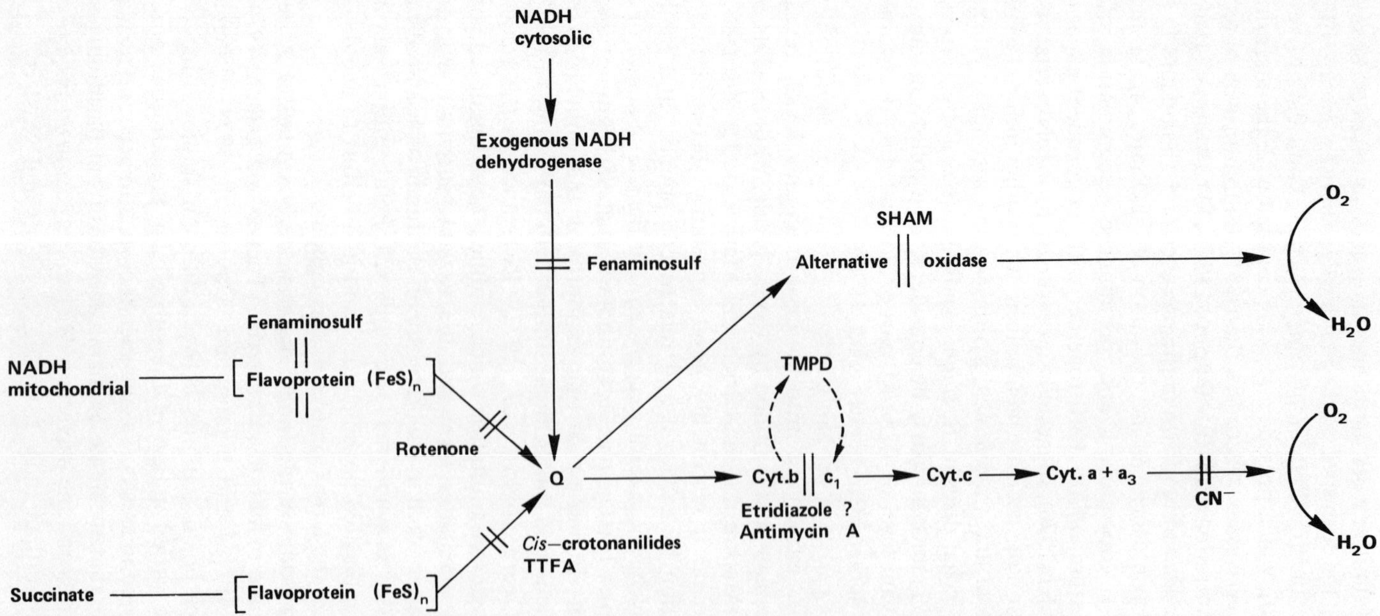

Figure 6 Components of the electron transport system in mitochondria including the alternative pathway and the external NADH dehydrogenase. The sites of action of fungicides and other inhibitors are illustrated, as is the TMPD shunt

Table 2 Inhibition of succinate oxidation by carboxin in preparations from fungal, mammalian, bacterial and higher plant systems

Source	Preparation	Assay	Inhibitory activity of carboxin	Reference
Ustilago maydis	Mitochondria[b]	O_2 uptake	K_i 0.32 μM	Mathre (1971a)
Ustilago maydis	Mitochondria	Succinate-DCPIP[c] reductase	K_i 0.5 μM	White (1971)
Ustilago maydis	Mitochondria	Succinate-DCPIP reductase	I_{50}[d] 0.36 μM	White *et al.* (1978)
Ustilago maydis (carboxin-resistant mutant no. 92)[a]	Mitochondria	Succinate-DCPIP reductase	I_{50} 37.0 μM	White *et al.* (1978)
Rat liver	Mitochondria	O_2 uptake	K_i 6.0 μM	Day *et al.* (1978)
Beef heart	Electron transport particles—non-phosphorylating inner membrane preparations	Succinate-PMS[c] reductase	K_i 0.78 μM	Mowery *et al.* (1977)
	Complex II	Succinate-DCPIP reductase	K_i 0.7–1.0 μM	Mowery *et al.* (1977)
		Succinate-DPB[c] reductase		Mowery *et al.* (1977)
	Soluble	Succinate-PMS reductase	No inhibitory effect	Mowery *et al.* (1977)
Micrococcus denitrificans	Membrane	O_2 uptake	K_i 16.0 μM	Tucker and Lillich (1974)
Saccharomyces cerevisiae	Mitochondria	O_2 uptake	K_i 270.0 μM	Mathre (1971a)
Phaseolus vulgaris	Mitochondria	Succinate-DCPIP reductase	26% inhibition at 100 μM	Ulrich and Mathre (1972)
Solanum tuberosum	Mitochondria	O_2 uptake	K_i 200.0 μM	Day *et al.* (1978)

[a] This mutant was highly resistant to carboxin, having an ED_{50} value of *c.* 37.0 μM.

[b] The term 'mitochondria' is used to indicate preparations containing intact organelles, as opposed to submitochondrial fractions. It does not imply the use of uncontaminated, homogeneous preparations.

[c] Abbreviations used: DCPIP, 2,6-dichlorophenolindophenol; PMS, phenazine methosulphate; and DPB, 2,3-dimethoxy-5-methyl-6-pentyl-1,4-benzoquinone.

[d] The I_{50} value is the concentration which inhibits the rate of DCPIP reduction by 50% as determined from a semilog plot of percentage inhibition against inhibitor concentration.

Mathre (1968) attributed the selectivity of carboxin and oxycarboxin in fungi to differential penetration between sensitive and resistant species. This is not, however, in accord with the data of Lyr *et al.* (1971) or Ben-Yephet *et al.* (1975a) and it would appear, therefore, that some other factors must be involved. Selectivity might reflect differential uptake into mitochondria and/or variations in the make-up and hence susceptibility of Complex II in different sources.

Although there is overwhelming agreement on the fact that carboxamides inhibit succinate oxidation, the exact nature of the site at the molecular level, and of the mechanism involved, have been subject to some debate. In one study employing inner membrane and soluble preparations of succinate dehydrogenase (SDH) from beef heart, it was found that while thenoyltrifluoroacetone (TTFA) inhibited the oxidation of succinate by the artificial electron acceptor phenazine methosulphate (PMS) carboxin did not (Schewe *et al.*, 1973). It was concluded, therefore, that the two inhibitors had different sites of action (Lyr, 1977; Lyr and Schewe, 1975). The results of Schewe *et al.* (1973) appear to be at odds with the majority of reports in the literature which show that like TTFA (Ziegler, 1961), carboxin has no effect on the solubilized enzyme (Georgopoulos *et al.*, 1975a; Ulrich and Mathre, 1972). Indeed, subsequent studies revealed no qualitative differences at all between the action of carboxins and TTFA although the former were more potent with K_i values of 0.002–3.0 μM compared with 9.0–9.7 μM for TTFA (Mowery *et al.*, 1976; 1977). The reduced sensitivity to TTFA of mitochondria from carboxin-resistant mutants of *U. maydis* is yet further evidence in support of a common site of action (Georgopoulos and Ziogas, 1977).

One of the methods used in attempts to pinpoint the site of action of carboxin has been electron paramagnetic resonance (e.p.r.) spectroscopy. This technique can detect and distinguish between the various iron–sulphur centres within Complex II and involved in electron transfer (Figure 6). Data from kinetic e.p.r. experiments suggest that both TTFA and carboxamides act by inhibiting the oxidation of one of these centres, the high-potential (HiPIP) Fe–S cluster (centre 3), thereby blocking electron transport to the Q pool (Ackrell *et al.*, 1977). Although 3′-methylcarboxin caused some line broadening in the e.p.r. signal of the HiPIP centre, this was considered to be of insufficient magnitude to indicate chelation with the carboxamide. Carboxamides might rather cause a change in the environment of the cluster as proposed by White *et al.* (1978).

Further details regarding the site of action of carboxamides have been obtained by way of binding studies using Complex II or inner membrane preparations from beef heart mitochondria, and the carboxin analogue 2,4,5-trimethyl-3-carboxanilinofuran (TCF) (Coles *et al.*, 1978). These showed that TCF could be displaced from specific binding sites not only by TTFA but also by 4′-hexoxycarboxin, one of the most active carboxin derivatives (Mowery *et al.*, 1977). In addition, a good relationship was observed between specific binding by TCF (K_D 1.0 ± 0.1 × 10^{-7} at 4 °C) and inhibition of succinate-2,6-dichlorophenolindophenol (DCPIP) reductase activity (K_i 1.1 × 10^{-7} at 25 °C). It is noteworthy that no binding was observed when TCF was incubated with a soluble preparation of SDH, an observation consistent with the lack of carboxin activity when assayed against this form of the enzyme; it was suggested that this possibly reflected the absence from soluble preparations of a component, perhaps a lipid, which is in some way critical to specific binding. Treatment of Complex II preparations with perchlorate to remove SDH resulted in a concomitant loss of specific binding, implying that the dehydrogenase may itself be instrumental in the binding of TCF.

Apart from the effect of carboxin on succinate oxidation, the fungicide was early on reported to inhibit the cyanide- and antimycin A-insensitive alternative respiratory pathway in sporidia of *U. maydis* (Sherald and Sisler, 1972). Lyr and Schewe (1975) extended this concept in studies with whole cells of the yeast, *Rhodotorula mucilaginosa*, and concluded that an iron–sulphur component of Complex II was itself the alternative oxidase. At the time, this proposal was consistent with the apparent loss of the alternative pathway in a carboxin-resistant mutant of *U. maydis* (Georgopoulos and Sisler, 1970). A recent study, however, using mitochondria from a wild-type strain and a highly carboxin-resistant mutant of the same fungus, revealed no specific effect of carboxin on the alternative pathway (Ziogas and Georgopoulos, 1979). On the other hand, evidence has been obtained using higher plant mitochondria (CN-resistant and CN-sensitive) which shows that carboxins inhibit both succinate oxidation and the alternative respiratory pathway (Day *et al.*, 1978).

The suggestion that the Fe–S cluster centre 3 is the terminal oxidase in the alternative respiratory pathway (Lyr and Schewe, 1975) seems unlikely in view of the fact that the alternative pathway inhibitor, salicyl-hydroxamic acid (SHAM) does not alter the e.p.r. signal from this component (Rich and Bonner, 1978). Furthermore, mitochondrial particles from *Sauromatum guttatum* which have been treated to remove SDH, retain alternative oxidase activity with NADH substrates (Rich and Bonner, 1977). With regard to the branch point between the cytochrome and alternative pathways, current thinking favours a site at the level of ubiquinone (Figure 6) (Hanssens *et al.*, 1978; Moore and Rich, 1980; Storey, 1976).

Aspects of the molecular structure of carboxin and related carboxamides which are required for activity have been elucidated via structure–activity relationships (SAR) determined both *in vitro* and *in vivo*, and at the biochemical level. The early observations of Snel *et al.* (1970) regarding the importance of the carboxamido group and the methyl on the β-carbon were confirmed by ten Haken and Dunn (1971). From the results of *in vivo* studies versus broad bean rust (*Uromyces fabae*) and wheat rust (*Puccinia recondita*), the latter authors concluded that the basic structure needed for activity was (**34**), hence the designation '*cis*-crotonanilides' to describe this group of compounds.

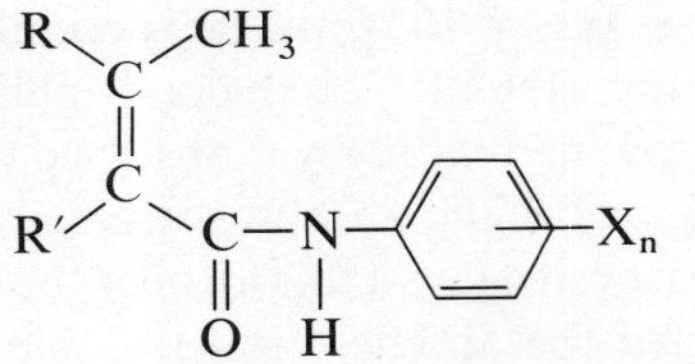

(34)

R = CH_3, R′ = H, or R = R′ + double bond forming a ring system.
X_n = one or more electron donating groups, or unsubstituted.

These conclusions were consolidated by the SAR studies of White and Thorn (1975) which examined the activities of 93 carboxamides against mitochondria isolated from two sensitive organisms, *U. maydis* and *Cryptococcus laurentii*. Although the conclusions drawn from the data were largely consistent with those of ten Haken and Dunn (1971) some modifications were indicated. For example, replacement of the methyl on the β-carbon and the phenyl attached to the nitrogen atom with ethyl and cyclohexyl groups, respectively, does not markedly affect activity. In addition, the effect of substituents on the anilino ring appear to be more complex than was originally described. The observation by Mathre (1971b) that the activity of carboxin and five related compounds could be correlated with their partition coefficients was not substantiated in the more extensive investigation of White and Thorn (1975).

A more recent SAR study has considered, among other aspects, the effects of substitution in the *N*-phenyl ring with alkyl and alkoxy groups, and of replacement of this ring with straight chain hydrocarbon groups (C_2–C_{14}) to give *N*-alkyl derivatives (White *et al.*, 1978). The activities exhibited by a large number of thiophene carboxamides have also recently been determined (White and Thorn, 1980). In addition to providing useful information regarding the molecular requirements for activity, the studies of White *et al.* (1978) and White and Thorn (1980) identified compounds which may be of practical value in situations where resistance to carboxin or oxycarboxin becomes established in the field. This latter aspect is considered more fully below.

As noted earlier some of the most compelling evidence indicating Complex II as the site of action of the *cis*-crotonanilides has come from studies employing fungicide-resistant mutants (Georgopoulos *et al.*, 1972; Georgopoulos and Ziogas, 1977). Resistance to carboxamides has been demonstrated in the laboratory with *U. maydis* (Georgopoulos and Sisler, 1970; Georgopoulos *et al.*, 1975a), *U. hordei* (Ben-Yephet *et al.*, 1974; 1975b), *R. solani*, *R. bataticola* (Grover and Chopra, 1970), *Aspergillus nidulans* (Gunatilleke *et al.*, 1976; van Tuyl, 1975), *Alternaria triticina* (Sankhla *et al.*, 1970), *Sclerotinia sclerotiorum*, *Fusarium solani* f.sp. *pisi* (Luc *et al.*, 1971) and *Neurospora crassa* (Sisler, cited by Dekker, 1972). Furthermore, resistance to oxycarboxin has been reported under natural conditions (Abiko *et al.*, 1975; Dirkse *et al.*, 1982).

In view of the fact that alterations in Complex II have been implicated as the basis of tolerance, the report (White *et al.*, 1978) that certain oxathiin carboxamides are nearly equally potent inhibitors of succinate oxidation in mitochondrial preparations from wild-type and mutant strains of *U. maydis* is noteworthy and may potentially be of practical significance. Indeed, some of the compounds tested actually showed a negative correlation to carboxin with respect to cross-resistance; for instance 4′-phenylcarboxin proved to be 13 times less active than carboxin versus succinate oxidation in mitochondria from a wild-type strain but 19–59 times more toxic towards preparations from moderately resistant mutants. Similar results have been obtained in a separate study (White and Thorn, 1980), again using mitochondria from *U. maydis* but employing a range of thiophene rather than oxathiin derivatives. White and Thorn (1980) concluded that 'with all the structural combinations available, it appears quite possible, in terms of inhibition, to overcome any type of mutation in a fungal SDC which arises through selection by carboxin or other carboxamide compounds'. Such a statement is highly encouraging at a time when fungicide resistance is viewed as potentially one of the most crucial barriers to the continued efficient use of systemic fungicides.

Fenaminosulf Fenaminosulf (**35**, 'Lesan'; 'Dexon') is a seed and soil fungicide for the control of diseases caused by *Pythium*, *Aphanomyces* and *Phytophthora* (Urbschat, 1960). The sensitivity of these Phycomycetes is considered to reflect the absence of barriers to fungicide penetration which exist in higher fungi such as *Fusarium* and *Rhizoctonia* (Tolmsoff, 1966). Selectivity might also depend on differential metabolism and detoxification of the fungicide (Tolmsoff, 1962).

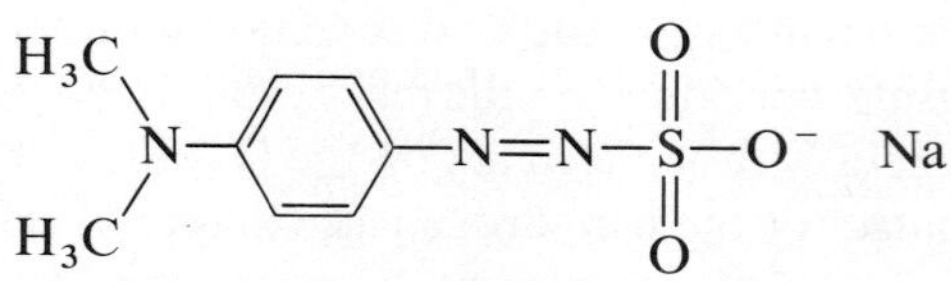

(**35**) Fenaminosulf

In contrast to the *cis*-crotonanilides, fenaminosulf interferes with the Complex I region of the respiratory chain by which the oxidation of metabolically generated NADH is linked to the terminal electron transport system (Figure 6). The first evidence to this effect was obtained by Tolmsoff (1962) who showed that fenaminosulf was a strong inhibitor of NADH (DPNH)-cytochrome c reductase in *Pythium*. Later experiments (Halangk and Schewe, 1975; Schewe *et al.*, 1975) using electron transport particles from beef heart mitochondria confirmed the potent inhibitory action of fenaminosulf against the oxidation of NADH, and also the lack of effect on the oxidation of succinate; the activity of NADH oxidase was inhibited by 50% in the presence of 1.4 μM fenaminosulf, whereas comparable inhibition of succinate oxidation required a fungicide concentration of 2 mM.

Although fenaminosulf affects NADH oxidation in broadly the same region as rotenone, i.e. Complex I, the precise sites of action of the two inhibitors within the dehydrogenase are apparently different. This is indicated by the fact that as well as inhibiting the endogenous NADH dehydrogenase, fenaminosulf is also inhibitory towards the externally facing NADH dehydrogenase located on the inner membrane of plant mitochondria (Halangk and Schewe, 1975; Schewe *et al.*, 1975). The exogenous enzyme is thought to donate reducing equivalents into the respiratory chain at the level of ubiquinone (Figure 6) but more importantly in the present context, the oxidative process is insensitive to Complex I inhibitors such as rotenone (Day and Wiskich, 1974). A further difference between the two inhibitors is the ability of fenaminosulf but not rotenone to inhibit the low molecular weight (type II) form of soluble NADH dehydrogenase. This form of the enzyme, which was prepared according to the method of Mahler *et al.* (1952), exhibits considerably modified molecular and catalytic properties compared to the native dehydrogenase (Singer and Gutman, 1970) and, therefore, results obtained using it should be interpreted with caution.

It has been suggested that the site of action of fenaminosulf within Complex I involves interference with the flavin part of the dehydrogenase (Halangk and Schewe, 1975; Schewe *et al.*, 1975). This proposal relies on the observation that the fungicide is not only inhibitory to NADH–CoQ_1 reductase, but also inhibits oxidation of NADH by menadione, and ferricyanide. Oxidation by the latter two acceptors was considered not to involve the participation of the iron–sulphur centres, implying that fenaminosulf probably interacts in some way with the FMN component of the enzyme.

Assignment of the flavin as the site of interaction with fenaminosulf should be considered alongside work done on the more established inhibitor, rotenone. The latter is now generally thought to act on the O_2-side of both the flavin and the iron–sulphur centres of the dehydrogenase (Figure 6) (Hatefi and Stiggall, 1976). This concerted view was arrived at, however, only after several years of intensive research during which a number of theories were put forward, each at variance with one another (Singer and Gutman, 1970). Ultimate resolution of the problem involved the application of several sophisticated techniques including binding studies and e.p.r. spectroscopy. It would be useful to examine fenaminosulf using similar approaches and hopefully thereby confirm its putative site of action.

Interpretation may be facilitated by the advances which have been made over recent years regarding both the structure (Ragan, 1980) and the functioning (Hatefi and Stiggall, 1976) of NADH dehydrogenase. It is perhaps surprising that so little attention has been given to fenaminosulf in view of its potential value as a biochemical probe for the fundamental study of NADH dehydrogenase.

Interference with the formation or function of membrane components

Organophosphorus compounds

An extensive range of organophosphorus compounds have been reported to exhibit antifungal activity, but the principal fungicides in this group are the phosphorothiolate (PTL) compounds, namely edifenphos (**36**, 'Hinosan'), *S*-benzyl *O*,*O*-di-isopropyl phosphorothioate (**37**; IBP; 'Kitazin P'), and the *O*,*O*-diethyl analogue of the latter, 'Kitazin' (**38**, EBP). Although the precise mode of action of another organophosphorus fungicide, pyrazophos (**39**, 'Afugan', 'Curamil'), has yet to be firmly established, this compound will be considered alongside the PTL fungicides in the present context. Its inclusion here is reasonable in view of structural considerations, though fungicide resistance data which will be referred to later might be cited as further justification. The PTL fungicides are effective against the blast disease of rice caused by *Pyricularia oryzae*, while pyrazophos finds principal use in the control of powdery mildews on a wide range of crops.

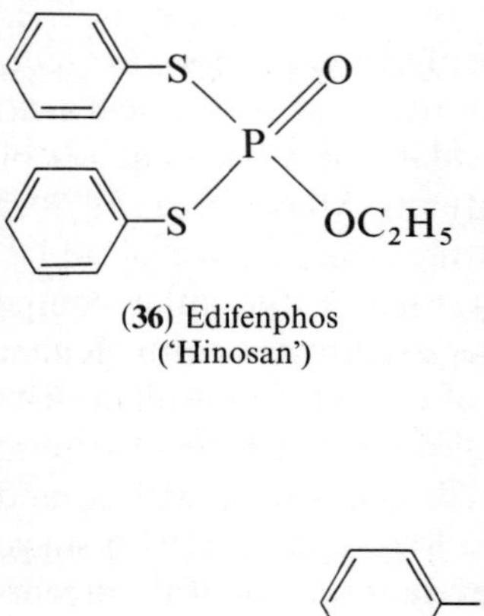

(**36**) Edifenphos
('Hinosan')

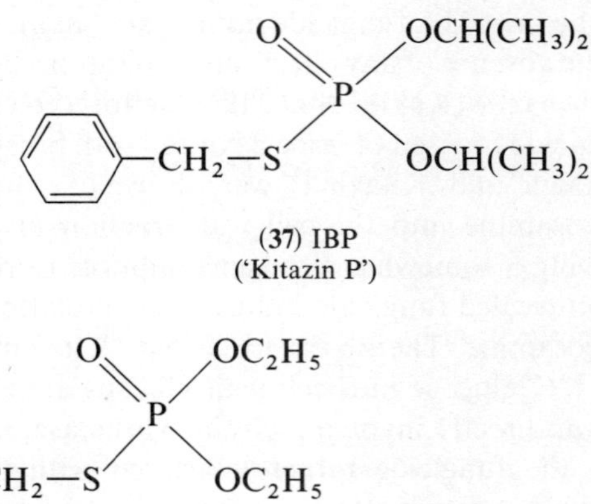

(**37**) IBP
('Kitazin P')

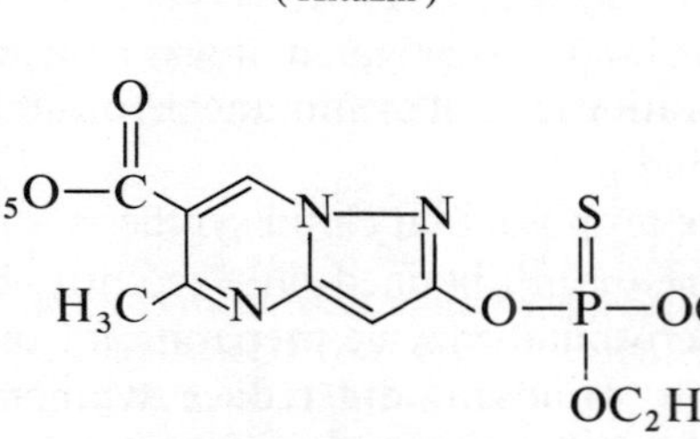

(**38**) EBP
('Kitazin')

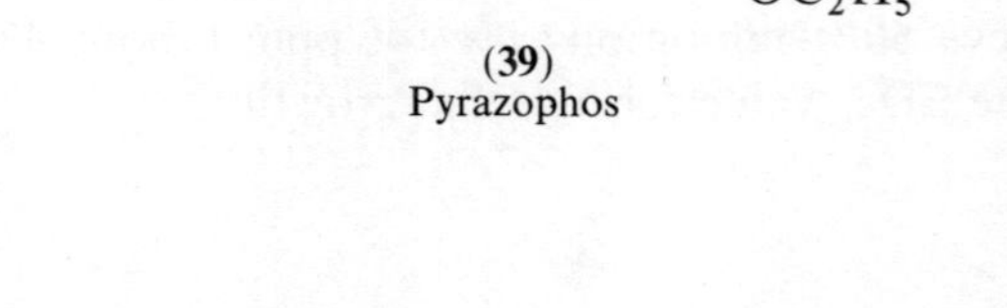

(**39**)
Pyrazophos

PTL fungicides Common structural features between the various PTL fungicides would suggest that they might possess similar mechanisms of action. Circumstantial support for this conjecture is provided by a number of studies on fungicide resistance employing mutant strains produced in the laboratory (Uesugi, 1981; Uesugi and Katagiri, 1977; Uesugi *et al.*, 1969; 1974). Observations include:

1. Cross-resistance between different PTL compounds—accordingly, a genetic change which confers resistance to one PTL compound also brings about resistance to other related fungicides.
2. Negatively-correlated cross-resistance between PTL compounds and phosphoramidate fungicides—resistance to the PTL fungicides, therefore, goes hand-in-hand with increased sensitivity towards phosphoramidates.
3. Synergism between PTL compounds and phosphoramidates versus wild-type representatives of *P. oryzae*—it has been shown that the combined inhibitory activity of PTL and phosphoramidate fungicides on the growth of *Pyricularia* wild-type strains exceeds the sum of their individual effects.

While these data appear consistent with a common site of action for PTL fungicides, they do not provide any indication as to the nature of the site.

The first PTL fungicide investigated in mode of action studies was 'Kitazin'. In the absence of any effects on respiration, or protein and nucleic acid synthesis in *P. oryzae* (Kakiki *et al.*, 1969), efforts to elucidate the mode of action of this fungicide focused on aspects of cell wall biosynthesis (Maeda *et al.*, 1970). Both 'Kitazin' and 'Kitazin P' were shown to reduce the incorporation of D-[1-^{14}C]-glucosamine into the cell wall fraction of *Pyricularia*, the latter compound proving a somewhat stronger inhibitor. Further experiments with 'Kitazin P' also revealed fungicide-induced accumulation of uridine diphosphate *N*-acetyl glucosamine. The observation that 'Kitazin' also inhibited the incorporation of [^{14}C]glucose into cell wall glucans appeared inconsistent with a mode of action directly involving chitin synthetase, and Maeda *et al.* (1970) suggested that the fungicide rather interfered with the transport of wall precursors, possibly through effects on the cell membrane; this was perhaps the first hint that PTL fungicides might interfere with some facet of lipid biosynthesis. In corresponding studies by de Waard (1972; 1974) in which the activities of both 'Kitazin' and 'Hinosan' were investigated, it was concluded that their toxicity probably reflected alterations in cell membrane permeability with any effects on chitin synthesis being indirect.

Confirmation for the proposal that chitin synthesis is not the primary site of action of 'Hinosan' has been obtained only recently, by direct assay of the fungicide against a microsomal enzyme preparation from *P. oryzae* (Kodama *et al.*, 1980). Although 'Hinosan' did reduce synthetase activity, the level required to achieve 50% inhibition was 146 ppm (about 470 μM), which is about 11 times the ED_{50} value for mycelial growth.

As well as looking at the effects of 'Hinosan' on chitin synthetase, Kodama *et al.* (1980) also investigated fungicide-induced changes in lipid biosynthesis. The latter aspect proved more fruitful, and at 10 ppm the fungicide inhibited the incorporation of L-[methyl-^{14}C]methionine, particularly into compound lipids. The product most affected was phosphatidyl choline (**XVII**), and the incorporation of other precursors (*S*-adenosyl-L-[methyl-^{14}C]methionine and [1-^{14}C]glycerol) into this phospholipid was similarly inhibited by 'Hinosan'. In assays of the enzymes involved in the biosynthesis of phosphatidyl choline from glycerol, 'Hinosan' specifically inhibited phosphatidyl ethanolamine (**XVI**): adenosylmethionine methyltransferase, implying that the primary mode of action of the fungicide is interference with the methylation reactions by which phosphatidyl ethanolamine is converted into phosphatidyl choline (Figure 7). 'Kitazin P' inhibits the formation of phosphatidyl choline at the same biochemical level as 'Hinosan' (Akatsuka *et al.*, 1977; Kodama *et al.*, 1979) and this common site of action probably forms the basis for cross-resistance between the two fungicides.

Using a mammalian cultured cell line (mineral oil-induced mouse plasmacytoma, MOPC-31C cells) Maeda *et al.* (1981) recently demonstrated

(XVI)

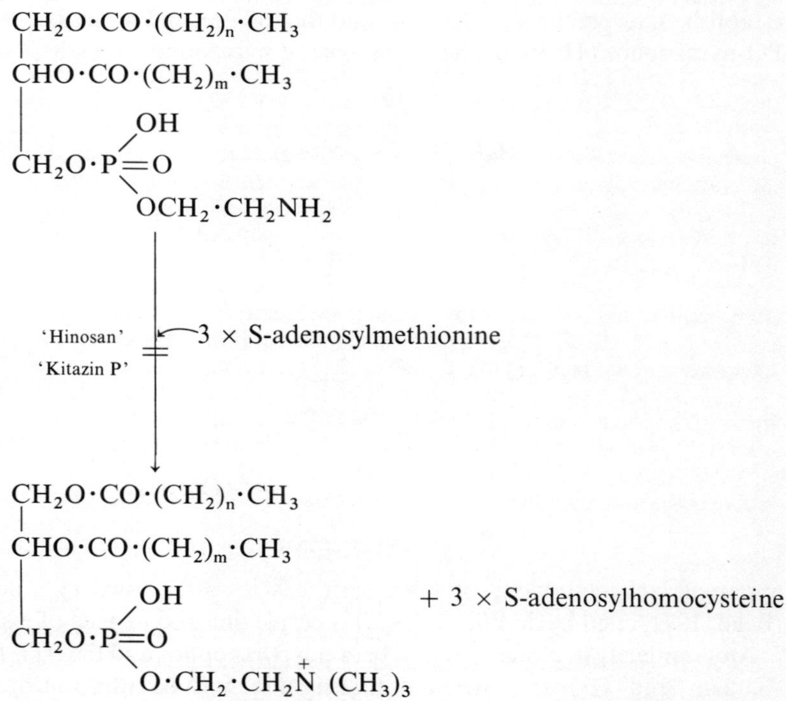

(XVII)

Figure 7 The methylation of phosphatidyl ethanolamine (**XVI**) to give phosphatidyl choline (**XVII**). This process in *P. oryzae* is inhibited by 'Hinosan' and 'Kitazin P'

that inhibition of phospholipid methylation by 'Kitazin P' was irreversible. This contrasted with cycloleucine where inhibitory effects were reversed after removing the toxicant by washing.

Pyrazophos Although pyrazophos is principally active against powdery mildews, it is also inhibitory to *P. oryzae*, and as the latter can be cultured *in vitro*, it has been the organism most extensively used in mode of action studies (de Waard, 1974; 1975). Pyrazophos was found by de Waard to have no effect on the synthesis of DNA, RNA, or protein in *P. oryzae*, but did partially inhibit O_2 uptake. At pH 4.0, however, both mycelial growth in liquid culture and O_2 uptake were somewhat more inhibited by the pyrazophos metabolite, 2-hydroxy-5-methyl-6-ethoxycarbonylpyrazolo [1,5-a]pyrimidine (PP) (**40**), and it was suggested, therefore, that this might in fact be the active principle. The toxicity of pyrazophos towards *Ustilago maydis* was likewise considered to reflect conversion into this metabolite (Georgopoulos *et al.*, 1975b). Conversely, lack of activity against *Pythium debaryanum*, *Saccharomyces cerevisiae* (de Waard, 1974) and pyrazophos-resistant strains of *P. oryzae* (de Waard and van Nistelrooy, 1980b), have been correlated with an inability to metabolize the fungicide to PP. While the relevance of PP to selectivity would appear to be established, its precise site of action, and that of a second fungitoxic metabolite, PO-pyrazophos (**41**) (phosphate analogue of pyrazophos) remain obscure. The

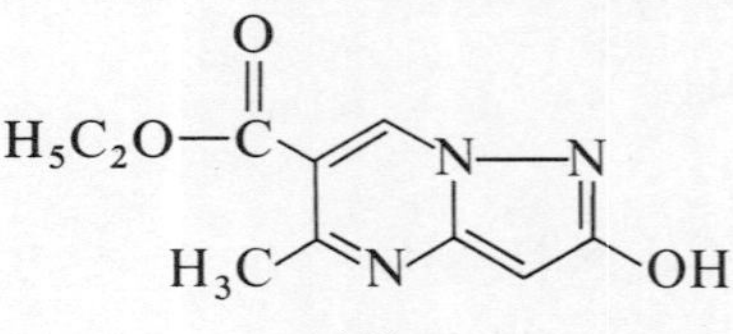

(**40**)
PP

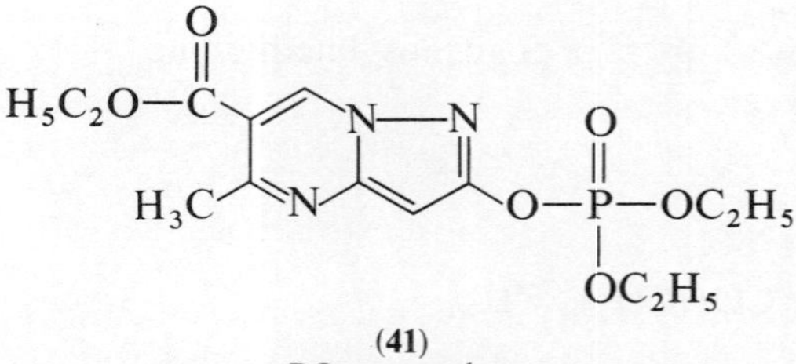

(**41**)
PO-pyrazophos

apparent lack of toxicity of both materials towards powdery mildew fungi (Belal, 1971, cited by de Waard, 1975) is perplexing and in need of explanation.

Any similarity in mode of action between pyrazophos and the PTL fungicides 'Kitazin' and 'Hinosan', would at first sight seem to be ruled out by a number of factors including the absence of effect of pyrazophos on membrane permeability and, as a consequence, on cell wall synthesis, and the apparently multisite inhibitory nature of PP (de Waard, 1974). At the same time, however, a recent study has shown pyrazophos-resistant strains of *P. oryzae* to be cross-resistant to these PTL compounds (de Waard and van Nistelrooy, 1980b). Several tentative explanations for the latter results were proffered, but the basis for cross-resistance is still unknown. Although the possibility of a common site of action was not raised explicitly, it would seem worthwhile to explore this possibility by investigating the effects of pyrazophos, and metabolites thereof, on lipid biosynthesis and in particular on the formation of phosphatidyl choline.

Isoprothiolane

The ketone dithioacetal, isoprothiolane (**42**, 'IPT'; 'Fuji-one') is a systemic fungicide applied to paddy water for the protection of rice. As well as controlling blast caused by *P. oryzae*, it also exhibits a useful 'insectistatic' action against planthoppers.

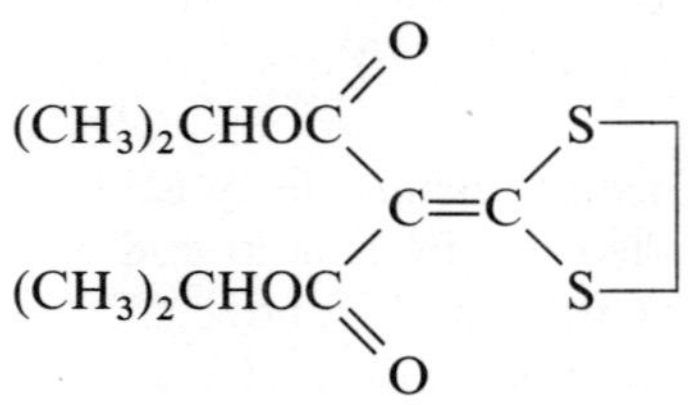

(**42**) Isoprothiolane

Interestingly, Katagiri and Uesugi (1977) found that strains of *P. oryzae* resistant to isoprothiolane were also resistant to PTL fungicides, despite the lack of a phosphorus atom in the former and also the presence of sulphur atoms in a thio-ether rather than thio-ester form. Furthermore, isoprothiolane exhibited identical properties with regard to interactions with phosphoramidates as had previously been observed with PTL compounds (see points 2 and 3 under PTL fungicides, p. 40). It appears, therefore, that isoprothiolane may act in the same manner against *Pyricularia* as the organophosphorus rice blast fungicides.

There is certainly good evidence that isoprothiolane has pronounced effects on lipid biosynthesis (Kakiki and Misato, 1979b). Incorporation of [^{14}C]acetate into fatty acids and tryglycerides is markedly decreased in the presence of the fungicide, as is the incorporation of L-[methyl-^{14}C]methionine into phospholipids. Collectively, these data would be consistent with a mechanism of action involving interference with the biosynthesis of fatty acids, though in assays against a preparation of fatty acid synthetase from *P. oryzae* the fungicide was without effect. A possible alternative site is the enzyme acetyl-CoA carboxylase which catalyses the formation of malonyl-CoA.

Fungicide-induced accumulation of *S*-adenosylmethionine might be indicative of an additional site of action, and in view of the results with the PTL fungicides it is tempting to suggest that this second site is the enzyme phosphatidyl ethanolamine:adenosylmethionine methyltransferase (Kaars Sijpesteijn, 1982). While such a conclusion is consistent with the cross-resistance mentioned earlier, it should be borne in mind that *S*-adenosylmethionine is used as a methyl donor in several other reactions, in addition to the formation of phosphatidyl choline. It is important, therefore, to eliminate other possible explanations for the build-up of *S*-adenosylmethionine, and to assay the fungicide against the putative target enzyme. If isoprothiolane proves not to inhibit the transferase, then the accumulation could merely be due to a lack of substrate as proposed by Kakiki and Misato (1979b).

In addition to effects on lipid biosynthesis, isoprothiolane has also been shown to inhibit the uptake and hence the incorporation of various monosaccharides into the cell wall (Kakiki and Misato, 1979a). With the PTL fungicides, alterations in sugar transport/cell wall biosynthesis were considered likely to be secondary effects resulting from impairment of membrane function (de Waard, 1972; Maeda *et al.*, 1970) and this might also be the case with isoprothiolane.

With the exception of assays against fatty acid synthetase, all of the biochemical experiments carried out by Kakiki and Misato (1979a; b) involved the use of isoprothiolane at 50 $\mu g\ ml^{-1}$ (about 172 μM), a concentration equal to the ED_{95} for appressorium formation and in excess of the ED_{95} for mycelial growth (Araki and Miyagi, 1976). In order to establish unequivocally the primary site of action of the fungicide it will be necessary to examine effects using lower concentrations. In this way it should be possible to unravel the apparently complex effects of isoprothiolane, and discriminate between those which are direct and those which are consequential. It should, indeed, be a general principle applied to all mode of action studies, namely that one localizes and examines the earliest physiological or biochemical effect which is induced by the lowest concentration of the test compound.

Prothiocarb and Propamocarb

While the precise mode of action of prothiocarb (**43**) remains uncertain, an effect on the permeability of fungal membranes is implicated. Propamocarb (**44**), a closely related derivative, probably acts in the same way. Both compounds are soil-applied systemic fungicides with specific action against diseases

$$(CH_3)_2\ N\cdot CH_2\cdot CH_2\cdot CH_2\cdot NH\cdot \overset{\overset{\displaystyle O}{\|}}{C}\cdot S\cdot C_2H_5$$

(**43**) Prothiocarb

$$(CH_3)_2N\cdot CH_2\cdot CH_2\cdot CH_2\cdot NH\cdot \overset{\overset{\displaystyle O}{\|}}{C}\cdot O\cdot Pr$$

(**44**) Propamocarb

caused by Oomycete fungi, e.g. species of *Pythium* and *Phytophthora*, mainly as prophylactics. Kaars Sijpesteijn *et al.* (1974) found prothiocarb to be similarly selective in its action *in vitro* against Oomycete fungi. Sensitivity was dependent on pH, it being very active at neutral values but less so at lower pH values. This depends to some extent, however, on the fungal species used for assay (Papavizas *et al.*, 1978). Activity *in vitro* is also influenced by temperature. In the case of prothiocarb, two types of temperature response were found. With fungi such as *Pythium splendens*, sensitivity increased with decrease in temperature, while with *Achlya radiosa*, sensitivity increased with increase in temperature. This was explained by suggesting that there are two modes of action for prothiocarb. In the case of *P. splendens*, toxicity depends on the whole molecule. This is presumably the case with propamocarb also. However, prothiocarb, and other ethyl thioester analogues, can break down to the volatile compound ethyl mercaptan and this is the fungitoxic principle for the *A. radiosa*-type fungi, but not for the *P. splendens*-type fungi. Ethyl mercaptan is presumed to be transformed metabolically into L-ethionine, an antimetabolite of L-methionine, and the fungitoxicity of prothiocarb for only the *A. radiosa*-type fungi was reversed by addition of L-methionine (Kerkenaar and Kaars Sijpesteijn, 1977). The relative importance of these two modes of action is not recorded but since prothiocarb and propamocarb (which cannot generate ethyl mercaptan) have similar activities in their use as fungicides, the generation of ethyl mercaptan is of doubtful practical importance.

It is interesting that *Phytophthora* species are generally very insensitive to prothiocarb *in vitro*, while in practice, diseases caused by *Phytophthora* are controlled well.

Mycelium of *Pythium ultimum* grown in media containing propamocarb leaked a variety of cell constituents at a rate exceeding that of control mycelium. This leakage was reduced when the sterols cholesterol, cholestanol or sitosterol (but not the methylated sterol lanosterol) were added to the growth medium. The sterols also reduced the fungitoxicity of propamocarb (Papavizas *et al.*, 1978). This had been found previously for prothiocarb (Kaars Sijpesteijn *et al.*, 1974).

These interesting results clearly require further work. In the studies of Papavizas *et al.* (1978), the basal growth medium used did not contain any sterols, and since these fungi are unable to synthesize sterols (e.g. Hendrix, 1970) a direct interaction of the fungicides with membrane-bound sterols, as with polyene antibiotics (Kinsky *et al.*, 1968), is unlikely. The fact that the induced leakage occurred only when mycelium was grown in the presence of propamocarb and not when the fungicide was added to pre-grown mycelium (Papavizas *et al.*, 1978) might suggest an effect on the biosynthesis of some membrane component, leading to an impairment of permeability that can be at least partly restored by addition of sterols.

Interference with nucleic acid metabolism and cell division

Pyrimidines

Ethirimol (**45**, 'Milstem') and its dimethylamino analogue, dimethirimol (**46**, 'Milcurb') are pyrimidine fungicides specifically active against certain powdery mildews (Bebbington *et al.*, 1969; Elias *et al.*, 1968); they have been recommended for use in cereals and cucurbits, respectively. Bupirimate (**47**, 'Nimrod'), which is the dimethylsulphamate ester of ethirimol, has been introduced more recently and is active against powdery mildews of apple and glasshouse rose (Finney *et al.*, 1975). It has been less intensively studied than its two predecessors but can reasonably be assumed to possess a similar or identical mode of action.

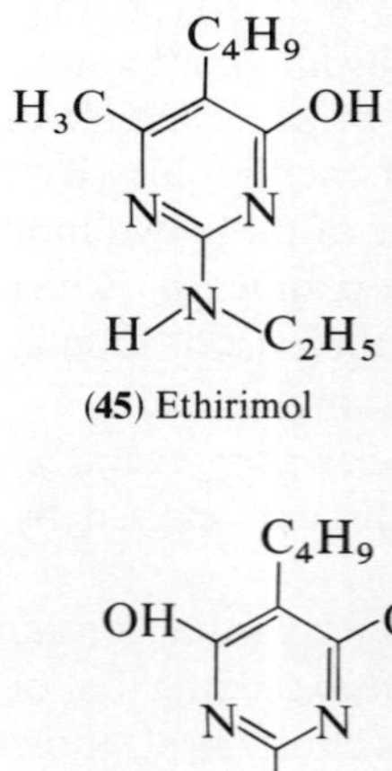

(**45**) Ethirimol

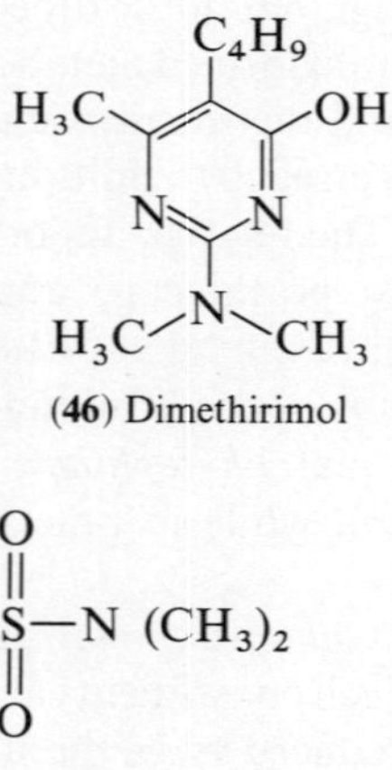

(**46**) Dimethirimol

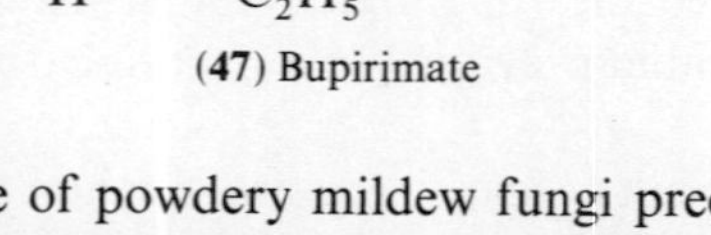

(**47**) Bupirimate

The biotrophic nature of powdery mildew fungi precludes mode of action studies on axenically-cultured mycelia, and consequently, the vast majority of work on ethirimol and dimethirimol has been carried out using artificially infected plant material. Considerable use has been made of 'reversal' experiments, which assess the ability of various biochemicals to antagonize systemic activity and fungitoxicity, so indicating likely sites of action.

In early reversal experiments (Bent, 1970; Slade *et al.*, 1972), riboflavin consistently proved to be the most active antagonist tested though its effect was attributed to photo-inactivation of the fungicides. Folic acid, pyridoxal-5-phosphate, adenine, and adenosine were also shown to partially alleviate fungitoxicity, collectively suggesting that the pyrimidine fungicides might act by interfering with tetrahydrofolate (THFA)-directed C-1 metabolism, and thereby purine biosynthesis (Sampson, 1969; Slade *et al.*, 1972). Dimethirimol, however, had no effect on purine biosynthesis in a cell-free system isolated from mildew-infected cucumber, nor did it significantly inhibit a number of enzymes involved in THFA/C-1 metabolism (Calderbank, 1971).

Contrary to the results of the reversal experiments cited above, Hollomon (1979a) found that riboflavin, folic acid and pyridoxal phosphate did not antagonize the activity of ethirimol. Other data suggested that *de novo* purine biosynthesis was unimportant during the primary stages of infection when ethirimol is thought to exert its effect, implying that interference with this process is probably not the mechanism of action of the fungicide. Ethirimol did, however, reduce the incorporation of [8-^{3}H] adenine into RNA, leading to the conclusion that the fungicide may affect adenine metabolism at a site subsequent to its synthesis.

A recent study (Hollomon, 1979b) has indicated that the site of inhibition is probably the enzyme adenosine deaminase, which catalyses the hydrolytic deamination of adenosine (**XVIII**) to inosine (**XIX**) (Figure 8); deoxyadenosine also serves as a substrate for the enzyme. The conversion of adenosine into inosine is one reaction in the salvage pathways of purine metabolism, whereby the reutilization of preformed bases, nucleosides, and nucleotides is mediated. Hollomon (1981) suggested that this enzyme is important during the early stages of infection by the ethirimol-sensitive fungus *Erysiphe graminis* f.sp. *hordei*, but is not involved in purine metabolism in two other pathogens of barley, *Rhynchosporium secalis* and *Puccinia hordei*; selectivity at this or a similar level would certainly appear necessary since differential uptake may not occur (Slade *et al.*, 1972). In general, specificity is considered to reflect the presence in susceptible powdery mildews of an unusual form of the enzyme which is absent from other fungi (Hollomon, 1979b).

Although overall the arguments put forward by Hollomon (1979a; b; 1981) appear sound there are several inconsistencies which are worthy of note. Firstly, in view of the specified mode of action, it is not immediately clear why adenosine should antagonize the fungitoxicity of ethirimol since the inhibition induced by the fungicide is of a non-competitive type. Similarly, it is difficult to rationalize the lack of effect of inosine in comparable experiments, though this could perhaps reflect a failure to penetrate tissues in effective concentrations. Finally, there is the absence of a correlation between fungal growth ED_{50} and enzyme K_I values for genotypes of *E. graminis* f.sp. *hordei* widely differing in their sensitivity to ethirimol. Hollomon (1979b) suggested that the

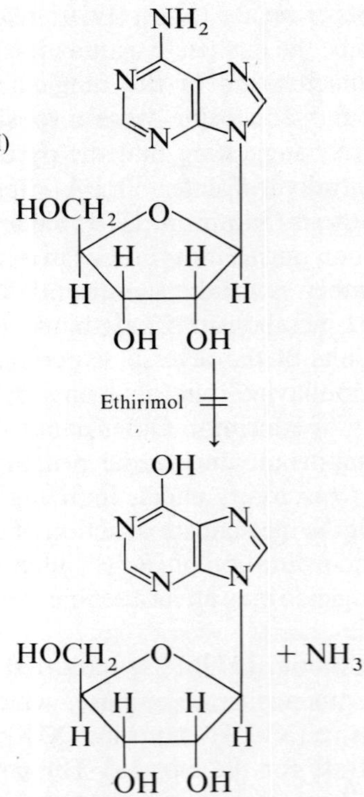

(XVIII)

(XIX)

Figure 8 The deamination of adenosine (**XVIII**) to give inosine (**XIX**). This reaction in *E. graminis* f.sp. *hordei* is inhibited by ethirimol

latter aspect could be rationalized if the activities of other enzymes involved in purine salvage became altered in resistant strains, such that the inhibitory effects of ethirimol were compensated for or circumvented.

In the most recent paper on the hydroxypyrimidine fungicides it was concluded that adenosine deaminase may be one site of action of these compounds (Hollomon and Chamberlain, 1981).

Acylalanines and related compounds

Metalaxyl and Furalaxyl The acylalanines, metalaxyl (**48**, 'Ridomil') and furalaxyl (**49**, 'Fongarid') were first described in 1977 (Schwinn *et al.*, 1977; Smith *et al.*, 1977; Urech *et al.*, 1977; Wiertsema and Wissink, 1977) and are systemic fungicides effective against Oomycetes of the order Peronosporales. Activity has been demonstrated against late blight of potato, downy mildews

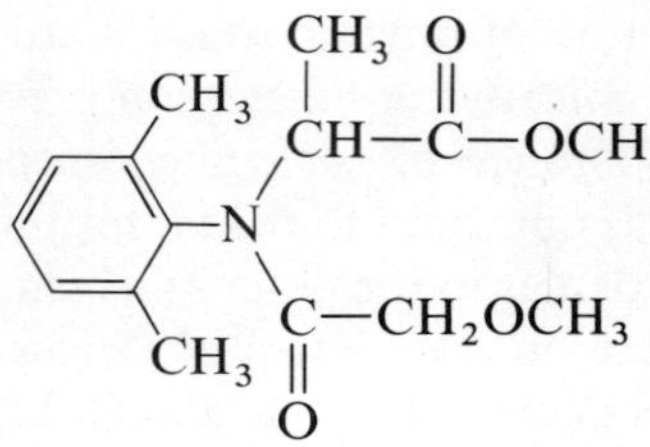

(**48**) Metalaxyl

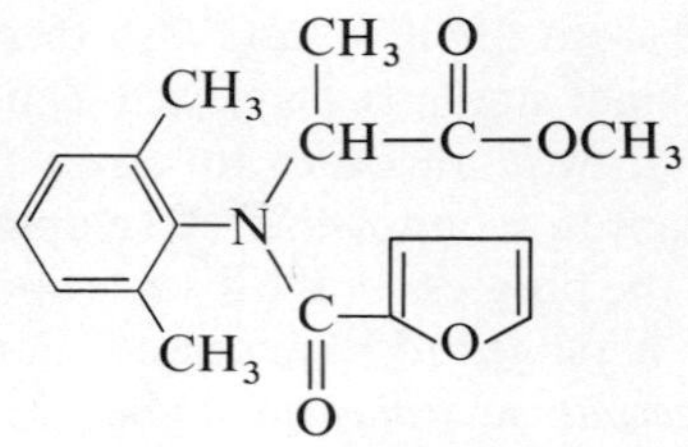

(**49**) Furalaxyl

and soil-borne diseases caused by species of *Pythium* and *Phytophthora* (refs 1–27 in Kerkenaar and Kaars Sijpesteijn, 1981). The fungitoxicity of metalaxyl, and presumably furalaxyl, is a property of the ester, since the free acid is not active (unpublished data of Pinkas and Edgington cited by Edgington, 1981).

The selectivity of both fungicides towards members of the Peronosporales has recently been confirmed in an extensive survey *in vitro* (Kerkenaar and Kaars Sijpesteijn, 1981), where the majority of representatives from this group proved to have ED_{50} values of 10 μg ml^{-1} or less; in general, metalaxyl was the more active of the two. *Pythium splendens* with an ED_{50} value for metalaxyl of 0.1 μg ml^{-1} (*c.* 0.36 μM) was employed in subsequent mode of action studies at the biochemical level (Kerkenaar, 1981a; b).

Of the biosynthetic processes investigated by Kerkenaar (1981a) metalaxyl exerted its most profound effects on the synthesis of nucleic acids. Inhibition of both DNA and RNA synthesis in *P. splendens* was apparent after exposure of mycelia to metalaxyl (0.1 μg ml^{-1}) for 0.5 h. Increases in fungal dry weight were also severely inhibited despite the extremely limited uptake of metalaxyl from bathing solutions into mycelia. The potency of the fungicide is further emphasized, when it is considered that biological activity is predominantly associated with one enantiomer of the racemic mixture (Moser and Vogel, 1978). Although the precise level at which metalaxyl interferes with nucleic acid synthesis was not established, the absence of abnormal nuclear staining patterns and the lack of any effect on the movement of nuclei into germ tubes would argue against a mode of action similar to that of carbendazim.

Inhibition of nucleic acid synthesis has also been observed in *Phytophthora megasperma* f.sp. *medicaginis* (Davidse, 1981). In this case effects on RNA were much more pronounced and consistent than those on DNA, suggesting that inhibition of RNA synthesis may be the primary site of action of the fungicide.

Although only direct effects of metalaxyl on fungi have been considered so far, it is conceivable that indirect components might also contribute to disease control. Localized host reactions may, for example, play a role in the restriction of *Phytophthora infestans* and *Plasmopara viticola* on tomato and grapevine, respectively (Staub *et al.*, 1980) and of *Bremia lactucae* on lettuce (Crute, 1978). The formation of encasements around haustoria (Hickey and Coffey, 1980) if derived from the host, and also the production of phytoalexins (Ward *et al.*, 1980) might similarly be viewed as possible contributory factors in preventing fungal growth. In order to assess the significance of such phenomena it is important to monitor their development with time, and to relate this information to the progress of the disease and in particular to the point at which fungal growth is judged to have ceased. For instance, in the control of *Phytophthora megasperma* var. *sojae* on soybean, it was concluded that metalaxyl functions principally via a direct action on the fungus, with necrosis and phytoalexin production being secondary effects (Stössel *et al.*, 1982).

Milfuram In common with metalaxyl and furalaxyl, the acylanilinobutyrolactone milfuram (**50**, ofurace; Ortho 20615; Chevron RE 20615) exhibits activity against Oomycetes (Staub and Hubele, 1980).

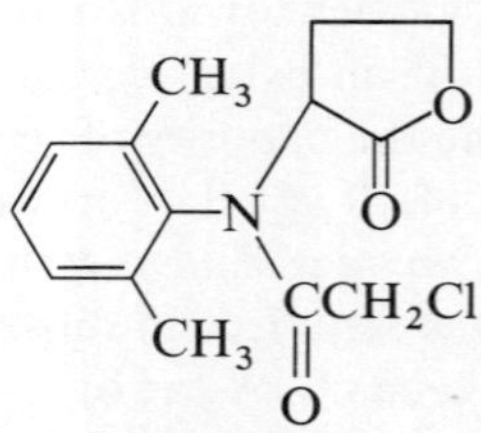

(**50**) Milfuram

Although milfuram is less potent than metalaxyl versus mycelial growth *in vitro* (Table 3) and also less inhibitory towards RNA synthesis (Table 4) (Fisher and Hayes, 1981), cross-resistance between the two compounds (Clerjeau and Simone, 1982) is suggestive of a common mode of action.

'*Galben*' The most recent addition to the acylalanine group of systemic fungicides is 'Galben' (**51**, proposed common name, benalaxyl; M 9834) (Bergamaschi *et al.*, 1981).

SAR studies have been reported (Gozzo *et al.*, 1981), as have investigations into uptake, translocation, and persistence (Garavaglia *et al.*, 1981). As yet, however, there is no information regarding the biochemical mode of action of

Table 3 ED_{50} values (μM) for furalaxyl, metalaxyl and 'Chevron RE 20615' against mycelial growth of *Pythium ultimum*, *Phytophthora nicotianae* and *Phytophthora palmivora*. (Fisher and Hayes, 1981)

	P. ultimum	*P. nicotianae*	*P. palmivora*
Furalaxyl	0.033	0.23	1.44
Metalaxyl	0.012	0.28	0.67
'Chevron RE 20615' (milfuram)	0.19	12.59	7.90

Table 4 Inhibition of RNA synthesis in *P. ultimum*, *P. nicotianae* and *P. palmivora* by furalaxyl, metalaxyl and 'Chevron RE 20615'. The table was compiled from the data of Fisher and Hayes (1981)

	P. ultimum	*P. nicotianae*	*P. palmivora*
Furalaxyl	42[a]	28	76
Metalaxyl	42	20	46
'Chevron RE 20615' (milfuram)	20	18	11

[a] The figures are given in terms of the percentage decrease in the incorporation of [^{14}C]uracil into RNA relative to controls.

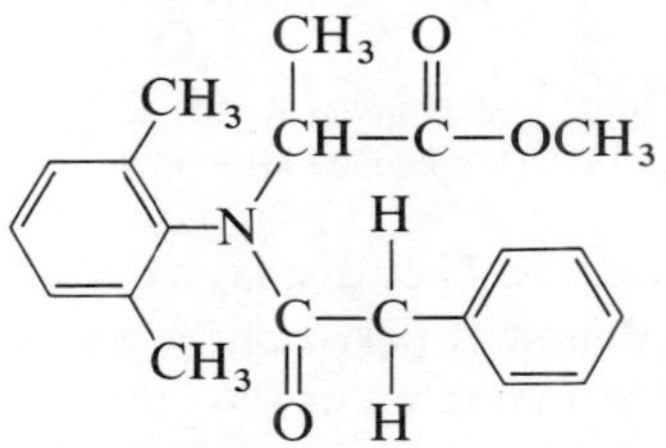

(**51**) 'Galben'

'Galben'. Notwithstanding this, cross-resistance data (Leroux and Gredt, 1982) are again consistent with a mode of action similar or identical to that of metalaxyl.

Benzimidazoles

The benzimidazoles collectively display a wide range of biocidal activities. They are used as anthelmintics in veterinary medicine (Baeder *et al.*, 1974; van den Bossche, 1972; Theodorides *et al.*, 1973; Actor *et al.*, 1967), antitumour agents (Atassi *et al.*, 1975; Atassi and Tagnon, 1975; De Brabander *et al.*, 1975) and most importantly as fungicides. The fungicidal properties were not recognized until three years after the first compound, thiabendazole (**52**) (TBZ) was introduced as an anthelmintic in 1961 (Brown *et al.*, 1961). Since then benomyl (**53**) (Delp and Klöpping, 1968), TBZ (Staron and Allard, 1964), fuberidazole (**54**) (Schumann, 1968), carbendazim (MBC) (**55**) (Hampel and Löcher, 1973) and the thiophanates (**56**, **57**) (Aelbers, 1971), have been introduced as fungicides and have proved highly effective in controlling a broad

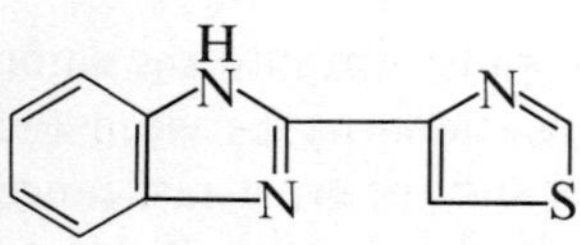

(**52**) Thiabendazole

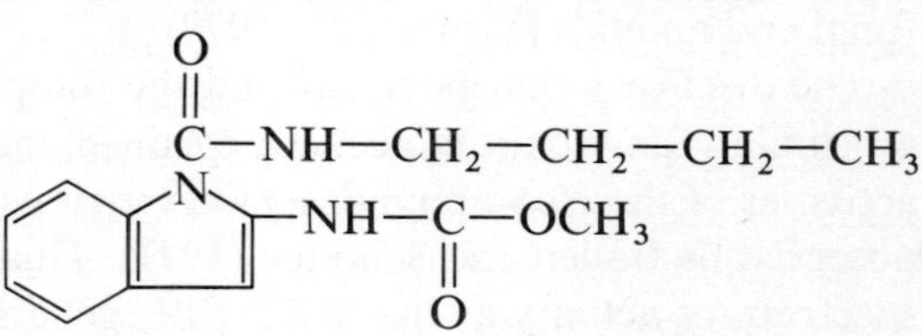

(**53**) Benomyl

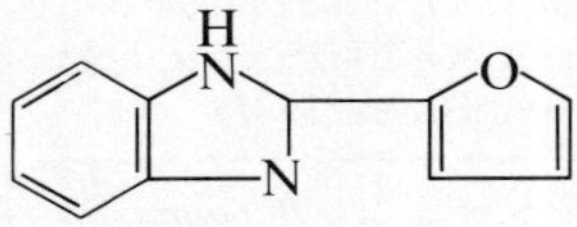

(**54**) Fuberidazole

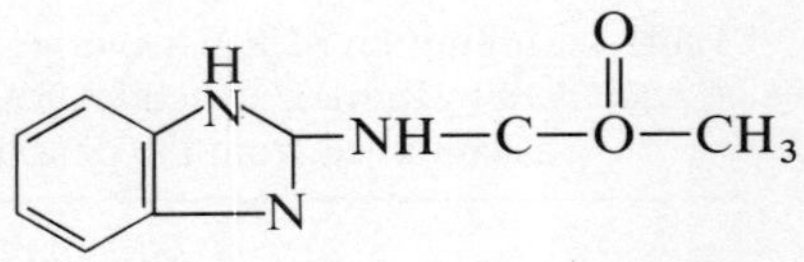

(**55**) MBC (Carbendazim)

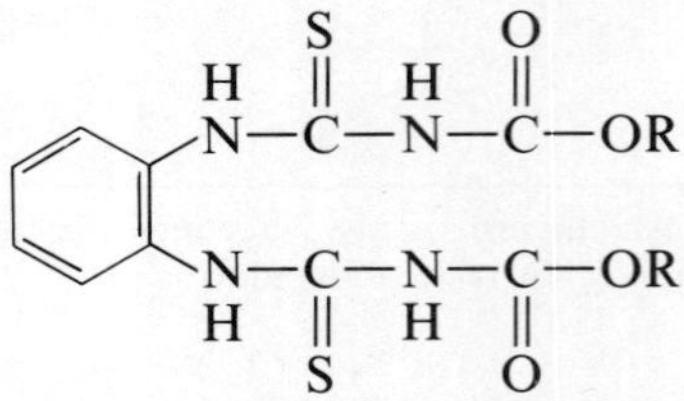

(**56**) Thiophanate methyl R = CH_3
(**57**) Thiophanate R = C_2H_5

range of important crop diseases. They display a considerable degree of systemic and therapeutic activity. Benomyl particularly was such an improvement on existing chemicals when first marketed in 1967 that it was used exclusively and intensively in many situations around the world. The intensity of use coupled with the specific mode of action of the benzimidazoles has unfortunately led to the development of severe resistance problems. There are now at least twenty-three fungal genera resistant to them (Delp, 1981).

Because of their remarkable properties and the problems associated with resistance the benzimidazoles have been extensively studied. Much is now known about their spectrum of activity (Aelbers, 1971; Bollen and Fuchs, 1970), symplastic movement in plants (Solel *et al.*, 1973), metabolism and mode of action.

Interestingly, the prime fungitoxic principles of benomyl and thiophanate-methyl have been ascribed not to the molecules themselves but to a common derivative MBC (**55**) (Clemons and Sisler, 1969; Selling *et al.*, 1970; Vonk and Kaars Sijpesteijn, 1971), itself, a powerful fungicide (Figure 9). Thiophanate is converted to the closely related ethyl benzimidazole-2-ylcarbamate (**58**) (EBC) (Vonk and Kaars Sijpesteijn, 1971). The conversions occur spontaneously in aqueous solution and proceed very rapidly in the case of benomyl (approximately 50% in 1 h). The thiophanates undergo conversion much more slowly in water but the rate can be greatly accelerated by metabolic activity in plants, fungi and animals (Vonk *et al.*, 1977).

The discovery that benomyl and the thiophanates transform into the same or similar fungitoxic molecule explained the striking similarity between the activities of these compounds against various diseases and the cross-resistance observed by Bollen and Scholten (1971). Thiabendazole also displays the same spectrum of activity as the MBC/EBC generators although it is generally less active (Edgington *et al.*, 1971). Moreover, strains of a fungal species resistant

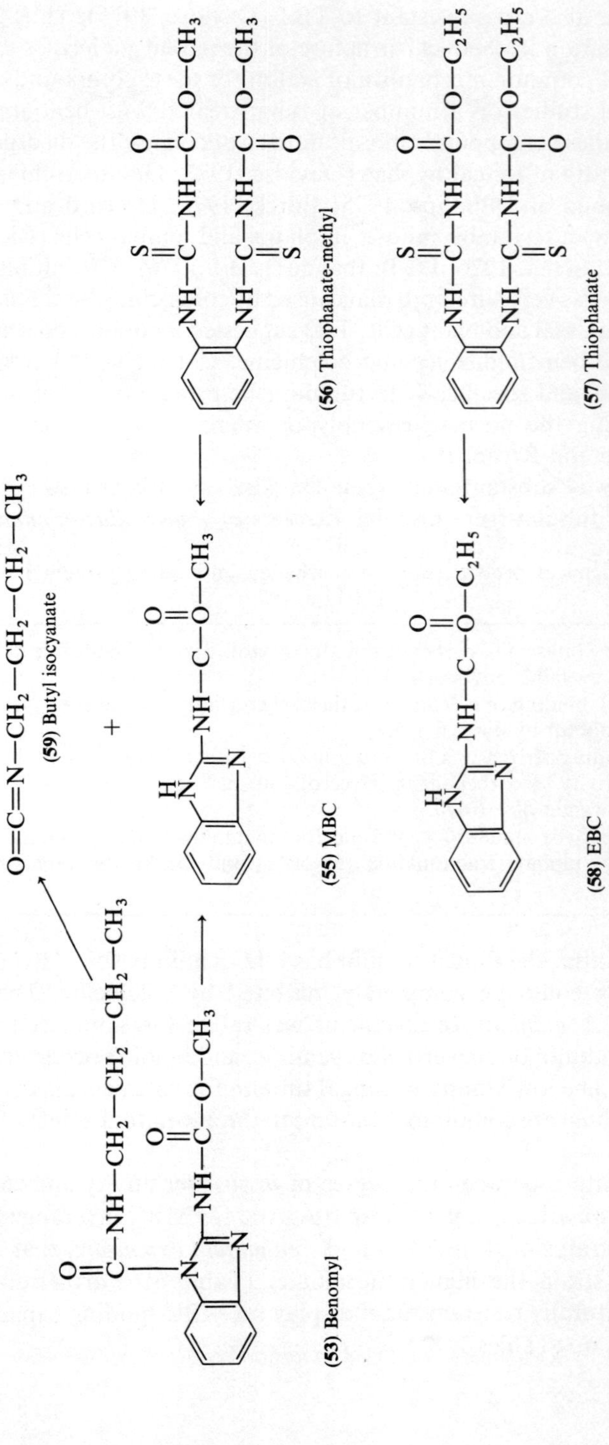

Figure 9 Conversion of benomyl and the thiophanates to MBC/EBC

to MBC were also cross-resistant to TBZ (Davidse, 1975). Thus despite considerable variation in chemical structure of the parent molecules it is apparent that there is a common mechanism of action for these compounds.

Cytological studies on a number of fungi treated with benomyl and MBC showed that these compounds inhibit mitosis and cause the disorganization of the fine structure of fungal hyphae (Davidse, 1973; Hammerschlag and Sisler, 1973; Richmond and Phillips, 1975; Borck, 1973; Howard and Aist, 1976). MBC was shown to inhibit mitosis in plants and animal cells (Richmond and Phillips, 1975; Seiler, 1976; De Brabander *et al.*, 1976). The inhibition in all of these systems was very similar to that caused by colchicine, the classical inhibitor of mitosis in animal and plant cells. This suggested a possible common mode of action for the benzimidazoles and colchicine. Colchicine is known to act by binding tightly and specifically to tubulin (the protein dimer of microtubules) thus preventing the normal assembly of microtubule subunits into spindle fibres (Wilson and Bryan, 1974).

This idea was substantiated when Davidse and Flach (1977) isolated and characterized tubulin from mycelial extracts of *Aspergillus nidulans* (Table 5).

Table 5 Biochemical properties of the carbendazim-binding protein in fungi. (From Davidse, 1979)

1. High affinity binding of carbendazim, oncodazole and thiabendazole at the same site. Binding is reversible and fast at 4 °C.
2. Low affinity binding of colchicine at the carbendazim-binding site.
3. Optimum pH for binding: 6.6–6.9.
4. Loss of binding activity in a first order manner with a $t^{1/2}$ of 6.5 h. Stabilization of the binding activity by carbendazim, glycerol and sucrose.
5. Molecular weight of 110 000.
6. Purification by a standard procedure for mammalian brain tubulin, consisting of ammonium sulphate fractionation (35–50 %) and DEAE-ion exchange chromatography.

They showed that the fungal tubulin has a high affinity for MBC and that the MBC binding could be completely inhibited by colchicine. The affinity of tubulin from *A. nidulans* for colchicine was rather low compared with that of mammalian tubulin but nevertheless genuine, and binding occurred at the same site as MBC. The low affinity of fungal tubulin for colchicine is consistent with the fact that fungi are commonly resistant to the antimitotic effect of this spindle poison.

This correlation between the degree of *in vivo* sensitivity and the affinity of tubulin for a spindle poison was also observed for MBC in a range of wild-type and mutant strains of *A. nidulans* and *Penicillium expansum*, that is, the more sensitive the strain the higher the affinity (Table 6). Furthermore, cell-free extracts of naturally resistant fungi display no MBC-binding capacity whereas sensitive ones do (Tables 7, 8).

Table 6 Carbendazim-binding activity in extracts of fungal, plant and animal cells. (From Davidse, 1979)

System	Binding activity dpm [^{14}C]carbendazim/ml extract
Aspergillus nidulans 003 (wild-type)	5640
Aspergillus nidulans 186 (extra sensitive)	12200
Aspergillus nidulans R (resistant)	320
Penicillium brevicompactum S	4140
Penicillium brevivompactum R	220
Penicillium corymbiferum S	2080
Penicillium corymbiferum R	140
Alternaria sp	370
Pythium irregulare	250
pea root tips	250
porcine brain	210

Table 7 Sensitivity of carbendazim and the dissociation constant of the carbendazim–tubulin complex in strains of *Aspergillus nidulans*. (From Davidse, 1979)

Strain	ED_{50} (radial growth on agar) (μM)	Dissociation constant (μM)
003	4.5	2.2
186	1.5	0.6
R	95	27

Table 8 Sensitivity of carbendazim and the dissociation constants of carbendazim–tubulin complex in strains of *Penicillium expansum*. (From Davidse, 1979)

Strain	ED_{50} (radial growth on agar) (μM)	Dissociation constant (μM)
S	0.4	0.8
SS	0.07	0.2
R	2500	—

The results strongly suggest that the selectivity of MBC within fungi is based on a differential affinity of fungal tubulin for this toxicant. Acquired resistance brought about by mutation, for example as demonstrated in *A. nidulans* and *P. expansum* that involve changes in the affinity of tubulin for MBC, do not affect the growth rate or sporulation of the strains, indicating normal functioning of the microtubular system (Davidse and Flach, 1977). This is a contributory factor in the development of resistance problems in practice.

Low affinities of tubulin from plant and animal cells may also account for the low toxicity of MBC for these organisms (Table 6) although an active detoxification mechanism may be a contributory factor in mammals (Gardiner *et al.*, 1974).

Davidse and Flach (1978) also showed that TBZ bound to fungal tubulin at the MBC-binding site although the TBZ-protein complex had a higher dissociation constant than that of MBC. On comparison of the dissociation constants with ED_{50} values (Table 9) it was concluded that the biochemical basis of the antifungal activities of TBZ is similar to that of MBC (Davidse, 1979), as predicted after the isolation of cross-resistant fungal strains (see above).

It is clear that benzimidazole fungicides share a capacity to act as powerful inhibitors of the microtubule assembly. This is an effective target site for a toxicant in organisms that possess tubulin having a high affinity for benzimidazoles and accounts for the potency of these fungicides. Microtubules play an essential role in the maintenance of cell shape and structure, cellular and nuclear division and organelle movements, etc.

Earlier reports on the mode of action of the benzimidazoles had regarded inhibition of respiration as being an important target site. Work by Hammerschlag and Sisler (1973) established that in the case of benomyl this effect was attributable to butyl isocyanate (**59**) (BIC) (Figure 9). BIC is formed during the conversion of benomyl to MBC. It is highly reactive and undoubtedly inhibits more processes in cells than does MBC. BIC gives rise to a very rapid inhibition of respiration and as a consequence DNA, RNA synthesis and growth etc., but MBC ultimately causes greater cell death by microtubule disarrangement. TBZ on the other hand, appears to be able to exert a direct effect on the terminal electron transport pathway between NADH and cytochrome c (Hammerschlag and Sisler, 1973; Allen and Gottlieb, 1970) albeit at relatively high concentrations. Nevertheless, it is likely that these secondary effects contribute to the overall field performance of these fungicides.

The biochemical activities of benzimidazole fungicides that have been investigated are summarized in Figure 10. Fuberidazole has received little or

Table 9 Sensitivity to carbendazim and thiabendazole and the dissociation constants of the thiabendazole- and carbendazim-tubulin complex in a wild-type strain of *Penicillium expansion*. (From Davidse, 1979)

Compound	ED_{50} (radial growth on agar) (μM)	Dissociation constant[a] (μM)
Carbendazim	0.4	0.8
Thiabendazole	7	10 ± 5[a]

[a] The dissociation constant of the thiabendazole–tubulin complex was measured as the inhibition constant for the ability of thiabendazole to inhibit carbendazim binding. Binding was determined with a charcoal binding assay (Davidse and Flach, 1977).

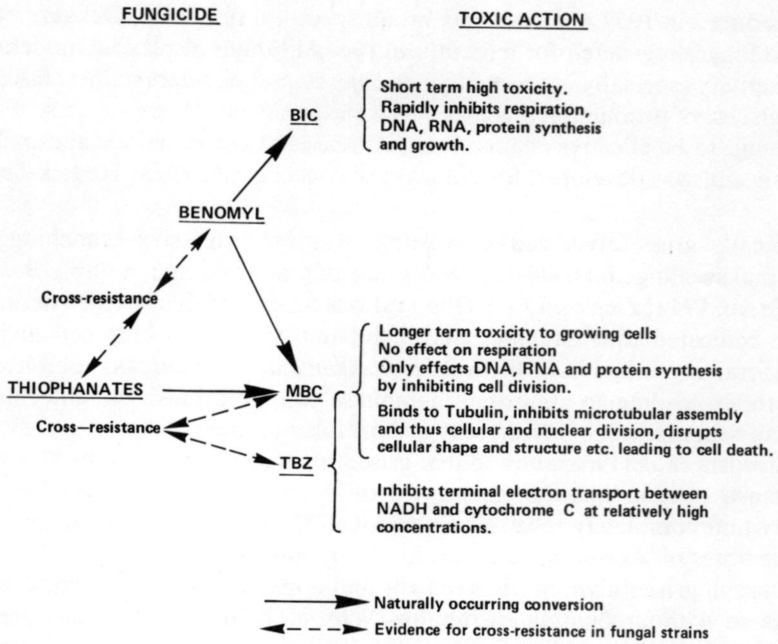

Figure 10 Biochemical activities and interelationships of the benzimidazole fungicides

no attention probably because of its much more limited use as a seed dressing for the control of *Fusarium* species. However, strains of fungi cross-resistant to benomyl and TBZ are also resistant to fuberidazole (Bollen 1971). This and the structural similarities between TBZ and fuberidazole suggests that fuberidazole has the same mechanism of action as TBZ.

Griseofulvin

Griseofulvin (**60**), an antibiotic, was first isolated from cultures of *Penicillium griseofulvum* in 1939 (Oxford *et al.*, 1939) and latterly from a variety of *Penicillium* species (Grove *et al.*, 1951). It is non-toxic to plants, animals, Oomycete fungi and certain yeasts and bacteria.

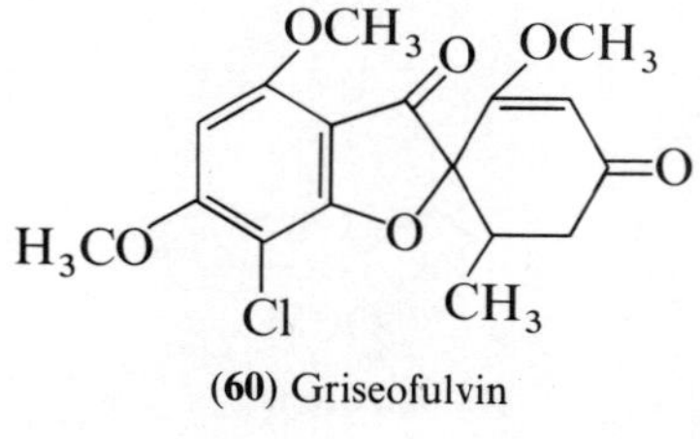

(**60**) Griseofulvin

Introduced in 1957 as a systemic, broad spectrum fungicide (Dekker, 1963), it is no longer registered for agricultural use. Although displaying moderately good activity especially in the control of *Botrytis* and powdery mildew diseases, the high cost of production precluded its widespread use. However, griseofulvin was found to be effective against fungal diseases of the skin, hair and nails of humans and was developed for clinical use (Allen *et al.*, 1973; Hildick-Smith, 1971).

Typically griseofulvin causes twisting, stunting, excessive branching and abnormal swelling and distortion of hyphae of sensitive fungi, notably *Botrytis allii* (Brian, 1949; Ziegler, 1963). Observations on griseofulvin-treated dermatophytes indicated that daughter nuclei did not separate during cell division (Thyagarajan *et al.*, 1963). Mitosis in *Basidiobolus ranarum* was inhibited by griseofulvin leading to abnormal metaphase configurations. Daughter nuclei failed to separate fully moving only a short distance apart.

Malawista *et al.* (1968) showed that griseofulvin at 10^{-5} M rapidly destroyed preformed mitotic spindles in oocytes of *Pectinaria gouldi*. The effect was rapidly and completely reversed on removal of the drug. A similar effect on spindle fibres of *Aspergillus nidulans* has been noted (Crackower, 1972).

Although griseofulvin inhibits mitosis and appears to damage spindle fibres it does so without binding to tubulin (Wilson, 1970) and does not prevent colchicine from doing so. Griseofulvin was found to be inactive in preventing microtubule polymerization *in vitro* (Wilson and Bryan, 1974). An MBC-resistant mutant of *A. nidulans* is as sensitive to griseofulvin as the wild strain. Thus it has been concluded (Kappas and Georgopoulos, 1974) that although MBC and griseofulvin interfere with chromosome segregation, cellular shape and structure, the sites of action on the mitotic apparatus are probably different. One possibility is that griseofulvin affects a process vital to the sliding of microtubules necessary for the separation of chromosomes. Alternatively, griseofulvin may not directly affect microtubule function at all but may act on other processes essential for cell division (Wilson and Bryan, 1974).

Chloroneb

Introduced in 1967, chloroneb (**61**) exhibits a degree of systemic activity. When applied to the soil it accumulates in the roots and lower parts of the plant

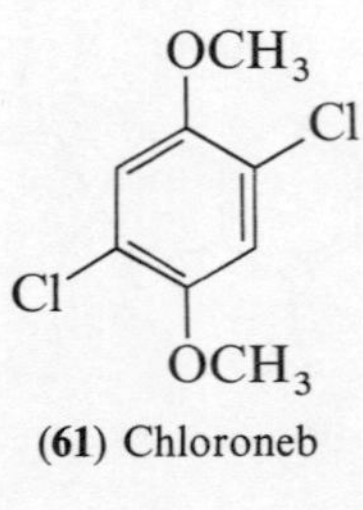

(**61**) Chloroneb

(Rhodes *et al.*, 1971). It gives good protection against a limited spectrum of soil-borne seedling diseases such as those caused by *Rhizoctonia*, *Pythium* and *Sclerotinium* species.

Chloroneb may be demethylated to the non-fungitoxic 2,5-dichloro-4-methoxyphenol (DCMP) in plants (Rhodes *et al.*, 1971) and in some fungi but not others (Hock and Sisler, 1969). Further metabolism was observed in bean plants where DCMP is glycosylated to form the β-D-glucoside (Thorn, 1973). Trace amounts of 2,5-dichloroquinone and 2,5-dichlorohydroquinone have also been detected (Rhodes *et al.*, 1971).

Chloroneb appears to inhibit cell division. Studies with *Ustilago maydis* showed that chloroneb inhibited the multiplication of spordia of this fungus (Tillman and Sisler, 1973). Furthermore, Georgopoulos *et al.* (1976) reported that chloroneb increases somatic segregation in diploid strains of *Aspergillus nidulans* indicating that this compound interferes with mitosis. Earlier reports (Hock and Sisler, 1969) suggested that chloroneb may inhibit DNA synthesis either directly or indirectly. This effect is now considered to have been a consequence of the prevention of cell division rather than the primary site of action. Many chloroneb-resistant fungal strains are also positively cross-resistant to a range of aromatic hydrocarbon fungicides, such as dichloran and biphenyl, and to the cyclic imides; vinclozolin, procymidone and iprodione (Leroux, 1981).

Pyroxychlor

Much of the interest in pyroxychlor (**62**) has centred on its systemic basipetal movement to control diseases caused by soil-borne Phycomycetes (Small *et al.*, 1977). Basipetal movement is a highly desirable feature, but one rarely found

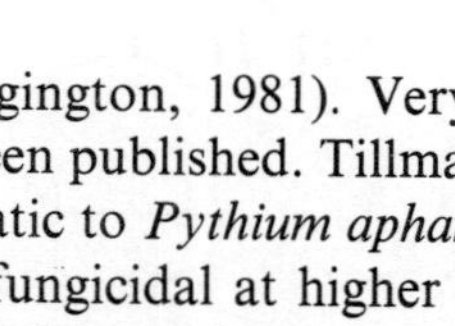

(**62**) Pyroxychlor

in pesticides in general (Edgington, 1981). Very little work on the mode of action of pyroxychlor has been published. Tillman and Ferguson (1980) found the compound to be fungistatic to *Pythium aphanidermatum* at low concentrations (below 50 μg/ml) but fungicidal at higher levels. Effects were obtained, in order of increasing sensitivity, against respiration, RNA synthesis, protein synthesis and DNA synthesis. Since the earliest and most pronounced effects were against the incorporation of ^{14}C thymidine into DNA, they concluded that DNA synthesis is the primary site of action.

Interference with chitin synthesis

Polyoxins

The mode of action of the polyoxins has been reviewed extensively (Misato and Kakiki, 1977; Woodcock, 1977; Kaars Sijpesteijn, 1977; Gooday, 1979a) and only a brief summary of the information is presented here.

The polyoxins are powerful nucleotide antibiotics active against fungi that contain chitin in their cell walls. They are ineffective against bacteria, Oomycetes, green plants and mammals. The polyoxins, isolated from culture fluids of *Streptomyces cacaoi* var *asensis* (Suzuki *et al.*, 1965; Isono *et al.*, 1967; Uramoto *et al.*, 1981), comprise a series of 14 compounds (polyoxin A to polyoxin N) (**63**, **64**) of similar physical and chemical characteristics but which show a marked specificity in their antifungal activity. Components C and I are of little use as fungicides whereas A is used to control brown spot of rice. Polyoxins B and G are useful against pear black spot fungus, apple cork spot and several *Botrytis* spp. while polyoxin D is most effective for the control of *Pellicularia sasakii* of rice.

Typically, polyoxins do not affect spore germination but they do induce distortion of germ tubes, bulbous hyphae susceptible to osmotic shock and inhibition of growth (Misato and Kakiki, 1977; Gooday, 1979a). Investigations with a variety of fungi have determined that chitin biosynthesis is the primary site of action of the polyoxins (Table 10). Chitin, a component of the primary wall of all filamentous fungi except the Oomycetes, is a homopolymer on *N*-acetylglucosamine units linked by β(1-4) glycosidic bonds (Figure 11). The polymer is formed from *N*-acetylglucosamine that is phosphorylated by *N*-acetylglucosamine kinase to *N*-acetylglucosamine-6-phosphate which in turn is

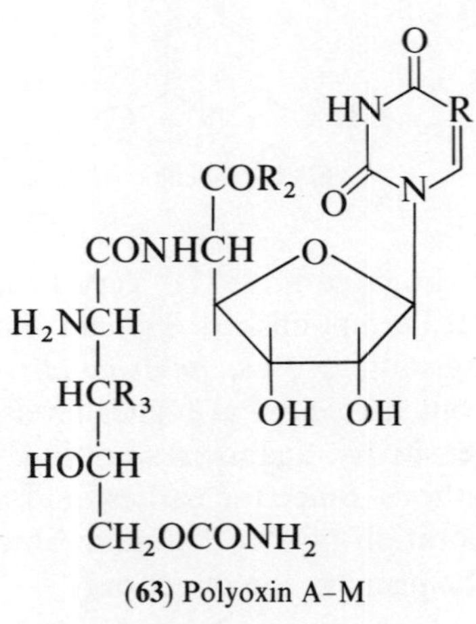

(**63**) Polyoxin A–M

	R_1	R_2	R_3
A	CH_2OH	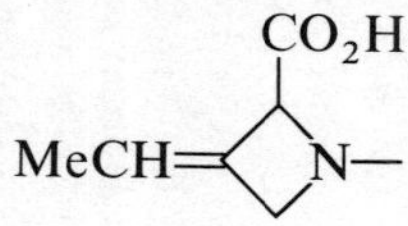	OH
B	CH_2OH	OH	OH
D	COOH	OH	OH
E	COOH	OH	H
F	COOH	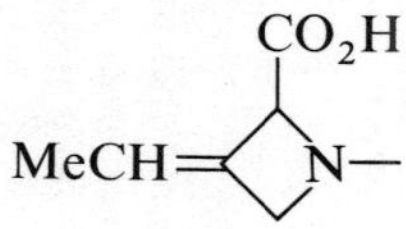	OH
G	CH_2OH	OH	H
H	Me	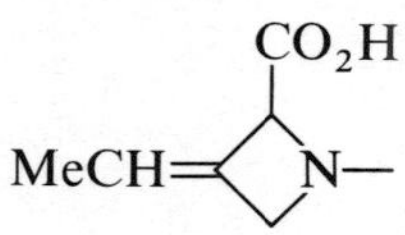	OH
K	H	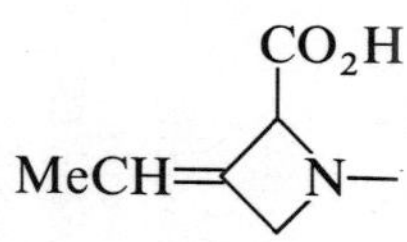	OH
L	H	OH	OH
M	H	OH	H

R substituents Polyoxin A-M

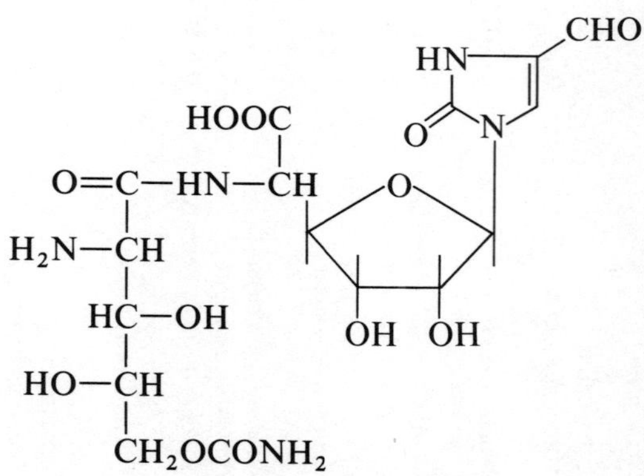

(**64**) Polyoxin N

Table 10 The action of polyoxin on fungi and insects (from Gooday, 1979a)

Organism	*K*m for UDP–GlcNAc (mM)[a]	K_i (μm)	Agent	Inhibitory conc (μm)	Effect	Reference
Chytridiomycetes						
Blastocladiella emersonii	1.8–4.1	—	Polyoxin B	50	—	Matsumae and Cantino, 1971
Zygomycetes						
Mucor hiemalis	0.35	—	Nikkomycin	10	Inhib. of growth and zygospores	Dähn *et al.*, 1976
Mucor rouxii	0.5–1.5	0.6	Polyoxin D	19	Apices burst	Bartnicki-Garcia and Lippman, 1972
Ascomycetes						
Alternaria kikuchiana	2.7–4.3	0.9–1.9	Polyoxin B	25	Bulbous hyphae	Hori *et al.*, and Ishizaki, *et al.*, 1974
Aspergillus flavus	20	6.5	Polyoxin D	—	—	Lopez-Romero and Ruiz-Herrera, 1976
Neurospora crassa	1–2	1.4	Polyoxin D	190	Germ tubes distorted	Endo *et al.*, 1970
Piricularia oryzae	3.3	3.4–33	Polyoxins A–M	5–40	Apices swell	Hori, *et al.*, 1971
Saccharomyces cerevisiae	0.6–0.9	0.5	Polyoxin D	2000	Bursting at buds	Keller and Cabib, 1971 and Bowers *et al.*, 1974
Trichoderma viride	—	—	Polyoxin D	100	Germination inhibited, hyphae bulge	Benitez, *et al.*, 1976
Basidiomycetes						
Agaricus bisporus	—	—	Polyoxin D	0.1	Growth inhib. autolysis	Wood and Hammond, 1977
Coprinus cinerus	0.9	3	Polyoxin D	0.5	No stipe elongation, autolysis	Gooday, *et al.*, 1976
Schizophyllum commune	—	—	Polyoxin D	10	No protoplast reversion	de Vries and Wessels, 1975
Insecta						
Leucophaea madera	—	—	Polyoxin D	3.5	Inhib. of regeneration of leg tissue	Sowa and Marks, 1975

[a] Measured for particulate preparations by Lineweaver–Burk plot, except for *C. cinereus* which is $(S)_{0\ 5}$ for solubilized enzyme. Measured under a variety of different conditions; not strictly comparable. Not including polyoxin C and I, which are essentially inactive.

converted into the immediate precursor uridine diphosphate *N*-acetylglucosamine (UDP–GLcNAc). Chitin synthase (EC 2.4.1.16) binds UDP–GLcNAc units into the polymer.

$$(\text{GlcNAc})_n + \text{UDP–GlcNAc} \xrightarrow{Mg^{2+}} (\text{GlcNAc})_{n+1} + \text{UDP}$$

(For a review of chitin in fungi see Gooday, 1979b; Stirling *et al.*, 1979; Bartnicki-Garcia *et al.*, 1979.)

The polyoxins have been found to be powerful inhibitors of chitin synthase from a variety of fungal sources (Table 10) indicating that the enzyme is the target site for the antibiotics. The kinetics of inhibition show that polyoxin competes with UDP–GLcNAc for an active site on the enzyme molecule (Gooday *et al.*, 1976). These results are consistent with the earlier observations

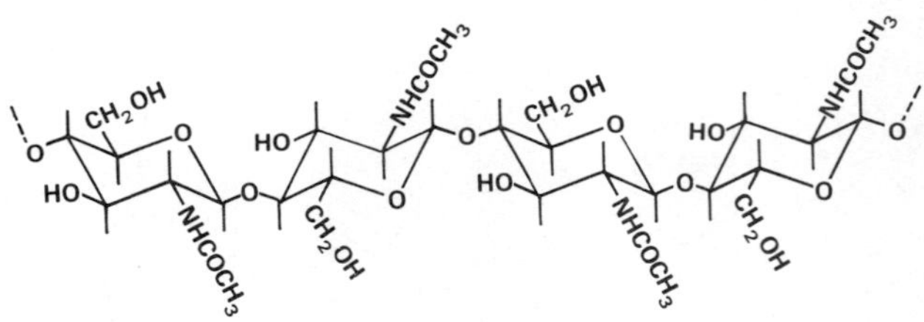

Figure 11 The covalent structure of chitin

that polyoxin D inhibits the incorporation of [^{14}C]glucosamine into chitin in a number of fungi and causes a concomitant accumulation of label in UDP–GlcNAc (Sasaki *et al.*, 1952; Endo *et al.*, 1970; Hori *et al.*, 1971). Detailed studies were made by Hori *et al.*, (1974b; c) on the interaction of a range of polyoxins and analogues on chitin synthase from *Pyricularia oryzae*. Their results suggest that the enzyme may have a specific binding site for the uridine moiety of UDP–GlcNAc, and for the pyrimidine nucleoside moiety of polyoxins (Figure 12).

Fungi vary widely in their sensitivities to polyoxins. This has been attributed not to intrinsic differences in the susceptibility of chitin synthase from different fungi but to the extents to which polyoxins can penetrate the cells and reach the enzyme. Evidence for this comes from two studies.

Firstly, Hori *et al.* (1974a) isolated chitin synthase from a number of polyoxin-resistant and sensitive strains of *Alternaria kikuchiana*. They showed that the enzymes from all the strains had similar K_i values for polyoxin B. Secondly, Mitani and Inoue (1968) found that simple peptides, such as glycyl-L-alanine, strongly antagonized the action of polyoxin *in vivo* but had no effect on its inhibitory activity on chitin synthetase *in vitro* (Hori *et al.*, 1974b).

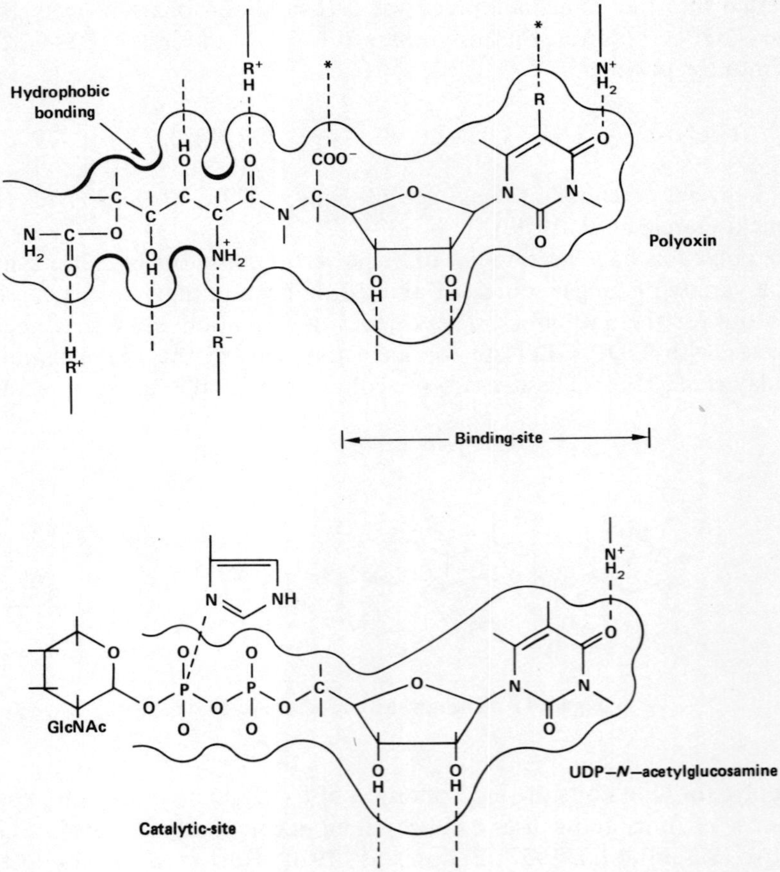

Figure 12 Proposed mechanism of interaction between Polyoxin or UDP-*N*-acetylglucosamine and the active centre of chitin synthetase. Reprinted with permission from Hori *et al.* (1974c). *The values of partial binding affinity for the —COOH and the —R(=CH_3) groups are 0.7 and 0.4 kcal/mole, respectively. (From Hori *et al.*, 1974b)

Interference with protein synthesis

It is notable that all of the commercially available fungicides known to inhibit protein synthesis are of natural origin or close synthetic analogues of naturally occurring compounds. The first ones, streptomycin (**65**) and cycloheximide (**66**), introduced as long ago as the late 1940s, represented an exciting new property in plant disease control agents at that time, namely systemicity, a property generally shared by other antibiotics.

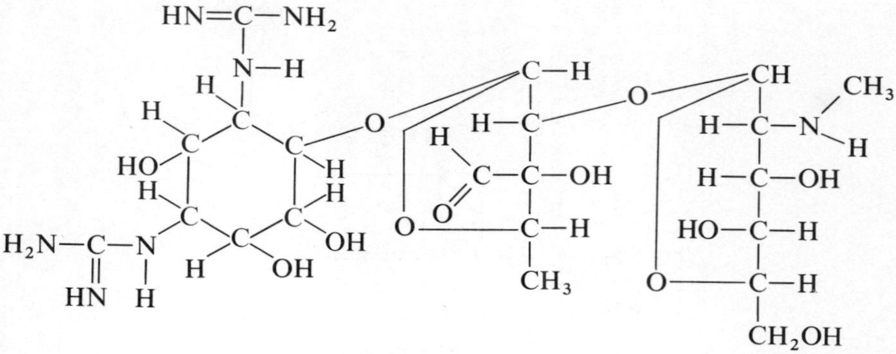

(**65**) Streptomycin

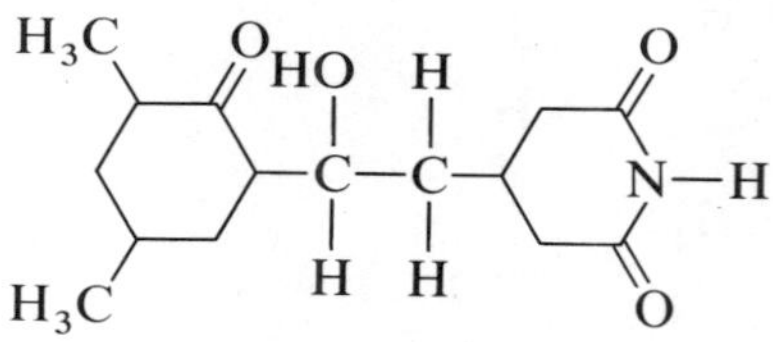

(**66**) Cycloheximide

Although many antibiotics active as inhibitors of protein synthesis have found extensive use in clinical medicine, comparatively few have proved effective as agricultural fungicides and with varying degrees of success.

Cycloheximide, although remarkably active against a wide range of fungi and yeasts, is inactive against bacteria. Its use against plant disease has been limited by phytotoxicity. Its main use has been for the control of many turf diseases, cherry leaf spot caused by *Coccomyces hiemalis* and powdery mildews of ornamentals.

Streptomycin shows activity against a large number of bacteria but is inactive on most fungi, protozoa, viruses and anaerobic bacteria. However, it has proved useful for the control of hop downy mildew and a broad range of bacterial plant pathogens.

Blasticidin S (**67**), developed in Japan, although active against both bacteria and fungi, proved particularly potent to *Pyricularia oryzae* and has been used extensively to control Blast disease of rice (Dekker, 1971). Phytotoxicity problems were reduced by applying blasticidin S as the monobenzylamine benzene sulphonate salt (Asakawa *et al.*, 1963).

Kasugamycin (**68**) rapidly superseded blasticidin for the control of Blast disease, when introduced in 1965, because of the wide margin between curvative and phytotoxic dosage. This systemic fungicide generally has weak bacteriostatic activity and has no effect on eukaryotic organisms except *P. oryzae*. Furthermore, its toxicity is only expressed in the rice plant and in extracts (Ishiyama *et al.*, 1965; Hamada *et al.*, 1965). The antifungal activity is only

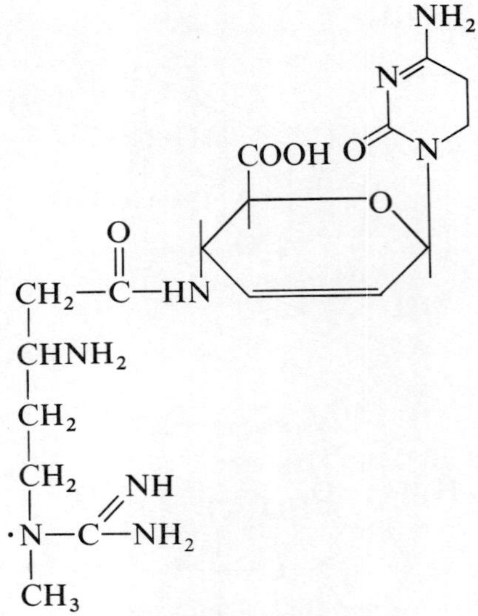

(**67**) Blasticidin S.

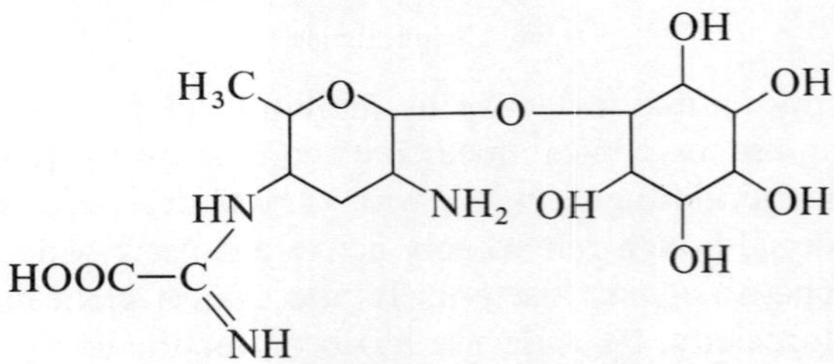

(**68**) Kasugamycin

apparent in plant juice at pH 6 and below. However, inhibition of protein synthesis does occur in a cell-free system from *P. oryzae* mycelia grown in the absence of plant extract (Masukawa *et al.*, 1968). It has been suggested (Siegel, 1977) that the rice extracts may effect the uptake and/or accumulation of kasugamycin by (a) causing changes in cell wall permeability, (b) affecting potential cellular detoxifying processes or (c) by modifying the structure of the toxicant.

The mode of action of antifungal protein synthesis inhibitors has been comprehensively reviewed by Siegel (1977). This account will only be concerned with the aforementioned antibiotics because they have been commercialized as fungicides. A brief summary of the processes of protein synthesis is included.

The biochemistry of protein synthesis

Protein synthesis in all organisms occurs in organelles known as ribosomes that comprise protein and RNA (ribosomal RNA, rRNA). The ribosomes are of high molecular weight, typically, 2.5 million for prokaryotes and of various sizes close to 5 million for eukaryotes, denoted 70-S and 80-S respectively by virtue of their sedimentation constants. Eukaryotic ribosomes are made up of 40-S and 60-S subunits whereas those from prokaryotes consist of 30-S and 50-S subunits.

Protein synthesis in fungi occurs in two parts of the cell; the cytoplasm and the mitochondria. Cytoplasmic protein synthesis occurs on 80-S ribosomes, typical of eukaryotes, but polypeptide assembly in mitochondria is associated with 55 to 80-S ribosomes and resembles protein synthesis in prokaryotic cells. In general though, the mechanisms of protein synthesis in prokaryotic and eukaryotes appear to be similar.

The ribosomes contain the machinery for assembling proteins (that is, for making the peptide bonds, etc. but not the instructions for determining the sequence of amino acids. This information is coded for in the base sequence of DNA and is communicated to the ribosomes by messenger RNA (mRNA). The sequence of ribonucleotides in mRNA is specified during the process of transcription by the base sequence of the DNA and thus mRNA is complementary to one of the strands of DNA. When mRNA binds to a ribosome it instructs it to synthesize the polypeptide corresponding to the transcribed gene(s), that is, the mRNA forms a template for setting up in exact sequence the amino acids of the protein being synthesized.

Amino acids for incorporation into the protein being synthesized are transferred to the ribosomal site by transfer RNA (tRNA). There is at least one specific tRNA for each of the twenty protein amino acids. Each tRNA carries a recognition sequence of three nucleotides (an anticodon). These nucleotides are complementary to the triplet codon on the mRNA ensuring that the amino acids are placed on the ribosome in the correct sequence.

The mechanism of protein synthesis is very complex and involves four distinct steps. Aminoacylation of tRNA, initiation, elongation and termination of polypeptide chains. The process in eukaryotes is summarized in Figure 13 and is outlined below.

The aminoacylation reaction In this reaction the amino acids become attached to tRNAs. Each specific amino acid/tRNA conjugation is catalysed by a specific aminoacyl synthase which esterifies the amino acids through their carboxyl groups to the 2′- or 3′-hydroxyl groups of the ribose moieties of the terminal adenylic acid residue of the tRNAs. The reaction takes place in two steps:

$$\text{ENZ} + \text{AA} + \text{ATP} \longrightarrow \text{ENZ.AA-AMP} + \text{PPi}$$

$$\text{ENZ.AA-AMP} + \text{tRNA} \longrightarrow \text{AA-tRNA} + \text{ENZ} + \text{AMP}$$

The ester linkage is a high energy bond whose energy is subsequently used during peptide bond formation on the ribosomes.

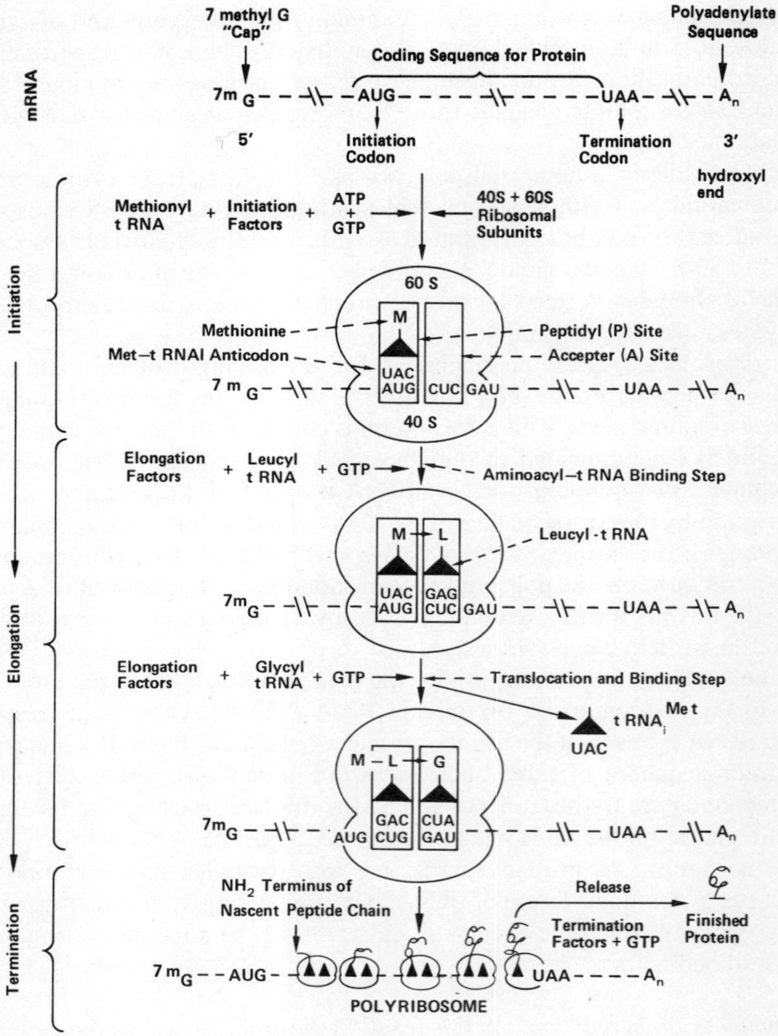

Figure 13 Schematic diagram of the major components and reactions involved in protein biosynthesis in eukaryotic organisms. (After Weeks, 1981)

Peptide chain initiation This process comprises four steps: (i) the formation of functional 40-S ribosomal subunits following the dissociation of 80-S ribosomes. A 40-S subunit is made functional by the addition of the soluble initiation factor eIF3 (eukaryotic initiation factor).

$$40S + eIF3 + mRNA \longrightarrow 40S\text{-}eIF3\text{-}mRNA$$

(ii) The binding of the initiator species of methionyl-tRNA (met-$tRNA_i$) to the 40-S subunit. In eukaryotes the initiator tRNA is methionyl-tRNA which recognizes specific codons of mRNA. Although there are two species of met-tRNA in the cytoplasm of eukaryotes, it is only met-$tRNA_f$ which serves as the initiator. The binding of met-$tRNA_f$ to the 40-S initiator site occurs in the presence of the initiation factors eIF1 and 2.

$$\text{40S-eIF3-mRNA} + (\text{eIF1,2}) \longrightarrow \text{eIF1,2-40S-eIF3-mRNA}$$

(iii) The binding and positioning of mRNA to form a 40-S ribosome pre-initiation complex;

$$\text{eIF1,2-40S-eIF3-mRNA} + \text{met-tRNA}_f \longrightarrow \text{40S-eIF3-mRNA-eIF1,2-met-tRNA}_f$$

and finally (iv) the coupling of a 60-S ribosomal subunit to this complex to form a complete 80-S ribosomal initiation complex ready for peptide bond formation. That is, methionine bound to met-$tRNA_i$ at the peptidyl site of the 80-S complex is in position to react with the amino acid moiety of an amino-acyl-tRNA entering the acceptor site of the ribosome.

Peptide chain elongation With initiation complete and the ribosomes complex primed, elongation can take place. This three-step process involves the 80-S ribosome, aminoacyl-tRNA, mRNA, peptidyltransferase and two elongation factors; EF1 and EF2. Firstly, aminoacyl-tRNA binds to the ribosomal acceptor (A) site as a complex of guanosine triphosphate (GTP) and EF1.

$$\text{AA-tRNA} + \text{GTP} + \text{EF1} \longrightarrow \text{AA-tRNA-GTP-EF1}$$

Binding of the complex to the triplet codons of the mRNA at the A site is governed by the codon–anticodon interaction (. . .). The peptidyl-(P) site of the mRNA is occupied with peptidyl-tRNA (pept-tRNA) and its codon (+ + +).

$$\text{AA-tRNA-GTP-EF1} + \frac{\overset{\text{pept-tRNA}}{+++} \qquad \cdots}{\text{P} \qquad\qquad \text{A}} \longrightarrow \frac{\overset{\text{pept-tRNA}}{+++} \qquad \overset{\text{AA-tRNA}}{\cdots}}{\text{P} \qquad\qquad \text{A}} + \text{GDP} + \text{EF1} + \text{Pi}$$

Secondly, peptide bond formation occurs between the amino group of the amino acid bound to the incoming tRNA and the carboxyl group of the terminal amino acid bound to the tRNA at the P site. This reaction is catalysed by peptidyltransferase (PT_f), a 60-S subunit protein. During this process the peptide chain is transferred from the tRNA at the P site to the tRNA at the A site which now contains an additional amino acid.

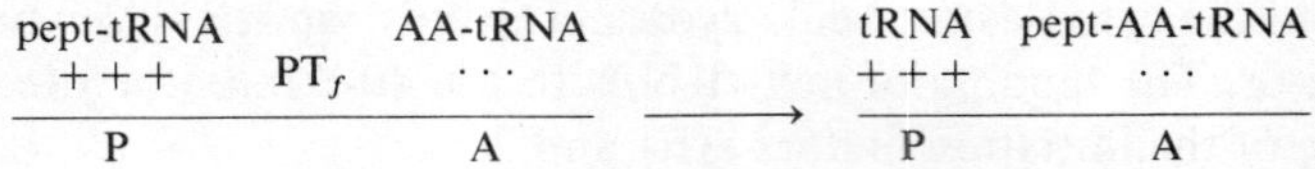

Thirdly, EF2 forms a stable complex with GTP. On binding to the ribosome the GTP of the complex is cleaved to GDP. The energy so derived enables EF2 to trigger (1) the movement of peptidyl-tRNA form the A site to the P site, (2) the movement of mRNA a distance of three nucleotides (× × ×) to the A site and (3) ejection of the deacylated tRNA, previously at the P site, from the ribosome.

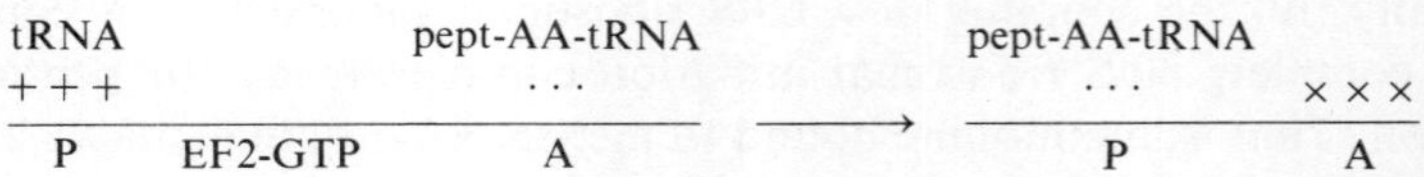

This process of elongation is cyclical with an amino acid being added each time. This continues until the protein being synthesized is complete and the mechanism terminated.

Peptide chain termination Special mRNA codons serve as signals for the termination of peptide chain elongation. When they reach the ribosomal A site they prevent binding of the aminoacyl-tRNA–GTP–EF1 complex. A peptide chain release factor (RF), with GTP attached, binds to the vacant A site and peptidyltransferase then cleaves the high energy ester bond and releases the completed protein.

The action of inhibitors of protein synthesis

Cycloheximide Cycloheximide is produced by *Streptomyces griseus* (Leach *et al.*, 1947) and is obtained as a by-product in the production of streptomycin. It is one of a group of compounds known as the glutarimide antibiotics which have a *β*-(2-hydroxyethyl)glutarimide moiety attached to a cyclic or acyclic ketone. Cycloheximide is the best known member of the group.

Cycloheximide was first recognized as a potent inhibitor of protein synthesis following studies with whole cells of *Saccharomyces carlsbergensis*, by Kerridge (1958). Since then its mode of action has been intensively studied in a variety of systems. It has been reported to act on a number of steps in cytoplasmic protein synthesis, principally as an inhibitor of translocation and transpeptidation of polypeptide chain elongation. However, there are some conflicting reports on its effect on chain initiation (Obrig *et al.*, 1971; Lodish *et al.*, 1971). Cycloheximide does not appear to inhibit the established steps in this process.

The binding site for cycloheximide was found to be on the 60-S subunit (Jimenez *et al.*, 1972; Battaner and Vasquez, 1971), following 'hybrid' ribosome reconstitution experiments. In this procedure the reactive subunit was determined by (a) mixing *in vitro* 40-S and 60-S subunits prepared from resistant and susceptible yeast strains and then incubating them in the presence or absence of the antibiotic (Jimenez *et al.*, 1972), and (b) preparing subunits from susceptible yeast and incubating them separately with or without cycloheximide prior to reconstitution. Inhibition of protein synthesis only occurred in reconstituted 'hybrid' ribosomes containing either resistant or antibiotic-treated 60-S subunits.

Evidence that cycloheximide, and other glutarimide antibiotics, inhibit chain elongation was deduced from their inhibitory effects on polysome breakdown in a variety of systems (e.g. Ennis and Lubin, 1964; Wettstein *et al.*, 1964; Trakatellis *et al.*, 1965; Godchaux *et al.*, 1967). Furthermore, cycloheximide at 4×10^{-6} M protects reticulocyte polysomes from NaF degradation when added prior to the salt (Godchaux *et al.*, 1967; Colombo *et al.*, 1965). In addition, the rate of initial reformation of the polysomes is slow in the presence of 4×10^{-6} M cycloheximide, a typical effect of an inhibitor of chain elongation. The rate at which reformation occurs is diagnostic for the type of inhibitor present (Wei *et al.*, 1974). Inhibitors of chain initiation completely inhibit the initial rate of polysome reformation while inhibitors of chain termination have no effect.

Cycloheximide inhibits elongation by preventing translocation and transpeptidation. The antibiotic blocks EF2 GTP dependent translocation of peptidyl-tRNA from the A site to the P site on the ribosome and prevents the release of deacylated tRNA (McKeehan and Hardesty, 1969; Baliga *et al.*, 1969; 1970; Obrig *et al.*, 1971). The inhibitory effect on translocation can be relieved by excess GTP, EF2 and sulphydryls.

Evidence that cycloheximide inhibits transpeptidation (peptide bond formation) comes from work by Pestka *et al.* (1972) and Schneider and Maxwell (1973). It was shown that the antibiotic behaves as a mixed inhibitor of the reaction of puromycin with nascent polypeptides on purified ribosomes in the absence of EF2. The puromycin reaction is an indicator of peptidyltransferase activity associated with peptide bond formation. Puromycin mimics the 3′-end of aminoacyl-tRNA and becomes peptidyl puromycin. This causes the release of the growing peptide chain from peptidyl-tRNA if the tRNA is at the P site but not if it is at the A site. Thus, inhibition of the puromycin reaction is indicative of an inhibitor of transpeptidation. Whether this is a major site of action of the antibiotic, however, has not been definitely established.

Inhibition of RNA synthesis has also been cited as a possibly important mode of action of cycloheximide. It has been shown to inhibit RNA polymerase

I of *Achyla bisexualis* and to reduce the capacity of this organism to synthesize RNA (Timberlake *et al.*, 1972).

Streptomycin Streptomycin, produced by *Streptomyces griseus*, is one of a group of compounds known as the aminoglycoside antibiotics. All have carbohydrate molecules containing one or more glycosidic linkages and a large number of hydrophilic groups including several amino or guanidino groups (see review by Umezawa, 1975).

The mode of action of streptomycin has received considerable attention, primarily because of its use in clinical medicine. Most of the work has been done on bacterial cell-free systems.

Streptomycin displays multiple inhibitory activity on protein synthesis in prokaryotes, but there are no reports that it can inhibit cytoplasmic protein synthesis in fungi or other eukaryotes. Its limited antifungal activity against certain species of Oomycetes and yeast (Gottlieb and Shaw, 1970; Tsao, 1970) may be due to increased uptake of the antibiotic by these organisms (Voros, 1965). In fungi the site of action is probably mitochondrial protein synthesis.

In prokaryotes it has been established that streptomycin binds to a 30-S ribosomal subunit protein, S-12 (Ozaki *et al.*, 1969; Chang and Flaks, 1972). The S-12 has to be associated with the 30-S subunit for binding to occur. Streptomycin inhibits chain elongation by 'locking' peptidyl-tRNA at the P site. This leads to distortion of the P and A sites and prevents the binding of aminoacyl-tRNA (Kaji and Tanaka, 1968; Modolell and Davis, 1969; 1970). Streptomycin irreversibly inactivates 70-S monosomes preventing them from forming new initiation complexes (Modolell and Davies, 1970; Luzzato *et al.*, 1969). Furthermore, as streptomycin acts on the ribosomal A site it blocks the release of peptides or f-met during chain termination (Scolnick *et al.*, 1968). Streptomycin also stimulates miscoding in cells and cell-free extracts (Pestka, 1971; Kaji, 1973). The precise causes of lethality have not been resolved but must be related to its irreversible binding to the ribosomes and the resultant effects described.

Blasticidin This antibiotic is produced by *Streptomyces griseochromogenes* (Fukunaga *et al.*, 1955) and its structure was established by Otake *et al.* (1965). It is a member of the 2-aminohexose pyrimidine nucleoside group that also includes gougerotin, amicetin and bamicetin. Although all these antibiotics are effective antibacterial agents only blasticidin S has antifungal activity.

Blasticidin was first implicated as an inhibitor of protein synthesis when it was shown to inhibit the incorporation of amino acids into protein in extracts of *Pyricularia oryzae* and *E. coli* (Huang *et al.*, 1964; Yamaguchi *et al.*, 1965). The activity was expressed at a site after the formation of aminoacyl-tRNA.

Blasticidin binds to a single site on the 50-S ribosomal subunit in prokaryotes (Kinoshita *et al.*, 1970) and to the 60-S subunit of eukaryotes (Battaner and

Vazquez, 1971). Competition experiments indicate that gougerotin and blasticidin both bind at the same sites (Kinoshita *et al.*, 1970; Barbacid and Vazquez, 1974).

Blasticidin S inhibits transpeptidation reactions of chain elongation as demonstrated by its effect on the puromycin reaction. In this way it is similar to cycloheximide. It exerts a two-phase reactivity. At low concentrations it acts as a competitive inhibitor and at high concentrations it is a non-competitive inhibitor of the puromycin reaction with ribosomal bound peptidyl-tRNA (Casky *et al.*, 1971; Pestka, 1972). Gougerotin was found to inhibit peptide chain elongation in bacterial and mammalian extracts without causing polysome breakdown (Casjens and Morris, 1965). Given that gougerotin and blasticidin S have the same binding site and similar mode of action this result may also be applicable to blasticidin. It has been postulated that these antibiotics react with the peptidyltransferase centre preventing the attachment of 3′ terminal CCA of aminoacyl-tRNA to the ribosomal A site (Siegel, 1977).

Kasugamycin Kasugamycin is a water-soluble base produced by *Streptomyces kasugaensis* (Umezawa *et al.*, 1965). Its structure was established by Suhara *et al.*, 1966). The antibiotic is classified as an aminoglycoside (streptomycin group) although the molecule lacks an inosamine moiety.

Most of the mode of action studies with kasugamycin have been with bacterial cell-free systems. Unlike streptomycin, kasugamycin does not cause miscoding (Tanaka *et al.*, 1966a; b), but it does bind to the 30-S ribosomal subunit of *E. coli* (Sparling, 1970). This was established by the 'hybrid' reconstituted ribosome procedure described earlier. Kasugamycin-resistant strains of *E. coli* were found to be deficient in methylation of the 16S-RNA. Methylation of 16S-RNA in cell-free extracts followed by reconstitution into 30S particles converts 16S-RNA from a kasugamycin-resistant organelle to a sensitive one (Helser *et al.*, 1972). It was concluded that the kasugamycin A locus involved in these mutations to resistance corresponds to an RNA methylase.

Kasugamycin inhibits initiation complex formation on 30-S subunits as well as on 70-S ribosomes (Okuyama *et al.*, 1971). Enzymatic and non-enzymatic binding of phe-tRNA is inhibited by kasugamycin but it is not as sensitive to the antibiotic as f-met-tRNA binding. This is inhibited by 2×10^{-5} M kasugamycin (Okuyama *et al.*, 1971).

Kasugamycin does not inhibit peptide bond formation or translocation of chain elongation as determined by the puromycin reaction on 70-S and 50-S subunits with peptidyl-tRNA (Pestka, 1972; Tanaka, 1972).

Studies on the activity of kasugamycin on fungal systems is very limited. In *P. oryzae* at high concentrations (1.5×10^{-3} M) kasugamycin was shown to inhibit partially poly-U-directed polyphenylalanine synthesis on ribosomes (Masukawa *et al.*, 1968). Binding of phe-tRNA to poly-U and ribosomes was also partially inhibited.

Mode of action not defined or equivocal

Cymoxanil

In common with the acylalanines and related compounds, cymoxanil (**69**, 'Curzate', DPX-3217) is one of the new generation of fungicides with activity against Oomycetes (Serres and Carraro, 1976). Cymoxanil has been shown to

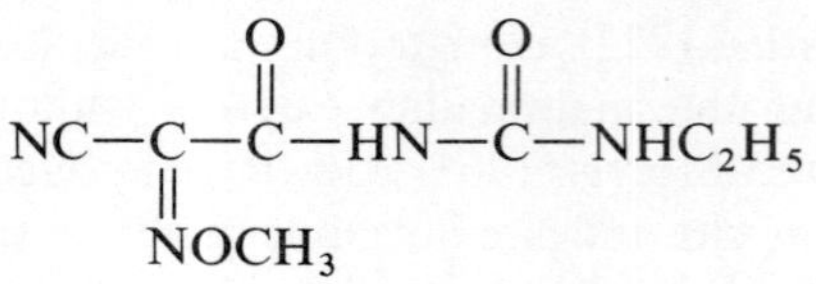

(**69**) Cymoxanil

control vine downy mildew, and late blight on tomato and potato (Absi, 1979; Douchet *et al.*, 1977; Garibaldi and Tamietti, 1980; Klöpping and Delp, 1980), but is comparatively ineffective against tobacco blue mould (O'Brien, 1978; Schiltz *et al.*, 1977), and soil-borne species of *Phytophthora* (Serres and Carraro, 1976). The fungicide exhibits good therapeutic effects, though apparently only local systemic movement (Clerjeau *et al.*, 1982; Roussel, 1978). Nevertheless, uptake and acropetal transport have been reported in tomato following application to the root (Kluge, 1978) and shoot (Klöpping and Delp, 1980) systems. Cymoxanil is rapidly metabolized in plants and loses activity within a few days (Conce and Duvauchelle, 1982; Douchet *et al.*, 1977). Consequently, it is formulated in combination with classical protectant fungicides, e.g. mancozeb or folpet, in order to prolong its protective effects.

As would be anticipated, the inhibitory activity of cymoxanil *in vitro* is limited to Phycomycetes. *Pythium debaryanum*, *Phytophthora cactorum* and *Mucor mucedo* were all susceptible to the fungicide, and had ED_{50} values for radial mycelial growth of 3, 20 and 40 $\mu g\ ml^{-1}$, respectively (Kluge, 1978). One *in vitro* study has revealed a remarkable differential sensitivity to cymoxanil among species of *Phytophthora*, with *P. infestans* proving more than 100 times more sensitive than other representatives (Richardson, L. T., 1976, *Pestic. Res. Rep.*, 385, cited by Staub and Hubele, 1980).

Information on the mode of action of cymoxanil is confined to a preliminary report by Despreaux *et al.* (1982). Interestingly these workers did not employ a Phycomycete fungus, but rather made use of a strain of *Botrytis cinerea* which was resistant to benzimidazoles but sensitive to cymoxanil. The most profound effects of cymoxanil were reductions in the uptake and incorporation of [^{14}C]uridine, suggesting that the fungicide might interfere with the synthesis of RNA. It should be stressed, however, that effects were also observed with other precursors, and although these were less pronounced than those found with uridine, they may nevertheless be significant.

Although early indications suggest that cymoxanil may interfere in some way with RNA synthesis, a recent study (Leroux and Gredt, 1982) found no cross-resistance between this fungicide and metalaxyl or related compounds.

Etridiazole

The fungicidal properties of etridiazole (**70**, 'Terrazole') were first described by Thomas *et al.* (1962). It is used principally against soil-borne Oomycetes such as *Pythium* and *Phytophthora*, but can also be employed in mixtures for the control of diseases caused by higher fungi. In addition to its activity *in vitro* towards Oomycetes, etridiazole also inhibits the Zygomycete, *Mucor mucedo* (Kluge, 1978). This contrasts with several other Oomycete fungicides (metalaxyl, furalaxyl, fenaminosulf and pyroxychlor) which are not active against *Mucor* (Kerkenaar and Kaars Sijpesteijn, 1981; Kluge, 1978).

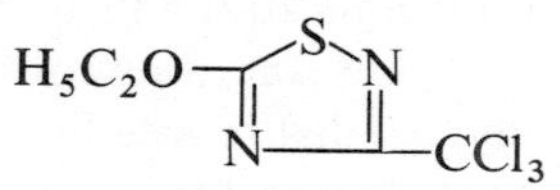

(**70**) Etridiazole

A difference of opinion exists regarding the mode of action of etridiazole. In essence, Halos and Huisman (1974; 1976a; b), suggest that the fungicide acts as an inhibitor of respiration, whereas Lyr and co-workers (Lyr, 1977; Lyr *et al.*, 1975; 1977; Radzuhn and Casperson, 1979) propose a totally different mechanism involving activation of phospholipase activity.

Using various species of *Pythium*, Halos and Huisman (1976b) showed that treatment with etridiazole produced inhibitory effects on respiration. Inhibition of O_2 uptake could be released, however, by addition of the artificial electron mediator, *N*,*N*,*N*′,*N*′-tetramethyl-*p*-phenylenediamine (TMPD). This suggested that the site of action of etridiazole was analogous to that of antimycin A, i.e. in the cyt.–bc_1 complex (Figure 6). Such a conclusion is based on the fact that TMPD acts as a bypass by which electrons are shunted around the antimycin A sensitive site (Figure 6) (Lee *et al.*, 1967).

In contrast to the data obtained using mitochondria from species of *Pythium*, oxidation of succinate and NADH by preparations from the insensitive fungus *Fusarium oxysporum* f.sp. *vasinfectum* was not inhibited by etridiazole. These results are in accord with the selectivity of the fungicide. A further point of circumstantial evidence consistent with the proposed site of action is the demonstration that tolerance to the fungicide relies on increased use of an alternative pathway of electron transport mediated by a component which precedes the cyt.–bc_1 complex, namely ubiquinone (Halos and Huisman, 1974; 1976a).

Although the evidence put forward by Halos and Huisman (1976b) appears incontrovertible, it should nevertheless by interpreted with caution, principally because of the rather high levels of etridiazole required to induce effects on respiration. For example, 50% inhibition of O_2 uptake in *P. debaryanum* required a fungicide concentration of about 1 mM, whereas the ED_{50} value for the same fungus was 13 μM. Admittedly this particular experiment was conducted over a short time period using intact mycelium, but nevertheless the discrepancy is not easily reconciled. Even in isolated mitochondria, etridiazole at 180 μM inhibited the oxidation of succinate and NADH by only 45 and 54% respectively. In such isolated systems one would normally anticipate a considerably enhanced susceptibility to the toxicant as a result of the absence of potential barriers to transport posed by the cell wall, plasmalemma and cytoplasm.

The alternative theory on the mechanism of action of etridiazole, which implicates fungicide-induced changes in phospholipase activity, is derived largely from studies using *M. mucedo*. Investigation by electron microscopy revealed marked ultrastructural alterations in hyphae exposed to etridiazole (Lyr, 1977; Lyr *et al.*, 1975). These were most pronounced in mitochondria, where treatment with the fungicide led to swelling and vacuolation of cristae, and enlargement of the outer membrane. Invagination of the plasmalemma and thickening of the cell wall were also observed.

Comparison of lipid compositions between control and fungicide-treated mitochondria from *Mucor* suggested that these effects were due to enhanced phospholipase activity (Radzuhn and Casperson, 1979). A rapid formation of lysolecithin in treated mitochondria was indicative of increased phospholipase activity. Lysolecithin is the product formed from phosphatidyl choline by the hydrolysis of one fatty acid ester bond, and it can be concluded, therefore, that the principal phospholipase activity stimulated by etridiazole is probably phosphatidase A. Abnormally high levels of phosphatidase A and other phospholipases would be expected to be detrimental to membrane integrity (Reman *et al.*, 1969; van Zutphen and van Deenen, 1967) and could, therefore, account for the ultrastructural changes noted earlier. Further evidence in favour of a role for phospholipases in the action of etridiazole is provided by the demonstration that the toxicity of the fungicide is partially alleviated by procaine (Lyr *et al.*, 1977), a reported antagonist of phospholipase activity (Scherphof *et al.*, 1972).

The microscopic studies on the effects of etridiazole were carried out using a fungicide concentration of 10 $\mu g\ ml^{-1}$ (about 40 μM) and are, therefore, quite valid since the ED_{50} value for *M. mucedo* is 30 μM (Radzuhn and Casperson, 1979). Unfortunately, however, some of the biochemical studies upon which the phospholipase theory is based, involved the use of rather higher levels of toxicant. The latter experiments can, therefore, be criticized in the same way as those of Halos and Huisman (1976b). Furthermore, there is still no explanation for the way in which etridiazole brings about increases in phospholipase activity. The basis for fungicide selectivity is similarly unresolved.

Cyclic imides (dicarboximides)

The 3,5-dichlorophenyl-*N*-cyclic imide group of fungicides comprises procymidone (**71**), iprodione (**72**) and vinclozolin (**73**). These compounds were introduced in 1969, 1971 and 1975 respectively and are also referred to as the dicarboximides. Their biological activity has been described (Hisada *et al.*, 1976; Lacroix *et al.*, 1974; Pommer and Mangold, 1975; Hess and Löcher, 1975). They are particularly effective in controlling diseases caused by *Botrytis*, *Sclerotinia* and *Monilinia* species and are used to protect a wide variety of vegetable and fruit crops. Most important in this respect is the control of *Botrytis cinerea* on vines. The cyclic imides are used mainly as protectant fungicides but they also display a curative action and a moderate degree of systemicity, particularly procymidone.

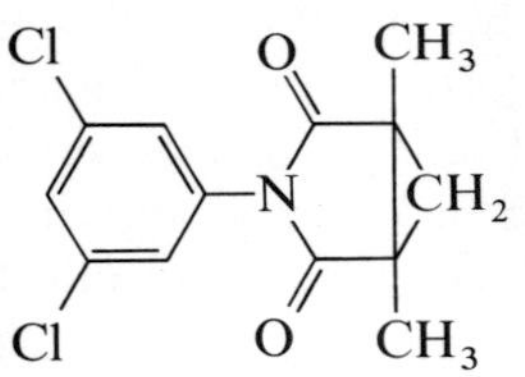

(**71**)

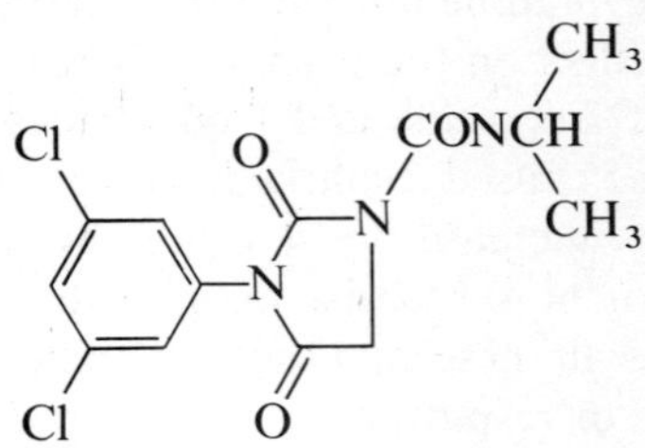

(**72**)

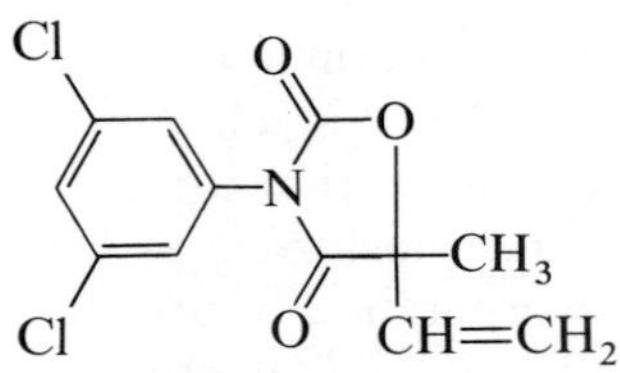

(**73**)

There have been several investigations into the mode of action of the dicarboximides but as yet no definitive evidence for a primary target site has been produced. It has been suggested (Fritz *et al.*, 1977) that they have similar modes of action because of the similarities in their chemical structure and spectrum of activity. Furthermore, there are many examples of mutant fungal strains displaying positive cross-resistance to the three cyclic imides, for example, *Botrytis cinerea*, *Aspergillus nidulans*, *Penicillium expansum*, *Rhizopus nigricans* and *Ustilago maydis* (Leroux *et al.*, 1981). Interestingly, this cross-resistance also extends to a considerable number of aromatic hydrocarbon compounds including the systemic fungicide chloroneb and the protectants biphenyl and dichloran (Leroux *et al.*, 1981). The mechanism of resistance has not been established.

There have been many suggestions as to the site of action of the dicarboximides following studies on a variety of fungal species. However, the reports are often contradictory. Procymidone toxicity towards *Botrytis cinerea* was thought to have been caused by effects on membrane functions required for transport processes (Hisada *et al.*, 1978). Pappas and Fisher (1979) detected no effects on membrane permeability of *B. cinerea*. Vinclozolin was thought to inhibit cell wall biosynthesis in *B. cinerea* (Eichhorn and Lorenz, 1978). Pappas and Fisher (1979) were only able to demonstrate an inhibitory effect on chitin synthesis with concentrations of the dicarboximides above those that cause a 50% reduction in growth rate. The slight effect of iprodione on the incorporation of [^{14}C]glucosamine into the cell wall of *S. sclerotorium* was considered to be a secondary effect (Reilly and Lamoureux, 1981).

Vinclozolin was shown to alter the triglyceride fraction of *Ustilago avenae* mycelia (Buchenauer, 1976). Pappas and Fisher (1979) noted that vinclozolin and procymidone inhibited triglyceride production in *B. cinerea* but iprodione had no effect on the process, whereas Fritz *et al.* (1977) observed an accumulation of free acids and triglycerides induced by iprodione and vinclozolin. Iprodione caused a marked decrease in 4-methylsterol production and a concomitant increase in 4,4-dimethylsterols indicating an interference with ergosterol biosynthesis. Despite these contradictions, there appears to be a consensus in these studies that the dicarboximides do not directly affect protein synthesis or respiration.

A number of observations indicate that the cyclic imides interact with nuclear processes. A 65% inhibition of growth and the number of nuclei in the germ tubes of *B. cinerea* exposed to procymidone for 2 h was reported by Hisada *et al.* (1978). Vinclozolin caused a 63% inhibition of cell multiplication and a 16% decrease in dry weight of sporidia of *U. avenae* (Buchenauer, 1976). Fritz *et al.* (1977) reported that vinclozolin and iprodione at 10 ppm rapidly and strongly inhibits the incorporation of uridine-2-^{14}C into both RNA and DNA. Pappas and Fisher (1979) noted that iprodione reduced the incorporation of [^{14}C]thymine into DNA although uracil incorporation into RNA was unaffected. This fungicide has a similar effect on *S. sclerotiorum* (Reilly and Lamoureux, 1981). Iprodione, procymidone and vinclozolin all cause a dose-dependent mitotic instability in *A. nidulans*, indicative of disruption of nuclear processes. Georgopoulos *et al.* (1979) conclude that the effect on the chromosomes and possibly on the mitotic spindle was a main cause of the fungitoxicity. Further elucidation of the mechanism of action of the cyclic imides is being actively pursued (Reilly and Lamoureux, 1981).

FUNGICIDES WITH AN INDIRECT MODE OF ACTION

Several fungicides require an activation process, usually metabolic, to convert them into chemicals that have a direct fungitoxic action against plant pathogens within their hosts. These have already been considered in the preceding sections. This section concerns fungicides that act either against the processes of secondary metabolism that are essential for fungi to be pathogenic, or on the host defence mechanisms to render a normally susceptible host resistant to a pathogen. Under the guise of 'alternative' chemical agents for disease control, these have been reviewed recently by Langcake (1981). That review attempted to show the approaches that might be taken to the discovery of alternative chemical agents, illustrating the principles with examples of experimental compounds that may act indirectly. It was stressed, however, that few of the examples quoted had found commercial application so far. This review is restricted to considering some recent examples of commercialized compounds that appear to have indirect action.

The problems associated with the elucidation of the modes of action of 'conventional' fungicides are compounded in the case of indirect-acting fungicides. A major problem lies in distinguishing between secondary or consequential effects and those that are of primary importance. This is made more difficult by our very incomplete understanding of the secondary metabolism essential for fungal pathogenesis and the mechanisms of host resistance. The problems are even greater in the case of compounds that control diseases caused by obligately biotrophic pathogens, those that cannot be grown in culture.

Nevertheless, a common feature of indirect-acting compounds is that they do not affect the growth of fungi *in vitro*. However, this is not an adequate criterion alone for an indirect action since it is never certain that the sensitivity of a fungus *in vitro* adequately reflects its sensitivity *in vivo* where it clearly has such different metabolic properties. In addition, even when a fungitoxic action *in vitro* can be shown, host resistant responses may yet occur *in vivo* and confuse the issue (see discussions of triadimefon and acylalanines above and in Langcake (1981)).

Three recently introduced commercial compounds, tricyclazole, TEPA (Aliette) and probenazole will be considered.

Tricyclazole

With the recent publications of Sisler and co-workers, there can be little doubt that tricyclazole (**74**) represents the best documented example of a fungicide with indirect action. The compound was introduced in the mid 1970s for the control of the rice blast disease (caused by *Piricularia oryzae*) against which it is systemic and highly specific. In an early publication, Froyd *et al.* (1976) had

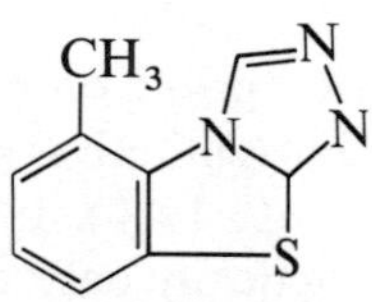

(**74**) Tricyclazole

shown that its activity *in vitro* was some 25–35 times less than that *in vivo* and this was confirmed by Tokousbalides and Sisler (1978). There was no evidence for the conversion of tricyclazole into a fungitoxic derivative in extracts of treated rice plants. The latter authors also noted that the fungicide inhibited the production by the fungus of the normal grey-black melanin pigment and that a brown soluble pigment tended to accumulate in the cultures instead. Although there was no clear reason at the time to associate this with the disease-suppressing properties of tricyclazole, their interest in the phenomenon was aroused by both the extreme sensitivity of melanin formation to this inhibition and the fact that mutants of *P. oryzae* unable to produce melanin were known to be non-pathogenic.

The melanins of many fungi are polyketide derivatives formed by the oxidative polymerization of 1,8-dihydroxynaphthalene (DHN, **XX**) (Figure 14). A putative toxin of the blast fungus, pyriculol (Iwasaki *et al.*, 1969), is also a polyketide derivative and this led Tokousbalides and Sisler (1978) to suggest that tricyclazole inhibits polyketide secondary metabolism that is essential for pathogenicity. It was not immediately supported, however, by the finding that production of pyriculol was actually enhanced by tricyclazole (Tokousbalides and Sisler, 1978), although production of another polyketide, 3,4-dihydro-4,8-dihydroxy-1(2H)-naphthalene (DDN, **XXI**), was strongly inhibited.

Melanin biosynthesis *per se* in fungi has been more completely studied in *Verticillium dahliae*. Since the production of melanin in *V. dahliae* was also inhibited by tricyclazole, this fungus was used to locate the precise site of inhibition (Tokousbalides and Sisler, 1979). Based on the pattern of accumulation of intermediates of melanin biosynthesis in tricyclazole-treated wild-type strains, they concluded that tricyclazole inhibits primarily the conversion of 1,3,8-trihydroxynaphthalene (**XXII**) into vermelone (**XXIII**). It resulted in an enhanced accumulation of scytalone (**XXIV**), 2-hydroxyjuglone (**XXV**) and DDN (**XXI**). At higher concentrations, other effects on sites of secondary sensitivity, e.g. that between 1,3,6,8-tetrahydroxynaphthalene (**XXVI**) and scytalone, caused a different accumulation of intermediates. In other experiments, intermediates of melanin biosynthesis were supplied to melanin-deficient (albino) mutants of *V. dahliae*. Tricyclazole inhibited the conversion of scytalone into melanin by these mutants, but not the conversion of vermelone or DHN (**XX**) into melanin (see Figure 14). Alterations in the melanin biosynthetic pathway were also induced in the fungus *Thielaviopsis basicola* by tricyclazole (Stipanovic and Wheeler, 1980). Although melanin biosynthesis is inhibited in a number of other fungi, disease control by tricyclazole appears to be restricted to *P. oryzae* on rice.

In *P. oryzae*, the action of tricyclazole on melanin biosynthesis is identical with that in *V. dahliae*. Melanin biosynthesis is sensitive to as little as 0.01 μg/ml of the compound (Woloshuk *et al.*, 1980). In addition, three mutants of *P. oryzae* were found that were genetically defective for melanin biosynthesis at

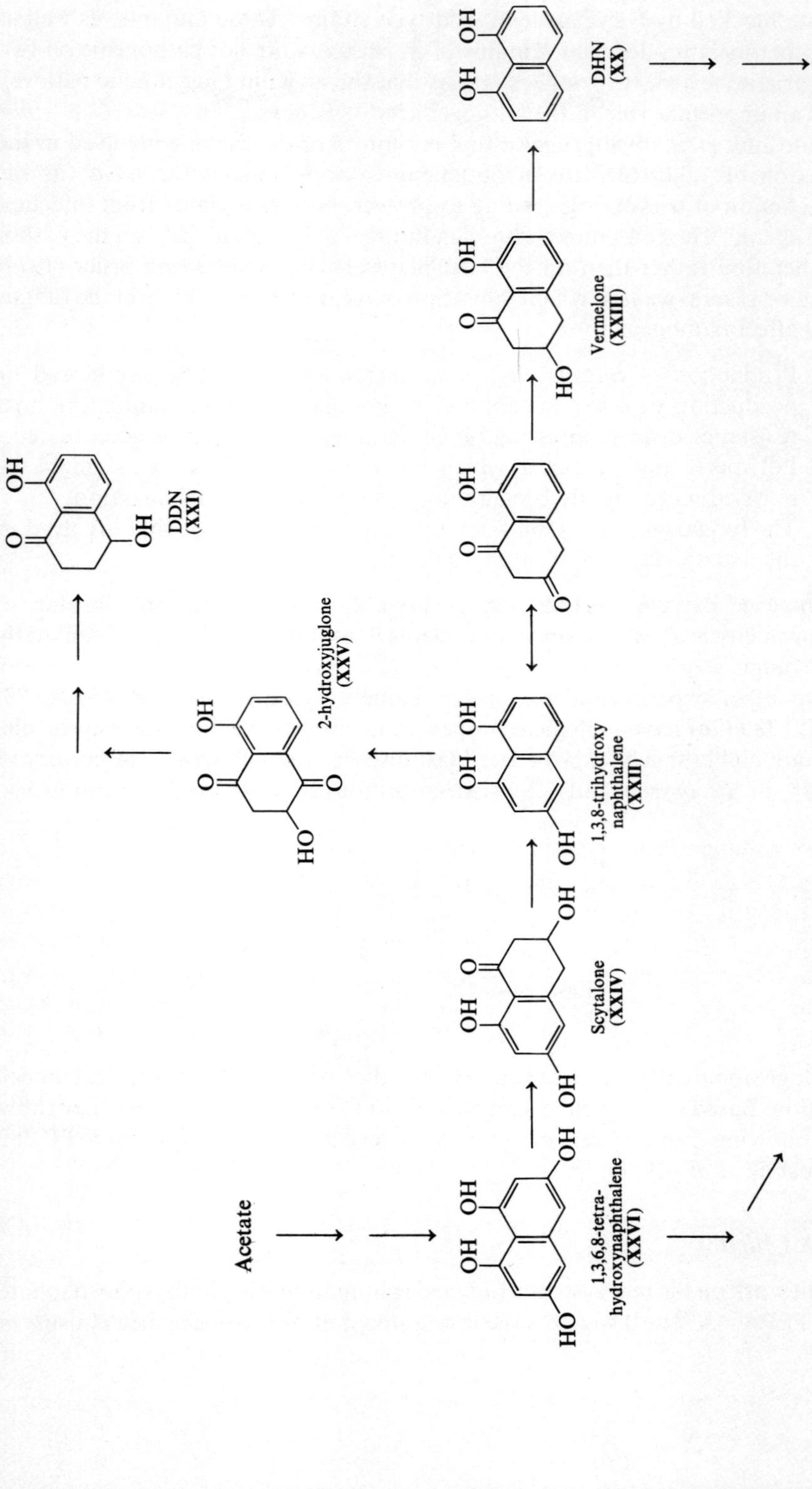

Figure 14 Proposed pathway of melanin biosynthesis in fungi

the site blocked by tricyclazole in wild-type strains. These mutants, as well as two other melanin-deficient mutants of *P. oryzae*, were not pathogenic on two rice varieties tested. It strongly suggests that the melanin biosynthetic pathway plays an important role in the pathogenicity of *P. oryzae*.

Woloshuk *et al.* (1980) suggest that inhibition of the enzyme involved in the reduction of 1,3,8-trihydroxynaphthalene to vermelone is the basis for the initial action of tricyclazole leading to protection of rice plants from infection by *P. oryzae*. The compound, although indirect in its action, acts on the pathogen therefore rather than on the host plant. Tokousbalides and Sisler (1979) discussed several ways in which inhibition of melanin biosynthesis in the fungus could affect pathogenicity.

(a) Production of certain phytotoxic metabolites might be suppressed or production of other metabolites which play a role in stimulating host resistance or in suppressing fungal pathogenicity might be accentuated.
(b) Polymeric wall material which elicit host defence systems might be released more rapidly because it is not stabilized by melanization.
(c) The fungal enzyme(s) inhibited by tricyclazole might be that involved in the detoxification of host phytoalexins.

Whatever its role melanization presumably has a unique relationship to pathogenicity in *P. oryzae* since tricyclazole does not control diseases caused by other fungi.

Two other experimental fungicides, namely pyroquilon (CGA 49104, **75**) and PP 389 (**76**) have biological properties identical with those of tricyclazole. They are highly specific for the rice blast disease, are only weakly inhibitory to growth of *P. oryzae* and affect melanization at very low concentrations.

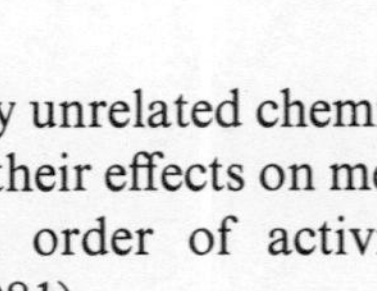

(**75**) Pyroquilon

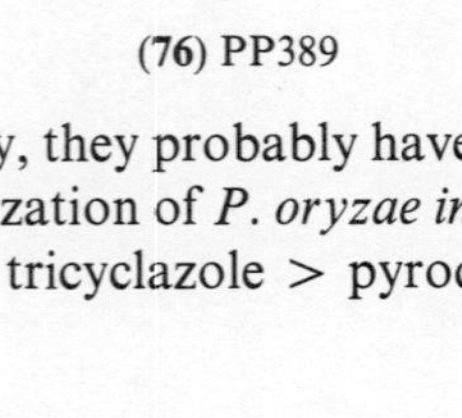

(**76**) PP389

Although apparently unrelated chemically, they probably have identical modes of action. Based on their effects on melanization of *P. oryzae in vitro*, they show the following rank order of activity: tricyclazole > pyroquilon > PP 389 (Woloshuk *et al.*, 1981).

TEPA (*Aliette*)

Recent work on the new systemic fungicide aluminium tris-*O*-ethyl phosphonate (**77**) (TEPA, 'Aliette') suggests the involvement of host defence mechanisms in fungicidal mode of action. The compound is representative of several metal salts of ethyl phosphonates that are principally active against diseases caused by Phycomycete fungi (e.g. downy mildews and *Phytophthora* spp.) on a range of crops. It also has some effects against diseases caused by fungi from other taxonomic groups, e.g. excoriose (*Phomopsis viticola*) of vine (Bertrand *et al.*, 1977). Most reports (Smith, 1979; Vo-Thi-Hai *et al.*, 1979; Raynal *et al.*, 1980; Bompeix *et al.*, 1980a) indicate that TEPA has very low fungicitoxicity *in vitro* whereas in host plants it strongly suppresses mycelial growth through the tissues. For example, at 500 μg/ml TEPA had no effects on growth of *P. capsici* on agar medium although when tomato leaves were floated on a solution containing 100 μg/ml, lesion development was totally suppressed (Vo-Thi-Hai *et al.*, 1979). This was accompanied by the development of a hypersensitive reaction characteristic of natural resistance (Bompeix *et al.*, 1980a; Durand and Salle, 1981). In uninfected leaves of tomato (Vo-Thi-Hai *et al.*, 1979) or grapevine (Raynal *et al.*, 1980), TEPA did not affect the normal contents of antifungal phenols and phytoalexin-like substances (e.g. stilbene-derived phytoalexins in vine (cf. Langcake and Pryce, 1977)). Inoculation of untreated plants increased the amounts of these substances while inoculation of treated plants gave the greatest concentrations. Treatment of plants with inhibitors of phenolic biosynthesis (glyphosate, α-aminooxy acetic acid, α-aminooxy phenyl propionic acid) annulled the effects of TEPA (Bompeix, 1982; Fettouche *et al.*, 1981).

Bompeix *et al.* (1980b) proposed two hypotheses to account for these results. Either the compound promotes elicitation of the defence response in conjunction with fungal components that can do this, or it interferes with pathogen products that normally suppress the defence response. However, more direct effects of TEPA against the fungus should not be totally discounted since Raynal *et al.* (1980) found that with *Plasmopara viticola* (the downy mildew pathogen of vine) zoospore liberation and motility *in vitro* were very sensitive at concentrations of 20 and 10 ppm respectively. An interesting and unexplained feature of TEPA is that while it protects a range of crops (e.g. tomato) from various *Phytophthora* spp., it is unable to do so on potato (Bompeix *et al.*, 1980a). It has also been suggested that the activity of TEPA is due to its degradation product, phosphorous acid, and that its action can be annulled by phosphate ions.

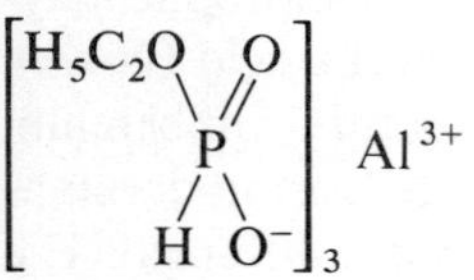

(77) Aliette

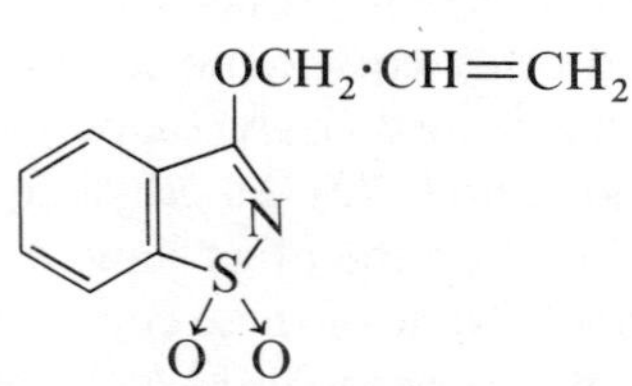

(78) Probenazole

$ClCH_2 \cdot CH_2—PO_3H$

(79) Ethrel

Probenazole

Probenazole (**78**, Oryzemate) is a new fungicide that is active, in common with tricyclazole, against the rice blast disease caused by *Piricularia oryzae*. It works mainly as a prophylactic and can be applied to the foliage directly or through the roots, in which case some curative properties are seen (Watanabe *et al.*, 1977). It is also active against the bacterium *Xanthomonas oryzae* on rice. Probenazole rapidly breaks down to allyl-*o*-sulphamoyl benzoate when applied to paddy water and on uptake into the rice plant is further converted into benzisothiazole-1,1-dioxide (saccharin) and *N*-D-glucopyranosyl saccharin. These latter two compounds are the major ones found in treated plants and may be the active principles (Uchiyama *et al.*, 1973). Probenazole and its breakdown products were not inhibitory to the growth on agar of 10 species of bacteria and 16 species of fungi, among them *P. oryzae* (Watanabe, 1977). With *P. oryzae in vitro*, probenazole inhibited appressorium formation and the penetration of a cellophane membrane. However, in the host plant, the effects of the compound against mycelial growth within lesions and of the subsequent sporulation were much more pronounced and this suggested the mediation of the host plant in the action of probenazole. In support of this, blast lesions on treated plants had some similarities with the hypersensitive-type lesions seen on genetically resistant cultivars, and the chemically induced resistance was eliminated by heat shock (Watanabe *et al.*, 1979). Moreover, probenazole enhanced the activities of host enzymes (peroxidase and phenylalanine ammonia lyase) that may be involved in the formation of physical barriers (e.g. lignin) to fungal invasion. Their activities were further increased by infection of treated plants (Iwata *et al.*, 1980). At the same time, an anti-conidial germination activity in the leaves was induced by probenazole treatment. Its activity was greatest in inoculated-treated plants but its presence in untreated-inoculated plants showed that it is a normal product of the host–pathogen interaction (Shimura *et al.*, 1981). Four antimicrobial factors (they also had antibacterial properties) were found and the major component (substance B) was identified as a mixture of fatty acids, principally linolenic acid (Sekizawa *et al.*, 1981). There was apparently no relationship of these antimicrobial substances with the momilactones, putative phytoalexins from rice (Cartwright *et al.*, 1980).

The disease-suppressing activity of probenazole is very similar to that of certain plant hormones and other compounds, namely 2,4-dichlorophenoxyacetic acid (2,4-D), indole acetic acid (IAA), naphthalene acetic acid (NAA), tryptophan, methionine and Ethrel (**79**) (an ethylene generator). Gibberellin, kinetin and abscisic acid (ABA) on the other hand stimulated lesion development (Matsumoto *et al.*, 1980). None of these compounds was antifungal. The disease-suppressing effects of probenazole, IAA, tryptophan, methionine and

Ethrel were all counteracted by application of ABA, as were the activities of these compounds, and fungal infection, in increasing peroxidase activity (Sekizawa and Watanabe, 1981; Sekizawa and Mase, 1981). Probenazole itself apparently does not have hormonal properties in plants.

It would seem therefore that the action of probenazole can be explained in several ways, possibly acting additively. It inhibits appressorium formation and penetration of cellophane membranes by the fungus in the absence of host tissue. This action may be reinforced by an activation of host defence mechanisms that are expressed as the formation of physical barriers (possibly lignin from the action of peroxidase on phenylalanine ammonia lyase-produced phenols) and chemical barriers (antimicrobial substances, among them linolenic acid). Probenazole has some characteristics in common with the effects of plant hormones on disease and this evidence has been developed into a working hypothesis (the signal transmission hypothesis; Sekizawa and Mase, 1980). This invokes auxins and ethylene as chemical mediators between a receptor that recognizes fungal products, and biochemical processes that result in the expression of host resistance. It is proposed that probenazole acts on the recognition stage of this process. Considerably more work, both on the molecular events in plant disease resistance and on probenazole, are required to substantiate this hypothesis.

CONCLUSIONS

The emergence of specific site-directed fungicides has greatly stimulated interest in the way in which these compounds act. It is appropriate therefore to give some of the reasons why such studies are of potential value. Firstly, mode of action studies on specific site-directed biocides in general have been of importance in increasing our knowledge of the target organisms studied. Just as these inhibitors have been useful tools for the biochemists in the past, so the elucidation of the mode of action of novel fungicides provides possible new tools for the biochemist in the future. Secondly, fungicide mode of action studies can reveal target sites in fungi that may be attacked with other, chemically unrelated compounds that may be more effective in practical use than the original compound. Although the rational design of pesticides from first principles based on an understanding of a target site has had little impact as yet on the process of pesticide discovery, the greater the number of potential target sites known, the greater the chance that this approach can be successfully implemented. Thirdly, an understanding of the mode of action of a fungicide may enable the full biological properties of that compound to be predicted and exploited. For example, it could be predicted that many of the sterol biosynthesis-inhibiting fungicides might be of value as hypocholesteremics for medical use and no doubt many agrochemical companies with pharmaceutical interests have

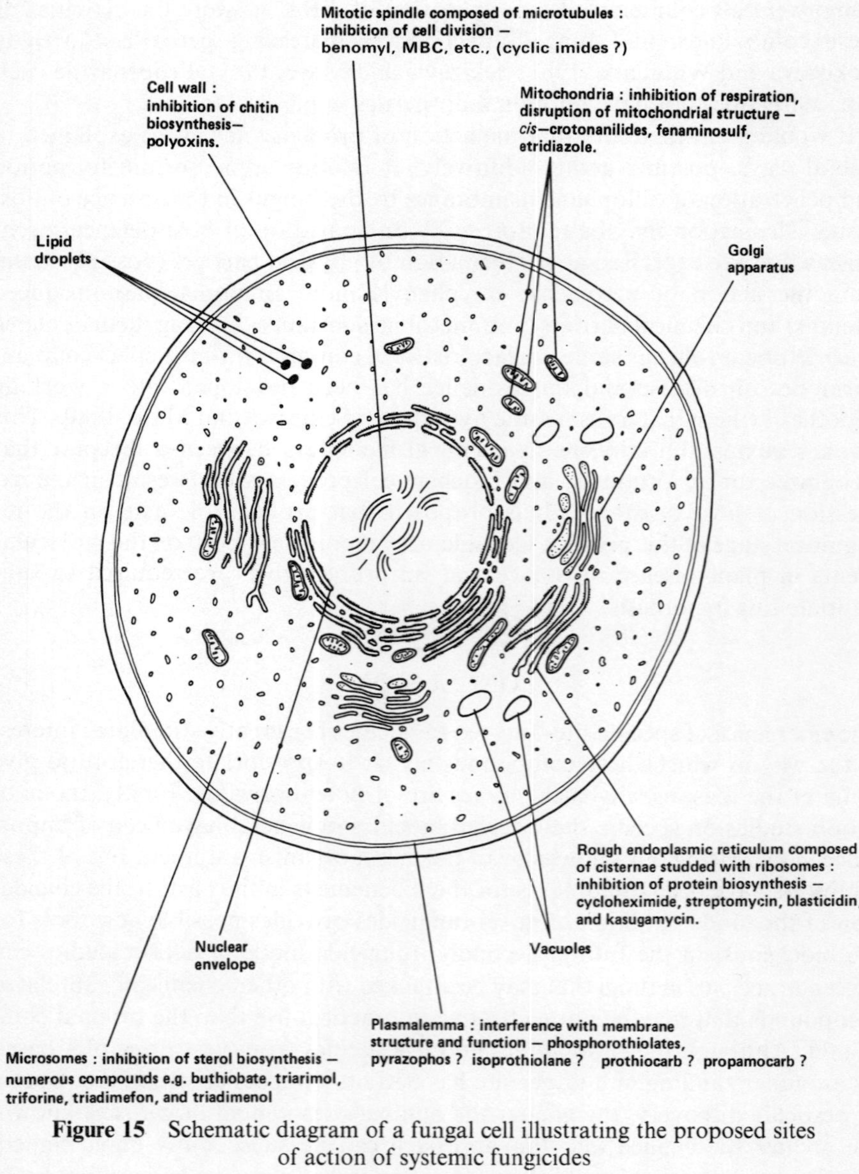

Figure 15 Schematic diagram of a fungal cell illustrating the proposed sites of action of systemic fungicides

examined this already. By the same token, mode of action studies may alert us to potential toxicological or environmental problems where the site of action (e.g. cell division) is common to a wide range of organisms including man and his benefactors. Finally, and no less important, there is the problem of resistance and cross-resistance. Since the discovery of therapeutic fungicides, the threat of fungal tolerance has been with us and in several cases that threat has materialized. Some fungicide target sites may be less prone to resistance problems than others while fungicides with the same mode of action, even if the compounds are apparently unrelated chemically, are liable to exhibit cross-resistance (e.g. cross resistance between sterol biosynthesis inhibitors, Barug and Kerkenaar, 1979). Clearly, a knowledge of mode of action is a sound basis for rational attempts to deal with the fungicide resistance problem.

In addition to the reasons outlined above for undertaking fungicide mode of action studies, those involved have been rewarded with the realization that systemic/therapeutic fungicides have very diverse modes of action and that to embark on such studies is a voyage of adventure. Whereas it can be said that a large proportion of all commercialized herbicides inhibit photosynthesis in one way or another, and that the majority of all insecticides act as nerve poisons, few generalizations about the modes of action of fungicides can as yet be made. Their sites of action range from energy-yielding metabolism through a range of biosynthetic processes to complex and as yet unexplained interactions with the balance between host plants and their highly specialized fungal parasites. Those with direct modes of action are summarized in Figure 15.

At least one generalization is possible, however, with regard to the mode of action of systemic fungicides that act against the growth of pathogens. These systemic fungicides, with few exceptions, inhibit metabolic processes (e.g. nuclear division, respiration, protein synthesis, sterol biosynthesis) that are common to fungi, their hosts and other non-target organisms (including man). Despite this, they normally show considerable selectivity towards their target fungi. It must be presumed therefore that this selectivity is based on: (a) small differences in the nature of the target site (e.g. an enzyme) in host and parasite resulting in differences in their sensitivities, (b) differential uptake or translocation, or (c) differential metabolism or other form of inactivation. The biochemical bases for selectivity of systemic fungicides, in addition to the structural requirements for systemic movement of compounds in plants (Edgington, 1981), are currently the major areas of ignorance in fungicide research. It is to be hoped that further progress in these areas, as well as more critical studies of fungicide mode of action, will be forthcoming.

Acknowledgements

The authors wish to thank A. Kerkenaar, P. Leroux, G. Sallese and M. R. Siegel for kindly providing manuscripts in press. Thanks are also due to several of our colleagues for their critical appraisal of various parts of this review. Finally, we would like to acknowledge the artwork of A. J. Reynolds and R. D. Hall, and the skilled and patient preparation of the typescript by Sheila Marshall and Susan Doherty.

REFERENCES

Aaronson, S. (1971). 'Inhibition of yeast multiplication by hypocholesteremic compounds and its annulment by lipids', *Proc. Soc. Exp. Biol. Med.*, **136**, 61–63.

Abiko, K., Kishi, K., and Yoshioka, A. (1975). 'Emergence of Plantvax-resistant strains of chrysanthemum rust fungus *Puccinia horiana*', *Ann. Phytopathol. Soc. Japan*, **41**, 100 (Abstr.).

Absi, M. (1979). '2 Cyano-*N*-[(ethylamino)carbonyl]-2-(methoxyimino) acetamide (DPX 3217), a new fungicide for the control of tomato and potato late blight, hop downy mildew and other Peronosporales', *Proc. 1979 Br. Crop Prot. Conf.—Pests and Diseases*, **2**, 311–318.

Ackrell, B. A. C., Kearney, E. B., Coles, C. J., Singer, T. P., Beinert, H., Wan, Y., and Folkers, K. (1977). 'Kinetics of the reoxidation of succinate dehydrogenase', *Arch. Biochem. Biophys.*, **182**, 107–117.

Ackrell, B. A. C., Kearney, E. B., Mowery, P., Singer, T. P., Beinert, H., Vinogradov, A. D., and White, G. A. (1976). 'Factors controlling the turnover number of succinate dehydrogenase: a new look at an old problem', in *Iron and Copper Proteins* (eds K. T. Yasunoba, H. F. Mower, and O. Hayaishi), pp. 161–181, Plenum, N.Y.

Actor, P., Anderson, E. L., DiCuollo, D. J., Ferlauto, R. J., Hoover, J. R. E., Pagano, J. F., Ravin, L. R., Scheidy, S. F., Stedman, R. J., and Theodorides, V. J. (1967). 'New broad spectrum anthelmintic, methyl 5(6)-butyl-2-benzimidazole carbamate', *Nature* (*Lond.*), **215**, 321–322.

Adams, B. G., and Parks, L. W. (1968). 'Isolation from yeast of a metabolically active water-soluble form of ergosterol', *J. Lipid Res.*, **9**, 8–11.

Aelbers, E., (1971). 'Thiophanate and thiophanate methyl, two new fungicides with systemic action', *Meded. Fac. Landbouwwet. Rijksuniv. Gent*, **36**, 126–134.

Akatsuka, T., Kodama, O., and Yamada, H. (1977). 'A novel mode of action of Kitazin P in *Pyricularia oryzae*', *Agric. Biol. Chem.*, **41**, 2111–2112.

Allen, A. M., Reinhardt, J. H., Akers, W. A., and Gunnison, D. (1973). 'Griseofulvin in the prevention of experimental human dermatophytosis', *Arch. Dermatol.*, **108**, 233–236.

Allen, P. M., and Gottlieb, D. (1970). 'Mechanism of action of the fungicide thiabendazole, 2-(4′-thiazoyl)benzimidazole', *Applied Microbiol.*, **20**(6), 919–926.

Anon. (1975). 'World pesticide markets', *Farm Chemicals*, **138**, 45.

Araki, F., and Miyagi, Y. (1976). 'Effect of isoprothiolane on the infection process of *Pyricularia oryzae*', *Ann. Phytopath. Soc. Japan*, **42**, 401–406.

Asakawa, M., Misato, T., and Fukunaga, K. (1963). "Studies on the prevention of the phytotoxicity of blasticidin S', *Pesticide and Technique* (*Tokyo*), **8**, 24.

Atassi, G., Schauss, C., and Tagnon, H. J. (1975). 'R 17934-NSC 238159: A new antitumour drug. II. Effect on mitotic cycle of L1210 leukemia cells *in vivo* and synergism with cytosine arabinoside (NSC 63 878)', *Eur. J. Cancer*, **11**, 609–614.

Atassi, G., and Tagnon, H. J. (1975). 'R 17934-NSC 238159. A new antitumour drug. I. Effect of experimental tumours and factors influencing effectiveness', *Eur. J. Cancer*, **11**, 599–607.

Baeder, C., Bähr, H., Christ, O., Düwel, D., Kellner, H. M., Kirsch, R., Loewe, H., Schultes, E., Schutz, E., and Westen, H. (1974). 'Fenbendazole: a new, highly effective anthelmintic', *Experientia* (*Basel*), **30**, 753–754.

Baliga, B. S., Cohen, S. A., and Munro, H. N. (1970). 'Effect of cycloheximide on the reaction of puromycin with polysome-bound peptidyl-tRNA', *FEBS Lett.*, **8**, 249–252.

Baliga, B. S., Pronczuk, A. W., and Munro, H. N. (1969). 'Mechanism of cycloheximide inhibition of protein synthesis in a cell-free system prepared from rat liver', *J. Biol. Chem.*, **244**, 4480–4489.

Barbacid, M., and Vazquez, D. (1974). '[G-^{3}H]Gougerotin binding to ribosomes. Heterogeneity of eukaryotic ribosomes', *Eur. J. Biochem.*, **44**, 445–453.

Barksdale, A. W. (1969). 'Sexual hormones of *Achlya* and other fungi', *Science*, **66**, 831–837.

Bartnicki-Garcia, S. E., and Lippman, E. (1972). 'Inhibition of *Mucor rouxii* by polyoxin D: effects on chitin synthetase and morphological development', *J. Gen. Microbiol.*, **71**, 301–309.

Bartnicki-Garcia, S. E., Ruiz-Herrera, J., and Bracker, C. E. (1979). 'Chitosomes and chitin synthesis' in *Fungal Walls and Hyphal Growth* (eds. J. H. Burnett and A. P. J. Trinci), pp. 149–168, Cambridge University Press.

Barton, D. H. R., Corrie, J. E. T., Marshall, P. J., and Widdowson, D. A. (1973). 'Biosynthesis of terpenes and sterols VII. Unified scheme for the biosynthesis of ergosterol in *Saccharomyces cerevisiae*', *Bio-org. Chem.*, **2**, 363–373.

Barton, D. H. R., Corrie, J. E. T., Widdowson, D. A., Bard, M., and Woods, R. A. (1974). 'Biosynthetic implications of the sterol content of ergosterol-deficient mutants of yeast', *J.C.S. Chem. Comm.*, 30–31.

Barug, D., and Kerkenaar, A. (1979). 'Cross-resistance of UV-induced mutants of *Ustilago maydis* to various fungicides which interfere with ergosterol biosynthesis', *Meded. Fac. Landbouwwet. Rijksuniv. Gent*, **44**, 421–427.

Battaner, E., and Vazquez, D. (1971). 'Inhibitors of protein synthesis by ribosomes of the 80S type', *Biochim. Biophys. Acta*, **254**, 316–330.

Bebbington, R. M., Brooks, D. H., Geoghegan, M. J., and Snell, B. K. (1969). 'Ethirimol: a new systemic fungicide for the control of cereal powdery mildews', *Chem. Ind.*, 1512.

Belal, M. H. (1971). 'Studien zur chemischen Bekämpfung des echten Mehltaues', Dissertation der Georg-August Universität, Göttingen.

Benitez, T., Villa, T. G., and Garcia Acha, I. (1976). 'Effect of polyoxin D on germination, morphological development and biosynthesis of the cell wall of *Trichoderma viride*', *Arch. Microbiol.*, **108**, 183–188.

Bent, K. J. (1970). 'Fungitoxic action of dimethirimol and ethirimol', *Ann. Appl. Biol.*, **66**, 103–113.

Bent, K. J. (1979). 'Fungicides in perspective: 1979', *Endeavour, N.S.*, **3**, 7–14.

Bent, K. J., and Skidmore, A. M. (1979). 'Dichlobutrazol; a new systemic fungicide', *Proc. 1979 Br. Crop. Prot. Conf.—Pests and Diseases*, **2**, 477–484.

Ben-Yephet, Y., Dinoor, A., and Henis, Y. (1975a). 'The physiological basis of carboxin sensitivity and tolerance in *Ustilago hordei*', *Phytopathology*, **65**, 936–942.

Ben-Yephet, Y., Henis, Y., and Dinoor, A. (1974). 'Genetic studies on tolerance of carboxin and benomyl at the asexual phase of *Ustilago hordei*', *Phytopathology*, **64**, 51–56.

Ben-Yephet, Y., Henis, Y., and Dinoor, A. (1975b). 'Inheritance of tolerance to carboxin and benomyl in *Ustilago hordei*', *Phytopathology*, **65**, 563–567.

Bergamaschi, P., Borsari, T., Garavaglia, C., and Mirenna, L. (1981). 'Methyl *N*-phenylacetyl-*N*-2,6-xylyl-DL-alaninate (M 9834), a new systemic fungicide controlling downy mildew and other diseases caused by Peronosporales', *Proc. 1981 Br. Crop Prot. Conf.—Pests and Diseases*, **1**, 11–18.

Bertrand, A., Ducret, J., Debourge, J.-C., and Horriere, D. (1977). 'Etude des propriétés d'une nouvelle famille de fongicides: les monoéthyl phosphites métalliques caractéristiques physico-chimiques et propriétés biologiques', *Phytiatrie-Phytopharmacie*, **26**, 3–18.

Bollen, G. J. (1971). 'Resistance to benomyl and some chemically related compounds in strains of *Penicillium* species', *Neth. J. Pl. Pathol.*, **77**, 187–193.

Bollen, G. J., and Fuchs, A. (1970). 'On the specificity of the *in vitro* and *in vivo* antifungal activity of benomyl', *Neth. J. Pl. Pathol.*, **76**, 299–312.

Bollen, G. J., and Scholten, G. (1971). 'Acquired resistance to benomyl and some other systemic fungicides in a strain of *Botrytis cinerea* in cyclamen', *Neth. J. Pl. Pathol.*, **77**, 83–90.

Bompeix, G. (1982). 'Mode d'action du Phosethyl Al', *Phytiatrie-Phytopharmacie* (in press).

Bompeix, G., Ravise, A., Fettuche, F., and Durand, M. C. (1980a). 'Modalités de l'obtention des nécroses bloquantes sur feuilles detachées de tomate par l'action du tris-*O*-éthyl phosphonate d'aluminium (phosethyl d'aluminium), hypotheses sur son mode d'action *in vivo*', *Ann. Phytopathol.*, **12**, 337–351.

Bompeix, G., Ravise, A., Raynal, G., Fettuche, F., and Durand, M. C. (1980b). 'Stimulation des réactions de défense et de synthèse de phytoalexines chez la vigne et la tomate sous l'influence du tris-*O*-ethyl phosphonate d'aluminium', in *Abstracts of 18 eme Colloque de la Société Francaise de Phytopathologie*, Toulouse, April 1980.

Borck, M. K. (1973). 'Studies on the mode of action of benomyl in *Neurospora crassa*', Ph.D. Thesis, Louisiana State University, 1–171.

van den Bossche, H. (1972). 'Biochemicals effects of the anthelmintic drug mebendazole', in *Comparative Biochemistry of Parasites* (ed. H. van den Bossche), pp. 139–157, Academic Press, New York.

van den Bossche, H., Willemsens, G., Cools, W., Lauwers, W. F. J., and LeJeune, L. (1978). 'Biochemical effects of miconazole on fungi. II. Inhibition of ergosterol biosynthesis in *Candida albicans*', *Chem. Biol. Interactions*, **21**, 59–78.

Bowers, B., Levin, G., and Cabib, E. (1974). 'Effect of polyoxin D on chitin synthesis and septum formation in *Saccharomyces cerevisiae*', *J. Bact.*, **119**, 564–575.

Brian, P. W. (1949). 'Studies on the biological activity of Griseofulvin', *Ann. Botany*, **13**, 59–77.

Brown, H. D., Matzuk, A. R., Ilvues, I. R., Peterson, L. H., Harris, S. A., Sarett, L. H., Edgerton, J. R., Yaktis, J. J., Campbell, W. C., and Cuckler, A. C. (1961). 'Antiparasitic drugs. IV. 2-(4′-thiazolyl)benzimidazole, a new anthelmintic', *J. Am. Chem. Soc.*, **83**, 1764–1765.

Brown, I. F. (1970). 'Biological activity of a new broad spectrum fungicide', *Proc. 7th Int. Congr. Plant Prot.*, 206.

Buchenauer, H. (1975). 'Systemisch-fungizide Wirkung und Wirkungsmechanismus von triadimefon (MEB 6447)', *Mitt. Biol. Bundesanst. Land-Forstwirtsch. Berlin-Dahlem*, **165**, 154.

Buchenauer, H. (1976). 'Preliminary studies on the mode of action of vinclozolin', *Meded. Fac. Landbouwwet, Rijksuniv. Gent*, **41**, 1509–1519.

Buchenauer, H. (1977a). 'Mode of action of triadimefon in *Ustilago avenae*', *Pestic. Biochem. and Physiol.*, **7**, 309–320.

Buchenauer, H. (1977b). 'Biochemical effects of the systemic fungicides fenarimol and nuarimol in *Ustilago avenae*', *Z. Pflanzenkr. Pflanzenschutz*, **84**, 286–299.

Buchenauer, H. (1977c). 'Mode of action and selectivity of fungicides that interfere with ergosterol biosynthesis', *Proc. 1977 Br. Crop. Prot. Conf.—Pests and Diseases*, **3**, 699–711.

Buchenauer, H. (1977d). 'Mechanism of action of the fungicide imazalil in *Usilago avenae*', *Z. Pflanzenkr. Pflanzenschutz*, **84**, 440–450.

Buchenauer, H. (1978a). 'Inhibition of ergosterol biosynthesis by triadimenol in *Ustilago avenae*', *Pestic. Sci.*, **9**, 507–512.

Buchenauer, H. (1978b). 'Analogy in the mode of action of fluotrimazole and clotrimazole in *Ustilago avenae*', *Pestic. Biochem. and Physiol.*, **8**, 15–25.

Buchenauer, H. (1979). 'Comparative studies on the antifungal activity of triadimefon, triadimenol, fenarimol, nuarimol, imazalil and fluotrimazole *in vitro*', *Z. Pflanzenkr. Pflanzenschutz*, **86**, 341–354.

Buchenauer, H. (1980). 'Interaction of different lipid components with various fungicides', *Z. Pflanzenkr. Pflanzenschutz*, **87**, 335–345.

Buchenauer, H., and Grossman, F. (1977). 'Triadimefon: mode of action in plants and fungi', *Neth. J. Pl. Pathol.*, **83**, (Suppl. 1), 93–103.

Buchenauer, H., and Kemper, K. (1981). 'Wirkungsweise von Propiconazol (Desmel®) in verschiedenen Pilzen'. *Meded. Fac. Landbouwwet. Rijksuniv. Gent*, **46**, 909–921.

Buchenauer, H., and Rohner, E. (1981). 'Effect of triadimefon and triadimenol on growth of various plant species as well as on gibberellin content and sterol metabolism in shoots of barley seedlings', *Pestic. Biochem. Physiol.*, **15**, 58–70.

Calderbank, A. (1971). 'Metabolism and mode of action of dimethirimol and ethirimol', *Acta Pyhtopathol. Acad. Sci. Hung.*, **6**, 355–363.

Cartwright, D. W., Langcake, P., and Ride, J. P. (1980). 'Phytoalexin production in rice and its enhancement by a dichlorocyclopropane fungicide', *Physiol. Plant Pathol.*, **17**, 259–267.

Casjens, S. R., and Morris, A. J. (1965). 'The selective inhibition of protein assembly by gougerotin', *Biochim. Biophys. Acta* **108**, 677–686.

Caskey, C. T., Beaudet, A. L., Scolnick, E. M., and Rosman, M. (1971). 'Hydrolysis of fmet-tRNA by peptidyl transferase', *Proc. Nat. Acad. Sci. U.S.*, **68**, 3163–3167.

Chang, F. N., and Flaks, J. G. (1972). 'Binding of dihydrostreptomycin to *Escherichia coli* ribosomes: characteristics and equilibrium of the reaction', *Antimicrob. Agr. Chemotherap.*, **2**, 294–307.

Chang, T. Y., Telakowski, C., Vanden Heuvel, W., Alberts, A. W., and Vagelos, P. R. (1977). 'Isolation and partial characterization of a cholesterol-requiring mutant of Chinese hamster ovary cells', *Proc. Nat. Acad. Sci.*, **74**, 832–836.

Child, J. J., Defago, G., and Haskins, R. H. (1969). 'The effect of cholesterol and polyene antibiotics on the permeability of the protoplasmic membrane of *Pythium* PRL 2142', *Can. J. Microbiol.*, **15**, 599–603.

Clark, T., Clifford, D. R., Deas, A. H. B., Gendle, P., and Watkins, D. A. M. (1978). 'Photolysis, metabolism and other factors influencing the performance of triadimefon as a powdery mildew fungicide', *Pestic. Sci.*, **9**, 497–506.

Clemons, G. P., and Sisler, H. D. (1969). 'Formation of a fungitoxic derivative from benlate', *Phytopathology*, **59**, 705.

Clerjeau, M., and Simone, J. (1982). 'Apparition en France de souches de mildiou (*Plasmopara viticola*) résistantes aux fongicides de la famille des anilides (métalaxyl, milfurame)', *Le Progrès Agricole et Viticole*, **3**, 59–61.

Clerjeau, M., Lafon, R., Bugaret, Y., and Simone, J. (1982). 'Mode d'action des nouveaux fongicides anti-mildiou chez la vigne', *Phytiatrie-Phytopharmacie*, in press.

Clifford, C. R., and Hislop, E. C. (1971). "The activities of some systemic fungicides against powdery mildew fungi', *Proc. 6th Br. Insectic. Fungic. Conf.*, 438–444.

Coles, C. J., Singer, T. P., White, G. A., and Thorn, G. D. (1978). 'Studies on the binding of carboxin analogs to succinate dehydrogenase', *J. Biol. Chem.*, **253**, 5573–5578.

Colombo, B., Felicetti, L., and Baglioni, C. (1965). 'Inhibition of protein synthesis by cycloheximide in rabbit reticulocytes', *Biochim. Biophys. Res. Commun.*, **18**, 389–395.

Conce, G., and Duvauchelle, S. (1982). 'Le mildiou biologice et méthodes de lutte', *Phytoma*, **337**, 31–34.

Coolbaugh, R. C., and Hamilton, R. (1976). 'Inhibition of *ent*-kaurene oxidation and growth by α-cyclopropyl-α-(*p*-methoxyphenyl)-5-pyrimidine methyl alcohol', *Pl. Physiol.*, **57**, 245–248.

Cope, J. E. (1980). 'Mode of action of miconazole on *Candida albicans*: effect on growth, viability and K+ release', *J. Gen. Microbiol.*, **119**, 245–251.

Crackower, S. H. B. (1972). 'The effects of griseofulvin on mitosis in *Aspergillus nidulans*', *Can. J. Microbiol.*, **18**, 683–687.

Crute, I. R. (1978). 'Studies on new systemic fungicides active against *Bremia lactucae*', in *Abstr. 3rd Int. Congr. Pl. Pathol.*, München, 1978, 019–384.

van Cutsem, J., and Thienpont, D. (1972). 'Miconazole, a broad-spectrum antibiotic agent with antibacterial activity', *Chemotherapy*, **17**, 392–404.

Dähn,U., Hagenmaier, H., Höhne, H., König, W. A., Wolf, G., and Zähner, H. (1976). 'Stoffwechselprodukte von Mikroorganismen 154 Mitteilung. Nikkomycin, ein neuer Hemmstoff der Chitinsynthese bei Pilzen', *Arch. Microbiol.*, **107**, 143–160.

Davidse, L. C., (1973). 'Antimitotic activity of methyl benzimidazol-2-yl-carbamate (MBC) in *Aspergillus nidulans*', *Pest. Biochem. Physiol.*, **3**, 317–325.

Davidse, L. C. (1975). 'Mode of action of methyl benzimidazol-2-yl carbamate (MBC) and some biochemical aspects of acquired resistance against this fungicide in *Aspergillus nidulans*', in *Systemfungizide IV. Int. Symp. Reinhardsbrunn, 1974* (eds. H. Lyr and C. Polter), pp. 137–143, Akademie-Verlag, Berlin.

Davidse, L. C. (1979). 'Mode of action and selectivity of benzimidazole compounds', in *Systemfungizide V. Int. Symp. Reinhardsbrunn, 1977* (eds. H. Lyr and C. Polter), pp. 276–286, Akademie-Verlag, Berlin.

Davidse, L. C. (1981). 'Mechanism of action of metalaxyl in *Phytophthora megasperma f.sp. medicaginis*', *Neth. J. Pl. Pathol.*, **87**, 254 (Abstr.).

Davidse, L. C., and Flach, W. (1977). 'Differential binding of methyl benzimidazol-2-yl carbamate to fungal tubulin as a mechanism of resistance to this antimitotic agent in mutant strains of *Aspergillus nidulans*', *J. Cell. Biol.*, **72**, 174–193.

Davidse, L. C., and Flach, W. (1978). 'Interaction of thiabendazole with fungal tubulin', *Biochim. Biophys. Acta*, **543**, 82–90.

Day, D. A., Arron, G. P., and Laties, G. G. (1978). 'The effect of carboxins on higher plant mitochondria', *FEBS Lett.*, **85**, 99–102.

Day, D. A., and Wiskich, J. T. (1974). 'The oxidation of malate and exogenous reduced nicotinamide adenine dinucleotide by isolated plant mitochondria', *Pl. Physiol.*, **53**, 104–109.

De Brabander, M., van der Veire, R., Aerts, F., Geuens, G., Borgers, M., Desplenter, L., and de Crée, J. (1975). 'A new anti-cancer drug interfering with microtubules. Effects on neoplastic cells cultured *in vitro* and *in vivo*', in *Microtubules and Microtubule Inhibitors* (eds. M. Borgers and M. De Brabander), pp. 509–521, North Holland Publ. Comp., Amsterdam.

De Brabander, M., van der Veire, R., Aerts, F., Geuens, S., and Hoebeke, J. (1976). 'A new culture model facilitating rapid quantitative testing of mitotic spindle inhibition in mammalian cells', *J. Nat. Cancer Inst.*, **56**, 357–363.

Dekker, J. (1963). 'Antibiotics in the control of plant diseases', *Ann. Rev. Microbiol.*, **17**, 243–262.

Dekker, J. (1971). 'Agricultural use of antibiotics', *World Rev. Pest. Control*, **10**, 9–23.

Dekker, J. (1972). 'Resistance', in *Systemic Fungicides* (ed. R. W. Marsh), pp. 156–174, Longman, London and New York.

Delp, C. J. (1981). 'Coping with resistance to plant disease control agents', *Pl. Disease*, **64**, 652–659.

Delp, C. J., and Klöpping, H. L. (1968). 'Performance attributes of a new fungicide and mite ovicide candidate', *Pl. Disease Rep.*, **52**, 95–99.

Despreaux, D., Fritz, R., and Leroux, P. (1982). 'Mode d'action biochimique du cymoxanil', *Phytiatrie-Phytopharmacie*, in press.

Dirske, F. B., Dil, M., Linders, R., and Rietstra, I. (1982). 'Resistance in white rust (*Puccinia horiana*) of chrysanthemum to oxycarboxin and benodanil in the Netherlands, *Meded. Fac. Landbouwwet. Rijksuniv. Gent*, **47**, in press.

Douchet, J. P., Absi, M., Hay, S. J. B., Muntan, L., and Villani, A. (1977). 'European results with DPX 3217 a new fungicide for the control of grape downy mildew and potato late blight', *Proc. 1977 Br. Crop Prot. Conf.—Pests and Diseases*, **2**, 535–540.

Durand, M. C., and Salle, G. (1981). 'Effect of aluminium tris-*O*-ethyl phosphonate (TEPA) on *Lycopersicum esculentum* Mill leaves infected with *Phytophthora capsici* Leon. Cytological and cytochemical study', *Agronomie*, **1**, 723–732.

Edgington, L. V. (1981). 'Structural requirements of systemic fungicides', *Ann. Rev. Phytopathol.*, **19**, 107–124.

Edgington, L. V., and Barron, G. L. (1967). 'Fungitoxic spectrum of oxathiin compounds', Phytopathology, **57**, 1256–1257.

Edgington, L. V., Khew, K. L., and Barron, G. L. (1971). 'Fungitoxic spectrum of Benzimidazole compounds', *Phytopathology*, **61**, 42–44.

Edgington, L. V., Walton, G. S., and Miller, P. M. (1966). 'Fungicide selective for Basidiomycetes', *Science* (*N.Y.*), **153**, 307–308.

Eichhorn, K. W., and Lorenz, D. H. (1978). 'Untersuchungen über die wirkung von vinclozolin gegenüber *Botrytis cinerea in vitro*', *Pflanzenkr. Pflanzenschutz*, **85**, 449–460.

Elias, R. S., Shephard, M. C., Snell, B. K., and Stubbs, J. (1968). '5-*n*-Butyl-2-dimethylamino-4-hydroxy-6-methylpyrimidine: a systemic fungicide', *Nature* (*Lond.*)., **219**, 1160.

Elliott, C. G. (1969). 'Effects of inhibitors of sterol synthesis on growth of *Sordaria* and *Phytophthora*', *J. Gen. Microbiol.*, **56**, 331–343.

Endo, A., Kakiki, K., and Misato, T. (1970). 'Mechanism of action of the antifungal agent polyoxin D', *J. Bact.*, **104**, 189–196.

Ennis, H. L., and Lubin, M. (1964). 'Cycloheximide: aspects of inhibition of protein synthesis in mammalian cells', *Science*, **146**, 1474–1476.

Fettouche, F., Ravise, A., and Bompeix, G. (1981). 'Suppression de la résistance induite-phosethyl Al-chez la tomate a *Phytophthora capsici* avec deux inhibiteurs-glyphosate et acide α-aminooxyacetique', *Agronomie*, **1**, 821–826.

Finney, J. R., Farrell, G. M., and Bent, K. J. (1975). 'Bupirimate—a new fungicide for the control of powdery mildews on apples and other crops', *Proc. 8th Br. Insectic. Fungic. Conf.*, **2**, 667–673.

Fisher, D. J. (1974). 'A note on the mode of action of the systemic fungicide tridemorph', *Pestic. Sci.*, **5**, 219.

Fisher, D. J., and Hayes, A. L. (1981). 'Fungicides, and mode of action Phycomycete fungicides', *Rep. Long Ashton Res. Stn. for 1979*, 137.

Forster, H., Buchenauer, H., and Grossmann, F. (1980a). 'Side effects of the systemic fungicides triadimefon and triadimenol on barley plants. I. Influence on growth and yield', *Z. Pflanzenkr. Pflanzenschutz*, **87**, 473–492.

Forster, H., Buchenauer, H., and Grossmann, F. (1980b). 'Side effects of the systemic fungicides triadimefon and triadimenol on barley plants. II. Cytokinin-like effects', *Z. Pflanzenkr. Pflanzenschutz*, **87**, 640–653.

Forster, H., Buchenauer, H., and Grossmann, F. (1980c). 'Side effects of the systemic fungicides triadimefon and triadimenol on barley plants. III. Further effects on metabolism', *Z. Pflanzenkr. Pflanzenschutz*, **87**, 717–730.

Frasinel, C., Patterson, G. W., and Dutky, S. R. (1978). 'Effect of triarimol on sterol and fatty acid composition of three species of *Chlorella*', *Phytochemistry*, **17**, 1567–1570.

Fritz, R., Leroux, P., and Gredt, M. (1977). 'Mechanism de l'action fongetoxique de la promidone (26019 RP ou Glycophene), de la vinchlozoline et du dicloran sur *Botrytis cinerea* Pers', *Phytopathol. Z.*, **90**, 152–163.

Froyd, J. D., Paget, C. J., Guse, L. R., Dreikorn, B. A., and Pafford, J. L. (1976). 'Tricyclazole: a new systemic fungicide for control of *Pyricularia oryzae* on rice', *Phytopathology*, **66**, 1135–1139.

Fryberg, M., Oehlschlager, A. C., and Unrau, A. M. (1974). 'Sterol biosynthesis in antibiotic-resistant yeast: Nystatin', *Arch. Biochem. Biophys.*, **160**, 83–89.

Fuchs, A., and Drandarevski, Ch. A. (1973). 'Fungitoxic spectrum and effectiveness of triforine *in vitro* and *in vivo*', *Z. Pflanzenkr. Pflanzenschutz*, **80**, 403–417.

Fuchs, A., and Drandarevski, Ch. A. (1976). 'The likelihood of development of resistance to systemic fungicides which inhibit sterol biosynthesis', *Neth. J. Pl. Pathol.*, **82**, 85–87.

Fuchs, A., and Ost, W. (1976). 'Translocation, distribution and metabolism of triforine in plants', *Archs. Envir. Contam. Toxicol.*, **4**, 30–43.

Fuchs, A., de Ruig, S. P., van Tuyl, J. M., and de Vries, F. W. (1977). 'Resistance to triforine: a nonexistent problem?', *Neth. J. Pl. Pathol.*, **83** (Suppl. 1), 189–205.

Fukunaga, K., Misato, T., and Asakawa, M. (1955). 'Blasticidin, a new antiphytopathogenic fungal substance', *Bull. Agr. Chem. Soc. Japan*, **19**, 181.

Garavaglia, C., Lorusso, S., Mirenna, L., Pizzingrilli, G. F., and Santi, R. (1981). 'Modalità dell'azione fungicida dell'antiperonosporico sistemico Galben[R]', *III Simposio chimica degli antiparassitari—Modi di azione*. Piacenza, 26–27.2.81. In press.

Gardiner, J. A., Kirkland, J. J., Klöpping, H. L., Sherman, H. (1974). 'Fate of benomyl in animals', *J. Agric. Fd. Chem.*, **22**, 419–427.

Garg, D. K., and Mehrotra, R. S. (1975). 'Effect of some fungicides on growth, respiration and enzyme activity of *Fusarium solani* f.sp. *pisi*', *Z. Pflanzenschutz*, **10**, 570.

Garibaldi, A., and Tamietti, G. (1980). 'Efficacia dei nouvi fungicidi sistemici nella lotta contro la peronospora del pomodoro', *Inf. Fitopatol.*, **30**, 41–44.

Gasztonyi, M. (1981). 'The diastereomeric ratio in the triadimenol produced by fungal metabolism of triadimefon and its role in fungicide selectivity', *Pestic. Sci.*, **12**, 433–438.

Gasztonyi, M., and Josepovits, G. (1979). 'The activation of triadimefon and its role in the selectivity of fungicide action', *Pestic. Sci.*, **10**, 57–65.

Gaylor, J. L., and Mason, H. S. (1968). 'Investigation of the component reactions of oxidative sterol demethylation. Evidence against participation of cytochrome P-450', *J. Biol. Chem.*, **243**, 4966–4972.

Georgopoulos, S. G., Alexandri, E., and Chrysayi, M. (1972). 'Genetic evidence for the action of oxathiin and thiazole derivatives on the succinate dehydrogenase system of *Ustilago maydis*', *J. Bact.*, **110**, 809–817.

Georgopoulos, S. G., Chrysayi, M., and White, G. A. (1975a). 'Carboxin resistance in the haploid, the heterozygous diploid, and the plant-parasitic dicaryotic phase of *Ustilago maydis*', *Pestic. Biochem. Physiol.*, **5**, 543–551.

Georgopoulos, S. G., Geerligs, J. W. G., and Dekker, J. (1975b). 'Sensitivity of *Ustilago maydis* to pyrazophos and one of its conversion products, and failure to induce resistance with UV-treatment', *Neth. J. Pl. Pathol.*, **81**, 38–41.

Georgopoulos, S. G., Kappas, A., and Hastie, A. C. (1976). 'Induced sectoring in diploid *Aspergillus nidulans* as a criterion of fungitoxicity by interference with hereditary processes', *Phytophthol.*, **66**, 217–220.

Georgopoulos, S. G., Sarris, M., and Ziogas, B. N. (1979). 'Mitotic instability in *Aspergillus nidulans* caused by the fungicides iprodione, procymidone and vinclozolin', *Pestic. Sci.*, **10**, 389–392.

Georgopoulos, S. G., and Sisler, H. D. (1970). 'Gene mutation eliminating antimycin A-tolerant electron transport in *Ustilago maydis*', *J. Bact.*, **103**, 745–750.

Georgopoulos, S. G., and Ziogas, B. N. (1977). 'A new class of carboxin-resistant mutants of *Ustilago maydis*', *Neth. J. Pl. Pathol.*, **83** (Suppl. 1), 235–242.

Gibbons, G. F., and Mitropoulos, K. A. (1972). 'Inhibition of cholesterol biosynthesis by carbon monoxide: accumulation of lanosterol and 24,25-dihydrolanosterol', *Biochem. J.*, **127**, 315–317.

Gibbons, G. F., and Mitropoulos, K. A. (1975). 'Effects of trans-1,4-bis(2-chlorobenzylaminomethyl)cyclohexane dihydrochloride and carbon monoxide on hepatic cholesterol biosynthesis from 4,4-dimethyl sterols *in vitro*', *Biochim. Biophys. Acta*, **380**, 270–281.
Girardet, F., and Buchenauer, H. (1980). 'Mode of action of phenapronil (Sisthane) on *Ustilago avenae* sporidia and *Erisyphe graminis* f.sp. *hordei*', *Meded. Fac. Landouwwet. Rijksuniv. Gent*, **45**, 435–443.
Godchaux, W., Adamson, S. D. and Herbert, E. (1967). 'Effects of cycloheximide on polyribosome function in reticulocytes', *J. Mol. Biol.*, **27**, 57–72.
Gooday, G. W. (1979a). 'The action of polyoxin in fungi', in *Systemfungizide V. Int. Symp.* Reinhardsbrum, 1977 (eds. H. Lyr and C. Polter), pp. 159–168. Akademie-Verlag, Berlin.
Gooday, G. W. (1979b). 'Chitin synthesis and differentiation in *Coprinus cinereus*' in *Fungal Walls and Hyphal Growth* (eds. J. H. Burnett and A. P. J. Trinci), pp. 203–224, Cambridge University Press.
Gooday, G. W., de Rousset-Hall, A., and Hunsley, D. (1976). 'Effect of polyoxin D on chitin synthesis in *Coprinus cinereus*', *Trans. Br. Mycol. Soc.*, **67**, 193–200.
Gottlieb, D., and Shaw, P. D. (1970). 'Mechanism of action of antifungal antibiotics', *Ann. Rev. Phytopathol.*, **8**, 371–397.
Gozzo, F., Masoero, M., and Zagni, A. (1981). 'Recenti sviluppi sulle relazioni attività—struttura di fungicidi sistemici antiperonosporici', *III Simposio chimica degli antiparassitari—Modi di azione*, Piacenza, 26–27.2.1981. In press.
Grove, J. F., Ismay, D., MacMillan, J., Mulholland, T. P. C., and Rogers, M. A. (1951). 'The structure of griseofulvin', *Chem. Ind.*, 219–220.
Grover, R. K., and Chopra, B. L. (1970). 'Adaptation of *Rhizoctonia* species to two oxathiin compounds and manifestations of the adapted isolates', *Acta Phytopathol. Acad. Sci. Hung.*, **5**, 113–121.
Gunatilleke, I. A. U. N., Arst, H. N., and Scazzocchio, C. (1976). 'Three genes determine the carboxin sensitivity of mitochondrial succinate oxidation in *Aspergillus nidulans*', *Genet. Res.*, **26**, 297–305.
ten Haken, P., and Dunn, C. L. (1971). 'Structure-activity relationships in a group of carboxanilides systemically active against broad bean rust (*Uromyces fabae*) and wheat rust (*Puccinia recondita*)', *Proc. 6th Br. Insectic. Fungic. Conf.*, **2**, 453–462.
Halangk, W., and Schewe, T. (1975). 'Untersuchungen zum biochemischen Wirkungsmechanismus von Dexon', in *Systemfungizide, IV Int. Symp. Reinhardsbrunn, 1974* (eds. H. Lyr and C. Polter), pp. 177–182, Akademie-Verlag, Berlin.
Halos, P. M., and Huisman, O. C. (1974). 'Reversal of ethazol inhibition by mycelial extracts from *Pythium* species', *Proc. Am. Phytopathol. Soc.*, **1**, 44 (Abstr).
Halos, P. M., and Huisman, O. C. (1976a). 'Mechanism of tolerance of *Pythium* species to ethazol', *Phytopathology*, **66**, 152–157.
Halos, P. M., and Huisman, O. C. (1976b). 'Inhibition of respiration in *Pythium* species by ethazol', *Phytopathology*, **66**, 158–164.
Hamada, M., Hashimoto, T., Takahashi, T., Yokayoma, S., Miyaka, M., Takeuchi, T., Okami, Y., and Umezawa, H. (1965). 'Antimicrobial activity of kasugamycin', *J. Antibiotics*, **18A**, 104–106.
Hammerschlag, R. S., and Sisler, H. D. (1973). 'Benomyl and methyl-2-benzimidazole carbamate (MBC): Biochemical cytological and chemical aspects of toxicity to *Ustilago maydis* and *Saccharomyces cerevisae*', *Pest. Biochem. Physiol.*, **3**, 42–54.
Hampel, H., and Löcher, F. (1973). 'Control of cereal diseases with carbendazim'. *Proc. 7th Br. Insectic. Fungic. Conf.*, 127–134.
Hanssens, L., Verachtert, H., and von Jagow, G. (1978). 'Participation of ubiquinone in the cyanide-insensitive respiration of *Moniliella tomentosa*', in *Functions of Alternative Terminal Oxidases*, Vol. 49 Proc. 11th FEBS Meeting, Copenhagen, 1977 (eds. H. Degn, D. Lloyd, and G. C. Hill), pp. 47–53, Pergamon Press, Oxford.

Hatefi, Y., and Stiggall, D. L. (1976). 'Metal-containing flavoprotein dehydrogenases', in *The Enzymes* Vol. XIII (ed. P. D. Boyer), pp. 175–297, Academic Press, London.

Hays, P. R., Parks, L. W., Pierce, H. D. Jr., and Oehlschlager, A. C. (1977). 'Accumulation of ergosta-8,4-dien-3β-ol by *Saccharomyces cerevisae* cultured with an azasterol antimycotic agent', *Lipids*, **12**, 666.

Helser, T. L., Davies, J. E., and Dahlberg, J. E. (1972). 'Mechanism of kasugamycin resistance in *Escherichia coli*', *Nature* (*New Biol.*)., **235**, 6–9.

Hendrix, J. W. (1970). 'Sterols in growth and reproduction of fungi', *Ann. Rev. Phytopathol.*, **8**, 111–130.

Henry, M. J., and Sisler, H. D. (1979). 'Effects of Miconazole and Dodecylimidazole on sterol biosynthesis in *Ustilago maydis*', *Antimicrob. Agents Chemother.*, **15**, 603–607.

Henry, M. J., and Sisler, H. D. (1981). 'Inhibition of ergosterol biosynthesis in *Ustilago maydis* by the fungicide 1-[2-(2,4-dichlorphenyl)-4-ethyl-1,3-dioxolan-2-ylmethyl]-1H-1,2,4-triazole', *Pestic. Sci.*, **12**, 98–102.

Hess, C., and Löcher, F. (1975). 'Experiences with vinclozolin in the control of *Botrytis cinerea* in strawberries', *Proc. Br. Insectic. Fungic. Conf., 8th, 1975*, **2**, 693–696.

Hickey, E. L., and Coffey, M. D. (1980). 'The effects of Ridomil on *Peronospora pisi* parasitizing *Pisum sativum*: an ultrastructural investigation', *Physiol. Plant Pathol.*, **17**, 199–204.

Hildick-Smith, G. (1971). 'Chemotherapy of fungal infections', in *Drills Pharmacology in Medicine* (ed. J. R. DiPalma) 4th ed., pp. 1755–1769, McGraw-Hill, New York.

Hippe, S., Buchenauer, H., and Grossmann, F. (1980). 'Influence of triadimefon on ultrastructure of sporidia of *Ustilago avenae*', *Z. Pflanzenkr. Pflanzenschutz*, **87**, 423–426.

Hisada, Y., Maeda, K., Tottori, N., and Kawase, Y. (1976). 'Plant disease control by *N*-(3′,5′-dichlorophenyl)-1,2-dimethylcyclopropane-1,2-dicarboximide', *J. Pesticide Sci.*, **1**, 145–149.

Hisada, Y., Kato, T., and Kawase, Y. (1978). 'The mechanisms of antifungal action of procymidone in *Botrytis cinerea*', *Ann. Phytopathol. Soc. Japan*, **44**, 509–518.

Hock, W. K., and Sisler, H. D. (1969). 'Metabolism of chloroneb by *Rhizoctonia solani* and other fungi', *J. Agr. Food Chem.*, **17**, 123–128.

Hollomon, D. W. (1979a). 'Evidence that ethirimol may interfere with adenine metabolism during primary infection of barley powdery mildew', *Pestic. Biochem. Physiol.*, **10**, 181–189.

Hollomon, D. W. (1979b). 'Specificity of ethirimol in relation to inhibition of the enzyme adenosine deaminase', *Proc. 1979 Br. Crop Prot. Conf.—Pests and Diseases*, **1**, 251–256.

Hollomon, D. W. (1981). 'Mode of action of hydroxypyrimidine fungicides', *Neth. J. Pl. Pathol.*, **87**, 255 (Abstr.).

Hollomon, D. W., and Chamberlain, K. (1981). 'Hydroxypyrimidine fungicides inhibit adenosine deaminase in barley powdery mildew', *Pestic. Biochem. Physiol.*, **16**, 158–169.

Hori, M., Eguchi, J., Kakiki, K., and Misato, T. (1974a). 'Studies on the mode of action of polyoxins VI. Effect of polyoxin B on chitin synthesis in polyoxin-sensitive and resistant strains of *Alternaria kikuchiana*', *J. Antibiotics*, **27**, 260–266.

Hori, M., Kakiki, K., and Misato, T. (1974b). 'Further study on the relation of polyoxin structure to chitin synthetase inhibition', *Agric. Biol. Chem.*, **38**, 691–698.

Hori, M., Kakiki, K., and Misato, T. (1974c). 'Interaction between polyoxin and active center of chitin synthetase', *Agric. Biol. Chem.*, **38**, 699–705.

Hori, M., Kakiki, K., Suzuki, S., and Misato, T. (1971). 'Studies on the mode of action of polyoxins III. Relation of polyoxin structure to chitin synthetase inhibition', *Agric. Biol. Chem.*, **35**, 1280–1291.

Howard, R. J., and Aist, J. R. (1976). 'Evidence for an antitubulin mode of action of MBC from antagonistic effects of heavy water', *Proc. Am. Phytopathol. Soc.*, **3**, 306.

Huang, T. K., Misatom, T., and Asuyama, H. (1964). 'Effect of blasticidin S on protein synthesis of *Piricularia oryzae*', *J. Antibiotics*, **17A**, 65–70.
Ishiyama, T., Hara, I., Matsuoka, M., Sato, K., Shimada, S., Izawa, R., Hashimoto, T., Hamada, M., Okami, Y., Takeuchi, T., and Umezawa, H. (1965). 'Studies on the preventive effect of kasugamycin on rice blast', *J. Antibiotics*, **18A**, 115–119.
Ishizaka, H., Mitsuoka, K., and Kunoh, H. (1974). 'Effect of polyoxin on fungi I. Optical microscopic observations of mycelia of *Alternaria kikuchiana*', *Ann. Phytopathol. Soc., Japan.*, **40**, 433–438.
Isono, K., Nagatsu, J., Kobinata, K., Sasaki, K., and Suzuki, S. (1967). 'Studies on polyoxins, antifungal antibiotics II. Isolation and characterisation of polyoxins C, D, E, F, G, H and I', *Agric. Biol. Chem.*, **31**, 190.
Iwasaki, S., Nozoe, S., and Okuda, S. (1969). 'Isolation and structural elucidation of a phytotoxic substance produced by *Pyricularia oryzae* Cav.', *Tetrahedron Lett.*, **45**, 3977–3980.
Iwata, M., Suzuki, Y., Watanabe, T., Mase, S., and Sekizawa, Y. (1980). 'Effect of probenazole on the activities of enzymes related to the resistant reaction in rice plant', *Ann. Phytopathol. Soc. Japan*, **46**, 297–306.
Jank, B., and Grossmann, F. (1971). '2-Methyl-5,6-dihydro-4-H-pyran-3-carboxylic acid anilide: a new systemic fungicide against smut diseases', *Pestic. Sci.*, **2**, 43–44.
Jimenez, A., Littlewood, B., and Davies, J. (1972). 'Inhibition of protein synthesis in yeast', in *Molecular Mechanisms of Antibiotic Action on Protein Bisosynthesis and Membranes* (eds. E. Munoz, F. Garcia-Ferrandiz, and D. Vazquez), Elsevier, New York, pp. 292–306.
Johnson, D., and Subden, R. (1977). 'Polyene antibiotic affinities for the sterols of resistant and sensitive strains of *Neurospora crassa*', *Can. J. Microbiol.*, **23**, 113–115.
Kaars Sijpesteijn, A. (1977). 'Effects on fungal pathogens', in *Systemic Fungicides* (ed. R. W. Marsh), pp. 131–159, Longman, London and New York.
Kaars Sijpesteijn, A. (1982). 'Mechanism of action of fungicides' in *Fungicide Resistance in Crop Protection* (eds. J. Dekker and S. G. Georgopoulos), Pudoc, Wageningen.
Kaars Sijpesteijn, A., Kerkenaar, A., and Overeem, J. C. (1974). 'Observations on selectivity and mode of action of prothiocarb (SN 41703)', *Meded. Fac. Landbouwwet. Rijksuniv. Gent*, **39**, 1027–34.
Kaji, A. (1973). 'Mechanism of protein synthesis and the use of inhibitors in the study of protein synthesis', in *Progress in Molecular and Subcellular Biology* (ed. F. E. Hahn) Vol. 3, Springer-Verlag, New York, pp. 85–143.
Kaji, H., and Tanaka, Y. (1968). 'Binding of dihydrostreptomycin to ribosomal subunits', *J. Mol. Biol.*, **32**, 221–230.
Kakiki, K., Maeda, T., Abe, H., and Misato, T. (1969). 'Studies on mode of action of organo phosphorus fungicide, Kitazin Part I. Effect on respiration, protein synthesis, necleic acid synthesis, cell wall synthesis, and leakage of intracellular substances', *Nippon Nogeikagaku Kaishi*, **43**, 37–44.
Kakiki, K., and Misato, T. (1979a). 'Effects of isoprothiolane on the uptake of sugars and fungal morphology', *J. Pestic. Sci.*, **4**, 129–135.
Kakiki, K., and Misato, T. (1979b). 'Effect of isoprothiolane on fatty acid synthesis', *J. Pestic. Sci.*, **4**, 305–313.
Kappas, A., and Georgopoulos, S. G. (1974). 'Interference of griseofulvin with the segregation of chromosomes at mitosis in diploid *Aspergillus nidulans*', *J. Bact.*, **119**, 334–335.
Katagiri, M., and Uesugi, Y. (1977). 'Similarities between the fungicidal action of isoprothiolane and organophosphorus thiolate fungicides', *Phytopathology*, **67**, 1415–1417.
Kato, T. (1977). 'Mechanism of action of Denmert, an inhibitor of sterol biosynthesis', *Neth. J. Pl. Pathol.*, **83**, (Suppl. 1), 113–120.

Kato, T., and Kawase, Y. (1976). 'Selective inhibition of the demethylation at C-14 in ergosterol biosynthesis by the fungicide Denmert (S-1358)', *Agric. Biol. Chem.*, **40**, 2379–2388.

Kato, T., Shoami, M., and Kawase, Y. (1980). 'Comparison of tridemorph with buthiobate in antifungal mode of action', *J. Pestic. Sci.*, **5**, 69–79.

Kato, T., Tanaka, S., Ueda, M., and Kawase, Y. (1974). 'Effects of the fungicide, S-1358, on general metabolism and lipid biosynthesis in *Monilinia fructigena*', *Agric. Biol. Chem.*, **38**, 2377–2384.

Kato, T., Tanaka, S., Ueda, M., and Kawase, Y. (1975a). 'Inhibition of sterol biosynthesis in *Monilinia fructigena* by the fungicide, S-1358', *Agric. Biol. Chem.*, **39**, 169–174.

Kato, T., Tanaka, S., Yamamoto, S., Kawase, Y., and Ueda, M. (1975b). 'Fungitoxic properties of a *N*-3-pyridylimidodithiocarbonate derivative', *Ann. Phytopathol. Soc. Japan*, **41**, 1–8.

Keller, F. A., and Cabib, E. (1971). 'Chitin and yeast budding. Properties of chitin synthetase from *Saccharomyces carlsbergensis*', *J. Biol. Chem.*, **246**, 160–166.

Kerkenaar, A. (1981a). 'On the antifungal mode of action of metalaxyl, an inhibitor of nucleic acid synthesis in *Pythium splendens*', *Pestic. Biochem. Physiol.*, **16**, 1–13.

Kerkenaar, A. (1981b). 'On the mode of action of metalaxyl in *Pythium splendens*', *Neth. J. Pl. Pathol.*, **87**, 254 (Abstr.).

Kerkenaar, A., Barug, D., and Kaars Sijpesteijn, A. (1979). 'On the antifungal mode of action of tridemorph', *Pestic. Biochem. Physiol.*, **12**, 195–204.

Kerkenaar, A., and Kaars Sijpesteijn, A. (1977). 'On the mode of action of prothiocarb', *Neth. J. Pl. Pathol.*, **83**, (Suppl. 1), 145–152.

Kerkenaar, A., and Kaars Sijpesteijn, A. (1979). 'On a difference in the antifungal activity of tridemorph and its formulated product, Calixin', *Pestic. Biochem. Physiol.*, **12**, 124–129.

Kerkenaar, A., and Kaars Sijpesteijn, A. (1981). 'Antifungal activity of metalaxyl and furalaxyl', *Pestic. Biochem. Physiol.*, **15**, 71–78.

Kerkenaar, A., Uchiyama, M., and Versluis, G. G. (1981). 'Specific effects of tridemorph on sterol biosynthesis in *Ustilago maydis*', *Pestic. Biochem. Physiol.*, **16**, 97–104.

Kerridge, D. (1958). 'The effect of Actidone and other antifungal agents on nucleic acid and protein synthesis in *Saccharomyces carlsbergensis*', *J. Gen. Microbiol.*, **19**, 497–506.

Kinoshita, T., Tanaka, N., and Umezawa, H. (1970). 'Binding of blasticidin S to ribosomes', *J. Antibiotics*, **23A**, 288–290.

Kinsky, S. C., Haxley, J., Kinsky, C. B., Damel, R. A., and van Deenen, L. L. M. (1968). 'Effect of cholesterol incorporation on sensitivity of liposomes to the polyene antibiotic filipin', *Biochim. Biophys. Acta*, **152**, 174–185.

Klöpping, H. L., and Delp, C. J. (1980). '2-Cyano-*N*-[(ethylamino)carbonyl]-2-(methoxyimino)acetamide, a new fungicide', *J. Agric. Fd. Chem.*, **28**, 467–468.

Kluge, E. (1978). 'Vergleichende Untersuchungen über die Wirksamkeit von Systemfungiziden gegen Oomyzeten', *Arch. Phytopathol. Pflanzenschutz*, **14**, 115–122.

Kodama, O., Yamada, H., and Akatsuka, T. (1979). 'Kitazin, P, inhibitor of phosphatidylcholine biosynthesis in *Pyricularia oryzae*', *Agric. Biol. Chem.*, **43**, 1719–1725.

Kodama, O., Yamashita, K., and Akatsuka, T. (1980). 'Edifenphos, inhibitor of phosphatidylcholine biosynthesis in *Pyricularia oryzae*', *Agric. Biol. Chem.*, **44**, 1015–1021.

Kraus, P. (1979). 'Studies on the mechanism of action of Baycor'. *Pflanzenschutz-Nachr. Bayer*, **32**, 17–30.

Lacroix, L., Bic, G., Burqaud, L., Guillot, M., Leblanc, R., Riottot, R., and Sauli, M. (1974). 'Etude des propriétés antifongiques d'une nouvelle famille de dérivés de l'hydantione et un particulier du 26019 R.P.', *Phytiatr. Phytopharm.*, **23**, 165–174.

Lala, A. K., Lin, H. K., and Bloch, K. (1978). 'The effect of some alkyl derivatives of cholesterol on the permeability properties and microviscosites of model membranes', *Bioorg. Chem.*, **7**, 437–445.

Langcake, P. (1974a). 'Sterols in potato leaves and their effects on growth and sporulation of *Phytophthora infestans*', *Trans. Brit. Mycol. Soc.*, **63**, 573–586.

Langcake, P. (1974b). 'Uptake of sterols by *Phytophthora infestans*, their intracellular distribution and metabolism', *Trans. Brit. Mycol. Soc.*, **64**, 55–65.

Langcake, P. (1981). 'Alternative chemical agents for controlling plant disease', *Phil. Trans. R. Soc. Lond. B.*, **295**, 83–101.

Langcake, P., and Pryce, R. J. (1977). 'A new class of phytoalexins from grapevine', *Experientia*, **33**, 151–152.

Leach, B. E., Ford, J. H., and Whiffen, A. J. (1947). 'Actidione, an antibiotic from *Streptomyces griseus*', *J. Am. Chem. Soc.*, **69**, 474.

Lee, C., Sottocasa, G. L., and Ernster, L. (1967). 'Use of artificial electron acceptors for abbreviated phosphorylating electron transport: flavin-cytochrome *c*', in *Methods in Enzymology* Vol. X (eds. R. W. Estabrook and M. E. Pullman), pp. 33–37, Academic Press, New York.

Leroux, P. (1981). 'Les modes d'action des substances antifongiques à usage agricoles', *La Defence des Vegetaux*, **207**, 59–83.

Leroux, P., and Gredt, M. (1976). 'Comparaisons des modes d'action fongitoxique du triadimefon (MEB 6447), du triarimol et de la triforine', *Phytopathol. Z.*, **86**, 276–279.

Leroux, P., and Gredt, M. (1978a). 'Etude de l'action d'acides gras, de sterols, et de divers dérives isopreniques vis à vis de quelques fongicides', *Phytopathol. Z.*, **91**, 177–181.

Leroux, P., and Gredt, M. (1978b). 'Effects of different systemic fungicides on ergosterol biosynthesis in *Botrytis cinerea*, *Penicillium expansum* and *Ustilago maydis*', *Ann. Phytopathol.*, **10**, 45–60.

Leroux, P., and Gredt, M. (1978c). 'Effet de l'imazalil [1-(2-(2,4-dichlorophényl)-2-(2-propenyloxy)éthyl)-1H-imidazole] sur la biosynthese de l'ergosterol chez *Penicillium expansum* Link'. *C.R. Acad. Sci. Ser. D.*, **286**, 427–9.

Leroux, P., and Gredt, M. (1982). 'Résistance croisée entre les fongicides anti-mildiou: quelques résultats de laboratoire', *Phytiatrie-Phytopharmacie*, in press.

Leroux, P., Gredt, M., and Fritz, R. (1976). 'Similitudes et différences entre les modes d'action de l'imazalil, du triadimefon, du triarimol et de la triforine', *Phytiatrie-Phytopharmacie*, **25**, 317–334.

Leroux, P., Gredt, M., and Fritz, R. (1981). 'Resistance to 3,5-dichlorophenyl-*N*-cyclic imide fungicides', *Neth. J. Pl. Pathol.*, **87**, 244–245.

Lodish, H. F., Houseman, D., and Jacobson, M. (1971). 'Initiation of hemoglobin synthesis. Specific inhibition by antibiotics and bacteriophage ribonucleic acid', *Biochemistry*, **10**, 2348–2356.

Lopez-Romero, E., and Ruiz-Herrera, J. (1976). 'Synthesis of chitin by particulate preparations from *Aspergillus flavus*', *Antonie van Leeuwenhoek*, **42**, 261–276.

Luc, L. H., Sung, P. Y., and Bochow, H. (1971). 'Untersuchungen über das Verhalten systemfungizidtoleranter Laborkulturen von *Sclerotinia sclerotiorum* und *Fusarium solani* in Infektions- und Bekämpfungsversuchen', *Arch. Pflanzenschutz*, **7**, 91–102.

Luzzatto, L., Apirion, D., and Schlessinger, D. (1969). 'Polyribosome depletion and blockage of the ribosome cycle by Streptomycin in *Escherichia coli*', *J. Mol. Biol.*, **42**, 315–335.

Lyr, H. (1977). 'Effect of fungicides on energy production and intermediary metabolism', in *Antifungal Compounds* Vol. 2 (eds. M. R. Siegel and H. D. Sisler), pp. 301–332, Marcel Dekker, New York.

Lyr, H., Casperson, G., and Laussmann, B. (1977). 'Wirkungsmechanismus von Terrazol bei *Mucor mucedo*', *Z. Allg. Mikrobiol.*, **17**, 117–129.

Lyr, H., Laussmann, B., and Casperson, G. (1975). 'Wirkungsmechanismus von Terrazol', *Z. Allg. Mikrobiol.*, **15**, 345–355.

Lyr, H., Luthardt, W., and Ritter, G. (1971). 'Wirkungsweise von Oxathiin-Derivaten auf die Physiologie sensitiver und insensitiver Hefearten', *Z. Allg. Mikrobiol.*, **11**, 373–385.

Lyr, H., and Schewe, T. (1975). 'On the mechanism of the cyanide-insensitive alternative pathway of respiration in fungi and higher plants and the nature of the alternative terminal oxidase', *Acta. Biol. Med. Ger.*, **34**, 1631–1641.

McCorkindale, N. J. (1976). 'The biosynthesis of terpenes and steroids', in *The Filamentous Fungi. Vol. II. Biosynthesis and Metabolism*' (eds. J. E. Smith and D. R. Berry), pp. 369–422, Edward Arnold, London.

McKeehan, W., and Hardesty, B. (1969). 'Mechanism of cycloheximide inhibition of protein synthesis in rabbit reticulocytes', *Biochem. Biophys. Res. Commun.*, **36**, 625–630.

Maeda, M., Tanaka, Y., Doi, O., Nakano, A., Nishijima, M., and Akamatsu, Y. (1981). 'Two types of inhibitor of phospholipid methylation in MOPC-31C cells with modified membrane phospholipids', *Biochim. Biophys. Acta*, **666**, 275–283.

Maeda, T., Abe, H., Kakiki, K., and Misato, T. (1970). 'Studies on the mode of action of organophosphorus fungicide, Kitazin Part II. Accumulation of an amino sugar derivative on Kitazin-treated mycelia of *Pyricularia oryzae*', *Agric. Biol. Chem.*, **34**, 700–709.

Mahler, H. R., Sarkar, N. K., Vernon, L. P., and Alberty, R. A. (1952). 'Studies on diphosphopyridine nucleotide-cytochrome *c* reductase', *J. Biol. Chem.*, **199**, 585–597.

Malawista, S. E., Sato, H., and Bensch, K. G. (1968). 'Vimblastine and Griseofulvin reversibly disrupt the living mitotic spindle', *Science*, **160**, 770–772.

Marriott, M. S. (1980). 'Inhibition of sterol biosynthesis in *Candida albicans* by imidazole-containing antifungals', *J. Gen. Microbiol.*, **117**, 253–255.

Masukawa, H., Tanaka, N., and Umezawa, H. (1968). 'Inhibition by kasugamycin of protein synthesis in *Piricularia oryzae*', *J. Antibiotics*, **21A**, 73–74.

Mathre, D. E. (1968). 'Uptake and binding of oxathiin systemic fungicides by resistant and sensitive fungi', *Phytopathology*, **58**, 1464–1469.

Mathre, D. E. (1970). 'Mode of action of oxathiin systemic fungicides. I. Effect of carboxin and oxycarboxin on the general metabolism of several Basidiomycetes', *Phytopathology*, **60**, 671–676.

Mathre, D. E. (1971a). 'Mode of action of oxathiin systemic fungicides III. Effect on mitochondrial activities', *Pestic. Biochem. Physiol.*, **1**, 216–224.

Mathre, D. E. (1971b). 'Mode of action of oxathiin systemic fungicides structure-activity relationships', *J. Agric. Fd. Chem.*, **19**, 872–874.

Matolcsy, G., Hamran, M., and Bordas, B. (1973). 'Assay of hypocholesteemic agents for antifungal activity', *Pestic. Sci.*, **4**, 267–271.

Matsumae, A., and Cantino, E. C. (1971). 'Sensitivity of the spores of *Blastocladiella emersonii* and related fungi to antibiotics and some other drugs', *J. Antibiotics*, **24**, 77–84.

Matsumoto, K., Suzuki, Y., Mase, S., Watanabe, T., and Sekizawa, Y. (1980). 'On the relationship between plant hormones and rice blast resistance', *Ann. Phytopathol. Soc. Japan*, **46**, 307–314.

Misato, T., and Kakiki, K. (1977). 'Inhibition of fungal cell wall synthesis and cell membrane function', in *Antifungal Compounds Vol. 2* (eds. M. R. Siegel and H. D. Sisler), pp. 277–300, Marcel Dekker, New York.

Mitani, M., and Inoue, Y. (1968). 'Antagonists of antifungal substance polyoxin', *J. Antibiotics*, **21**, 492–496.

Mitropoulos, K. A., and Gibbons, G. F. (1976). 'Effect of triarimol on cholesterol biosynthesis in rat-liver subcellular fractions', *Biochem. Biophys. Res. Comm.*, **71**, 892–900.

Modolell, J., and Davis, B. D. (1969). 'Mechanism of inhibition of ribosomes by streptomycin', *Nature*, **224**, 345–348.

Modolell, J., and Davis, B. D. (1970). 'Breakdown by streptomycin of initiation complexes formed on ribosomes of *Escherichia coli*', *Proc. Nat. Acad. Sci. U.S.*, **67**, 1148–1155.

Moore, A. L., and Rich, P. R. (1980). 'The bioenergetics of plant mitochondria', *TIBS*, **5**, 284–288.

Moser, H., and Vogel, Ch. (1978). 'Preparation and biological activity of the enantiomers of CGA 48988, a new systemic fungicide', in *Abstr. 4th Int. Congr. Pestic. Chem. (IUPAC)*, Zürich, 1978, II–310.

Mowery, P. C., Ackrell, B. A. C., Singer, T. P., White, G. A., and Thorn, G. D. (1976). 'Carboxins: powerful selective inhibitors of succinate oxidation in animal tissues', *Biochem. Biophys. Res. Comm.*, **71**, 354–361.

Mowery, P. C., Steenkamp, D. J., Ackrell, B. A. C., Singer, T. P., and White, G. A. (1977). 'Inhibition of mammalian succinate dehydrogenase by carboxins', *Arch. Biochem. Biophys.*, **178**, 495–506.

Nelson, R. R., Huisingh, D., and Webster, R. K. (1967). 'Sexual differentiation in *Cochliobolus carbonum* as influenced by inhibition and repair of steroid biosynthesis', *Phytopathology*, **57**, 1081–1085.

Nes, W. R. (1974). 'Role of sterols in membranes', *Lipids*, **9**, 596–612.

de Nollin, S., and Borgers, M. (1976). 'An ultrastructural and cytochemical study of *Candida albicans* after *in vitro* treatment with imidazoles', *Mycosen*, **19**, 317–328.

O'Brien, R. G. (1978). 'Systemic chemicals for tobacco blue mould control', *Pl. Disease Rep.*, **62**, 277–279.

Obrig, T. G., Culp, W., McKeenan, W., and Hardesty, B. (1971). 'The mechanism by which cycloheximide and related glutarimide antibiotics inhibit peptide synthesis on reticulocyte ribosomes', *J. Biol. Chem.*, **246**, 174–181.

Odds, F. C. (1980). 'Laboratory evaluation of antifungal agents; a comparative study of five imidazole derivatives of clinical importance', *J. Antimicrob. Agents Chemotherap.*, **6**, 749–761.

Okuyama, A., Machiyama, N., Konishita, T., and Tanaka, N. (1971). 'Inhibition by kasugamycin of initiation complex formation on 30S ribosomes', *Biochem. Biophys. Res. Commun.*, **43**, 196–199.

Otake, N., Takeuchi, S., Endo, T., and Yonehara, H. (1965). 'The structure of blasticidin S', *Tetrahedron Letters*, **19**, 1411.

Oxford, A. E., Raistrick, H., and Simonart, P. (1939). 'Griseofulvin, $C_{17}H_{17}O_6Cl$, a metabolic product of *Penicillium griseofulvum* dierckx', *Biochem. J.*, **33**, 240–248.

Ozaki, M., Mizushima, S., and Nomura, N. (1969). 'Identification and function characterisation of the protein controlled by the streptomycin-resistant locus in *E. coli*', *Nature*, **222**, 333–339.

Papavizas, G. C., O'Neill, N. R., and Lewis, J. A. (1978). 'Fungistatic activity of propyl-*N*-(γ-dimethylaminopropyl)carbamate on *Pythium* spp. and its reversal by sterols', *Phytopathology*, **68**, 1667–1671.

Pappas, A. C., and Fisher, D. J. (1979). 'A comparison of the modes of action of vinclozolin, procymidone, iprodione and prochloraz against *Botrytis cinerea*', *Pestic. Sci.*, **10**, 239–246.

Pearson, A. J. A. (1977). 'Fungicides—The new generation', *Span*, 129–132.

Pestka, S. (1971). 'Inhibitors of ribosome functions', *Annu. Rev. Microbiol.*, **25**, 487–562.

Pestka, S. (1972). 'Studies on transfer ribonucleic acid-ribosomes complexes XIX. Effect of antibiotics on peptidyl puromycin synthesis on polyribosomes from *Escherichia coli*', *J. Biol. Chem.*, **247**, 4669–4678.

Pestka, S., Rosenfeld, H., Harris, R., and Hintikka. (1972). 'Studies on transfer ribonucleic acid-ribosome complexes XXI. Effect of antibiotics on peptidyl-puromycin synthesis by mammalian polyribosomes', *J. Biol. Chem.*, **247**, 6895–6900.

Polter, C., and Casperson, G. (1979). 'The influence of tridemorph on the lipid composition and the ultrastructure of *Botrytis cinerea*', in *Systemfungizide, V Int. Symp. Reinhhardsbrunn, 1977* (eds. H. Lyr and C. Polter) Akademie-Verlag, Berlin.

Pommer, E. H., and Kradel, J. (1969). 'Mebenil (BAS 3050 F), a new compound with specific action against some Basidiomycetes', *Proc. 5th Br. Insectic. Fungic. Conf.*, **2**, 563–568.

Pommer, E.-H., and Mangold, D. (1975). 'Vinchlozolin (BAS 352F), ein neuer wirkstoff zur bekampflung von *Botrytis cinerea*', *Meded. Fac. Landbouwwet. Rijksuniv. Gent.*, **40**, 713–722.

Radzuhn, B., and Casperson, G. (1979). 'Zum Wirkungsmechanismus von Terrazol', in *Systemfungizide, V Int. Symp. Reinhardsbrunn, 1977* (eds. H. Lyr and C. Polter), pp. 195–206, Akademie-Verlag, Berlin.

Ragan, C. I. (1980). 'The molecular organization of NADH dehydrogenase', in *Subcellular Biochemistry*, Vol. 7 (ed. D. B. Roodyn), pp. 267–307, Plenum Press, New York.

Ragsdale, N. N. (1975). 'Specific effects of triarimol on sterol biosynthesis in *Ustilago maydis*', *Biochim. Biophys. Acta*, **380**, 81–96.

Ragsdale, N. N., and Sisler, H. D. (1970). 'Metabolic effects related to fungitoxicity of carboxin', *Phytopathology*, **60**, 1422–1427.

Ragsdale, N. N., and Sisler, H. D. (1972). 'Inhibition of ergosterol synthesis in *Ustilago maydis* by the fungicide triarimol', *Biochem. Biophys. Res. Comm.*, **46**, 2048–2053.

Ragsdale, N. N., and Sisler, H. D. (1973). 'Mode of action of triarimol in *Ustilago maydis*', *Pestic. Biochem. Physiol.*, **3**, 20–29.

Ragsdale, N. N., and Sisler, H. D. (1975). 'Origin of excess fatty acids in *Ustilago maydis* treated with sterol inhibitors', *Proc. Am. Phytopathol. Soc.*, **2**, 55 (Abstr.).

Ragsdale, N. N., and de Waard, M. A. (1977). 'Sterol patterns in fenarimol-susceptible and tolerant strains of *Aspergillus nidulans*', *Proc. Am. Phytopathol. Soc.*, **4**, 218 (Abstr.).

Raynal, G., Ravise, A., and Bompeix, G. (1980). 'Action of Aluminium tris-*O*-ethyl phosphonate on *Plasmopara viticola* pathogenicity and on stimulation of defense reactions of grapevine', *Ann. Phytopathol.*, **12**, 163–175.

Reilly, C. C., and Lamoureux, G. L. (1981). 'The effects of the fungicide, iprodione, on the mycelium of *Sclerotinia sclerotiorum*', *Phytopathol.*, **71**, 722–727.

Reman, F. C., Demel, R. A., de Gier, J., van Deenen, L. L. M., Eibl, H., and Westphal, O. (1969). 'Studies on the lysis of red cells and bimolecular lipid leaflets by synthetic lysolecithins, lecithins and structural analogs', *Chem. Phys. Lipids*, **3**, 221–233.

Rhodes, R. C., Pease, H. L., and Brantley, R. K. (1971), 'Fate of C^{14}-labelled chloroneb in plants and soils', *J. Agric. Fd Chem.*, **19**, 745–749.

Rich, P. R., and Bonner, W. D., jr. (1977). 'Epr studies of higher plant mitochondria in relation to alternative respiratory oxidations', *Pl. Physiol.*, **59** (*Suppl.*), 59 (Abstr.).

Rich, P. R., and Bonner, W. D., jr. (1978). 'The nature and location of cyanide and antimycin resistant respiration in higher plants', in *Functions of Alternative Terminal Oxidases*, Vol. 49 Proc. 11th FEBS Meeting, Copenhagen, 1977 (eds. H. Degn, D. Lloyd, and G. C. Hill), pp. 149–158, Pergamon Press, Oxford.

Richmond, D. V., and Phillips, A. (1975). 'The effect of benomyl and carbendazim on mitosis in hyphae of *Botrytis cinerea*. Pers. ex. Fr. and roots of *Allium cepa L.*', *Pest. Biochem. Physiol.*, **5**, 367–379.

Rose, A. H. (1976). 'Chemical nature of membrane components', in *The Filamentous Fungi, Vol. II Biosynthesis and Metabolism* (eds. J. E. Smith and D. R. Berry), p. 323, Edward Arnold, London.

Roussel, C. (1978). 'Nouveaux fongicides contre le mildou de la vigne. Des découvertes deja prometteuses, des mises au point encore nécessaires', *Phytoma*, **30**, 13–14.

Sampson, M. J. (1969). 'The mode of action of a new group of species specific pyrimidine fungicides', *Proc. 5th Br. Insectic. Fungic. Conf.*, **2**, 483–487.

Sankhla, B., Sankhla, H. C., and Delela, G. G. (1970). 'Adaptation of *Alternaria triticina* Prasada and Prabhu to fungicides', *Rajasthan J. Agric. Sci.*, **1**, 116–119.

Sasaski, S., Ohta, N., Yamaguchi, I., Kuroda, S., and Misato, T. (1952). 'Studies on polyoxin action. Part I. Effect on respiration and synthesis of protein, nucleic acids and cell-wall of fungi'. *J. Agric. Chem. Soc. Japan*, **42**, 633–638.

Scallen, T. J., Srikantaiah, M. V., Seetharam, B., Hansbury, E., and Gavey, K. L. (1974). 'Sterol Carrier Protein Hypothesis', *Fed. Proc.*, **33**, 1733–1746.

Schatz, A., Bugie, E., and Waksman, S. A. (1944). 'Streptomycin, a substance exhibiting antibiotic activity against Gram-positive and Gram-negative bacteria', *Proc. Soc. Exp. Biol. Med.*, **55**, 66–69.

Schenkman, J. B., Remmer, H., and Estabrook, R. W. (1967). 'Spectral studies of drug interaction with hepatic microsomal cytochrome', *Mol. Pharmacol.*, **3**, 113–123.

Scherphof, G. L., Scarpa, A., and van Toorenenbergen, A. (1972). 'The effect of local anesthetics on the hydrolysis of free and membrane-bound phospholipids catalyzed by various phospholipases', *Biochim. Biophys. Acta*, **270**, 226–240.

Schewe, T., Hiebsch, Ch., and Halangk, W. (1975). 'Zur Wirkung des Systemfungizids Dexon auf einige NADH-Dehydrogenasen', *Acta Biol. Med. Ger.*, **34**, 1767–1775.

Schewe, T., Rapoport, S., Böhme, G., and Kunz, W. (1973). 'Zum Angriffspunkt des Systemfungizids Carboxin in der Atmungskette', *Acta Biol. Med. Ger.*, **31**, 73–86.

Schiltz, P., Delon, R., Cazamajour, F., Podeur, G., and Boulogne, R. (1977). 'Comparaison de quelques fongistatiques et de produits endothérapiques pour la lutte contre le mildiou du tabac', *Ann. Tab. Sect. II.*, **14**, 127–154.

Schlüter, K., and Weltzien, H. C. (1971). 'Ein beitrag zur Wirkungsweise systemischer Fungizide auf *Erysiphe graminis*', *Meded. Fac. Landbouwwet. Rijksuniv. Gent*, **36** 1159–1164.

von Schmeling, B., and Kulka, M. (1966). 'Systemic fungicidal activity of 1,4-oxathiin derivatives', *Science (N.Y.)*, **152**, 659–660.

Schmitt, P., and Benveniste, P. (1979). 'Effect of fenarimol on sterol biosynthesis in suspension cultures of bramble cells', *Phytochemistry*, **18**, 1659–1665.

Schmitt, P., Benveniste, P., and Leroux, P. (1981). 'Accumulation of 9β,19-cyclopropyl sterols in suspension cultures of bramble cells cultured with tridemorph', *Phytochemistry*, **20**, 2153–2159.

Schmitt, P., Scheid, F., and Benveniste, P. (1980). 'Accumulation of Δ8,14-sterols in suspension cultures of bramble cells treated with an azasterol antimycotic (A25822B)', *Phytochemistry*, **19**, 525–530.

Schneider, J. A., and Maxwell, E. (1973). Peptidylpuromycin formation on mammalian polysomes. Studies on transpeptidation and translocation, *Biochemistry*, **12**, 475–481.

Schumann, G. (1968). 'Beitrag zur systemischen wirkung fungizider benzimidazol-derivatives als getreidebeizmittel.' *Nachrichtenbl, Dtsch, Pflzenschutzdienstes. (Braunschw)*, **20**, 1–5.

Schwinn, F. J., Staub, T., and Urech, P. A. (1977). 'A new type of fungicide against diseases caused by Oomycetes', *Meded. Fac. Landbouwwet. Rijksuniv. Gent*, **42**, 1181–1188.

Scolnick, E., Tompkins, R., Caskey, T., and Nirenberg, M. (1968). 'Release factors differing in specificity for terminator codons', *Proc. Nat. Acad. Sci. U.S.*, **61**, 768–774.

Seiler, J. P. (1976). 'The mutagenicity of benzimidazole and benzimidazole derivatives. VI. Cytogenetic affects of benzimidazole derivatives in the bone marrow of the mouse and the chinese hamster', *Mutat. Res.*, **40**, 339–348.

Sekizawa, Y., and Mase, S. (1980). 'Recent progress in studies on non-fungicidal controlling agent, probenazole, with reference to the induced resistance mechanism of rice plant,' *Rev. Plant. Prot. Res.*, **13**, 114–121.

Sekizawa, Y., and Mase, S. (1981). 'Mode of controlling action of probenazole against rice blast disease with reference to the induced resistance mechanism in rice plant', *J. Pestic. Sci.*, **6**, 91–94.

Sekizawa, Y., and Watanabe, T. (1981). 'On the mode of action of probenazole against rice blast', *J. Pestic. Sci.*, **6**, 247–255.

Sekizawa, Y., Shimura, M., Suzuki, A., and Iwata, M. (1981). 'Anticonidial germination factors induced in the presence of probenazole in infected host leaves. II. Structural elucidation of the major component (Substance B)', *Agric. Biol. Chem.*, **45**, 1437–1439.

Selling, H. A., Vonk, J. W., and Kaars Sijpesteijn, A. (1970). 'Transformation of the systemic fungicide methyl thiophanate into 2-benzimidazole carbamic acid methyl ester', *Chem. Ind.*, 1625–1626.

Serres, J. M., and Carraro, G. A. (1976). 'DPX-3217, a new fungicide for the control of grape downy mildew, potato late blight and other Peronosporales', *Meded. Fac. Landbouwwet. Rijksuniv. Gent*, **41**, 645–650.

Sherald, J. L., Ragsdale, N. N., and Sisler, H. D. (1973). 'Similarities between the systemic fungicides triforine and triarimol', *Pestic. Sci.*, **4**, 719–727.

Sherald, J. L., and Sisler, H. D. (1972). 'Selective inhibition of antimycin A-insensitive respiration in *Ustilago maydis* and *Ceratocystis ulmi*', *Pl. Cell Physiol.*, **13**, 1039–1052.

Sherald, J. L. and Sisler, H. D. (1975). 'Antifungal mode of action of triforine', *Pestic. Biochem. Physiol.*, **5**, 477–488.

Shimura, M., Iwata, M., Tashiro, N., Sekizawa, Y., Suzuki, Y., Mase, S., and Watanabe, T. (1981). 'Anticonidial germination factors induced in the presence of probenazole in infected host leaves. I. Isolation and properties of four active substances', *Agric. Biol. Chem.*, **45**, 1431–1435.

Shive, J. B., and Sisler, H. D. (1976). 'Effects of ancymidol (a growth retardant) and triarimol (a fungicide) on the growth, sterols and gibberellins of *Phaseolus vulgaris* (L)', *Pl. Physiol.*, **57**, 640–644.

Shively, O. D., and Mathre, D. E. (1971). 'Mode of action of oxathiin systemic fungicides. IV. Effect of carboxin on solute leakage from hypae of *Rhizoctonia solani*', *Can. J. Microbiol.*, **17**, 1465–1470.

Siegel, M. R. (1977). 'Effect of fungicides on protein synthesis', in *Antifungal Compounds* Vol. 2 (eds. M. R. Siegel and H. D. Sisler), pp. 399–438. Marcel Dekker, New York.

Siegel, M. R., Kerkenaar, A., and Kaars Sijpesteijn, A. (1977). 'Antifungal activity of the systemic fungizide imazalil', *Neth. J. Pl. Pathol.*, **83** (Suppl. 1), 121–133.

Siegel, M. R., and Ragsdale, N. N. (1978). 'Antifungal mode of action of imazalil', *Pestic. Biochem. Physiol.*, **9**, 48–56.

Siegel, M. R., and Sisler, H. D. (1977). *Antifungal Compounds* Vols. I, II, III. Marcel Dekker, New York.

Siegel, M. R. and Solel, Z. (1981). 'Effects of imazalil on a wild type and fungicide resistant strain of *Aspergillus nidulans*', *Pestic. Biochem. Physiol.*, **15**, 222–233.

Singer, T. P., and Gutman, M. (1970). 'The NADH dehydrogenase of the respiratory chain', in *Pyridine Nucleotide-Dependent Dehydrogenases*, Proc. Advanced Study Institute, Univ. of Konstanz, 1969 (ed. H. Sund), pp. 375–391, Springer-Verlag, Berlin, Heidelberg.

Sisler, H. D., and Ragsdale, N. N. (1977). 'Fungitoxicity and growth regulation involving aspects of lipid biosynthesis', *Neth. J. Pl. Pathol.*, **83**, (Suppl. 1), 81–92.

Slade, P., Cavell, B. D., Hemingway, R. J., and Sampson, M. J. (1972). 'Metabolism and mode of action of dimethirimol and ethirimol', *Pestic. Chem.*, **5**, 295–303.

Small, L. W., Martin, R. A., and Edgington, L. V. (1977). 'A comparison of the translocation within plants of pyroxychlor and a 6-amino analogue', *Neth. J. Pl. Pathol.*, **83** (Suppl. 1), 63–70.

Smith, J. M., Cartwright, J. H. M., and Smith, E. G. (1977). 'Furalaxyl (CGA 38140): a new fungicide for the control of downy mildew in lettuces and hops', *Proc. 1977 Br. Crop Prot. Conf.—Pests and Diseases*, **2**, 633–640.
Smith, P. M. (1979). 'A study of the effects of fungitoxic compounds on *Phytophthora cinnamomi* in water', *Ann. Appl. Biol.*, **93**, 149–157.
Snel, M., von Schmeling, B., and Edgington, L. V. (1970). 'Fungitoxicity and structure-activity relationships of some oxathiin and thiazole derivatives', *Phytopathology*, **60**, 1164–1169.
Sobus, M. T., Holmlund, C. E., and Whittaker, N. F. (1977). 'Effects of the hypocholesteremic agent trifluperidol on the sterol, steryl ester and fatty acid metabolism of *Saccharomyces cerevisiae*, *J. Bact.*, **130**, 1310–1316.
Solel, Z., Schooley, J. M., and Edgington, L. V. (1973). 'Uptake and translocation of benomyl and carbendazim (methyl benzimidazole-2-yl-carbamate) in the symplast', *Pestic. Sci.*, **4**, 713–718.
Sowa, B. A., and Marks, E. P. (1975). 'An *in vitro* system for the quantitative measurement of chitin synthesis in the cockroach: inhibition by TH 6040 and polyoxin D'. *Insect. Biochem.*, **5**, 855–859.
Sparling, P. F. (1970). 'Kasugamycin resistance: 30S ribosomal mutation with an unusual location on the *Escherichia coli* chromosome'. *Science*, **167**, 56–58.
Staron, T., and Allard, G. (1964). 'Propriétés antifongique du 2-(4′-thiazolyl)-benzimidazole ou thiabendazole', *Phytiatr.-Phytopharm. Rev. Fr. Med. Parm. Veg.*, **13**, 163–168.
Staub, T. H., Dahmen, H., and Schwinn, F. J. (1980). 'Effects of Ridomil^R on the development of *Plasmopara viticola* and *Phytophthora infestans* on their host plants', *Z. Pflanzenkr. Pflanzenschutz*, **87**, 83–91.
Staub, T. H., and Hubele, A. (1980). 'Recent advances in the chemical control of Oomycetes', in *Chemie der Pflanzenschutzund Schädlingsbekämpfungsmittel* (ed. R. Wegler), Vol. 6, pp. 389–422, Springer-Verlag, Berlin, Heidelberg.
Stipanovic, R. D., and Wheeler, M. H. (1980). 'Accumulation of 3,3′-biflaviolin, a melanin shunt product, by tricyclazole-treated *Thielaviopsis basicola*', *Pestic. Biochem. Physiol.*, **13**, 198–201.
Stirling, J. L., Cook, G. A., and Pope, A. M. S. (1979). 'Chitin and its degradation', in *Fungal Walls and Hyphal Growth* (eds. J. H. Burnett and A. P. J. Trinci), pp. 169–188, Cambridge University Press.
Storey, B. T. (1976). 'Respiratory chain of plant mitochondria XVIII. Point of interaction of the alternate oxidase with the respiratory chain', *Pl. Physiol.*, **58**, 521–525.
Stössel, P., Lazarovits, G., and Ward, E. W. B. (1982). 'Light and electron microscopy of *Phytophthora* rot in soybeans treated with metalaxyl', *Phytopathology*, **72**, 106–111.
Sud, I. J., Chou, D. L., and Feingold, D. S. (1979). 'Effect of free fatty acids on liposome susceptibility to imidazole antifungals', *Antimicrob. Agents Chemother.*, **16**, 660–663.
Sud, I. J., and Feingold, D. S. (1981a). 'Mechanisms of action of the antimycotic imidazoles', *J. Invest. Dermatol.*, **76**, 438–441.
Sud, I. J., and Feingold, D. S. (1981b). 'Heterogeneity of action mechanisms among antimycotic imidazoles', *Antimicrob. Agents Chemother.*, **20**, 71–74.
Suhara, Y., Maeda, K., and Umezawa, H. (1966). 'Chemical studies on kasugamycin V. The structure of kasugamycin'. *Tetrahedron Letters*, **12**, 1239–1244.
Suzuki, S., Isono, K., Nagutsu, J., Mizutani, T., Kawashima, Y., and Mizuno, T. (1965). 'A new antibiotic, polyoxin A', *J. Antibiotics*, **18**, 131–132.
Swamy, K. H. S., Sirsi, M., and Rao, G. R. (1974). 'Studies on the mechanism of action of miconazole: Effect of miconazole on respiration and cell permeability of *Candida albicans*', *Antimicrob. Agents Chemother.*, 420–425.

Tanaka, N. (1972). 'Inhibition by several antibiotics of initiation of protein synthesis and peptide chain elongation', in *Molecular Mechanisms of Antibiotic Action on Protein Biosynthesis and Membranes* (eds. E. Munoz, F. Garcia-Ferrandiz and D. Vazquez), pp. 279–291, Elsevier, New York.

Tanaka, N., Yamaguchi, H., and Umezawa, H. (1966a). 'Mechanism of Kasugamycin action on polypeptide synthesis', *J. Biochem.* (*Tokyo*), **60**, 429–434.

Tanaka, N., Yoshida, Y., Sashikata, K., Yamaguchi, H., and Umezawa, H. (1966b). 'Inhibition of polypeptide synthesis by kasugamycin, an aminoglycosidic antibiotic', *J. Antibiotics*, **19A**, 65–68.

Tanaka, S., Kato, T., Yamamoto, S., and Yoshioka, H. (1977). 'Structure modifications of *S-n*-butyl-*S′-p-tert*-butylbenzyl-*N*-3-pyridyldithiocarbonimidate (S-1358, Denmert®) and fungicidal activities', *Agric. Biol. Chem.*, **41**, 1953–1959.

Theodorides, V. J., Chang, J., DiCuollo, C. J., Grass, G. M., Parish, R. C., and Scott, G. C. (1973). 'Oxibendazole, a new broad spectrum anthelminthic effective against gastro-intestinal nematodes of domestic animals', *Br. Vet. J.*, **129**, XCVII–XCVIII.

Thomas, W. D., jr., Eastburg, P. H., and Bankuti, M. D. (1962). 'The comparative fungicidal properties of acidic nitrogen derivatives of polyhalo-alkyl sulfenyl chlorides', *Phytopathology*, **52**, 754–755 (Abstr.).

Thorn, G. D. (1973). 'Uptake and metabolism of chloroneb by *Phaseolus vulgaris*', *Pestic. Biochem. Physiol.*, **3**, 137–140.

Thyagarajan, T. R., Srivastava, O. P., and Vora, V. V. (1963). 'Some cytological observations on the effect of griseofulvin on dermatophytes', *Naturwiss.*, **50**, 524–525.

Tillman, R. W., and Ferguson, M. W. (1980). 'Toxicity of pyroxychlor to *Pythium aphanidermatum*', *Phytopathology*, **70**, 441–444.

Tillman, R. W., and Sisler, H. D. (1973). 'Effect of chloroneb on the growth and metabolism of *Ustilago maydis*', *Phytopathology*, **63**, 219–225.

Timberlake, W. E., McDowell, L., and Griffin, D. H. (1972). 'Cycloheximide inhibition of the DNA-dependent RNA-polymerase I of *Achlya bisexualis*', *Biochem. Biophys. Res. Commun.*, **46**, 942–947.

Tokousbalides, M. C., and Sisler, H. D. (1978). 'Effect of tricyclazole on growth and secondary metabolism of *Pyricularia oryzae*', *Pestic. Biochem. Physiol.*, **8**, 26–32.

Tokousbalides, M. C., and Sisler, H. D. (1979). 'Site of inhibition by tricyclazole in the melanin biosynthetic pathway by *Verticillium dahliae*', *Pestic. Biochem. Physiol.*, **11**, 64–73.

Tolmsoff, W. J. (1962). 'Biochemical basis for biological specificity of Dexon (*p*-dimethyl-amino-benzenediazo sodium sulfonate)', *Phytopathology*, **52**, 755 (Abstr.).

Tolmsoff, W. J. (1966). 'Biochemical basis for biological specificity of Dexon (*p*-dimethyl-amino-benzenediazo sodium sulfonate), a respiratory inhibitor', *Dissert. Abstr.*, **26**, 5707.

Tomiku, T., Fujikawa, T., Sato, S., and Ando, S. (1979). 'Occurrence of *Xanthomonas oryzae* infection of rice and time and number of applications of probenazole granules.' *Nogyo Oyobi Engei.*, **54**, 339–340.

Trakatellis, A. C., Montjar, M., and Axelrod, A. E. (1965). 'Effect of cycloheximide on polysomes and protein synthesis in the mouse liver', *Biochemistry*, **4**, 2065–2071.

Trocha, P. J., Jasne, S. J., and Sprinson, D. B. (1974). 'Novel sterols in ergosterol deficient yeast', *Biochem. Biophys. Res. Comm.*, **59**, 666–671.

Trocha, P. J., and Sprinson, D. B. (1976). 'Location and regulation of early enzymes of sterol biosynthesis in yeast', *Arch. Biochem. Biophys.*, **174**, 45–51.

Tsao, P. H. (1970). 'Selective media for isolation of pathogenic fungi', *Ann. Rev. Phytopathol.*, **8**, 157–186.

Tucker, A. N., and Lillich, T. T. (1974). 'Effect of the systemic fungicide carboxin on electron transport function in membranes of *Micrococcus denitrificans*', *Antimicrob. Agents Chemother.*, **6**, 572–578.

van Tuyl, J. M. (1975). 'Acquired resistance to carboxin in *Aspergillus nidulans*', *Neth. J. Pl. Pathol.*, **81**, 122–123.

van Tuyl, J. M. (1977). 'Genetic aspects of resistance to imazalil in *Aspergillus nidulans*', *Neth. J. Pl. Pathol.*, **83** (Suppl. 1), 169–176.

Uchiyama, M., Abe, H., Sato, R., Shimura, M., and Watanabe, T. (1973). 'Fate of 3-allyloxy-1,2-benzisothiazole-1,1-dioxide (Oryzemate) in rice plants', *Agric. Biol. Chem.*, **37**, 737–745.

Uesugi, Y. (1981). 'Resistance to fungicides in *Pyricularia oryzae*', *J. Pestic. Sci.*, **6**, 239–246.

Uesugi, Y., and Katagiri, M. (1977). 'Back-mutation of an organophosphorus thiolate-resistant mutant of *Pyricularia oryzae*', *Neth. J. Pl. Pathol.*, **83** (Suppl. 1), 243–252.

Uesugi, Y., Katagiri, M., and Fukunaga, K. (1969). 'Resistance in *Pyricularia oryzae* to antibiotics and organophosphorus fungicides', *Bull. Nat. Inst. Agric. Sci. Ser. C.* (*Plant Pathol. Entomol.*), **23**, 93–112.

Uesugi, Y., Katagiri, M., and Noda, O. (1974). 'Negatively correlated cross-resistance and synergism between phosphoramidates and phosphorothiolates in their fungicidal actions on rice blast fungi', *Agric. Biol. Chem.*, **38**, 907–912.

Ulrich, J. T., and Mathre, D. E. (1972). 'Mode of action of oxathiin systemic fungicides V. Effect on electron transport system of *Ustilago maydis* and *Saccharomyces cerevisiae*', *J. Bact.*, **110**, 628–632.

Umezawa, S. (1975). 'The chemistry and conformation of aminoglycoside antibiotics', in *Drug action and drug resistance in bacteria. Aminoglycoside antibiotics* (ed. S. Mitsuhashi), Vol. 2, University of Tokyo Press.

Umezawa, H., Okami, Y., Hashimoto, T., Suhara, Y., Hamada, M., and Takeuchi, T. (1965). 'A new antibiotic, kasugamycin', *J. Antibiotics*, **18a**, 101–103.

Uramoto, M., Matsuoka, M., Liehr, J. G., McCloskey, J. A., and Isono, K. (1981). 'Polyoxin N, a new member of the polyoxin family', *Agric. Biol. Chem.*, **48**(8), 1901–1902.

Urbschat, E. (1960). 'Gegen Boden- und Blattpilze hochwirksame Azoverbindungen', *Angew. Chem.*, **72**, 981–985.

Urech, P. A., Schwinn, F., and Staub, T. (1977). 'CGA 48988, a novel fungicide for the control of late blight, downy mildews and related soil-borne diseases', *Proc. 1977 Br. Crop Prot. Conf.—Pests and Diseases*, **2**, 623–631.

Vonk, J. W., and Dekhuijzen, H. M. (1979). 'Transport and metabolism of (3H) imazalil in barley and cucumber', *Med. Fac. Landbouwwet. Rijksuniv. Gent*, **44**, 927–934.

Vonk, J. W., and Kaars Sijpesteijn, A. (1971). 'Methyl benzimidazol-2-ylcarbamate, the fungitoxic principle of thiophanate-methyl', *Pest. Sci.*, **2**, 160–164.

Vonk, J. W., Mihanovic, B., and Kaars Sijpesteijn, A. (1977). 'A biochemical mechanism for the conversion of thiophanate-methyl into methyl benzimidazol-2-ylcarbamate (MBC) in plant tissue', *Neth. J. Pl. Pathol.*, **83** (Suppl. 1), 269–276.

Voros, J. (1965). 'Streptomycin sensitivity of Oomycetes due to increased absorption of streptomycin by their mycelia', *Phytopathol. Z.*, **54**, 249–257.

Vo-Thi-Hai, Bompeix, G., and Ravise, A. (1979). 'Rôle du tris-*O*-éthyl phosphonate d'aluminium dans la stimulation de défense des tissus de tomate', *C. R. Acad. Sci. Ser. D.*, **288**, 1171–1174.

de Vries, O. M. H., and Wessels, J. G. H. (1975). 'Chemical analysis of cell wall regeneration and reversion of protoplasts from *Schizophyllum commune*', *Arch. Microbiol.*, **102**, 209–218.

de Waard, M. A. (1972). 'On the mode of action of the organophosphorus fungicide Hinosan', *Neth. J. Pl. Path.*, **78**, 186–188.

de Waard, M. A. (1974). 'Mechanism of action of the organophosphorus fungicide pyrazophos', *Meded. LandbHogesch. Wageningen*, **74**(14), 1–98.

de Waard, M. A. (1975). 'Mode of action of the organophosphorus fungicide pyrazophos', in *Systemfungizide, IV Int. Symp. Reinhardsbrunn, 1974* (eds. H. Lyr, and C. Polter), pp. 197–202, Akademie-Verlag, Berlin.

de Waard, M. A., and Gieskes, S. A. (1977). 'Characterization of fenarimol-resistant mutants of *Aspergillus nidulans*', *Neth. J. Pl. Pathol.*, **83** (Suppl. 1), 177–188.

de Waard, M. A., and van Nistelrooy, J. G. M. (1979). 'Mechanism of resistance to fenarimol in *Aspergillus nidulans*', *Pestic. Biochem. Physiol.*, **10**, 219–229.

de Waard, M. A., and van Nistelrooy, J. G. M. (1980a). 'An energy-dependent efflux mechanism for fenarimol in a wild-type strain and fenarimol resistant mutants of *Aspergillus nidulans*', *Pestic. Biochem. Physiol.*, **13**, 255–266.

de Waard, M. A., and van Nistelrooy, J. G. M. (1980b). 'Mechanism of resistance to pyrazophos in *Pyricularia oryzae*', *Neth. J. Pl. Pathol.*, **86**, 251–258.

Wain, R. L., and Carter, G. A. (1977). 'Historical aspects' in *Systemic fungicides* (ed. R. W. Marsh), pp. 6–31, Longman, London and New York.

Ward, E. W. B., Lazarovits, G., Stössel, P., Barrie, S. D., and Unwin, C. H. (1980). 'Glyceollin production associated with control of *Phytophthora* rot of soybeans by the systemic fungicide, metalaxyl', *Phytopathology*, **70**, 738–740.

Watanabe, T. (1977). 'Effect of probenazole (Oryzemate) on each stage of rice blast fungus (*Piricularia oryzae* Cav) in its life cycle', *J. Pestic. Sci.*, **2**, 395–404.

Watanabe, T., Igarashi, H., Matsumoto, K., Seki, S., Mase, S., and Sekizawa, Y. (1977). 'The characteristics of probenazole (Oryzemate) for the control of rice blast', *J. Pestic. Sci.*, **2**, 291–296.

Watanabe, T., Sekizawa, Y., Shimura, M., Suzuki, Y., Matsumoto, M., Iwata, M., and Mase, S. (1979). 'Effects of probenazole (Oryzemate) on rice plants with reference to controlling rice blast', *J. Pestic. Sci.*, **4**, 53–59.

Weeks, D. P. (1981). 'Protein Biosynthesis: Mechanisms and Regulation', in *The Biochemistry of Plants. A Comprehensive Treatise* (eds. in chief P. K. Stumpf and E. E. Conn), Vol. 6, Proteins and Nucleic Acids (ed. A. Marcus), pp. 491–529.

Weete, J. D. (1973). 'Sterols of the fungi: distribution and biosynthesis', *Phytochemistry*, **12**, 1843–1864.

Wei, C.-M., Hansen, B. S., Vaughan, M. H. Jr., and McLaughlin, C. S. (1974). 'Mechanisms of action of the mycotoxin trichodermin, a 12,13-epoxytrichothecene', *Proc. Nat. Acad. Sci. U.S.*, **71**, 713–717, (1974).

Wettstein, F. O., Noll, H., and Penman, S. (1964). 'Effect of cycloheximide on ribosomal aggregates engaged in protein synthesis *in vitro. Biochem*', *Biophys. Acta*, **87**, 525.

White, G. A. (1971). 'A potent effect of 1,4-oxathiin systemic fungicides on succinate oxidation by a particulate preparation from *Ustilago maydis*', *Biochem. Biophys. Res. Comm.*, **44**, 1212–1219.

White, G. A., and Thorn, G. D. (1975). 'Structure-activity relationships of carboxamide fungicides and the succinic dehydrogenase complex of *Cryptococcus laurentii* and *Ustilago maydis*', *Pestic. Biochem. Physiol.*, **5**, 380–395.

White, G. A., and Thorn, G. D. (1980). 'Thiophene carboxamide fungicides: structure-activity relationships with the succinate dehydrogenase complex from wild-type and carboxin-resistant mutant strains of *Ustilago maydis*', *Pestic. Biochem. Physiol.*, **14**, 26–40.

White, G. A., Thorn, G. D., and Georgopoulos, S. G. (1978). 'Oxathiin carboxamides highly active against carboxin-resistant succinic dehydrogenase complexes from carboxin-selected mutants of *Ustilago maydis* and *Aspergillus nidulans*', *Pestic. Biochem. Physiol.*, **9**, 165–182.

Wiertsema, W. P., and Wissink, G. H. (1977). 'Fongarid[R] 50WP, a new fungicide for ornamentals', *Meded. Fac. Landbouwwet. Rijksuniv. Gent*, **42**, 1189–1194.

Wilson, L. (1970). 'Properties of colchicine binding protein from chick embryo brain. Interactions with vinca alkaloids and podophyllotoxin', *Biochemistry*, **9**, 4999–5007.
Wilson, L., and Bryan, J. (1974). 'Biochemical and pharmacological properties of microtubules', *Adv. Cell. Mol. Biol.*, **3**, 21–72.
Wolf, G. (1980). Paper presented at NATO conference on 'Active Defense Mechanisms in Plants', Cape Sounion, Greece, May 1980.
Woloshuk, C. P., Sisler, H. D., and Dutky, S. R. (1979). 'Mode of action of the azasteroid antibiotic 15-aza-24-methylene-D-homocholesta-8,14-dien-3β-ol in *Ustilago maydis*', *Antimicrob. Agents Chemother.*, **16**, 98–103.
Woloshuk, C. P., Sisler, H. D., Tokousbalides, M. C., and Dutky, S. R. (1980). 'Melanin biosynthesis in *Pyricularia oryzae:* site of tricyclazole inhibition and pathogenicity of melanin-deficient mutants', *Pestic. Biochem. Physiol.*, **14**, 256–264.
Woloshuk, C. P., Wolkow, P. M., and Sisler, H. D. (1981). 'The effect of three fungicides, specific for the control of rice blast disease, on the growth and melanin biosynthesis of *Pyricularia oryzae* Cav', *Pestic. Sci.*, **12**, 86–90.
Wood, D. A., and Hammond, J. B. W. (1977). 'Inhibition of growth and development of *Agaricus bisporus* by polyoxin D', *J. Gen. Microbiol.*, **98**, 625–628.
Woodcock, D. (1977). 'Structure-activity relationships' in *Systemic fungicides* (ed. R. W. Marsh), pp. 32–84, Longman, London and New York.
Yamaguchi, H., and Iwata, K. (1979). 'Effect of fatty acyl group and sterol composition on sensitivity of lecithin liposomes to imidazole antimycotics', *Antimicrob. Agents Chemother.*, **15**, 706–711.
Yamaguchi, H., Yamamoto, C., and Tanaka, N. (1965). 'Inhibition of protein synthesis by Blasticidin S. 1. Studies with cell-free systems from bacterial and mammalian cells', *J. Biochem.* (*Tokyo*), **57**, 667–677.
Yoshida, Y., Aoyama, Y., Kumaoka, H., and Kubota, S. (1977). 'A highly purified preparation of cytochrome P-450 from microsomes of anaerobically grown yeast', *Biochem. Biophys. Res. Comm.*, **78**, 1005–1010.
Ziegler, D. M. (1961). 'The mechanism of coenzyme Q reduction in heart mitochondria', in *Biological Structure and Function* (eds. T. W. Goodwin and O. Lindberg), pp. 253–264, Academic Press, New York.
Ziegler, H. (1963). 'Untersuchungen über die wirkung von Griesofulvin auf *Microsporum canis* (4. Mitteilung)', *Z. Allgem. Mikrobiol.*, **3**, 211–224.
Ziogas, B. N., and Georgopoulos, S. G. (1979). 'The effect of carboxin and of thenoyltrifluoroacetone on cyanide-sensitive and cyanide-resistant respiration of *Ustilago maydis* mitochondria', *Pestic. Biochem. Physiol.*, **11**, 208–217.
van Zutphen, H., and van Deenen, L. L. M. (1967). 'The effect of lysolecithin on the electrical resistance of lecithin bilayer membranes', *Chem. Phys. Lipids*, **1**, 389–391.

Progress in Pesticide Biochemistry and Toxicology, Volume 3
Edited by D. H. Hutson and T. R. Roberts

CHAPTER 2

Metabolism of fungicides in plants

J. W. Vonk

INTRODUCTION

Fungicides are applied directly to the above-ground parts of plants or they may enter plants via the soil and roots or by seed treatment. Once in contact with the plant, the fungicide will be subject to natural processes which may alter its chemical structure. Such conversions can be of an enzymatic nature, but they can also be caused by purely chemical processes. It is sometimes difficult to distinguish between these two processes and in this chapter the terms metabolism and metabolite refer to any conversion process and product in and on plants.

Knowledge of the metabolic fate of fungicides (and other pesticides) in crops is necessary to assess the nature of any residue of the compound. Such assessment is part of the toxicological evaluation of the pesticide, and therefore in most countries data on metabolism of pesticides in plants are required to obtain a registration. As a consequence, manufacturers are conducting pesticide metabolism studies, but only a minor part of the information is published. Although this is understandable from the point of view of protection of their products, it is a regrettable situation for our understanding of the fate of xenobiotics in plants.

Fungicides distinguish themselves from other chemicals only by the fact that they kill fungi or prevent their development on or in plants. This means that the interaction between plants and fungicides is atypical: the same type of reaction may occur with other chemicals e.g. herbicides and insecticides. When appropriate, metabolic features of other pesticides will also be discussed in this study.

When considering chemical and biological conversion of fungicides in plants one may distinguish between several types of reactions. The structure of the compound is sometimes enlarged by conjugation with plant products e.g. sugars or amino acids. These conversions are almost exclusively of enzymatic nature. However, pure chemical reactions of fungicides or their metabolites with plant constituents also occur.

Minor alterations of the molecule take place by hydroxylation, or by sulphoxidation. Other slight changes in the structure are: reduction of a nitro group to an amino group and oxidation of a hydroquinone to a quinone or a phosphorothionate group to the corresponding oxon.

Breakdown of fungicides takes place when the molecular structure is split into smaller parts. These may eventually be transformed into structures which enter the metabolic pool of the plant and can be either completely metabolized, e.g. into CO_2, or incorporated into high molecular weight plant constituents. However, survey of the literature shows that the changes taking place are often minor and that in fact the skeleton, especially of aromatics and heterocyclics, remains intact over extended periods.

METABOLITES AND PATHWAYS

In the last decade some reviews of the literature have been published which deal in detail with the metabolism of fungicides in plants (Kaars Sijpesteijn *et al.*, 1977; Vonk and Kaars Sijpesteijn, 1977). The present review will therefore only summarize data on metabolites and pathways, mostly in tabulated form. The literature since 1977 is discussed in more detail.

Inorganic Chemicals

The largest proportion of the fungicides used still consists of inorganic chemicals like elemental sulphur and copper compounds. Very little is known about transformation of elemental sulphur in plants. It may be speculated that it is oxidized to give finally sulphate which is taken up in the metabolic pool of the plant.

Dithiocarbamates

The dithiocarbamates are the most extensively used organic fungicides in agriculture. It is appropriate to distinguish between two classes of compounds which have very different types of metabolism in plants, namely the *N*-dimethyldithiocarbamates (thiram, ferbam, ziram) and the bisdithiocarbamates (nabam, maneb, zineb, mancozeb, propineb).

Table 1 and Figure 1 summarize the metabolites of dimethyldithiocarbamates. Most of our knowledge originates from the work of Dekhuijzen (1964) and Kaslander *et al.* (1961, 1962) (cf. Kaslander, 1966). The two most important metabolites found after application of sodium dimethyldithiocarbamate (**1**) to roots or leaves of several plant species were a *β*-*S*-glucoside (**2**) and an L-alanine derivative (**3**). Thiazolidine-2-thione-4-carboxylic acid (**4**) was formed non-enzymatically from (**3**). Thiram (**5**) undergoes exactly the same reactions after reduction to dimethyldithiocarbamate ions.

The presence of these metabolites in plants has been confirmed by other authors working with labelled compounds (Massaux, 1963; Hylin and Chin, 1967; Raghu *et al.*, 1975; Kumarasamy and Raghu, 1976). Work with ^{35}S-labelled compounds showed that at least a part of the applied fungicide finds its way into S-containing amino acids which are incorporated into proteins (Raghu *et al.*, 1975). Hylin and Chin (1967) found besides the above-mentioned compounds four unidentified metabolites in papaya treated with ^{35}S and ^{14}C-labelled dimethyldithiocarbamate.

Demethylation does not take place with dimethyldithiocarbamate-glucoside (**2**) or dimethylthiocarbamoylthioalanine (**3**).

Dekhuijzen (1964) showed that the glucose conjugate (**2**) and the alanine derivative (**3**) are interconvertible, probably via the dimethyldithiocarbamate ion.

The formation of glucose conjugates in plants is very well known, e.g. phenyl glucosides from phenols. A variety of natural thioglucosides is known to be present in plants, but to the best of our knowledge the formation of dimethyldithiocarbamate-*S*-glucoside is the only example of the formation of an *S*-glucoside from a pesticide in plants.

In addition to dimethylthiocarbamoylthioalanine several other alanine derivatives have been reported to occur in plants as metabolites of foreign

Table 1 Metabolites of dimethyldithiocarbamates in plants

Parent compound	Metabolite	Plant species[a] Way of treatment[b]	Label	Evidence for structure[c]	Reference
Sodium dimethyldithiocarbamate (**1**)	*N*,*N*-dimethylthiocarbamoyl-l-thio-β-D-glucopyranoside (**2**)	Pot. (slices); Fr. bean (slices)	—	Isolation	Kaslander (1966); Kaslander *et al.* (1961; 1962)
		Cuc. (R, L); Br. bean (L); Fr. bean (L); Tob. (L)	—	PC	Dekhuijzen (1964)
		Pap. (L)	^{14}C, ^{35}S	PC	Hylin and Chin (1967)
	β-(*N*,*N*-dimethylthiocarbamoylthio)-L-α-alanine (**3**)	Pot. (slices)	—	Isolation	Kaslander (1966); Kaslander *et al.* (1961; 1962)
		Cuc. (R, L); Br. bean (L); Fr. bean (L); Tob. (L)	—	PC	Dekhuijzen (1964)
		Pap. (L)	^{14}C, ^{35}S	PC	Hylin and Chin (1977)
	Thiazolidine-2-thione-4-carboxylic acid (**4**)	Cuc. (R); Pot. (slices)	—	Isolation, PC	Kaslander (1966); Kaslander *et al.* (1961; 1962) Dekhuijzen (1964)
	CS_2 and dimethylamine	Pap. (L)	^{14}C, ^{35}S	PC	Hylin and Chin (1967)
Thiram (**5**)	Metabolites (**2**) and (**3**)	Cuc. (L); Tob. (L)	—	PC	Dekhuijzen (1964)
		Cuc. (L)	^{35}S	PC	Massaux (1963)
	(**2**), (**3**) and (**4**)	Maize (L)	^{35}S	PC	Raghu *et al.* (1975)
Ziram	Metabolites (**2**), (**3**) and (**4**)	Rice (L)	^{35}S	PC	Kumarasamy and Raghu (1976)

[a] Abbreviations: Pot. = potato; Fr. bean = french bean; Cuc. = cucumber; Br. bean = broad bean; Tob. = tobacco; Pap. = papaya.
[b] R = root; L = leaf.
[c] PC = paper chromatographic comparison with reference compounds.

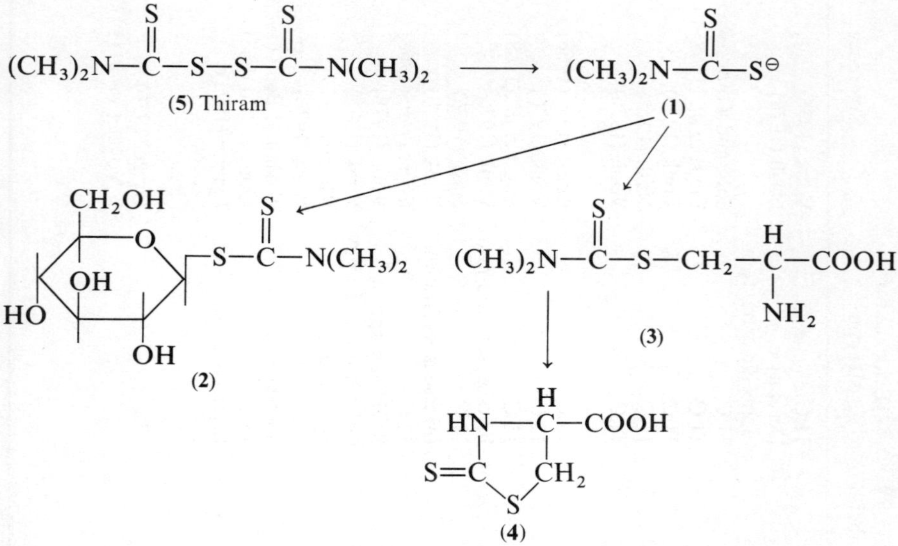

Figure 1 Pathways of metabolism of dimethyldithiocarbamates in plants

compounds. The herbicide 3-amino-1,2,4-triazole is metabolized in bean and tomato plants (Massini, 1963), the major metabolic product being β-(3-amino-1,2,4-triazolyl-1)-α-alanine (**6**).

Several examples exist of the formation of *S*-alanine-(cysteine-) derivatives of pesticides in plants. Lamoureux and Rusness (1980b) detected *S*-(pentachlorophenyl)cysteine (**105**) as a metabolite of pentachloronitrobenzene in peanuts, whereas the cysteine derivative of atrazine (**7**) was formed as one of the amino acid and peptide conjugates present in sorghum (Esser *et al.*, 1975).

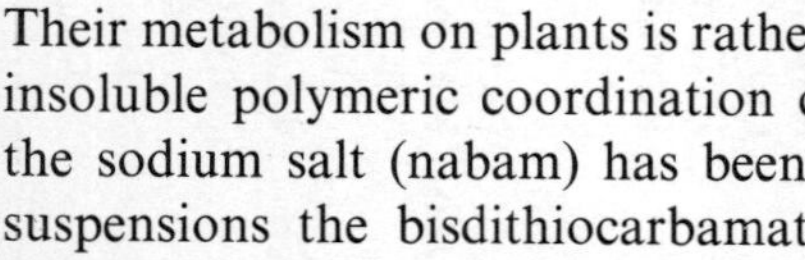

(**6**)

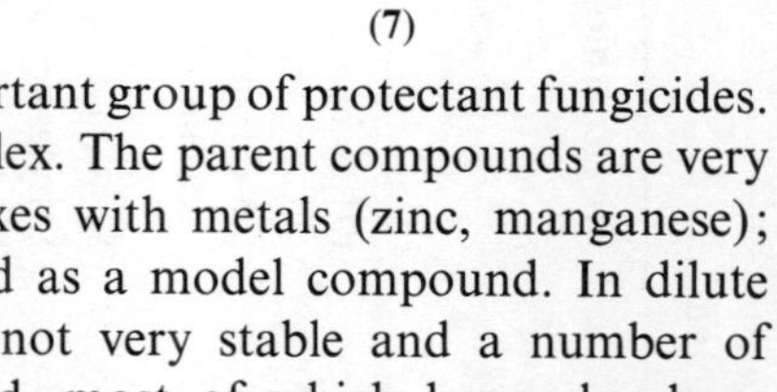

(**7**)

The bisdithiocarbamates are a very important group of protectant fungicides. Their metabolism on plants is rather complex. The parent compounds are very insoluble polymeric coordination complexes with metals (zinc, manganese); the sodium salt (nabam) has been studied as a model compound. In dilute suspensions the bisdithiocarbamates are not very stable and a number of degradation products have been observed, most of which have also been detected as metabolites in plants (Table 2, Figure 2). It is not clear whether at

Table 2 Metabolites of ethylenebisdithiocarbamates in plants

Parent compound	Metabolite	Plant species Way of application[a]	Label	Evidence for structure[b]	Reference
Zineb	Ethylenethiourea (**8**)	Lettuce (L)	^{14}C	TLC	Vonk (1976)
		Soybean (S, L)	^{14}C	Derivative, specific extraction	Nash (1976)
		Cucumber (L)	^{35}S	TLC	Sato and Tomizawa (1960)
		Tomato (L)	—	TLC	Engst *et al.* (1968)
	Ethyleneurea (**9**)	Lettuce (L)	^{14}C	TLC	Vonk (1976)
		Soybean (S)	^{14}C	TLC	Nash (1976)
	5,6-dihydro-3 *H*-imidazo[2,1-c]-1,2,4-dithiazole-3-thione (**14**)	Lettuce (L)	^{14}C	TLC	Vonk (1976)
		Tomato (L)	—	TLC	Engst *et al.* (1968)
	2-imidazoline (**10**)	Lettuce (L)	^{14}C	TLC	Vonk (1976)
	Ethylenediisothiocyanate (**15**)	Tomato	—	TLC	Engst *et al.* (1968)
	Hydantoine (**16**)	Lettuce (L)	^{14}C	Inverse isot. dilution	Vonk (1978
		Soybean (S)	^{14}C	TLC	Nash (1976)
	Ethylene diamine (**17**)	Lettuce (L)	^{14}C	Inverse isot. dilution	Vonk (1978)
	Sulphur	Tomato (L)	—	TLC	Engst *et al.* (1968)
	CO_2	Lettuce (L)	^{14}C	KOH adsorption	Vonk (1978)
	High molecular weight compounds	Lettuce (L)	^{14}C	Gel-Chromatography	Vonk (1978)
Maneb	Ethylenethiourea (**8**)	Tomato (L); Bean (L)	^{14}C	Specific analysis	Rhodes (1977)
		Soybean (S, L)	^{14}C	Specific analysis	Nash (1976)
		Tomato (L)	^{14}C	Specific analysis	Nash and Beall (1980)
	Metabolites (**14**), (**15**), sulphur	Tomato (L)	—	TLC	Engst *et al.* (1968)
Mancozeb	Metabolites (**8**), (**9**), (**10**), (**14**), (**17**), 3-(2-imidazolin-2-yl)-2-imidazolidinethione (**11**), sulphur, sulphate	Sugar beet (L)	^{14}C, ^{35}S, ^{3}H	Inverse isot. dilution	Lyman (1971)
Propineb	4-methylethylenethiourea (**12**), 5,6-dihydro-6-methyl-3 *H*-imidazo[2,1-c]-1,2,4-dithiazole-3-thione (**18**), 4-methylethyleneurea (**13**), 4-methylimidazoline (**19**)	Apple (F); Grape (F)	^{14}C	TLC	Drèze (1976) Vogeler *et al.* (1977)

[a] L = leaf; S = soil; F = fruit

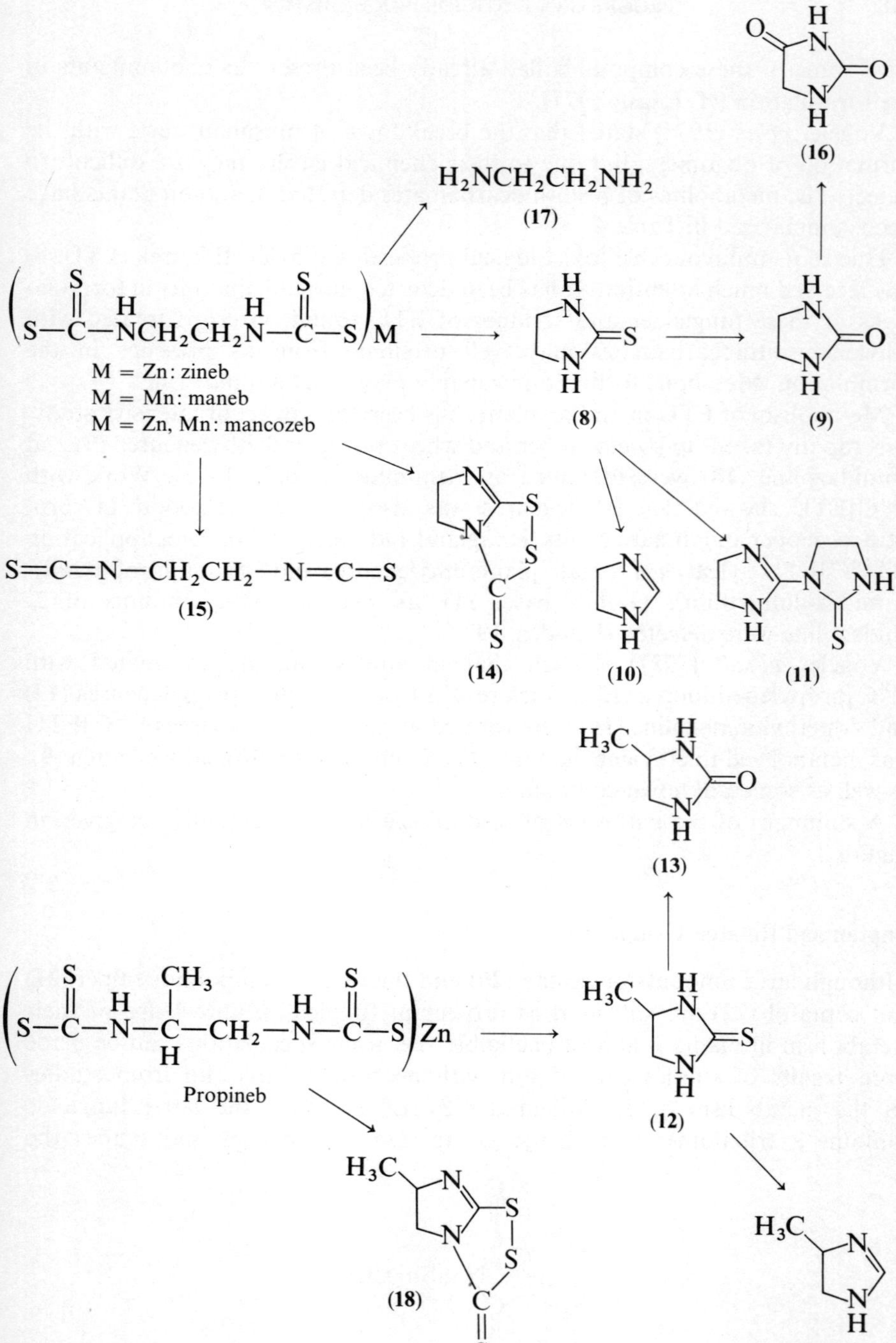

Figure 2 Pathways of metabolism of bisdithiocarbamates in plants

least some of these compounds had already been present as contaminants in the formulation (cf. Engst, 1977).

Vogeler *et al.* (1975) stated that the breakdown of propineb starts with the formation of oligomers, but due to their chemical nature they are difficult to detect. The metabolites of bisdithiocarbamates detected in and on plants have been summarized in Table 2.

Due to its unfavourable toxicological properties, ethylenethiourea (ETU, **8**) has received much attention. It has been detected in small amounts in formulations of these fungicides and residues of ETU found on crops treated with ethylenebisdithiocarbamates may well originate from its presence in the formulation. Metabolic formation probably plays only a minor role.

Metabolism of ETU in and on plants has been the subject of intensive study. It is rapidly taken up by cucumber and wheat plants and ethyleneurea (**9**) and 2-imidazoline (**10**) were identified as metabolites (Vonk, 1975). Work with [^{14}C]ETU showed that ethyleneurea was also a major metabolite in corn, lettuce, pepper and tomato plants (Hoagland and Frear, 1976). After application of [^{14}C]ETU to leaves of tomato plants ethyleneurea and 3-(2-imidazolin-2-yl)-2-imidazolidinethione (Jaffé's base, **11**), as well as small amounts of 2-imidazoline were detected (Rhodes, 1977).

Vogeler *et al.* (1977) showed that in apples and grapes treated with [^{14}C]propylenethiourea (**12**) (which results from propineb) propyleneurea (**13**) and 4-methylimidazoline (**19**) were formed as metabolites, whereas [^{14}C]ETU was metabolized to ethylene diamine (**17**), 2-imidazoline (**10**), ethyleneurea (**9**), as well as some unknown compounds.

A summary of the pathways of bisdithiocarbamate metabolism is given in Figure 2.

Captan and Related Compounds

Although large amounts of captan (**20**) and the related compounds folpet (**21**) and captafol (**22**) are still used as protectant fungicides, knowledge of their metabolism in plants is almost negligible. Yet some speculations can be made from results of studies carried out with microorganisms and from studies on the metabolism of dichlofluanid (**23**). Like captan, the latter fungicide contains a trihalomethylsulphenyl group. On strawberries and grapes the

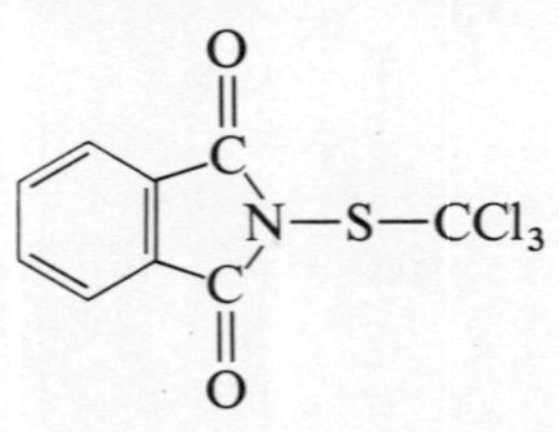

(**21**)

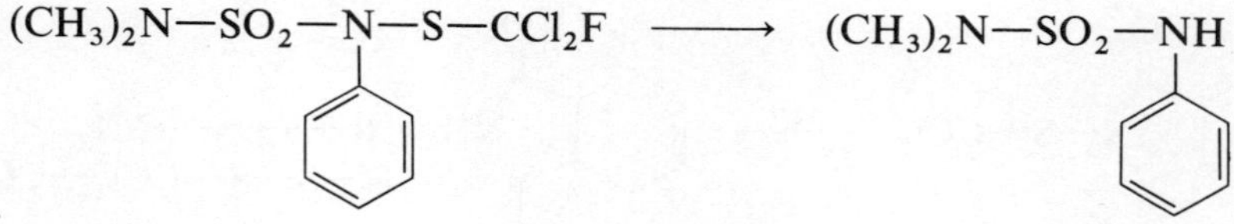

(23) (24)

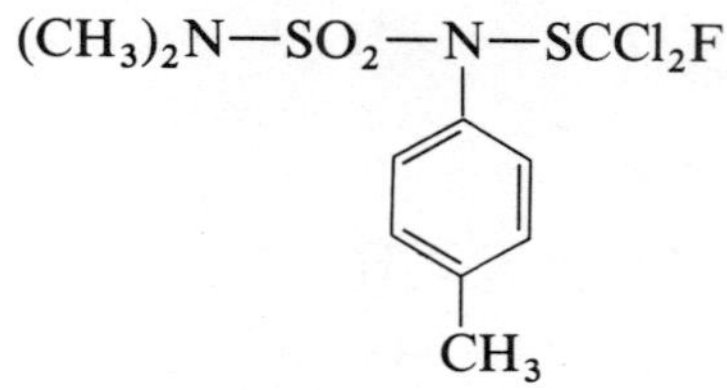

(25)

formation of *N,N*-dimethyl-*N'*-phenylsulphamide (**24**) has been observed (Vogeler and Niessen, 1967). A similar degradation pattern may be expected to occur with the related fungicide tolylfluanide (**25**).

By analogy, captan might be transformed into tetrahydrophthalimide (**26**). Captafol (**22**) seems to be hydrolysed rapidly to the corresponding dicarboximide (**26**), together with the acid amide and dicarboxylic acid (Casida and Lykken, 1969).

From *in vitro* reactions and studies with fungal spores it is known that the $—SCCl_3$ moiety can be transferred to the *S*-atoms of soluble and insoluble thiols (Richmond and Somers, 1966; 1968), e.g. cysteine and glutathione (Lukens and Sisler, 1958). Eventually the $—SCCl_3$ moiety is probably converted into gaseous products like thiophosgene (**27**), H_2S, HCl and COS (Figure 3). The $—SCCl_3$ group and thiophosgene are believed to be intermediates in the formation of thiazolidine-2-thione-4-carboxylic acid (**4**), an addition product with cysteine (Lukens and Sisler, 1958), and in the formation of a thiazolidine derivative of glutathione (**28**) (Richmond and Somers, 1968).

Similar conversions occur with folpet (Siegel, 1970). However, until now no studies have been undertaken to elucidate the metabolic fate of captan and related compounds on plants and the conversion products mentioned remain speculative. The uptake of captan by the leaves of plants may be restricted, thus impeding metabolic conversion. After application of [^{35}S]captan to the roots of broad beans only very small amounts of label were recovered in the leaves (Somers and Richmond, 1962).

Little is known of the effect of light on captan, folpet and captafol, although it is known that their fungicidal activity is destroyed by sunshine and UV light (Serra, 1964). Dichlofluanid was shown to be converted by irradiation under artificial conditions (Clark and Watkins, 1978) into a number of products including *N,N*-dimethyl-*N'*-phenylsulphamide, phenylisocyanate, phenylisothiocyanate and dimethylamidosulphonyl chloride.

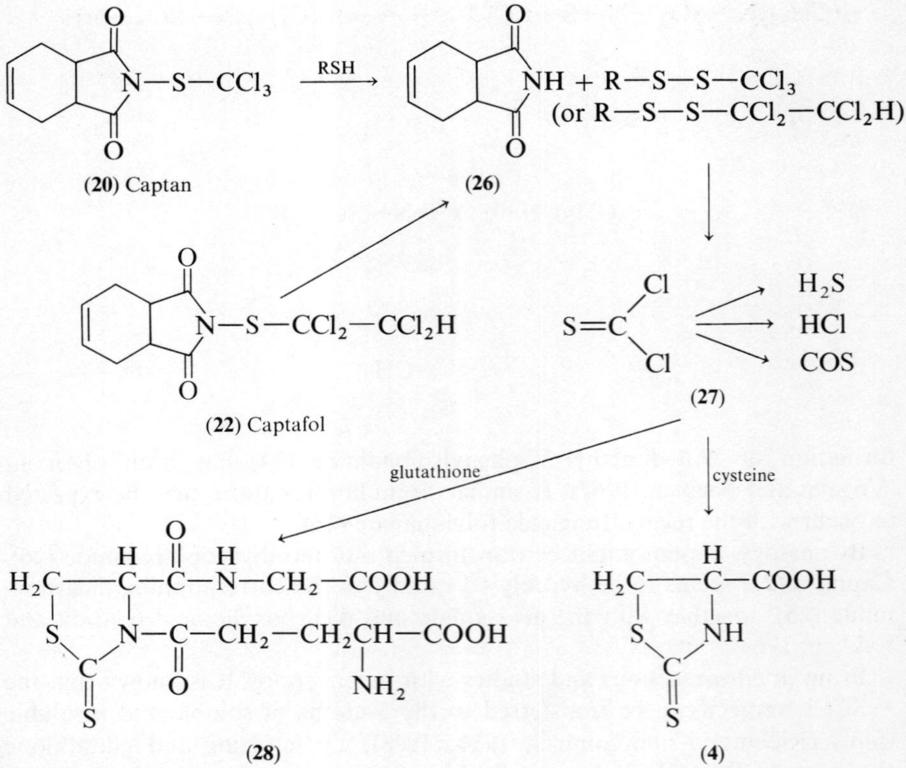

Figure 3 Possible pathways of conversion of captan and related compounds in plants

Benzimidazole Derivatives

Since the introduction of benomyl (**29**) this compound has attracted much attention due to its excellent systemic properties. In plants methyl benzimidazol-2-yl carbamate (carbendazim, **30**) is the most important metabolite. It is considered to be the active principle of benomyl and is also used as a systemic fungicide. Although benomyl rapidly loses the butylcarbamoyl group to give carbendazim in dilute aqueous solution and in organic solvents (Clemons and Sisler, 1969; Kilgore and White, 1970; Chiba and Doornbos, 1974), spray deposits of benomyl on various leaf surfaces remain largely intact for at least three weeks (Baude *et al.*, 1973). Contrary to earlier views, the conversion of benomyl into carbendazim is probably not hydrolytic but rather due to formation of butylisocyanate. The conversion of benomyl into carbendazim appears to be promoted by photolysis (Kilgore and White, 1970), and sunlight might therefore influence the rate of this conversion.

Although carbendazim seems to be rather resistant to degradation, several degradation products have been reported. One has the impression that the benzimidazole ring is rather stable in plants. Siegel (1973) found no label of ^{14}C-ring-labelled carbendazim in products of intermediary metabolism in strawberry plants after 88 days. Solel *et al.* (1973) found that during a period of 14 days only 1.56% of ^{14}C-ring-labelled carbendazim applied to cucumber leaves was respired as $^{14}CO_2$. Yet Rouchaud *et al.* (1974) found some metabolites resulting from cleavage of the benzimidazole nucleus. On the other hand, data on $^{14}CO_2$ formation from ^{14}C-labelled pesticides in plants are rare and therefore there is not much information concerning the stability of benzene derivatives in plants. In some cases this stability may be considerable.

A review of the conversion products of benomyl or carbendazim applied to plants is given in Table 3, and Figure 4 gives pathways of metabolism.

It is not known whether the *N*-benzimidazolyl carbamate bond is more stable in plants than the insecticidal *N*-methyl carbamate bond and the herbicidal *N*-phenyl carbamate bond. According to Still and Herrett (1975) the latter is not cleaved in plants. Also, for several methyl carbamate insecticides cleavage of the carbamate group seems to play a minor role (Kuhr, 1968). Upon prolonged incubation of carbendazim in plants the formation of 2-aminobenzimidazole (**31**) is observed (Siegel, 1973).

Hydroxylated derivatives of carbendazim have not been reported in plants, although evidence was obtained for the presence of *O* or *N* conjugates (Siegel, 1973). Upon prolonged contact between melon plants and carbendazim Rouchaud *et al.* (1974) were able to identify benzimidazole (**32**), 2-aminobenzonitrile (**33**) and aniline (**34**). It is not improbable that benzimidazole was formed by a photochemical reaction. A similar mechanism has been observed by Jacob *et al.* (1975). They found formation of small amounts of benzimidazole from thiabendazole (**35**) on the leaves of sugar beets only under direct sunlight, whereas no transformation was found under artificial light. Another compound formed under sunlight conditions was benzimidazole-2-carboximide (**36**).

Erwin *et al.* (1970) and Wang *et al.* (1971) studied the fate of ring-^{14}C-labelled thiabendazole in cotton plants after root uptake, and speculated that conjugates of metabolites were formed.

After application of thiophanate-methyl (**37**) to plants the major metabolite observed is carbendazim (**30**). It was first detected by Soeda *et al.* (1972a) in bean plants and later by several other investigators in other plant species. Carbendazim is responsible for the fungicidal activity of thiophanate-methyl (Vonk and Kaars Sijpesteijn, 1971).

Formation of carbendazim from thiophanate-methyl on plant leaves may be due to a photochemical conversion (Buchenauer *et al.*, 1973), which has also been observed on glass plates (Soeda *et al.*, 1972b). There is, however, also evidence that enzymatic processes are involved in the conversion of thiophanate-methyl into carbendazim (Vonk *et al.*, 1977). Besides carbendazim, small

Table 3 Metabolites of benzimidazole derivatives and related compounds in plants

Parent compound	Metabolite	Plant species Way of application[a]	Label	Evidence for structure[b]	Reference
Benomyl (**29**)	Methylbenzimidazole-2-yl carbamate (**30**, carbendazim)	Cotton (R)	—	TLC, Isolation	Sims *et al*, (1969)[c]
		Bean (R)	—	TLC	Peterson and Edgington (1969)
		Cucumber (R)	—	TLC	Fuchs *et al.* (1972)
		Apple, Cuc., Ban., Or., Gr. (L)	^{14}C	Specific analysis	Baude *et al.* (1973)
		Pea (R)	^{14}C	TLC	Siegel and Zabbia (1972)
	2-aminobenzimidazole (**31**)	Pea (R)	^{14}C	TLC	Siegel and Zabbia (1972)
		Pear (L)	—	Specific analysis	Ben-Aziz and Aharonson (1975)
Carbendazim (**30**)	2-aminobenzimidazole (**31**)	Strawberry (R)	^{14}C	TLC	Siegel (1973)
		Tom. (R, L); Pep. (R, L)	—	Specific analysis	Ben-Aziz and Aharonson (1974)
		Melon (R)	^{3}H	GLC	Rouchaud *et al.* (1974)
	Benzimidazole (**32**)	Melon (R)	^{3}H	GLC	Rouchaud *et al.* (1974)
	2-aminobenzonitrile (**33**)				
	Aniline (**34**)				
	O- and *N*-conjugates, ^{14}C-associated with hemicellulose	Strawberry	^{14}C	TLC	Siegel (1973)
Thiabendazole (**35**)	Benzimidazole (**32**)	Sugar beet (L)	^{14}C	Inverse isotope dilution	Jacob *et al.* (1975)
	Benzimidazole-2-carboximide (**36**)				
Thiophanatemethyl (**37**)	Carbendazim (**30**)	Several plant species	^{14}C	TLC	Soeda *et al.* (1972a)
	Dimethyl-4,4′-*o*-phenylene-bis-allophanate (**38**)	Apple (L); Grape (L)	^{14}C	TLC	Soeda *et al.* (1972b)

[a] Abbreviations: Cuc. = cucumber; Ban. = banana; Or. = orange; Gr. = grape; Tom. = tomato; Pep. = pepper; R = root; L = leaf.
[b] TLC: thin-layer chromatographic comparison with reference compounds; GLC: comparison with reference compounds by gas-chromatography.
[c] The transformation of benomyl into carbendazim is very much documented; only a few references are shown.

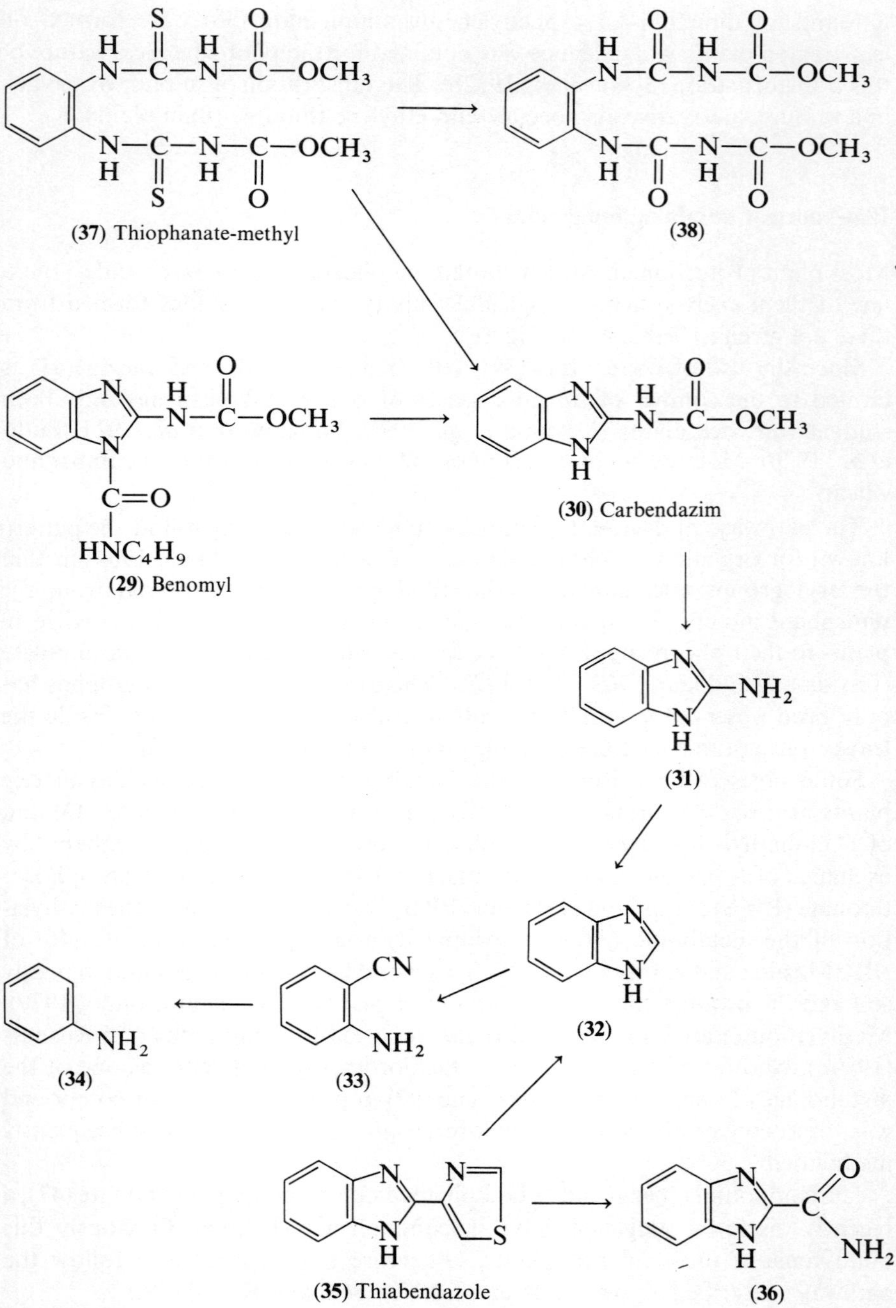

Figure 4 Pathways of metabolism of benzimidazole derivatives in plants

amounts of dimethyl-4,4′-*o*-phenylene-bis-allophanate (**38**) were formed on leaf-treated plants and evidence was obtained for the photochemical nature of this transformation (Soeda *et al.*, 1972b). The substitution of sulphur by oxygen in a thiourea derivative also occurs with ethylene thiourea (**8**) in plants.

Phosphorus-Containing Fungicides

Most plant protection chemicals containing phosphorus are insecticides, but a few of them exert systemic fungicidal activity. The metabolites formed from these are given in Table 4 and Figure 5.

Since the use of edifenphos (**39**), IBP (Kitazin-P) (**40**) and Inezin (**41**) is limited to the combat of fungal diseases of rice, metabolism has only been studied with rice plants (Ueyama *et al.*, 1973; Yamamoto *et al.*, 1973; Endo *et al.*, 1970). Metabolism of pyrazophos (**42**) has been studied in cucumber and wheat.

The pathways of degradation of these fungicides generally follow the pattern known for organic phosphate insecticides (Casida and Lykken, 1969) in that the aryl groups and sometimes the alkyl groups are hydrolysed from the phosphate moiety. Phosphorothionate insecticides (P═S) readily oxidize in plants to their phosphate (P═O) derivatives, which usually do not accumulate (Casida and Lykken, 1969; Bull, 1972). The oxon analogue of pyrazophos has only been observed in small amounts in leaf-treated wheat plants inside the leaves, but not in cucumber or apple plants (Gorbach *et al.*, 1975).

Some noteworthy features of the metabolism of these fungicides in rice plants are: (a) the formation of *S*,*S*,*S*-triphenylphosphorotrithioate (**43**) and of *O*,*O*-diethyl-*S*-phenylphosphorothioate (**44**) from edifenphos, probably by exchange of *S*-phenyl and *O*-ethyl radicals; (b) the formation of a phosphorothionate (P═S) compound (**45**) from IBP by isomerization and (c) the methylation of the metabolite *O*,*O*-di-isopropyl hydrogen phosphorothioate (**46**) of IBP (Masuda and Kanazawa, 1973). This methylation also takes place in paddy soil and the product can be taken up in rice plants (Masuda and Endo, 1979). Methylation of an SH compound is also reported by Lamoureux and Rusness (1980b), who found formation of pentachloromethylthiophenol as one of the metabolites of pentachloronitrobenzene (**92**) in peanuts. The latter compound was first converted into pentachlorothiophenol, which was subsequently methylated.

The compounds Inezin and *O*,*O*-diphenyldichloromethylphosphonate (**47**), a recently disclosed fungicide, have in common a C–P bond. Obviously this bond remains intact in rice plants. Otherwise these compounds follow the pathways described above (Endo *et al.*, 1970; Bedi and Roy, 1979).

Recently, a new phosphorus-containing fungicide was discovered, phosethyl-Al (aluminium tris(ethylphosphonate)). This compound is systemically active

Table 4 Metabolites of phosphorus-containing fungicides in plants

Parent compound	Metabolite	Plant species Way of application[a]	Label	Evidence for structure[b]	References
Pyrazophos (**42**)	Pyrazophos-oxon (*O*-pyrazophos) (**48**)	What (L)	^{14}C	TLC	Gorbach *et al.* (1975)
	β-glucoside of 6-ethoxycarboxyl-2-hydroxy-5-methylpyrazolo-[1,5-a]pyrimidine (**49**, **50**)	Cucumber (L); Wheat (L)	^{14}C	TLC	Gorbach *et al.* (1975)
Edifenphos (**39**)	*O*-ethyl-*S*-phenyl hydrogen phosphorothioate (**51**)	Rice (L)	^{35}S, ^{32}P	TLC, GLC	Ueyama *et al.* (1973)
	S-phenyl dihydrogen phosphorothioate (**52**)				
	S,*S*,*S*-triphenyl phosphorotrithioate (**43**)				
	S,*S*-diphenyl hydrogen phosphorodithioate (**53**)				
	O,*O*-diethyl-*S*-phenyl phosphorothioate (**44**)				
	Thiophenol (**54**)				
	Phenylsulphonic acid (**55**)				
Kitazin P (**40**)	*S*-benzyl-*O*-isopropyl hydrogen phosphorothioate (**56**)	Rice (L)	^{35}S, ^{32}P	TLC, GLC	Yamamoto *et al.* (1973)
	O,*O*-diisopropyl hydrogen phosphorothioate (**46**)				
	O,*O*-diisopropyl hydrogen phosphate (**57**)				
	O-isopropyl dihydrogen phosphate (**58**)				
	O-benzyl-*O*,*O*-diisopropyl phosphorothioate (**45**)				
	Benzyl mercaptane (**59**)				
	Benzyl disulphide (**60**)				
	O,*O*-diisopropyl-*S*-methyl phosphorothioate (**61**)	Rice (R)		GLC	Masuda and Kanazawa (1973)
Inesin (**41**)	*S*-benzyl hydrogen phenyl phosphonothioate (**62**)	Rice			Endo *et al.* (1970)
	Phenyl phosphonothioic acid (**63**)				
O,*O*-diphenyl-dichloromethyl phosphonate (**47**)	*O*-phenyl hydrogen dichloromethyl phosphonate	Rice (L)	^{32}P	TLC, PC	Bedi and Roy (1979)

[a] Abbreviations: L = le—f; R = root.

[b] TLC: thin-layer chromatographic comparison with reference compounds (GLC: gas-liquid chromatography).

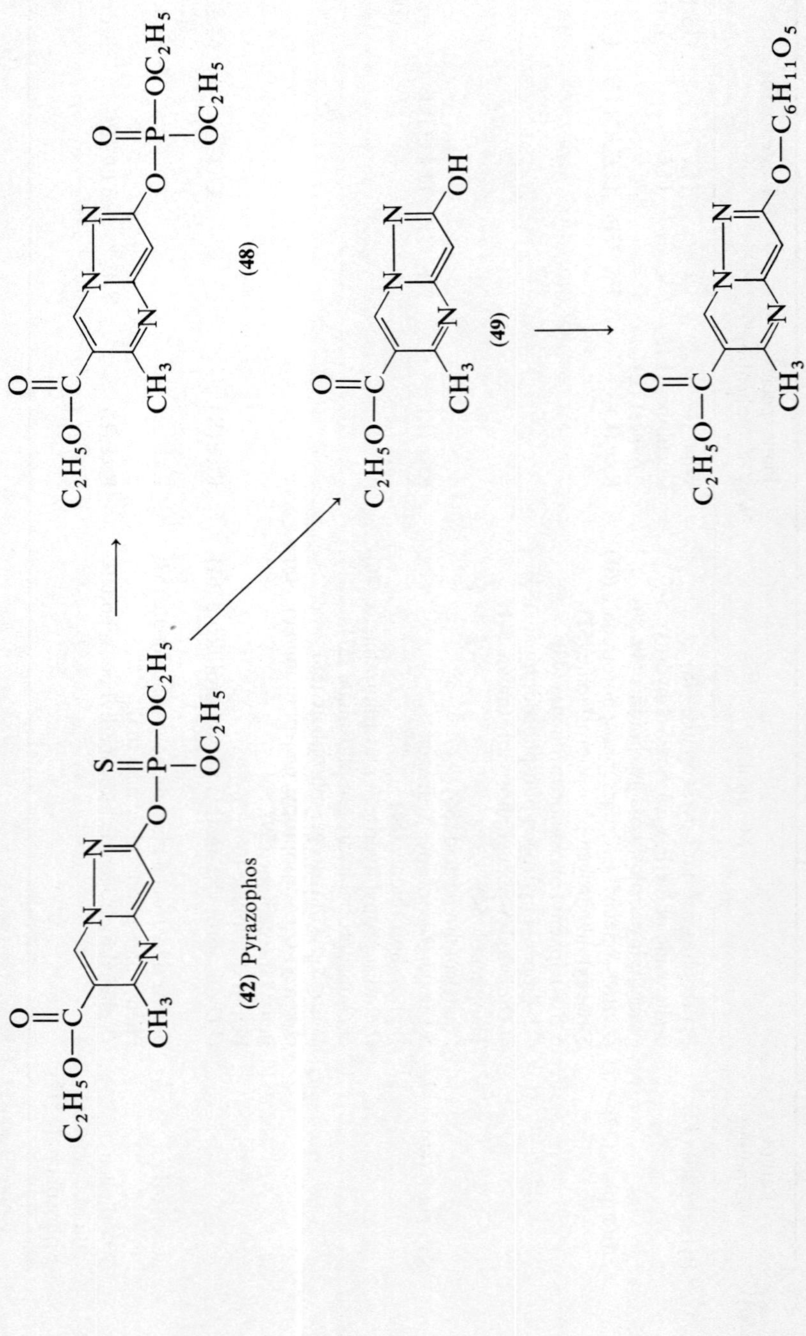

(42) Pyrazophos

(48)

(49)

(50)

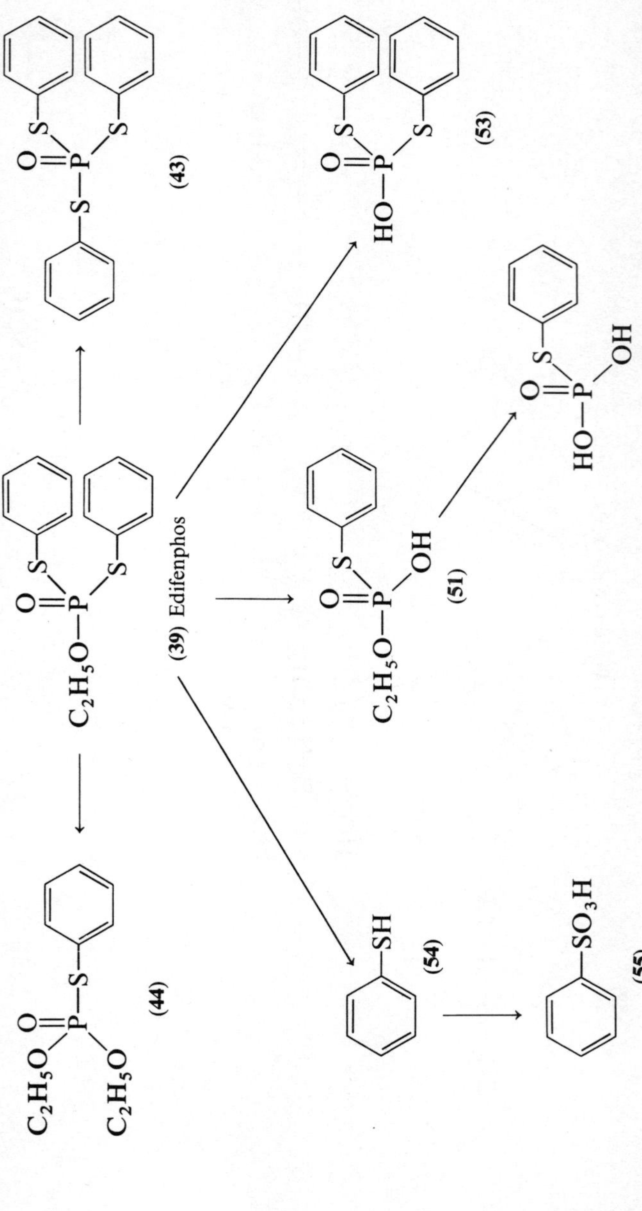

(continued)

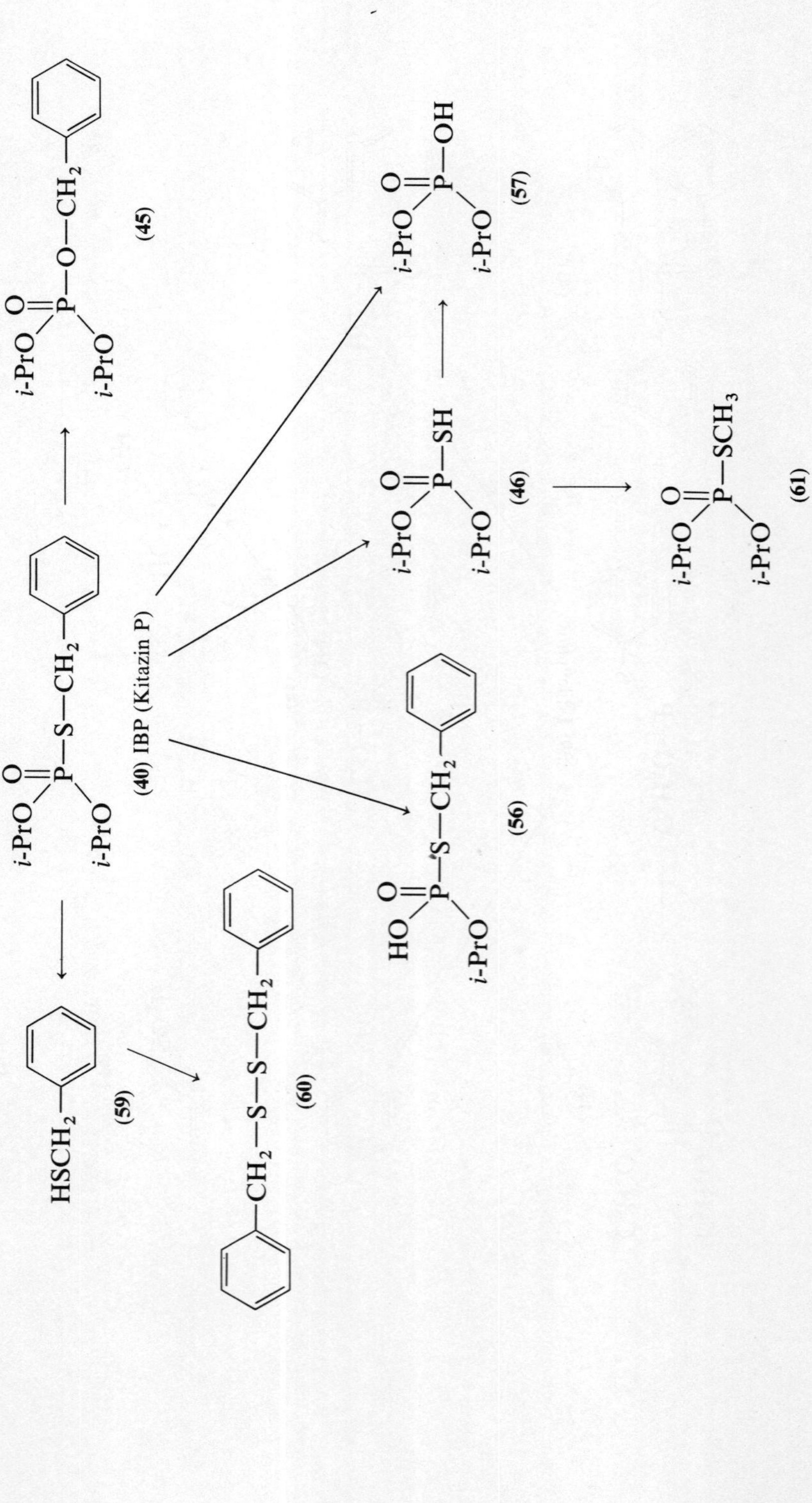
(40) IBP (Kitazin P)
(45)
(46)
(56)
(57)
(59)
(60)
(61)

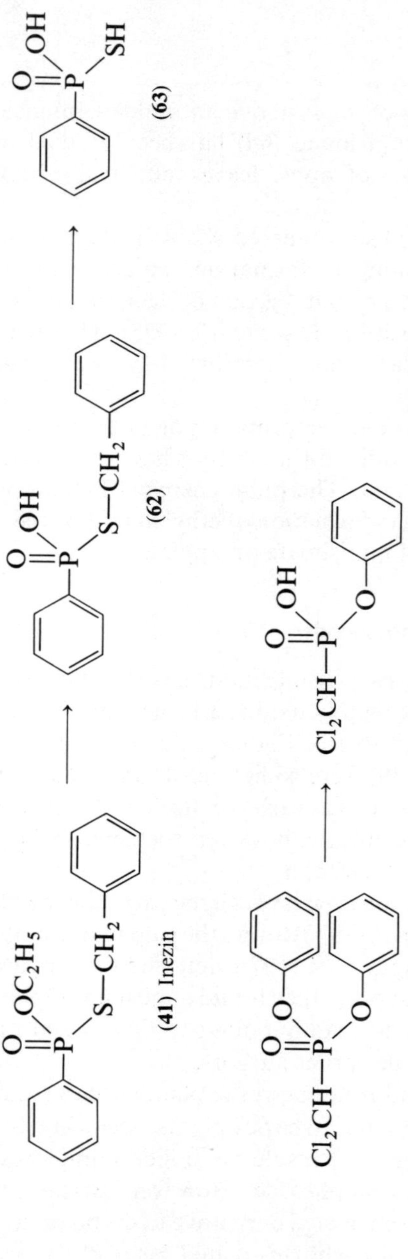

Figure 5 Pathways of metabolism of phosphorus-containing fungicides in plants

against *Phycomycete* fungi. Of its metabolism in plants, no more is known than that the major residue following foliar or soil application is phosphorous acid (Williams *et al.*, 1977).

Pyrimidines

The metabolism in plants of the systemic antimildew fungicides ethirimol (**64**), dimethirimol (**65**) and bupirimate (**66**) has been studied in detail (Table 5, Figure 6). On the surface of apple leaves and fruits bupirimate is readily converted into ethirimol.

Three routes of metabolism observed are *N*-dealkylation, hydroxylation of the butyl group and conjugate formation. Hydroxylation appears to be a relatively minor pathway. The butyl group of the herbicide secbumetone is also hydroxylated on the 2-position (Esser *et al.*, 1975). The structures of ethirimol and dimethirimol are related and therefore they have a common metabolite (**70**).

Cavell *et al.* (1976) reported the formation of a number of photodimerization products of ethirimol and dimethirimol which have, however, not been detected in crops exposed to sunlight. The photochemical breakdown of ethirimol in aqueous solution resulted in formation of ethylguanidine (**68**), a compound also observed as metabolite of bupirimate on apples.

Oxathiins and Related Compounds

Carboxin (**77**), used as a seed disinfectant, has systemic activity against smut infections of cereals. It is readily oxidized in several plant species to the non-fungitoxic *S*-oxide (**78**) (Table 6, Figure 7). Small amounts of the sulphone derivative (oxycarboxin, **79**) were sometimes found as well. Oxycarboxin is as active against the same diseases as carboxin itself and is also used as a fungicide. The amounts formed in plants are, however, too small to be of any importance for the fungicidal action of carboxin.

There seems to be some controversy with regard to the metabolites of carboxin in plants. Whereas Chin *et al.* (1970) found the sulphoxide as the main metabolite of carboxin in barley, Briggs *et al.* (1974) detected the *p*-hydroxyphenyl derivative (**80**) as the major metabolite; they found almost no sulphoxide. The hydroxy derivative was unstable and readily bound to lignin, and therefore may have escaped the attention of the earlier authors.

Carboxin formed bound residues in the plant species studied but the nature of the bound residue is obscure. In barley plants, according to Chin *et al.* (1970), carboxin is bound to lignin as insoluble anilide complexes, which probably contain mainly the intact sulphoxide. However, Briggs *et al.* (1974) found complexation of the hydroxyphenyl derivative of carboxin to lignin. Hydrolysis of the insoluble residue of carboxin-treated bean plants yielded aniline (**34**)

Table 5 Metabolites of pyrimidine fungicides in plants

Parent compound	Metabolite	Plant species Way of application[a]	Label	Evidence for structure[b]	Reference
Bupirimate (**66**)	Ethirimol (**64**)	Apple (F, L)	^{14}C	TLC	Teal and Cavell (1975)
	Ethirimol-*O*-glucoside (**69**)	Apple (F, L)			
	Ethylguanidine (**68**)	Apple (F, L)			
Ethirimol (**64**)	*O*-glucoside + possibly phosphate conjugate (**69**)	Barley (R); Wheat (R)	^{14}C	TLC	Calderbank (1971); Cavell *et al.* (1971)
	Desethylethirimol (**70**)				
	O-glucoside of (**70**), (**71**)				
	2-hydroxybutyl derivative of ethirimol (**72**)				
	4-butyl-3,5-dioxopyrazolidine (**73**)	Cereals			Hemingway (1976)
	5-butyryl-2-ethylamino-5-methyl-oxazolin-4-one (**74**)				
Dimethirimol (**65**)	*O*-glucoside + possibly phosphate conjugate (**75**)	Cucumber (R)	^{14}C	TLC	Calderbank (1971); Cavell *et al.* (1971)
	Des methyldimethirimol (**76**)				
	Glucoside of (**76**), (**67**)				
	Didesmethyldimethirimol (**70**) (2-amino-5-butyl-6-hydroxy-4-methyl-pyrimidine)				

[a] Abbreviations: F = fruit; L = leaf; R = root.
[b] TLC: thin-layer chromatographic comparison with reference compounds.

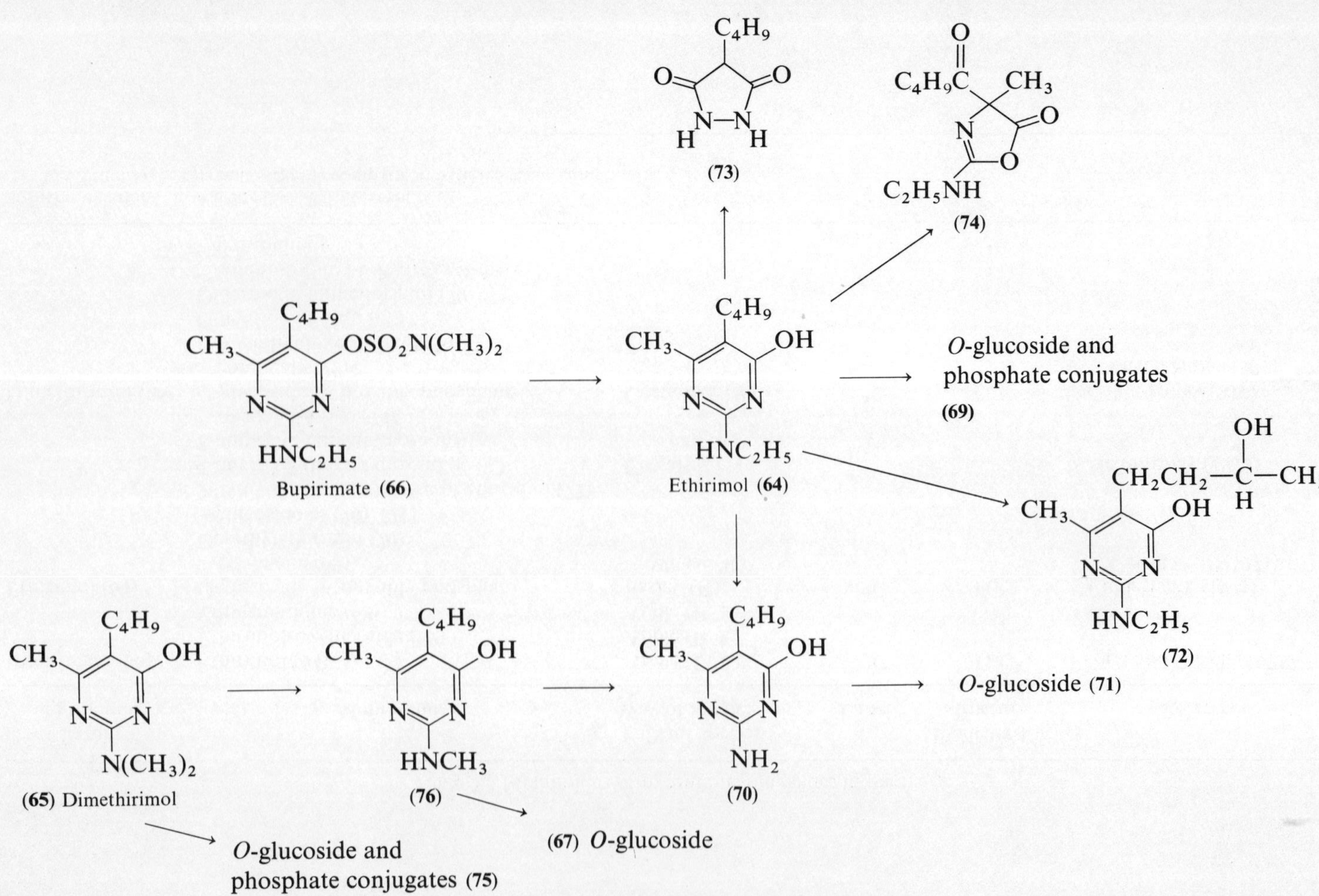

Figure 6 Pathways of metabolism of pyrimidine fungicides in plants

Table 6 Metabolites of oxathiins and related compounds in plants

Parent compound	Metabolite	Plant species Way of application[a]	Label	Evidence for structure[b]	Reference
Carboxin (**77**)	Sulphone of carboxin (oxycarboxin, **79**)	Cotton (Sd)	—	TLC	Allam and Sinclair (1969)
		Barley (Sd); Wheat (Sd)	^{14}C	TLC	Chin *et al.* (1970)
	Sulphoxide of carboxin (**78**)	Barley (Sd); Wheat (Sd)	^{14}C	TLC	Chin *et al.* (1970)
		Bean (R)	^{14}C	TLC	Snel and Edgington (1970)
	Insoluble complex with lignin	Barley (Sd); Wheat (Sd)	^{14}C	Special analysis	Chin *et al.* (1970)
		Bean (R)	^{14}C	Hydrolysis, TLC	Snel and Edgington (1970)
	p-hydroxyphenyl derivative of carboxin (**80**)	Barley (Sd)	^{14}C	MS	Briggs *et al.* (1974)
Pyracarbolid (**83**)	2-hydroxy-3-phenylaminocarbonyl-6-hydroxyhex-2-ene (**84**)	Wheat (L)	^{14}C	TLC	Oeser *et al.* (1975)

[a] Abbreviations: Sd = seed treatment; R = root; L = leaf.
[b] TLC = thin-layer chromatographic comparison with reference compounds; MS = mass spectrometry.

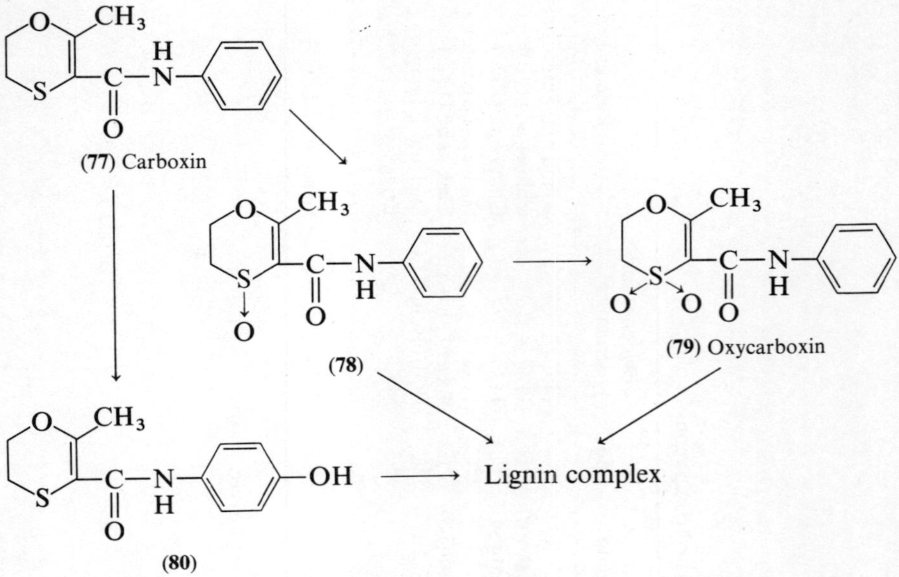

Figure 7 Pathways of metabolism of carboxin in plants

(Snel and Edgington, 1970). It was speculated that aniline had been formed by hydrolytic cleavage of carboxin and subsequently had been bound to plant polymers like lignin. It is known that aniline derivatives, e.g. 3,4-dichloroaniline (**81**) formed from the herbicide propanil (**82**), are bound to lignin in rice plants (Yih *et al.*, 1968; Still and Herrett 1975).

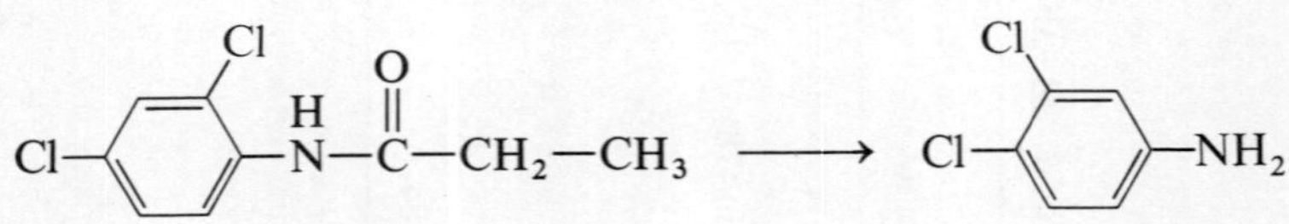

(**82**) (**81**)

Oxycarboxin, the sulphone derivative of carboxin, was more stable than carboxin after uptake in bean plants (Snel and Edgington, 1970). It was transformed into unknown polar metabolites. Also the ^{14}C-label of oxycarboxin was incorporated into acetone-insoluble plant residues, especially in the roots.

The formation of sulphoxides from sulphides is a commonly observed conversion in plants. Although a slow oxidation of carboxin in water was observed, sulphoxide formation in plants is obviously much faster. The mechanism of oxidation might be chemical rather than enzymatic, since Lyr *et al.* (1974) found rapid formation of carboxin-sulphoxide by H_2O_2. Moreover, photochemical decomposition of carboxin on leaves of bean plants by sunlight was found by Buchenauer (1975). After irradiating aqueous carboxin solutions with UV light most of the carboxin had been converted into the sulphoxide, but only a small amount could be detected as the sulphone.

Of other fungicides related to the oxathiins only the metabolism of pyracarbolid (**83**) in wheat has been reported (Oeser *et al.*, 1975). Small amounts of 2-hydroxy-3-phenylaminocarbonyl-6-hydroxyhex-2-ene (**84**) were found.

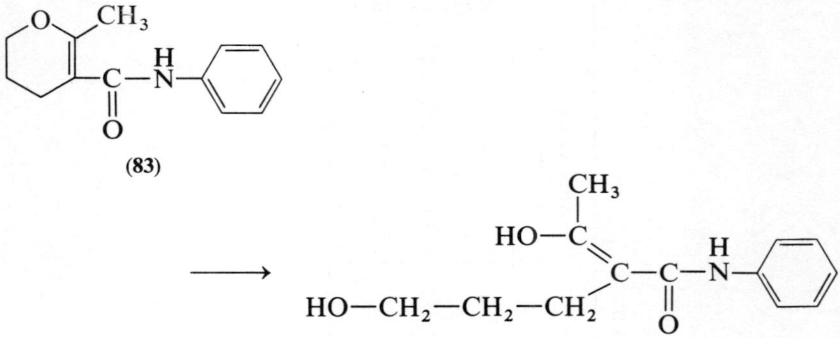

(83) **(84)**

Chlorinated Benzene Derivatives

Chloroneb (**85**), a systemic fungicide used for seed treatment, is demethylated in plants (Table 7, Figure 8). This observation was made with plants treated via the soil. Since chloroneb is very stable in soil (Rhodes *et al.*, 1971), it can be assumed that the demethylation had occurred in the plants. In bean plants the metabolite resulting from demethylation of one methoxy group, 2,5-dichloromethoxyphenol (**86**), is conjugated with glucose (Thorn, 1973).

Chlorothalonil (**90**) is a broad spectrum protectant fungicide. There is no detailed information about its metabolism in plants, but grapes have been investigated for residues of 4-hydroxy-2,5,6-trichloroisophthalonitrile (**91**), of which small amounts have been detected (Northover and Ripley, 1980).

It has been shown that reduced glutathione and protein-SH groups can react with chlorothalonil by nucleophilic substitution at the 4 and 6 position (Tillman *et al.*, 1973). Any penetrating chlorothalonil probably would result in such conversion products in plants.

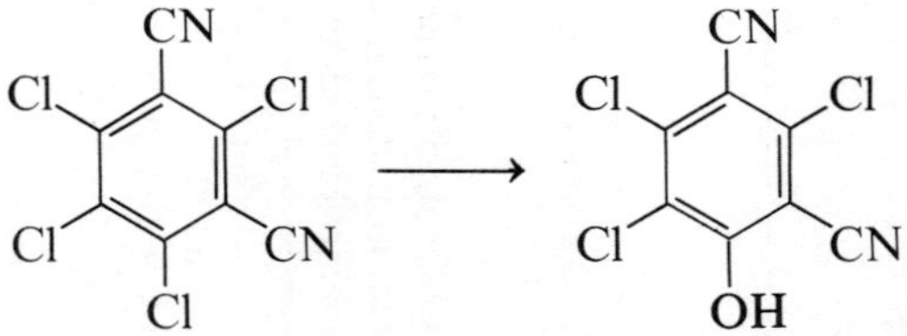

(90) **(91)**

Table 7 Metabolites of chloroneb (**85**) in plants

Metabolite	Plant species[a] Way of application[b]	Label	Evidence for structure[c]	Reference
2,5-dichloromethoxyphenol (**86**)	Sn. bn. (S); Cot. (S)	^{14}C	TLC, IR, MS	Rhodes *et al.* (1971)
	Bush bean (R)	^{14}C	TLC, Inverse isotope dilution	Thorn (1973)
2,5-dichloromethoxyphenyl-β-D-glucopyranoside (**87**)	Bush bean (R)	^{14}C	id	Thorn (1973)
2,5-dichlorohydroquinone (**88**)	Sn. bn. (S); Cot. (S)	^{14}C	TLC, IR, MS	Rhodes *et al.* (1971)
2,5-dichlorobenzoquinone (**89**)	id	^{14}C	id	Rhodes *et al.* (1971)

[a] Abbreviations: Sn. bn. = snap bean; Cot. = cotton.
[b] S = soil; R = root.
[c] TLC = thin-layer chromatography; IR = infra-red spectroscopy; MS = mass spectrometry.

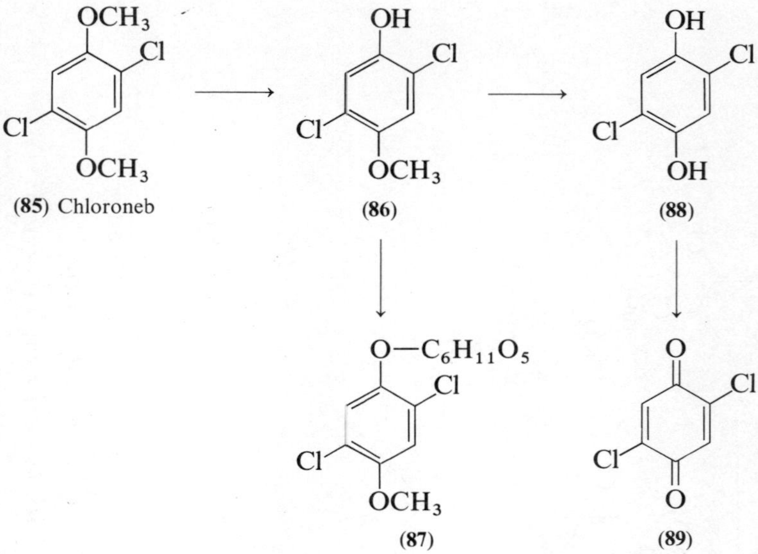

Figure 8 Pathways of metabolism of chloroneb in plants

PCNB (*quintozene*, *pentachloronitrobenzene*) (**92**) is a fungicide used mainly for soil treatment and for treatment of seeds and transplants, but also for some crop applications. Its metabolism has been studied in several plant species but always after application to the soil. Since some of the conversion products found in plants are also formed in soil, it is not clear whether these are really formed in the plant or are taken up as such from the soil. Thus pentachloroaniline (**93**) and pentachlorothioanisole (**94**) were identified as metabolites both in soil and in plants (cf. Kaars Sijpesteijn *et al.*, 1977). Since earlier studies were mostly performed with technical PCNB, impurities of PCNB like hexachlorobenzene, pentachlorobenzene (**95**) and tetrachloronitrobenzene (**96**) have also been detected in plants.

Recently a number of studies have been carried out (Begum *et al.*, 1979; Lamoureux and Rusness, 1980b; Rusness and Lamoureux, 1980) with pure ^{14}C-labelled PCNB added to soil or to nutrient solution. Plants grown in this soil or the nutrient solution contained a large number of products (Table 8, Figure 9). In onions the conversion products were present in shoots and roots but not inside the bulb (Begum *et al.*, 1979). Lamoureux and Rusness (1980b) studied the conversion of PCNB in peanut plants which had taken up the compound from nutrient solution; they found several metabolites which had already been detected earlier in plants grown in PCNB-treated soil. This suggested that they were true plant metabolites.

In addition to the compounds mentioned in Table 8, four minor conversion product structures were proposed on the basis of their mass spectra. These

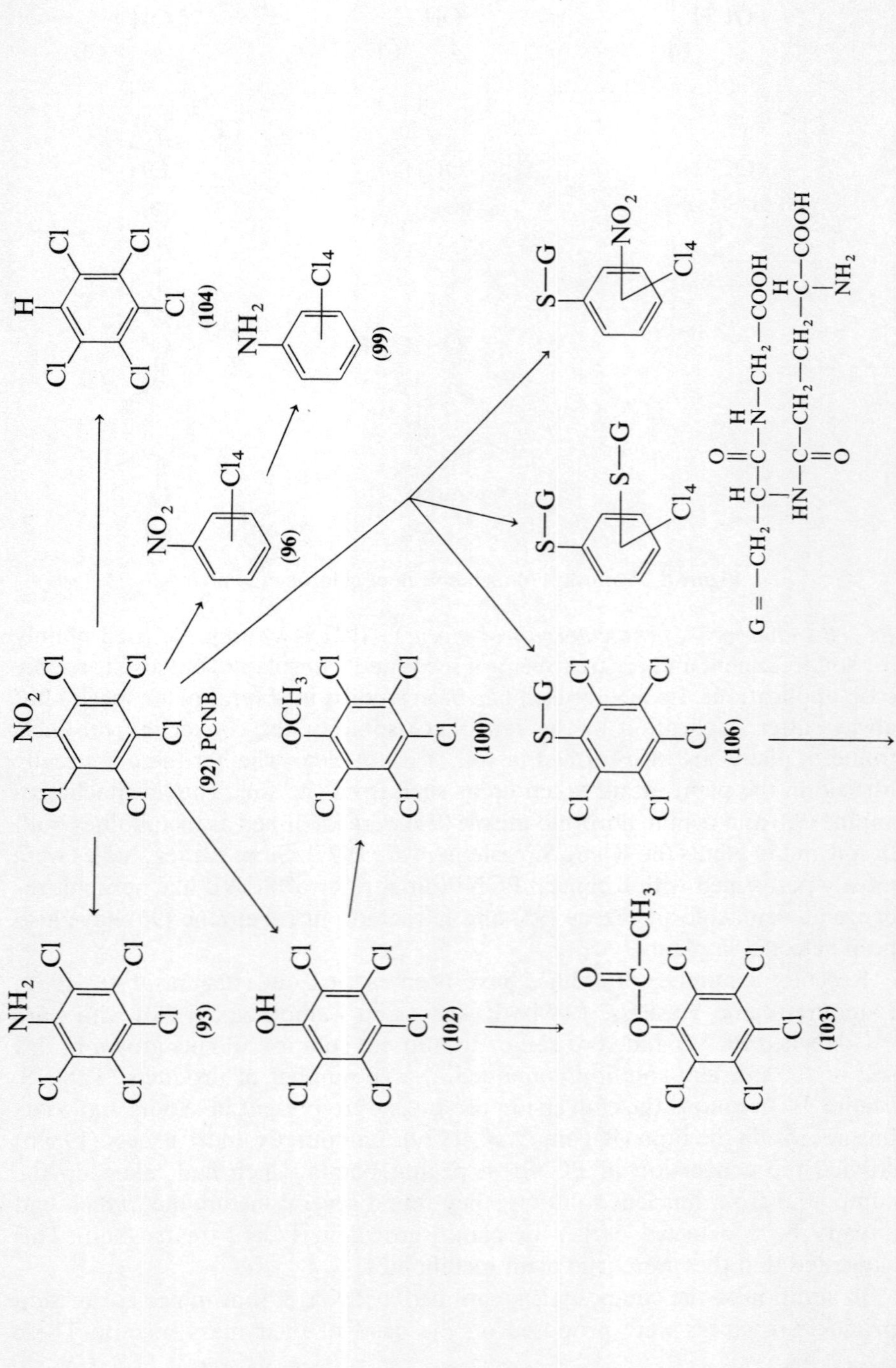
(92) PCNB
(93)
(96)
(99)
(100)
(102)
(103)
(104)
(106)
G = —CH₂—C—C—N—CH₂—COOH
HN—C—CH₂—CH₂—C—COOH
NH₂

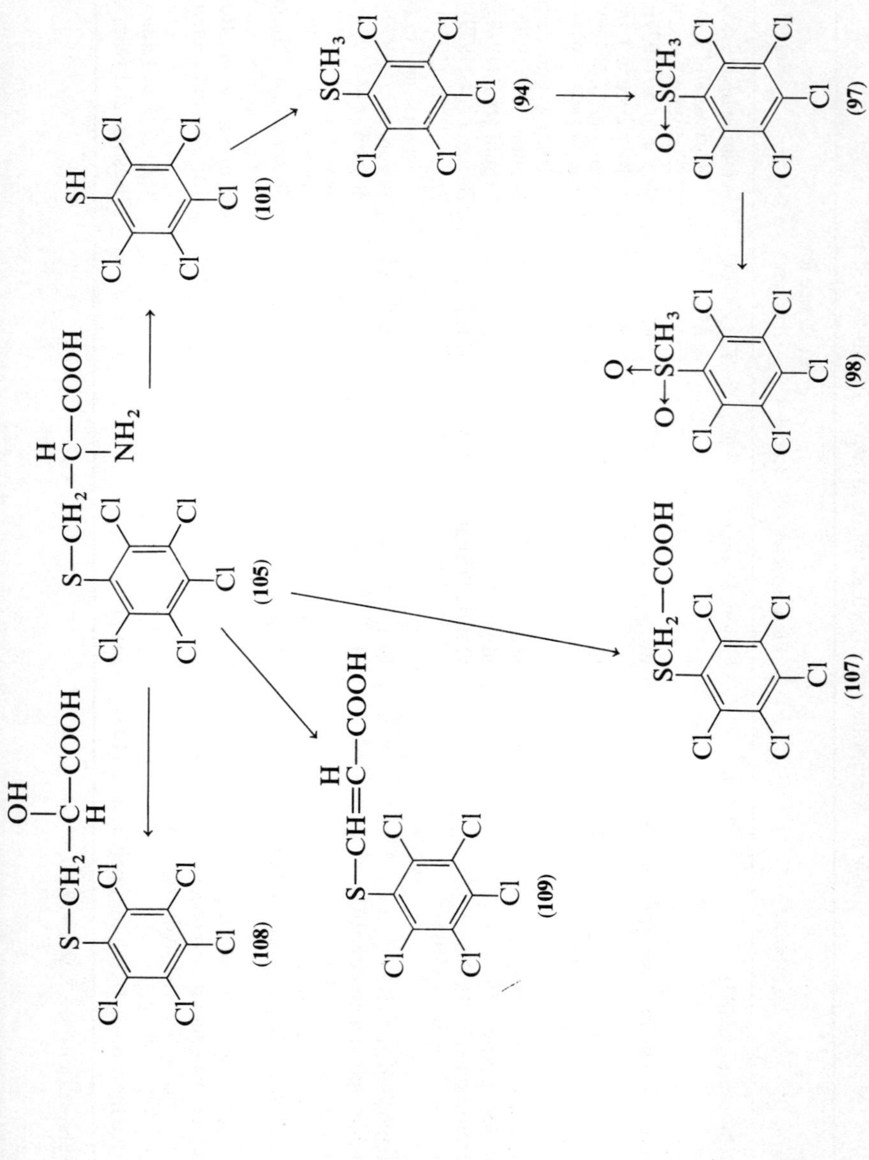

Figure 9 Pathways of metabolism of PCNB in plants

Table 8 Metabolites of PCNB in plants grown in PCNB-treated soil

Metabolite	Plant species	Label	Evidence for structure[a]	Reference
Pentachloroaniline (**93**)	Onion, Peanut	^{14}C	TLC, GLC, MS	Begum *et al.* (1979); Rusness and Lamoureux (1980); Lamoureux and Rusness (1980b)
Pentachlorophenylmethylsulphoxide (**97**)	id	^{14}C	id	Begum *et al.* (1979); Rusness and Lamoureux (1980)
Pentachlorophenylmethylsulphone (**98**)	Onion		id	Begum *et al.* (1979)
Pentachlorothioanisole (**94**)	id		id	Begum *et al.* (1979); Rusness and Lamoureux (1980)
Pentachlorobenzene (**95**)	id		id	Begum *et al.* (1979)
Tetrachloroaniline (**99**)	id		id	Begum *et al.* (1979)
Pentachloroanisole (**100**)	Onion, Peanut		id	Begum *et al.* (1979)
Pentachlorothiophenol (**101**)	id		id	Begum *et al.* (1979) Rusness and Lamoureux (1980)
Pentachlorophenol (**102**)	Onion		id	Begum *et al.* (1979)
Pentachlorophenyl acetate (**103**)	id		id	Begum *et al.* (1979)
Tetrachlorobenzene (**104**)	id		id	Begum *et al.* (1979)
Tetrachloronitrobenzene (**96**)	id		id	Begum *et al.* (1979)
S-(pentachlorophenyl)cysteine (**105**)	Peanut			Rusness and Lamoureux (1980); Lamoureux and Rusness (1980b)
S-(pentachlorophenyl)glutathione (**106**)	id			Lamoureux and Rusness (1980b)
S-(pentachlorophenyl)-2-thioacetic acid (**107**)	id		id	Rusness and Lamoureux (1980)
S-(pentachlorophenyl)-2-hydroxy-3-thiopropanoic acid (**108**)	id		id	Rusness and Lamoureux (1980)
S-(pentachlorophenyl)-3-thiopropenoic acid (**109**)	id		id	Rusness and Lamoureux (1980)

[a] Abbreviations: see Tables 3 and 7.

metabolites were derived from the above-mentioned compounds by dechlorination, formylation or introduction of the thiomethyl group (Begum *et al.*, 1979). Lamourex and Rusness (1980b) also found other glutathione conjugates and further metabolites which are not discussed here.

Dichloran The metabolism of dichloran (2,6-dichloro-4-nitroaniline, **110**), a protectant fungicide mainly used against *Botrytis*, has been studied in bean plants after uptake of the compound from a nutrient solution by the roots (Groves and Chough, 1970). Dichloran was found to disappear and the label was mainly incorporated in unextractable plant constituents. A small amount of $^{14}CO_2$ evolves from ^{14}C-dichloran-treated plants. The binding of dichloran to unextractable plant constituents is not surprising in view of studies carried out by Still and co-workers on the fate of herbicide-derived chloroanilines in rice plants (cf. Still and Herrett, 1975). The latter are rapidly bound to the plant lignin. Soybean plants were reported to metabolize dichloran to a compound tentatively identified as 4-amino-3,5-dichloromalonanilic acid (**112**) (USDA, 1968). 2,6-Dichlorophenylenediamine (**111**) is assumed to be an intermediate.

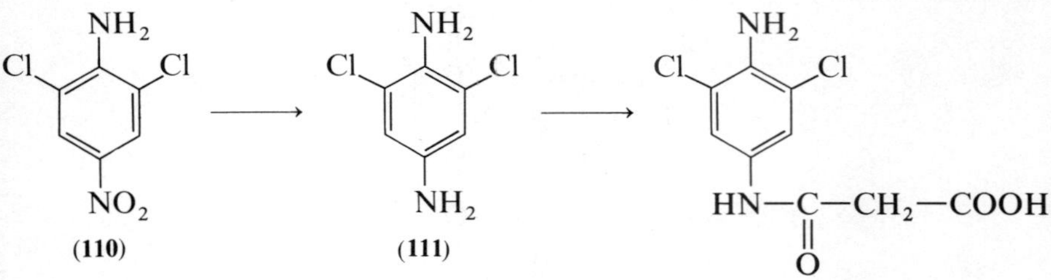

(**110**) (**111**) (**112**)

Dicarboximides

This group of systemic fungicides comprises the structurally related compounds dichlozoline (**113**), vinclozoline (**114**), procymidone (**115**) and iprodione (**116**). Their spectrum of activity is rather narrow: they act mainly against *Sclerotinia* and *Botrytis* species. Little has been published on their metabolism in plants.

Dichlozoline is no longer of practical importance. Its metabolism in bean and grape plants has been studied by Sumida *et al.* (1973). The metabolites found are given in Table 9 and Figure 10. Metabolites (**118**) and (**119**) were detected in root-treated bean plants. These two compounds were also formed in sterilized distilled water. Since they were not found in bean plants injected with dichlozoline, it can be assumed that they were taken up as such from the solution.

Procymidone (**115**), a compound related in structure to dichlozoline, was applied to the first leaf of a cucumber plant. Twelve days after application 80 % of the radioactivity of a methanolic extract obtained from all parts of the plants

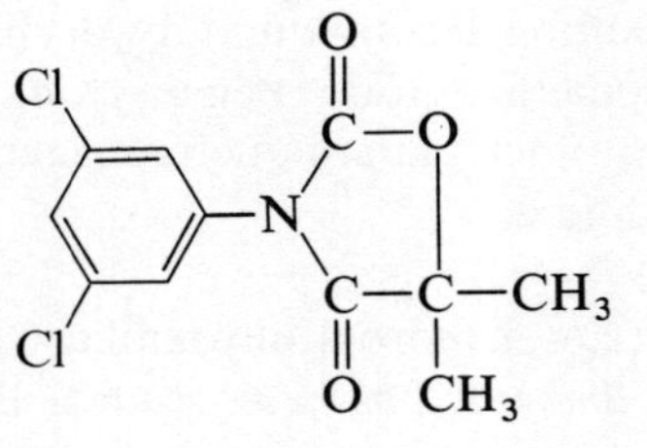

(113)

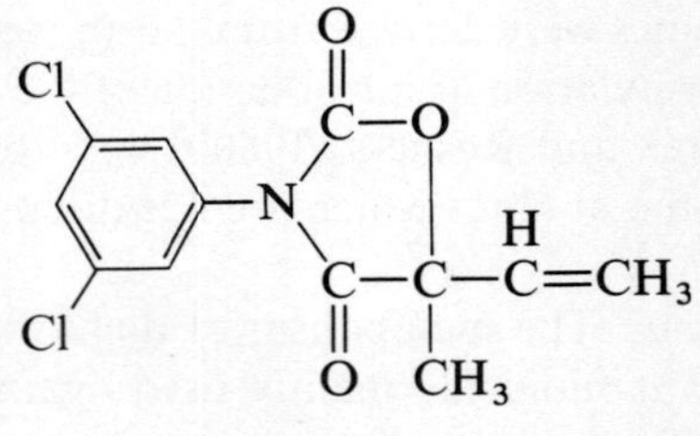

(114)

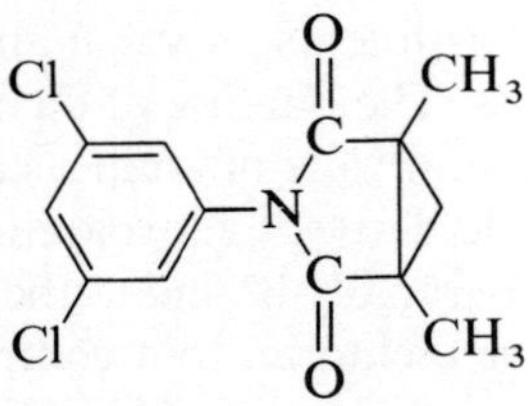

(115)

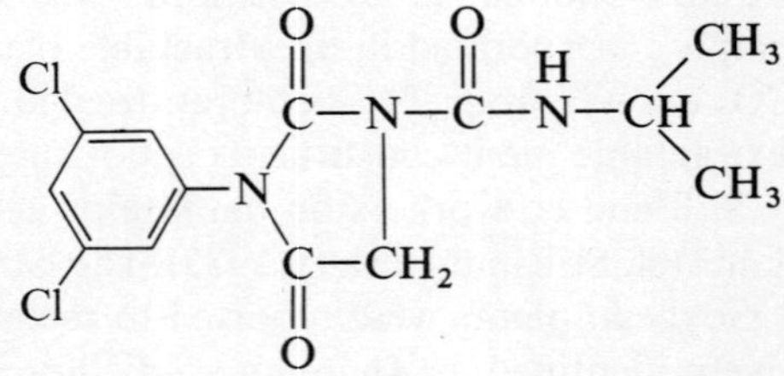

(116)

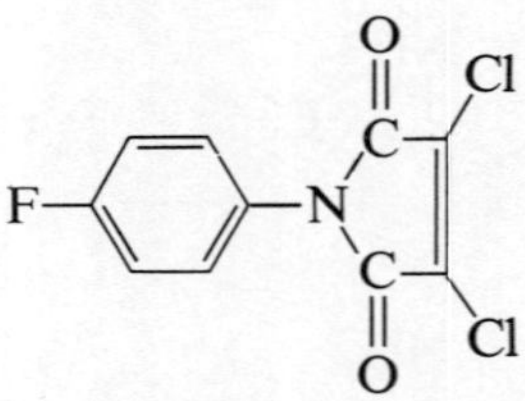

(117)

co-chromatographed with authentic procymidone (Hisada *et al.*, 1977), and a few unidentified metabolites were detected (Hisada *et al.*, 1976).

Iprodione (**116**) was shown to undergo a structural rearrangement in ethanol (Cooke *et al.*, 1979). It is not known whether the same reaction occurs on plant leaves.

Table 9 Metabolites of dichlozoline in plants. All metabolites were identified by TLC comparison with reference compounds (Sumida *et al.*, 1973)

Metabolite	Plant species (way of application)	Label
N-3,5-dichlorophenyl-*N*-α-hydroxyisobutyryl carbamic acid (**118**)	Bean (roots)	^{14}C
α-hydroxyisobutyryl-3,5-dichloroanilide (**119**)	Bean (roots)	^{14}C
	Grape (leaves, after stem injection)	^{14}C, ^{3}H
4-hydroxydichlozoline (**120**)	Grape	^{14}C, ^{3}H

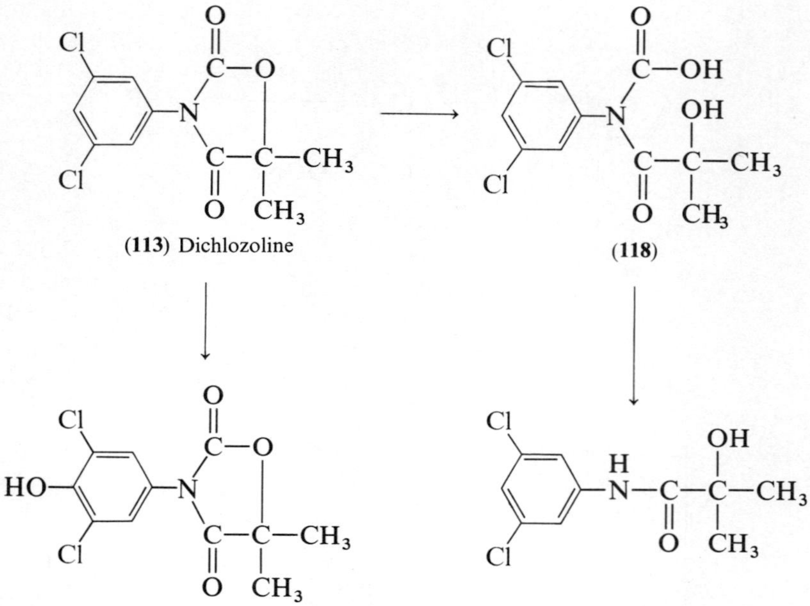

(113) Dichlozoline; (118); (120); (119)

Figure 10 Pathways of metabolism of dichlozoline in plants

Fluoroimide (**117**) belongs structurally to the dicarboximide group, but it is not systemic. It has a broader spectrum of antifungal activity than the previously mentioned compounds. Its fate on leaves and fruits of apple trees using ^{14}C-labelled compound has been reported by Ogawa *et al.*, (1980). A major metabolite was identified by MS, IR and NMR as sodium 4-fluorodichloromaleanilate (**121**), resulting from hydrolysis of the imide ring. Other minor metabolites found were 4-fluoroaniline (**125**), (E)-2,3-dichloro-*N*-(4-fluorophenyl)acrylamide (**122**), *N*-(4-fluorophenyl)maleimide (**123**) and *N*-(4-fluorophenyl)succinimide (**124**), which were identified by co-chromatography on TLC and MS. The latter two metabolites resulted from ring dechlorination and subsequent reduction. Formation of the acid form of (**121**) occurred rapidly on TLC plates and it is probable that on the surface of the leaves its formation is also mainly chemical.

Miscellaneous Compounds

Fenaminosulf (**126**), a soil fungicide with slight systemic activity, formed four unidentified compounds in sugar beets treated through their roots with fenaminosulf labelled with ^{14}C in the methyl groups (Hills, 1962). No data are available on the metabolism of fenaminosulf labelled in the benzene ring.

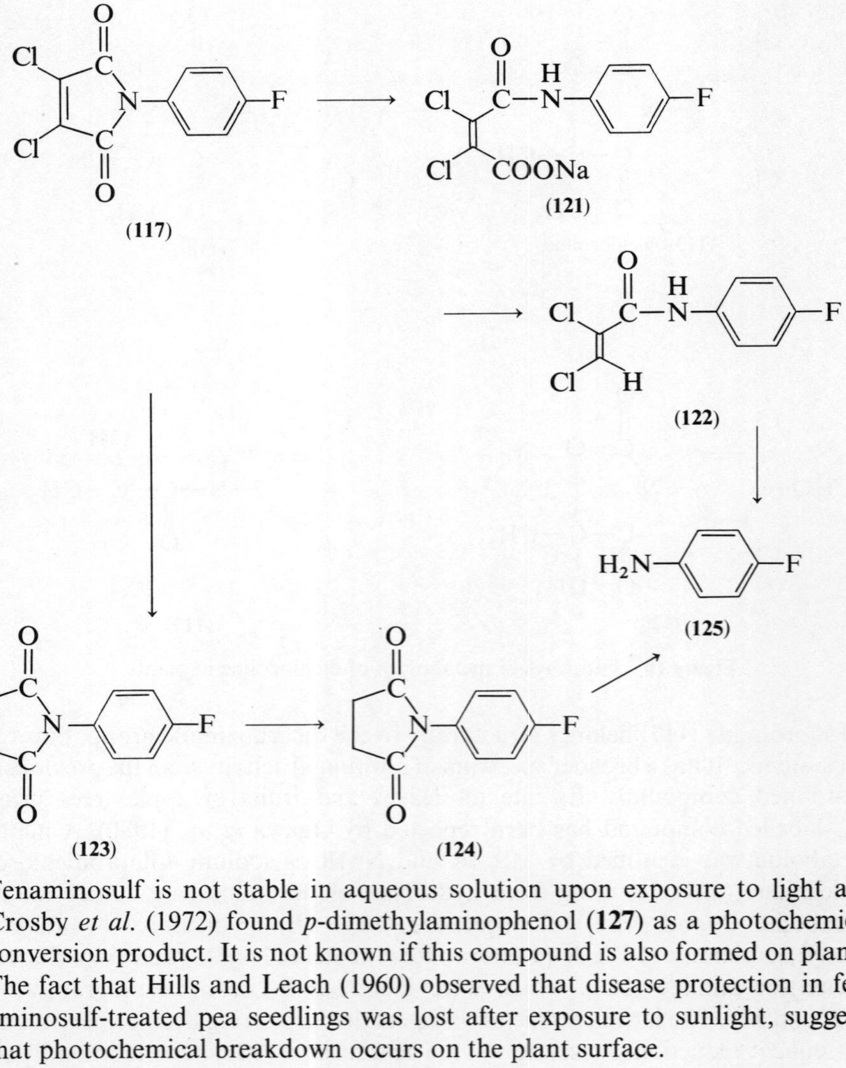

(117) (121) (122) (125) (123) (124)

Fenaminosulf is not stable in aqueous solution upon exposure to light and Crosby *et al.* (1972) found *p*-dimethylaminophenol (**127**) as a photochemical conversion product. It is not known if this compound is also formed on plants. The fact that Hills and Leach (1960) observed that disease protection in fenaminosulf-treated pea seedlings was lost after exposure to sunlight, suggests that photochemical breakdown occurs on the plant surface.

Dodine The fate of dodine (**128**) ^{14}C-labelled in the guanidine carbon, which was painted throughout the season on leaves of an apple tree, was studied by Curry

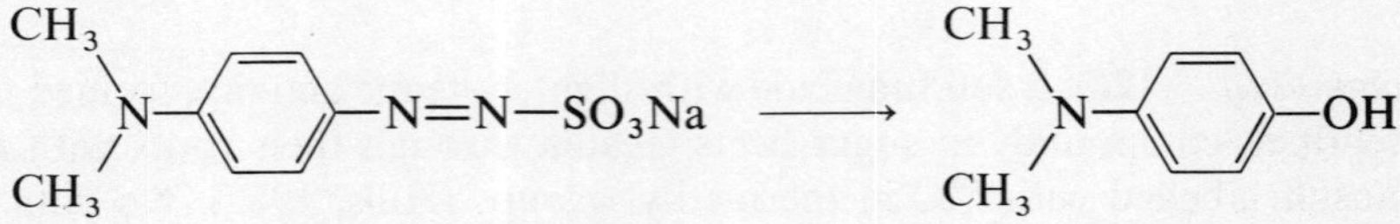

(126) (127)

(1962). Radioactivity could be detected after several weeks in various other compounds, including guanidine (**129**) or perhaps guanidine monosubstituted with a small group. After hydrolysis, labelled creatine (**130**) was found.

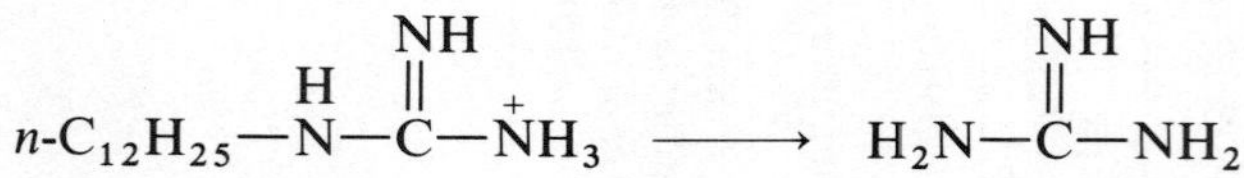

(**128**) (**129**)

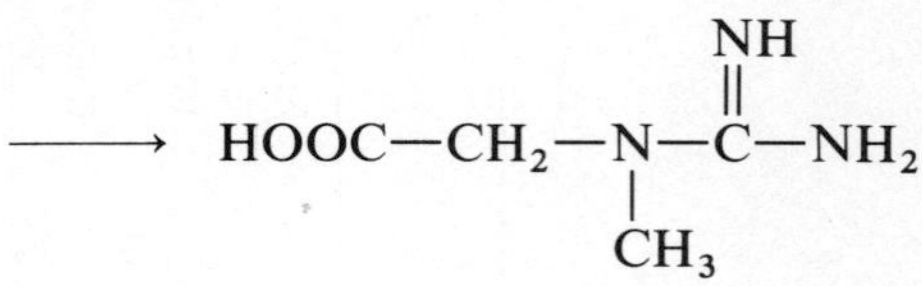

(**130**)

Tridemorph (**131**) is a systemic fungicide used as a foliar spray to control powdery mildew of cereals. It consists of two diastereoisomers in which the methyl groups have a different spatial arrangement. The tridecyl group is probably highly branched (Kost and Gurfinkel, 1975).

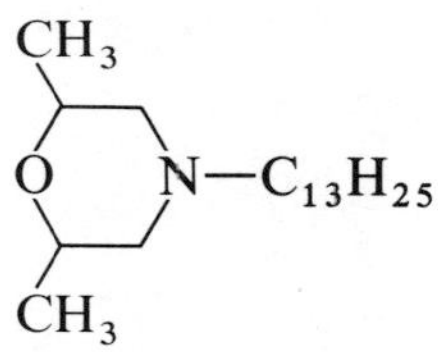

(**131**)

Analysis of barley plants sprayed under field conditions with unlabelled and ^{3}H-labelled tridemorph, and grown to maturity, showed the presence of unchanged tridemorph and a major metabolite in all parts of the plants. A top mass peak with $m/e = 313$ of this metabolite suggested that it may be a hydroxylated derivative of tridemorph. Small amounts of other, more polar metabolites also occurred (Waring and Wolfe, 1975).

Triforine (**132**) is effective as a systemic fungicide against powdery mildew and rusts. The metabolism of this compound in plants has been the subject of extensive studies mainly performed by Rouchaud and his co-workers (Rouchaud *et al.*, 1977; Rouchaud *et al.*, 1978a; b; 1979a; b) (Table 10).

Earlier work of Fuchs *et al.* (1976) showed that upon root-treatment of plants of several species a few metabolites were observed. The main metabolite, according to Fuchs, was different from any known non-enzymatic breakdown product of triforine. Von Bruchhausen and Stiasni (1973) found piperazine (**133**) in shoots of barley plants a short time after treatment with triforine by soil drench. Rouchaud *et al.* (1977) found that the major metabolite present in

Table 10 Metabolites of triforine in plants

Metabolite	Plant species Way of application[a]	Label	Evidence for structure[b]	Reference
Unknown major metabolite, not identical with breakdown product in water	Bean (R); Tomato (R); Barley (R)	^3H	TLC	Fuchs *et al.* (1976)
Piperazine (**133**)	Barley (S)	^3H	Inv. isot. dilution	Von Bruchhausen and Stiasni (1973)
N-[2,2,2-trichloro-1-(piperazin-1-yl)-ethyl]formamide (**134**)	Barley (S, L)	^3H	TLC	Rouchaud *et al.* (1977); Rouchaud *et al.* (1978a)
Iminodiacetic acid (**135**)	Barley (L)	^3H	TLC	Rouchaud *et al.* (1978a)
Glycine	Barley (L)	^3H	TLC	Rouchaud *et al.* (1978a)

[a] Abbreviations: R = root; S = soil drench; L = leaf.
[b] TLC = co-chromatography on thin-layer plates.

barley after soil drench was *N*-[2,2,2-trichloro-1-(piperazin-1-yl)ethyl]-formamide (**134**). After leaf treatment of barley with [^{3}H]triforine under field conditions this compound was also detected in very small amounts (Rouchaud *et al.*, 1978c). Remarkably, no piperazine was observed in this experiment and this compound had probably been degraded completely between the treatment and harvest of barley plants (2 months). Iminodiacetic acid (**135**) and glycine, two degradation products of piperazine (Rouchaud *et al.*, 1978b), were present in the grain and straw.

A fair amount of ^{3}H-label (the bound residue) remained in the solid plant residue of root-treated immature barley shoots after extraction with chloroform and methanol. This consisted partly of metabolite (**134**) and piperazine (**133**); both are extractable with methanol/HCl. The remainder of the label was probably incorporated into starch and cellulose and associated with lignin (Rouchaud *et al.*, 1978a). The bound residues (not methanol-soluble) of [^{3}H]triforine present in straw and grain of field-treated barley were also investigated (Rouchaud *et al.*, 1979a); the bound residue in straw contained 88 % of the total ^{3}H present, whereas in the grain 75 % was 'bound'. Methanol-HCl extracted small amounts of labelled iminodiacetic acid, serine, glycine and ethanol amine and ^{3}H label was also found in the cellulose and lignin fractions. The major part of ^{3}H was, however, incorporated into the hemicellulose fraction (as monosaccharides). In the grain the bound ^{3}H-label was mainly incorporated into the protein fraction (as radioactive amino acids) and into the starch fraction (as radioactive glucose). A hypothetical scheme of the conversions of precursors of the bound residues is given in Figure 11.

These pathways of breakdown resemble the proposed pathways of breakdown of the plant growth regulator chlormequat in wheat, via choline and sarcosine to serine, glycine and hydroxypyruvic acid (Dekhuijzen and Vonk, 1974).

Prothiocarb (**136**) exerts a systemic fungistatic effect on a variety of *Phycomycetes*. Its metabolism in tomato plants has been studied after root uptake from nutrient solution or soil drench (Iwan *et al.*, 1977). The propyl-^{14}C-labelled compound was used in this study. The main conversion product was *N*,*N*-dimethylpropane-1,3-diamine (**137**), obviously resulting from hydrolysis of the thiocarbamate moiety. Other metabolites were methyl-tetrahydro-1,3-diazine-2-one (**138**) and bis-(3-dimethylaminopropyl)urea (**139**). The latter two metabolites obviously result from the easy release of methyl mercaptane from prothiocarb (Kaars Sijpesteijn, 1977). The former metabolite (**138**) is formed by internal nucleophilic attack with demethylation, the second is possibly formed with an isocyanate as an intermediate which reacts with the *N*,*N*-dimethylpropanediamine (**137**). It is unclear whether the hydrolysis of prothiocarb is a biological process, since after three weeks in the nutrient solution 75 % was converted into (**137**). Small amounts of (**138**) and (**139**) were also present in the nutrient solution.

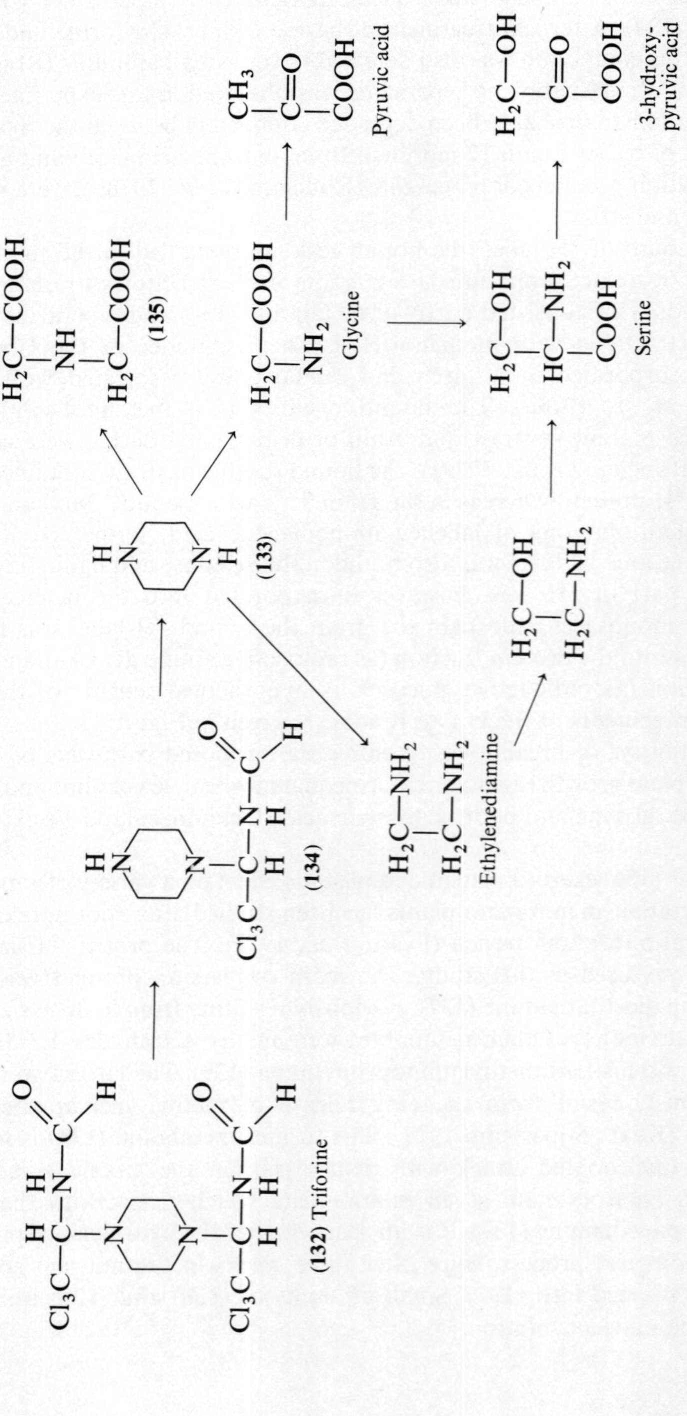

Figure 11 Pathways of metabolism of triforine in plants

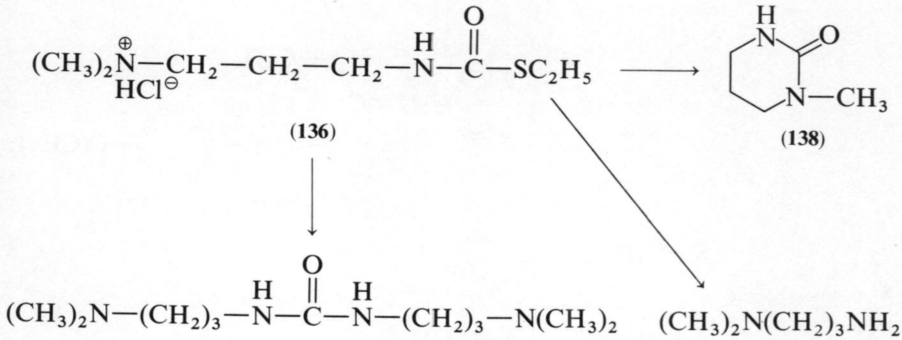

(136) (138) (139) (137)

Buthiobate (**140**) is a selective fungicide, highly effective in controlling various powdery mildew diseases. Metabolism of the ^{14}C-labelled compound in snap bean, cucumber and barley has been studied by Ohkawa *et al.* (1976), after leaf application. As metabolites the *S*-oxide (**141**), *S*-*p*-*tert*-butylbenzyl-*N*-3-pyridylthiolocarbamate (**142**), bis(*p*-*tert*-butylbenzyl)disulphide (**143**) and 4-*tert*-butylbenzaldehyde (**144**) were identified by co-chromatography. All these metabolites were also formed by influence of light on buthiobate, suggesting that their mechanism of formation was photochemical rather than biological. It is remarkable that only one of the *S*-atoms was oxidized to the *S*-oxide.

Triadimefon (**145**) is a recently developed systemic fungicide. Vogeler (1976) reported the reduction of triadimefon in barley to 1-(4-chlorophenoxy)-3,3-dimethyl-1-(1,2,4-triazol-1-yl)butan-2-ol (triadimenol, **146**). Later this metabolite was also found in other plant species (see Table 11). Evidence has been presented that triadimenol may be regarded as the active fungitoxic principle of triadimefon (Gasztonyi and Josepovits, 1979). Clark *et al.* (1978) observed that the metabolite triadimenol occurred in two diastereoisomeric forms which could be expected since triadimefon contains an asymmetric centre. The diastereoisomers are difficult to separate. 4-Chlorophenol (**147**) was detected by Rouchaud *et al.* (1981) in barley plants, 66 days after treatment with triadimefon. Most of it was in a conjugated form, which was liberated upon hydrolysis.

A photochemical mechanism may be responsible for the formation of 4-chlorophenol, since Clark *et al.* (1978) found this compound as one of the minor compounds after irradiation of triadimefon with light. On the other hand, an enzymatic hydrolysis of the ether bond seems equally possible. The hydrolysis of herbicidal chlorophenoxy acetic acids, which also contain an ether bond, to chlorophenols in several plant species is well documented (Loos, 1975). In a related fungicide, imazalil (**148**), the ether bond was possibly hydrolysed in cucumber plants after leaf treatment. ^{3}H-labelled imazalil was shown to form large amounts of bound residue (Vonk and Dekhuijzen, 1979).

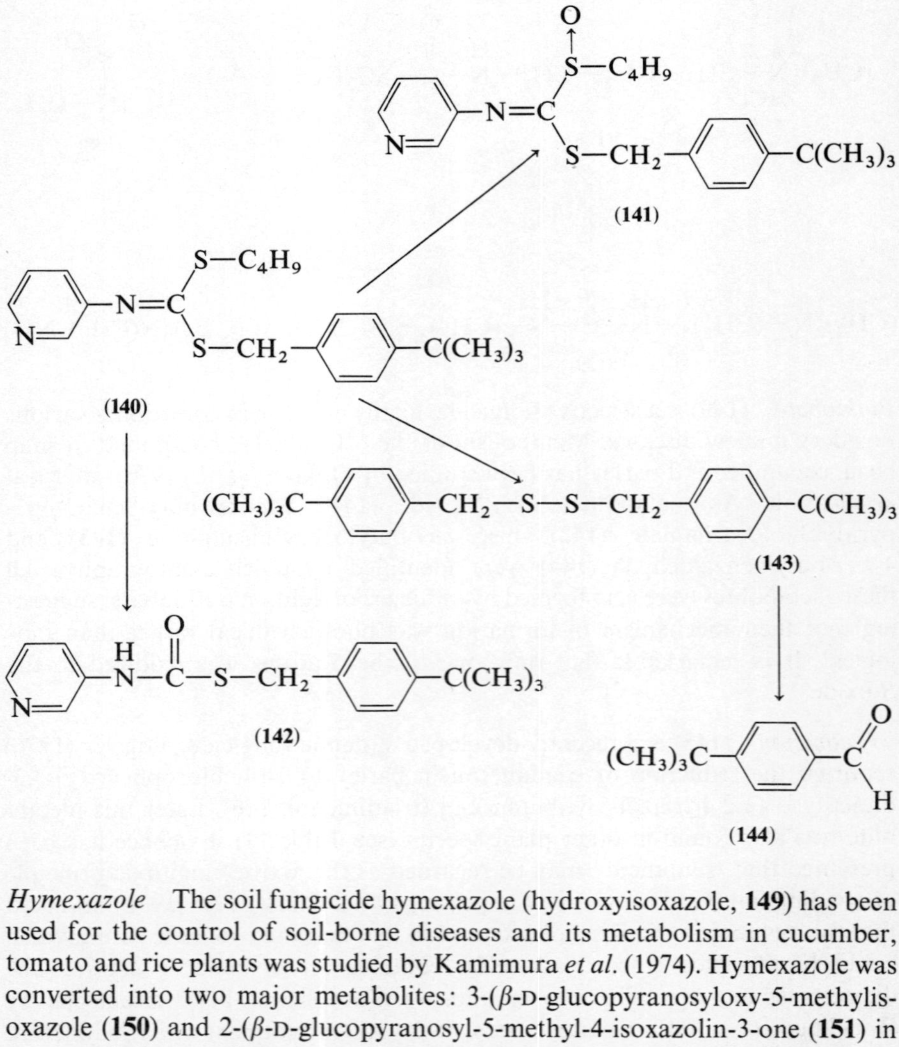

(140) (141) (142) (143) (144)

Hymexazole The soil fungicide hymexazole (hydroxyisoxazole, **149**) has been used for the control of soil-borne diseases and its metabolism in cucumber, tomato and rice plants was studied by Kamimura *et al.* (1974). Hymexazole was converted into two major metabolites: 3-(β-D-glucopyranosyloxy-5-methylisoxazole (**150**) and 2-(β-D-glucopyranosyl-5-methyl-4-isoxazolin-3-one (**151**) in roots and shoots of the treated plants. The compounds were isolated from cucumber plants and their structure elucidated by mass spectrometry, NMR, infra-red and UV spectroscopy, as well as by hydrolysis. Synthesized samples were identical with the metabolites. The *O*-glucoside (**150**) displayed toxic activity against *Pythium debaryanum*, but the *N*-glucoside (**151**) was inactive in this respect. Five days after application of the *O*-glucoside to cucumber seedlings, these contained large amounts of the *N*-glucoside and hymexazole. After application of the *N*-glucoside no *O*-glucoside was detected and only a trace of the parent compound. It may be concluded that the *O*-glucoside was more rapidly hydrolysed in plants.

Table 11 Metabolites of triadimefon in plants

Metabolite	Plant species Way of application[a]	Label	Evidence for structure[b]	Reference
Triadimenol (**146**)	Barley (L?)	^{14}C		Vogeler (1976)
	Cucumber (L); Wheat (L); Bean (L); Tomato (L)	—	TLC, GLC	Gasztonyi and Josepovits (1979)
	Marrow (L)	—	TLC, MS	Clark *et al.* (1978)
	Clover (S)	—	GLC	Fisher *et al.* (1979)
	Barley (L)	^{14}C	TLC	Rouchaud *et al.* (1981)
4-chlorophenol (**147**) (free and conjugated)	Barley (L)	^{14}C	TLC	Rouchaud *et al.* (1981)

[a] Abbreviations: L = leaf; S = soil.
[b] TLC = cochromatography with reference compounds; GLC = gas-liquid chromatography; MS = mass spectrometry.

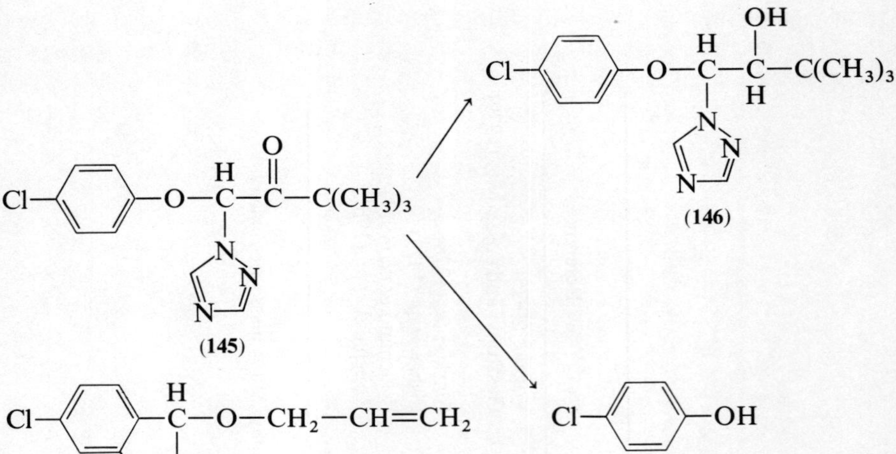

(145)

(146)

(147)

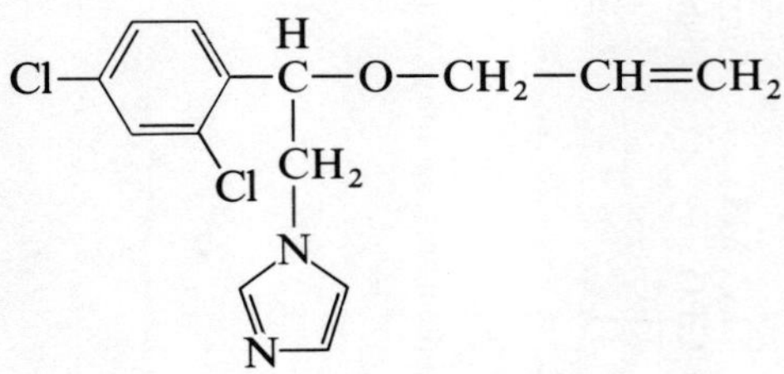

(148)

Cymoxanil (**152**) is highly effective against grape downy mildew as well as tomato and potato late blight. In grapes treated eight times by leaf spray, Belasco *et al.* (1981) found only small amounts of 1-ethyl-5,6-dihydro-5-methoxyimino-6-imino-(1*H*,3*H*)pyrimidine-2,4-dione (**153**), the major photolytic decomposition product of cymoxanil. Amino acid analysis by radio-GLC revealed that ^{14}C-labelled glycine accounted for most of the total radioactivity in grapes, potato tubers and tomatoes treated with [2-^{14}C]-cymoxanil. It has been demonstrated that [^{14}C]glycine present in potato and

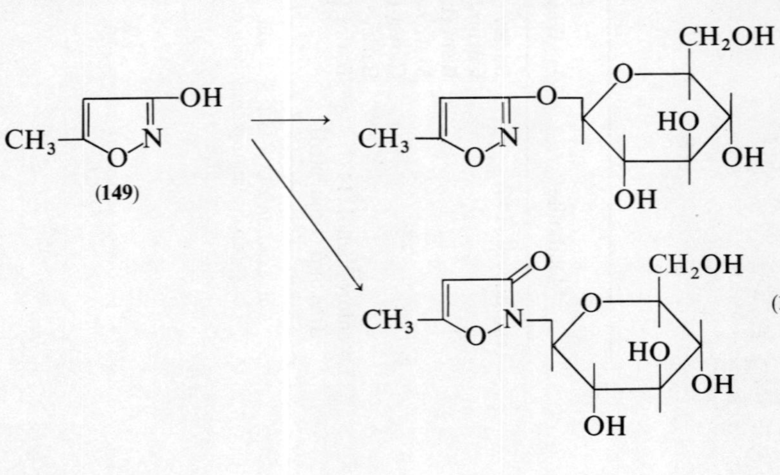

(149)

(150)

(151)

tomato plants was almost exclusively labelled in the 2-position, suggesting that the label in glycine was a result of direct metabolism rather than a form of incorporated label. Significant levels of ^{14}C were found in other amino acids, sugars, starch, fatty acids and lignin, indicating incorporation of ^{14}C via various metabolic pathways.

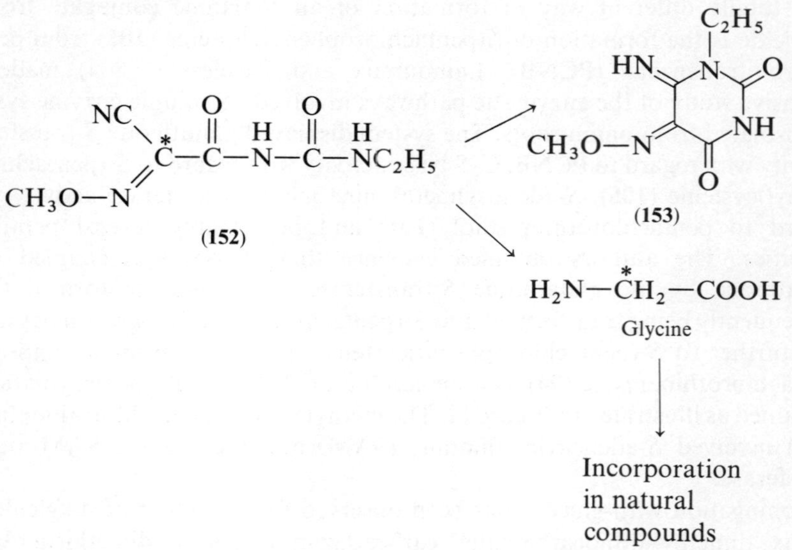

Triphenyltin compounds Freitag and Bock (1974) studied the degradation of [^{113}Sn]triphenyltin chloride on sugar beet leaves in a greenhouse experiment. Within 10 days, the amount of triphenyltin had decreased to about one-half; after only 3 days, 10% of the initially applied radioactivity already was present as diphenyltin species which disappeared gradually thereafter, and small amounts of monophenyltin (1%) could be detected. The main product, inorganic tin, accounted for 67% after 42 days, whereas 12% remained unextractable.

BIOCHEMICAL MECHANISMS

In the preceding section metabolites of fungicides in plants and the pathways of their formation have been discussed. It is remarkable that only limited information is available about the biochemical mechanisms of these conversions.

One of the mechanisms by which foreign compounds are converted in plants is the conjugation with amino acids, peptides and glucose. The mechanism of formation of an *S*-alanine derivative from dimethyldithiocarbamate (DDC-alanine, **3**) in plants has not been studied experimentally, but a very plausible biochemical explanation was given by Kaars Sijpesteijn and Van der Kerk (1965). In plants, cysteine can be synthesized from serine and homocysteine, with cystathionine as an intermediate. It is tempting to suggest that DDC-alanine is formed in plants by the enzyme which can couple homocysteine with serine to form cystathionine, the DDC ion (**1**) in this case replacing homocysteine. A similar enzymatic mechanism, viz. reaction with serine is probably responsible for the formation of aminotriazolyl alanine (**6**) (Massini, 1963).

A totally different way of formation of an *S*-alanine conjugate from a fungicide is the formation of *S*(pentachlorophenyl)cysteine (**105**) from pentachloronitrobenzene (PCNB). Lamoureux and Rusness (1980a) made an extensive study of the enzymatic pathways involved. A soluble enzyme system was isolated from onion roots. The system displayed glutathione *S*-transferase activity with regard to PCNB, C–S lyase activity with regard to *S*-(pentachlorophenyl)cysteine (**105**), *S*-adenosylmethionine methyl transferase activity with regard to pentachlorothiophenol (**101**) and presumably several peptidase activities. The authors obtained evidence that PCNB was coupled with glutathione by the glutathione *S*-transferase, the conjugate formed (**106**) subsequently being transformed into *S*-(pentachlorophenyl)-γ-glutamylcysteine and further to *S*-(pentachlorophenyl)cysteine. The pathway of formation of pentachlorothioanisole (**94**) as a metabolite of PCNB could be enzymatically explained as illustrated in Figure 12. The methylation of pentachlorothiophenol (**101**) involved *S*-adenosylmethionine (SAM) and the enzyme SAM-methyl transferase.

Conjugation with glucose has been observed for a number of fungicides in plants: dimethyldithiocarbamates, carbendazim, ethirimol, dimethirimol and hymexazole, or with metabolites of pyrazophos, ethirimol, dimethirimol, chloroneb and triadimefon. *O*-glucoside formation is most commonly observed, but with hymexazole an *N*-glucoside was detected as well, whereas with dimethyldithiocarbamate an *S*-glucoside is formed. In the formation of glucosides from natural and foreign phenolic compounds, the enzyme involved is UDP-glucosyl transferase; uridine diphosphoglucose (UDPG) functions as a glucose donor. Although no evidence is available it seems likely that UDPG is also involved in the formation of *S*-glucosides (Kaslander, 1966; Dutton and Illing, 1972).

An enzyme, arylamine *N*-glucosyltransferase, which catalyses the biotransformation of a variety of aromatic amines such as 3,4-dichloroaniline, has been isolated from soybeans (Still and Herrett, 1975).

A conjugate of dichloran was reported to be the *N*-malonyl derivative (**112**). No details, however, are known of this remarkable product. Roberts (1977) identified a glucosyl–malonyl derivative of the herbicide flamprop-methyl in plants. A cysteinyl–malonyl conjugate of pentachloronitrobenzene in peanut plants has been reported by Lamoureux and Rusness (1980b).

Several examples have been reported of oxidation of fungicides in plants, e.g. the replacement of sulphur by oxygen in pyrazophos, thiophanate-methyl and ethylenethiourea, and the sulphoxidation of carboxin and buthiobate. Not

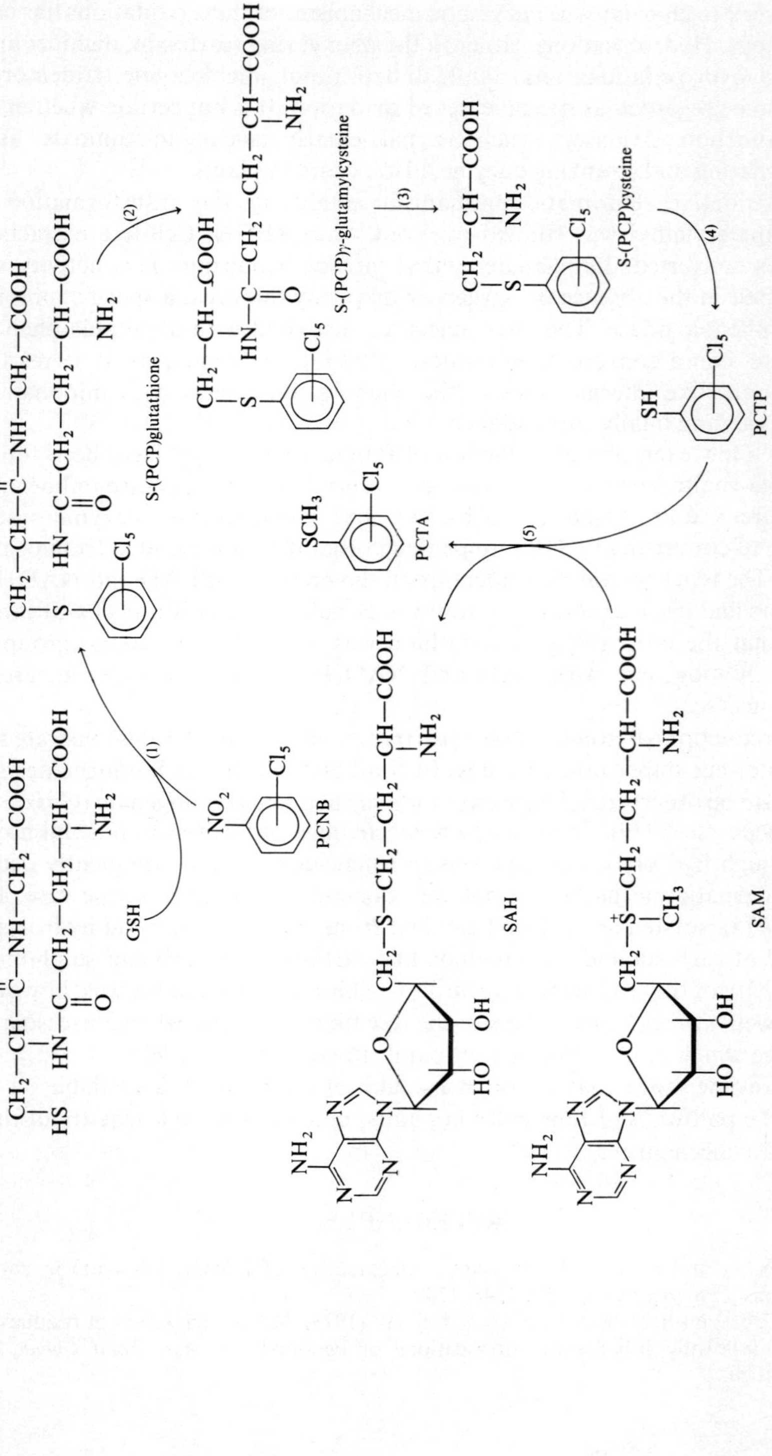

Figure 12 Pentachlorothioanisole formation in onion root. (1) GSH-S-transferase; (2) carboxy peptidase; (3) γ-glutamyl transpeptidase; (4) C–S lyase; (5) methyl transferase. (Reproduced from Lamoureux and Rusness (1980a) by permission of Academic Press)

much work to elucidate the enzymatic mechanisms of these oxidations has been carried out. Hydroxylations, either in the phenyl ring (carboxin, dichlozoline), or alkyl hydroxylations (ethirimol, dimethirimol, dichlozoline, tridemorph) may also be regarded as special cases of oxidation. It is not certain whether the mixed-function oxidase, which is particularly active in animals as a hydroxylating and oxidizing enzyme, also occurs in plants.

An oxidative, enzymatic mechanism catalysing the transformation of thiophanate-methyl was studied by Vonk *et al.* (1977). Cell-free extracts of potatoes converted thiophanate-methyl into carbendazim. This activity was suppressed in the absence of oxygen or by phenylthiourea, a specific inhibitor of polyphenoloxidase. The latter enzyme, incubated with a suitable phenolic substrate, could convert thiophanate-methyl into carbendazim. It seems that the quinone-like intermediates of the phenoloxidase react with thiophanate-methyl yielding finally carbendazim.

Only a few examples of reduction of fungicides in plants have been found. Pentachloronitrobenzene (PCNB) is reduced to pentachloroaniline (**93**) (Lamoureux and Rusness, 1980b). An aryl nitroreductase enzyme system capable of converting PCNB into pentachloroaniline was isolated from peanut plants. The reaction was dependent upon the presence of FAD and NADH as cofactors and upon anaerobic conditions. Suzuki and Uchiyama (1975) earlier found that the nitro group of parathion was reduced to an amino group by spinach homogenate with FAD and NADH as cofactors under anaerobic conditions.

The reduction of triadimefon into triadimenol could be demonstrated in filtrates of cucumber homogenates. In bean and wheat leaf homogenates this could also be observed, albeit slower than in cucumber homogenate (Gasztonyi and Josepovits, 1979). The enzymatic mechanism of this reaction is unknown.

Although hydrolytic degradations of fungicides in plants frequently occur, their enzymatic mechanism is not well documented, except in one case. Lyr *et al.* (1975) isolated an aryl-acyl amidase from barley which could hydrolyse a number of carboxanilide compounds like carboxin and carboxin sulphoxide. The activity of the enzyme with regard to *o*-chloropropionanilide was, however, about twenty times higher. The enzyme resembled an aryl-acyl amidase isolated from rice which also hydrolysed propanil (Frear and Still, 1968).

This review shows that although a wealth of information is available on the metabolic pathways of fungicides in plants, much has to be learnt about their enzymatic mechanisms.

REFERENCES

Allam, A. I., and Sinclair, J. B. (1969). 'Degradation of DMOC (Vitavax) in cotton seedlings', *Phytopathology*, **59**, 1548–1549.

Baude, F. J., Gardiner, J. A., and Han, J. C. Y. (1973). 'Characterization of residues on plants following foliar spray applications of benomyl', *J. Agr. Food Chem.*, **21**, 1084–1090.

Bedi, S., and Roy, N. K. (1979). 'Metabolic fate of dichloromethyl-*O*,*O*-diphenylphosphonate and dichloromethyl-*O*,*O*-di-(*p*-nitrophenyl)phosphonate in/on rice plants', *J. Environ. Sci. Health* **B14**, 443–458.

Begum, S., Scheunert, I., Haque, A., Klein, W., and Korte, F. (1979). 'Conversion of [^{14}C]pentachloronitrobenzene in onions', *Pestic. Biochem. Physiol.*, **11**, 189–200.

Belasco, I. J., Han, J. C-Y., Chrzanowski, R. L., and Baude, F. J. (1981). 'Metabolism of [^{14}C]cymoxanil in grapes, potatoes and tomatoes', *Pestic. Sci.*, **12**, 355–364.

Ben-Aziz, A., and Aharonson, N. (1974). 'Dynamics of uptake, translocation and disappearance of thiabendazole and methyl 2-benzimidazolecarbamate in pepper and tomato plants', *Pestic. Biochem. Physiol.*, **4**, 120–126.

Ben-Aziz, A., and Aharonson, N. (1975). 'Persistence of benzimidazole derivatives in plants', in *Pesticides* (eds F. Coulston and F. Korte), pp. 135–139, Thieme Verlag, Stuttgart.

Briggs, D. E., Waring, R. H., and Hackett, A. M. (1974). 'The metabolism of carboxin in growing barley', *Pestic. Sci.*, **5**, 599–607.

Buchenauer, H. (1975). 'Differences in light stability of some carboxylic acid anilide fungicides in relation to their applicability for seed and foliar treatment', *Pestic. Sci.*, **6**, 525–535.

Buchenauer, H., Edgington, L. V., and Grossmann, F. (1973). 'Photochemical transformation of thiophanate-methyl and thiophanate to alkyl benzimidazole-2-yl carbamates', *Pestic. Sci.*, **4**, 343–348.

Bull, D. L. (1972). 'Metabolism of organophosphorus insecticides in animals and plants', *Residue Rev.*, **43**, 1–22.

Calderbank, A. (1971). 'Metabolism and mode of action of dimethirimol and ethirimol', *Acta Phytopathol. Acad. Scient. Hung.*, **6**, 355–363.

Casida, J. E., and Lykken, L. (1969). 'Metabolism of organic pesticide chemicals in higher plants', *Ann. Rev. Plant Physiol.*, **20**, 607–636.

Cavell, B. D., Hemingway, R. J., and Teal, G. (1971). 'Some aspects of the metabolism and translocation of the pyrimidine fungicides', *Proc. 6th British Insectic. Fungic. Conf.*, 431–437.

Cavell, B. D., Pollard, S. J., and Wells, C. H. J. (1976). 'Photodimerisation of some systemic pyrimidine fungicides', *Chem. Ind.*, **1976**, 566–567.

Chiba, M., and Doornbos, F. (1974). 'Instability of benomyl in various conditions', *Bull. Environ. Contam. Toxicol.*, **11**, 273–274.

Chin, W-T., Stone, G. M., and Smith, A. E. (1970). 'Metabolism of carboxin (Vitavax) by barley and wheat plants', *J. Agr. Food Chem.*, **18**, 709–712.

Clark, T., Clifford, D. R., Deas, A. H. B., Gendle, P., and Watkins, D. A. M. (1978). 'Photolysis, metabolism and other factors influencing the performance of triadimefon as a powdery mildew fungicide', *Pestic. Sci.*, **9**, 497–506.

Clark, T., and Watkins, D. A. M. (1978). 'Photolysis of dichlofluanid', *Pestic. Sci.*, **9**, 225–228.

Clemons, G. P., and Sisler, H. D. (1969). 'Formation of a fungitoxic derivative from Benlate', *Phytopathology*, **59**, 705–706.

Cooke, B. K., Loeffler, R. S. T., and Pappas, A. C. (1979). 'The structural rearrangement of iprodione in ethanolic solution', *Pestic. Sci.*, **10**, 393–398.

Crosby, D. G., Moilanen, K. W., Nakagawa, M., and Wong, A. S. (1972). 'Photonucleophilic reactions of pesticides', in *Environmental Toxicology of Pesticides* (eds F. Matsumura, G. M. Boush and T. Misato), pp. 423–433, Academic Press, New York.

Curry, A. N. (1962). 'Translocation and metabolism of dodecylguanidine acetate (dodine) fungicide in apple trees, using C-14 radiotagged dodine', *J. Agr. Food Chem.*, **10**, 13–17.

Dekhuijzen, H. M. (1964). 'The systemic action of dimethyldithiocarbamates on cucumber scab caused by Cladosporium cucumerinum and the conversion of these compounds by plants', *Netherlands J. Plant Pathol.*, **70** (Suppl. 1), 1–75.

Dekhuijzen, H. M., and Vonk, C. R. (1974). 'The distribution and degradation of chlormequat in wheat plants', *Pestic. Biochem. Physiol.*, **4**, 346–355.

Drèze, Ph. (1976). 'Analyse de résidus et de métabolites de pesticides au moyen de molécules marquées', *Annales de Gembloux*, **82**, 67–70.

Dutton, G. J., and Illing, H. P. A. (1972). 'Mechanism of biosynthesis of thio-β-D-glucuronides and thio-β-D-glucosides', *Biochem. J.*, **129**, 539–550.

Endo, K., Mori, Y., Kakiki, K., and Tomomasa, M. (1970). 'Absorption, translocation and metabolic fate of radioactive Inegin in rice plant', *Nippon Nogei Kagaku Kaishi*, **44**, 356 (Chem. Abstr. 74, 12100).

Engst, R. (1977). 'Ethylenethiourea', *Pure Appl. Chem.*, **49**, 675–689.

Engst, R., Schnaak, W., and Rattba, H. (1968). 'Fungizide Wirkung und Rückstandsbildung von Abbauprodukten des Maneb und Zineb an freilandbehandelten Tomaten', *Nachrichtenblad Deutsch. Pflanzenschutzdienst*, **22**, 26–29.

Erwin, D. C., Wang, M. C., and Sims, J. J. (1970). 'Translocation of 2-(4-thiazolyl)benzimidazole in cotton', *Phytopathology*, **60**, 1291.

Esser, H. O., Dupuis, G., Ebert, E., Vogel, C., and Marco, G. J. (1975). 's-Triazines' in *Herbicides, Chemistry, Degradation and Mode of Action* (eds P. C. Kearney and D. D. Kaufman), Vol. 1, pp. 129–208, Marcel Dekker, New York.

Fisher, D. J., Pickard, J. A., and McKenzie, C. M. (1979). 'Uptake of the systemic fungicide triadimefon by clover and its effect on symbiotic nitrogen fixation', *Pestic. Sci.*, **10**, 75–82.

Frear, D. S., and Still, G. G. (1968). 'The metabolism of 3,4-dichloropropionanilide in plants. Partial purification and properties of an aryl acylamidase from rice', *Phytochemistry*, **7**, 913–920.

Freitag, K-D., and Bock, R. (1974). 'Degradation of triphenyltin chloride on sugar beet plants and in rats', *Pestic. Sci.*, **5**, 731–739.

Fuchs, A., De Vries, F. W., and Aalbers, A. M. J. (1976). 'Uptake, distribution and metabolic fate of ^{3}H-triforine in plants. I. Short-term experiments', *Pestic. Sci.*, **7**, 115–126.

Fuchs, A., Van den Berg, G. A., and Davidse, L. C. (1972). 'A comparison of benomyl and thiophanates with respect to some chemical and systemic fungitoxic characteristics', *Pestic. Biochem. Physiol.*, **2**, 191–205.

Gasztonyi, M., and Josepovits, G. (1979). 'The activation of triadimefon and its role in the selectivity of fungicide action', *Pestic. Sci.*, **10**, 57–65.

Gorbach, S., Thier, W., Kellner, H. M., Schulze, E. F., Kuenzler, K., and Fischer, H. (1975). 'Environmental impact of pyrazophos. 1. Degradation in plants and in the rat', in *Pesticides* (eds F. Coulston and F. Korte), pp. 840–844, Thieme Verlag, Stuttgart.

Groves, K., and Chough, K. S. (1970). 'Fate of the fungicide 2,6-dichloro-4-nitroaniline (DCNA) in plants and soils', *J. Agr. Food Chem.*, **18**, 1127–1128.

Hemingway, R. J. (1976). 'Persistence of pesticides in plants', in *The Persistence of Insecticides and Herbicides* (ed. K. I. Beynon), pp. 55–68, British Crop Protection Council (Monograph no. 17), Droitwich.

Hills, F. J. (1962). 'Uptake, translocation and chemotherapeutic effect of *p*-dimethylaminobenzene diazo sodium sulfonate (Dexon) in sugar beet seedlings', *Phytopathology*, **52**, 389–392.

Hills, F. J., and Leach, L. D. (1960). 'Effect of light on certain biological, physical and chemical properties of *p*-dimethylaminobenzene diazo sodium sulfonate (Dexon)', *Phytopathology*, **50**, 639.

Hisada, Y., Maeda, K., Kawase, Y., and Miyamoto, J. (1976). 'Uptake and translocation of a systemic fungicide, S-7131, in cucumber plants', *J. Pestic. Sci.*, **1**, 201–206.

Hisada, Y., Kato, T., and Kawase, Y. (1977). 'Systemic movement in cucumber plants and control of cucumber gray mould by a new fungicide, S-7131', *Neth. J. Plant Pathol.*, **83** Suppl. 1, 71–78.

Hoagland, R. E., and Frear, D. S. (1976). 'Behavior and fate of ethylenethiourea in plants', *J. Agr. Food Chem.*, **24**, 129–133.
Hylin, J. W., and Chin, B. H. (1967). 'Delayed phytotoxicities produced by dithiocarbamate fungicide residues', *Abstr. 6th Intern. Congr. Plant Protection*, pp. 614–615, Vienna.
Iwan, J., Hertel, B., and Radzuweit, J. (1977). 'Transformations of prothiocarb in plant and soil', *Neth. J. Plant Pathol.*, **83** Suppl. 1, 277–284.
Jacob, T. A., Carlin, J. R., Walker, R. W., Wolf, F. J., and VandenHeuvel, W. J. A. (1975). 'Photolysis of thiabendazole', *J. Agr. Food Chem.*, **23**, 704–709.
Kaars Sijpesteijn, A. (1977). 'Effects on fungal pathogens', in *Systemic Fungicides* (ed. R. W. Marsh), pp. 131–159, Second Edition, Longman, London.
Kaars Sijpesteijn, A., Dekhuijzen, H. M., and Vonk, J. W. (1977). 'Biological conversion of fungicides in plants and microorganisms', in *Antifungal Compounds* (eds M. R. Siegel and H. D. Sisler) Vol. 2, pp. 91–147, Marcel Dekker, New York.
Kaars Sijpesteijn, A., and Van der Kerk, G. J. M. (1965). 'Fate of fungicides in plants', *Ann. Rev. Phytopathol.*, **3**, 127–152.
Kamimura, S., Nishikawa, M., Saeki, H., and Takaki, Y. (1974). 'Absorption and metabolism of 3-hydroxy-5-methylisoxazole in plants and the biological activities of its metabolites', *Phytopathology*, **64**, 1273–1281.
Kàslander, J. (1966). *Metabolic fate of dithiocarbamates.* Thesis State University, Utrecht.
Kaslander, J., Kaars Sijpesteijn, A., and Van der Kerk, G. J. M. (1961). 'On the transformation of dimethyldithiocarbamate into its β-glucoside by plant tissues', *Biochim. Biophys. Acta*, **52**, 396–397.
Kaslander, J., Kaars Sijpesteijn, A., and Van der Kerk, G. J. M. (1962). 'On the transformation of the fungicide sodium dimethyldithiocarbamate into its alanine derivative by plant tissues', *Biochim. Biophys. Acta*, **60**, 417–419.
Kilgore, W. W., and White, E. R. (1970). 'Decomposition of the systemic fungicide 1991 (Benlate)', *Bull. Environ. Contam. Toxicol.*, **5**, 67–69.
Kost, D., and Gurfinkel, E. (1975). 'The isolation and structure determination of two diastereoisomers of *N*-tridecyl-2,6-dimethylmorpholine (Tridemorph)', *J. Chromatography*, **108**, 207–211.
Kuhr, R. J. (1968). 'Metabolism of methylcarbamate insecticide chemicals in plants', *J. Sci. Fd. Agric., Suppl. 1968*, 44–49.
Kumarasamy, R., and Raghu, K. (1976). 'Conversion of the fungicide ziram in rice plants', *Agr. Biol. Chem.*, **40**, 2297–2298.
Lamoureux, G. L., and Rusness, D. G. (1980a). 'In vitro metabolism of pentachloronitrobenzene to pentachloromethylthiobenzene by onion: characterization of glutathione S-transferase, cysteine C-S lyase, and S-adenosylmethionine methyl transferase activities', *Pestic. Biochem. Physiol.*, **14**, 50–61.
Lamoureux, G. L., and Rusness, D. G. (1980b). 'Pentachloronitrobenzene metabolism in peanut. Part I: Mass spectral characterization of seven glutathione-related conjugates produced in vivo or in vitro', *J. Agr. Food Chem.*, **28**, 1057–1070.
Loos, M. A. (1975). 'Phenoxyalkanoic acids', in *Herbicides, Chemistry, Degradation and Mode of Action* (eds P. C. Kearney and D. D. Kaufman) Vol. 1, pp. 1–128, Marcel Dekker, New York.
Lukens, R. J., and Sisler, H. D. (1958). 'Chemical reactions involved in the fungitoxicity of captan', *Phytopathology*, **48**, 235–244.
Lyman, W. R. (1971). 'The metabolic fate of Dithane M-45', *Pure Appl. Chem., Suppl. 1971*, 243–256.
Lyr, H., Ritter, G., and Banasiak, L. (1974). 'Detoxication of carboxin', *Zeitschr. Allg. Mikrobiol.*, **14**, 313–320.

Lyr, H., Ritter, G., and Polter, C. (1975). 'Detoxifikation des Carboxins', in *Systemfungizide* (eds H. Lyr and C. Polter), pp. 167–176, Akademie-Verlag, Berlin.

Massaux, F. (1963). 'Contribution à l'étude de l'action systémique du thiram (TMTD) chez les plantes à l'aide de thiram marqué au 35 S', *Meded. Landbouwhogeschool Opzoekingsstat. Staat Gent*, **28**, 590–596.

Massini, P. (1963). 'Aminotriazolylalanine: a metabolic product of aminotriazole from plants', *Acta Bot. Neerlandica*, **12**, 64–72.

Masuda, T., and Endo, S. (1979). 'Methylation of dialkyl hydrogen phosphorothioates and dithioates in soil and translocation of the methylated products to rice plants', *J. Pestic. Sci.*, **4**, 505–510.

Masuda, T., and Kanazawa, J. (1973). 'Residue of Kitazin P metabolite in rice plant and soil', *Agr. Biol. Chem.*, **37**, 2931.

Nash, R. G. (1976). 'Uptake of ethylenebis(dithiocarbamate) fungicides and ethylenethiourea by soybeans', *J. Agr. Food Chem.*, **24**, 596–601.

Nash, R. G., and Beall, M. L. (1980). 'Fate of maneb and zineb fungicides in microagroecosystem chambers', *J. Agr. Food Chem.*, **28**, 322–330.

Northover, J., and Ripley, B. D. (1980). 'Persistence of chlorothalonil on grapes and its effect on disease control and fruit quality', *J. Agr. Food Chem.*, **28**, 971–974.

Oeser, H., Gorbach, S., and Boerner, E. (1975). 'Translocation and metabolism of pyracarbolid in wheat', in *Pesticides* (eds F. Coulston and F. Korte), pp. 557–561, Thieme Verlag, Stuttgart.

Ogawa, K., Aizawa, H., and Yamauchi, F. (1980). 'Absorption, translocation and metabolism of fluoroimide in inu-apple trees, Malus prunifolia Borkhousen', *J. Pestic. Sci.*, **5**, 29–36.

Ohkawa, H., Shibaike, R., Okihara, Y., Morikawa, M., and Miyamoto, J. (1976). 'Degradation of the fungicide Denmert (S-*n*-butyl S′-*p*-*tert*-butylbenzyl *N*-3-pyridyldithiocarbonimidate, S-1358) by plants, soils and light', *Agr. Biol. Chem.*, **40**, 943–951.

Peterson, C. A., and Edgington, L. V. (1969). 'Quantitative estimation of the fungicide benomyl using a bioautograph technique', *J. Agr. Food Chem.*, **17**, 898–899.

Raghu, K., Murthy, N. B. K., Kumarasamy, R., Rao, S. R., and Sane, P. V. (1975). 'Fate and persistence of thiram in plants and soils', in *Origin and Fate of Chemical Residues in Food, Agriculture and Fisheries*, pp. 137–148, IAEA, Vienna.

Rhodes, R. C. (1977). 'Studies with manganese [^{14}C]ethylenebis(dithiocarbamate) ([^{14}C]maneb) fungicide and [^{14}C]ethylenethiourea ([^{14}C]ETU) in plants, soil and water', *J. Agr. Food Chem.*, **25**, 528–533.

Rhodes, R. C., Pease, H. L., and Brantley, R. K. (1971). 'Fate of C14-labelled chloroneb in plants and soils', *J. Agr. Food Chem.*, **19**, 745–749.

Richmond, D. V., and Somers, E. (1966). 'Studies on the fungitoxicity of captan. IV. Reactions of captan with cell thiols', *Ann. Appl. Biol.*, **57**, 231–240.

Richmond, D. V., and Somers, E. (1968). 'Studies on the fungitoxicity of captan. VI. Decomposition of ^{35}S-labelled captan by Neurospora crassa conidia', *Ann. Appl. Biol.*, **62**, 35–43.

Roberts, T. R. (1977). 'Metabolism of flamprop-methyl in wheat', *Pestic. Sci.*, **8**, 463–472.

Rouchaud, J. P., Decallonne, J. R., and Meyer, J. A. (1974). 'Metabolic fate of methyl-2-benzimidazole carbamate in melon plants', *Phytopathology*, **64**, 1513–1517.

Rouchaud, J. P., Decallonne, J. R., and Meyer, J. A. (1977). 'Metabolism of the fungicide triforine in barley plants', *Pestic. Sci.*, **8**, 65–70.

Rouchaud, J. P., Decallonne, J. R., and Meyer, J. A. (1978a). 'The nature of bound residues derived from triforine in barley plants', *Pestic. Sci.*, **9**, 47–78.

Rouchaud, J. P., Decallonne, J. R., and Meyer, J. A. (1978b). 'Metabolism of [2,5-^{14}C]piperazine in barley plants', *Pestic. Sci.*, **9**, 139–145.

Rouchaud, J. P., Moons, Ch., Decallonne, J. R., and Meyer, J. A. (1978c). 'Metabolism of [^{3}H]triforine in barley, grown in the field: the unbound radioactive residue', *Pestic. Sci.*, **9**, 587–594.

Rouchaud, J. P., Moons, Ch., and Meyer, J. A. (1979a). 'Characterisation of bound residues of [^{3}H]triforine in the straw of barley grown in the field', *Pestic. Sci.*, **10**, 438–446.

Rouchaud, J. P., Moons, Ch., and Meyer, J. A. (1979b). 'Characterisation of bound residues of [^{3}H]triforine in barley grain grown in the field', *Pestic. Sci.*, **10**, 509–518.

Rouchaud, J. P., Moons, Ch., and Meyer, J. A. (1981). 'Fate of [^{14}C]triadimefon in barley', *Med. Fac. Landbouww. Rijksuniv. Gent*, **33**, 331–336.

Rusness, D. G., and Lamoureux, G. L. (1980). 'Pentachloronitrobenzene metabolism in peanut. Part II: Characterization of chloroform-soluble metabolites produced in vivo', *J. Agr. Food Chem.*, **28**, 1070–1077.

Sato, T., and Tomizawa, C. (1960). 'Decomposition of zinc ethylenebisdithiocarbamate (zineb) applied to plants', *Bull. Nat. Inst. Agr. Sci., Ser. C*, **12**, 181–186.

Serra, G. (1964). 'The degradation of captan, Phaltan and difolatan by light', *Phytiatry Phytopharm.*, **13**, 107–110.

Siegel, M. R. (1970). 'Reactions of certain trichloromethyl sulfenyl fungicides with low molecular weight thiols', *J. Agr. Food Chem.*, **18**, 823–826.

Siegel, M. R. (1973). 'Distribution and metabolism of methyl 2-benzimidazolecarbamate, the fungitoxic derivative of benomyl, in strawberry plants', *Phytopathology*, **63**, 890–896.

Siegel, M. R., and Zabbia, A. J. (1972). 'Distribution and metabolic fate of the fungicide benomyl in dwarf pea', *Phytopathology*, **62**, 630–634.

Sims, J. J., Mee, H., and Erwin, D. C. (1969). 'Methyl 2-benzimidazolecarbamate, a fungitoxic compound isolated from cotton plants treated with methyl 1-(butyl carbamoyl)-2-benzimidazolecarbamate (benomyl)', *Phytopathology*, **59**, 1775–1776.

Snel, M., and Edgington, L. V. (1970). 'Uptake, translocation and decomposition of systemic oxathiin fungicides in bean', *Phytopathology*, **60**, 1708–1716.

Soeda, Y., Kosaka, S., and Noguchi, T. (1972a). 'Identification of alkyl 2-benzimidazolecarbamates as a major metabolite of thiophanate fungicides in/on bean plant', *Agr. Biol. Chem.*, **36**, 817–823.

Soeda, Y., Kosaka, S., and Noguchi, T. (1972b). 'The fate of thiophanate-methyl fungicide and its metabolites on plant leaves and glass plates', *Agr. Biol. Chem.*, **36**, 931–936.

Solel, Z., Schooley, J. M., and Edgington, L. V. (1973). 'Uptake and translocation of benomyl and carbendazim (methyl benzimidazol-2-yl carbamate) in the symplast', *Pestic. Sci.*, **4**, 713–718.

Somers, E., and Richmond, D. V. (1962). 'Translocation of captan by broad bean plants', *Nature*, **194**, 1194–1195.

Still, G. G., and Herrett, R. A. (1975). 'Methylcarbamates, carbanilates, and acylanilides', in *Herbicides, Chemistry, Degradation and Mode of Action* (eds. P. C. Kearney and D. D. Kaufman) Vol. 2, pp. 609–664, Marcel Dekker, New York.

Sumida, S., Yoshihara, R., and Miyamoto, J. (1973). 'Degradation of 3-(3′,5′-dichlorophenyl)-5,5-dimethyl oxazolidine-2,4-dione by plants, soil and light', *Agr. Biol. Chem.*, **37**, 2781–2790.

Suzuki, T., and Uchiyama, M. (1975). 'Pathway of nitro reduction of parathion by spinach homogenate', *J. Agr. Food Chem.*, **23**, 281–286.

Teal, G., and Cavell, B. D. (1975). 'Degradation of bupirimate fungicide on apples and in water', *Proc. 8th British Insectic. Fungic. Conf.*, 25–30.

Thorn, G. D. (1973). 'Uptake and metabolism of chloroneb by Phaseolus vulgaris', *Pestic. Biochem. Physiol.*, **3**, 137–140.

Tillman, R. W., Siegel, M. R., and Long, J. W. (1973). 'Mechanism of action and fate of the fungicide chlorothalonil (2,4,5,6-tetrachloroisophthalonitrile) in biological systems. 1. Reactions with cells and subcellular components of Saccharomyces pastorianus', *Pestic. Biochem. Physiol.*, **3**, 160–167.

Ueyama, I., Uesugi, Y., Tomizawa, C., and Murai, T. (1973). 'Metabolic fate of *O*-ethyl-*S,S*-diphenylphosphorodithiolate (Hinosan) in rice plant', *Agr. Biol. Chem.*, **37**, 1543–1551.

USDA, (1968). *Progress report on pesticides and related activities, 1967*, p. 194.

Vogeler, K. (1976). 'Der Metabolismus von Triadimefon, in Gerste unter Gewächshausbedingungen und im Boden', Bayer internal report RA 272, cited by W. Specht, *Bayer Nachrichten*, **30**, 55–71.

Vogeler, K., Drèze, Ph., Rapp, A., Steffan, H., and Ullemeyer, H. (1975). 'The distribution of radioactivity in fruits after application of ^{14}C-labelled Antracol, compared with the potential evolution of CS_2, in *Pesticides* (eds F. Coulston and F. Korte), pp. 183–188, Thieme Verlag, Stuttgart.

Vogeler, K., Drèze, P., Rapp, A., Steffan, H., and Ullemeyer, H. (1977). 'Distribution and metabolism of propineb in apples and grapes, and of its degradation products propylenethiourea and ethylenethiourea in apples', *Pflanzenschutz-Nachrichten Bayer*, **30**, 72–97.

Vogeler, K., and Niessen, H. (1967) 'Kolorimetrische und gaschromatographische Bestimmungen von Rückständen in Pflanzen nach Anwendung von Euparen', *Pflanzenschutz-Nachrichten Bayer*, **20**, 534–549.

Von Bruchhausen, V., and Stiasni, M. (1973). 'Transport of the systemic fungicide Cela W 524 (triforine) in barley plants. II. Uptake and metabolism', *Pestic. Sci.*, **4**, 767–773.

Vonk, J. W. (1975). *Chemical decomposition of bisdithiocarbamate fungicides and their metabolism by plants and microorganisms*. Thesis, State University of Utrecht.

Vonk, J. W. (1976). 'Metabolism of [^{14}C]zineb on lettuce', *Med. Fac. Landbouww. Rijksuniv. Gent*, **41**, 1383–1392.

Vonk, J. W. (1978). 'New aspects of the metabolism of [^{14}C]zineb on lettuce', *Abstr. 4th Intern. Congress of Pesticide Chemistry*, p. V–515, Zürich.

Vonk, J. W., and Dekhuijzen, H. M. (1979). 'Transport and metabolism of [^{3}H]imazalil in barley and cucumber', *Med. Fac. Landbouww. Rijksuniv. Gent*, **44**, 927–934.

Vonk, J. W., and Kaars Sijpesteijn, A. (1971). 'Methyl benzimidazol-2-yl carbamate, the fungitoxic principle of thiophanate-methyl', *Pestic. Sci.*, **2**, 160–164.

Vonk, J. W., and Kaars Sijpesteijn, A. (1977). 'Metabolism' in *Systemic Fungicides* (ed. R. W. Marsh) 2nd ed., pp. 160–175, Longman, London.

Vonk, J. W., Mihanović, B., and Kaars Sijpesteijn, A. (1977). 'A biochemical mechanism for the conversion of thiophanate-methyl into methyl benzimidazol-2-yl carbamate (MBC) in plant tissue', *Netherlands J. Plant Pathol.* **83**, Suppl. 1, 269–276.

Wang, M. C., Erwin, D. C., Sims, J. J., Keen, N. T., and Borum, D. E. (1971). 'Translocation of ^{14}C-labeled thiabendazole in cotton plants', *Pestic. Biochem. Physiol.*, **1**, 188–195.

Waring, R. M., and Wolfe, M. S. (1975). 'Distribution of tridemorph and its metabolites in barley grown to maturity', *Pestic. Sci.*, **6**, 169–172.

Williams, D. J., Beach, B. G. W., Horrière, D., and Marechal, G. (1977). 'LS 74-783, a new systemic fungicide with activity against Phycomycete disease', *Proc. 9th British Insectic. Fungic. Conf.*, 565–573.

Yamamoto, H., Tomizawa, C., Uesugi, Y., and Murai, T. (1973). 'Absorption, translocation and metabolism of *O,O*-diisopropyl S-benzyl phosphorothiolate (Kitazin P) in rice plant', *Agr. Biol. Chem.*, **37**, 1553–1561.

Yih, R. Y., McRae, D. H., and Wilson, H. F. (1968). 'Metabolism of 3′,4′-dichloropropionanilide: 3,4-dichloroaniline-lignin complex in rice plants', *Science*, **161**, 376–377.

Progress in Pesticide Biochemistry and Toxicology, Volume 3
Edited by D. H. Hutson and T. R. Roberts

CHAPTER 3

The mode of action of herbicides

A. D. Dodge

INTRODUCTION

The use of chemicals to kill plants has been practised to a limited extent for hundreds of years (Smith and Secoy, 1976). In addition to normal cultivation methods of weed control, the ability to poison plants by the direct application of chemicals proved attractive. The inherent toxicity of certain early herbicides not only to weeds but also to crops and farmer, was mitigated by the development from the early 1930s of organic herbicides which interacted at low concentrations with plant biochemical and physiological processes and led to plant death. Plants by definition are immobile and therefore a 'sitting target' for herbicide attack and ease of application. Although this characteristic is apparently advantageous to the farmer, it has also proved to be of benefit, unknowingly in general, to the agrochemical industry in the development of new herbicides. Immobility has necessitated the evolution of a number of protective mechanisms not only for resistance to predator or pathogen attack but to a changing physical environment. Many herbicides appear to operate by overtaxing or inhibiting certain of the biochemical protective mechanisms which deal with such problems as excess excitation energy, free radicals and toxic oxygen species. In addition, the ability of plants to detoxify foreign molecules by metabolism, conjugation and compartmentalization has proved in many instances to be the basis of selectivity.

During the last fifty or so years since the discovery of the dinitrophenol herbicides we have seen an increasing array of new compounds which possibly number up to 150 on a world scale, and also a considerable increase in potential potency of the herbicidal compound (Graham-Bryce, 1981). This, together with the improvement of application methods, means that considerably less active ingredient is used. For example, the new herbicide chlorsulfuron (DPX 4189) (**1**) shows growth inhibition at 2.8×10^{-9} M (Ray, 1980). In spite of the increasing activity of potential herbicides, only a relatively small amount of active ingredient will actually reach specific site(s) and accumulate in sufficient concentration to induce a primary lethal effect. In many instances, where herbicide selectivity is apparent, the lethal concentration is not reached because of metabolism, inactivation or lack of movement.

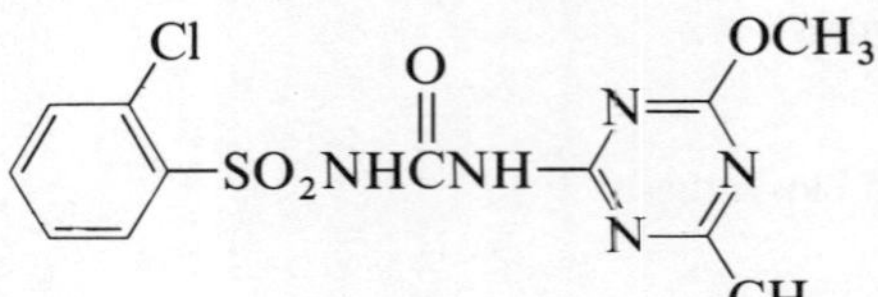

(**1**)

A cursory survey of the extensive literature on herbicide mode of action reveals that certain processes are target sites for herbicide action where a primary lethal effect is induced. In the case of almost 50% of all herbicides this is certainly due to interaction at a site on the chloroplast electron transport chain. Such interaction will lead to a cessation of photosynthesis and secondary damaging events. With most remaining herbicides we are unable to identify specific points of molecular interaction, but can describe other events such as the inhibition of biosynthetic events or cell division. This field of study is also confused by the range of investigations not only with varying plant material but with a wide range of concentrations. It is possible to induce changes in almost any biochemical or physiological processes if sufficient concentration of chemical is applied. In spite of these problems it is intended in this brief review to attempt to identify primary sites of action and to consider the subsequent events which lead to the premature death of plants.

PHOTOSYNTHETIC PROCESSES

Electron transport inhibitors

Photosynthetic inhibition is a good mode of herbicide action as compounds which work in this way are more likely to have fewer problems of animal toxicity. The electron transport system of chloroplasts leading to the reduction of $NADP^+$ and the generation of a proton gradient linked to an associated phosphorylation system, is located within chloroplast thylakoid membranes. The organization of the two photosystems and associated electron carriers is summarized in Figure 1. Since the initial work of Wessels and Van der Veen (1956) with urea herbicides, a large number of compounds (Table 1) have been shown to inhibit electron transport, normally demonstrated by using isolated chloroplasts (Hill Reaction). An extensive literature is available which indicates the specific chemical requirements for potential photosynthetic electron transport inhibitor herbicides. See recent papers by Trebst and Draber (1979), Van Assche and Carles (1982) and Shipman (1981; 1982).

The specific site of interaction of the herbicides mentioned in Table 1 is the reducing side of photosystem II at point A (Figure 1). Between photosystem II and plastoquinone two hypothetical components Q and B are suggested to operate in sequence. Present ideas visualize these components embedded in the lipoprotein thylakoid membrane and covered by a proteinaceous shield. All herbicides which operate at this site contain a hydrophobic component and this portion of the molecule will affect the relationship with the membrane lipid components. The polar portion of the molecule binds non-covalently to a protein component of around 32 K daltons (Pfister *et al.*, 1981). The interaction of the herbicidal molecule with the membrane possibly induces a change in the

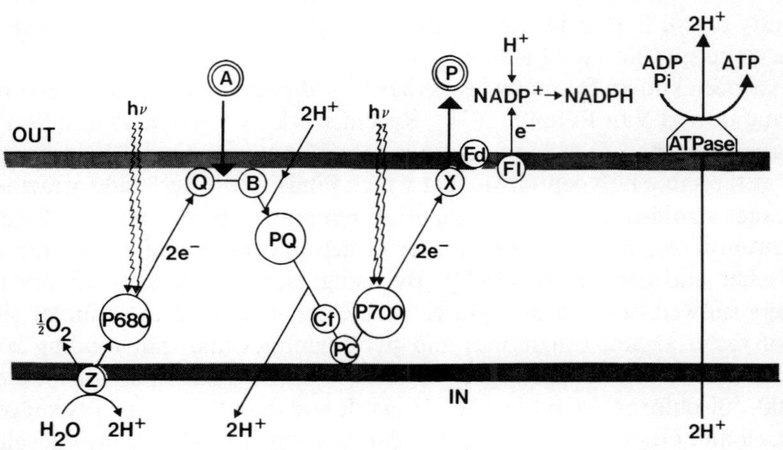

Figure 1 Schematic representation of electron transport and photosynthetic phosphorylation within a chloroplast thylakoid membrane (modified from Trebst, 1974). Electron flow is driven by photosystem II and photosystem I through the reaction centres P680 and P700. Abbreviations: PQ, plastoquinone; Cf. cytochrome f; PC, plastocyanin; Fd, ferredoxin; Fl, ferrodoxin $NADP^+$ reductase; Z, Q, B and X are hypothetical intermediates; A indicates the point of interaction of the majority of photosynthetic electron transport inhibitors; P the possible point of electron deviation caused by the bipyridyl herbicides e.g. paraquat

Table 1 Some herbicides which primarily affect chloroplast electron transport

Family	Examples
Anilide	Pentanochlor, propanil
Benzothiadiazinone	Bentazon(4)
Diphenylether	Flurodifen(12)
Hydroxybenzonitrile	Ioxynil(3), bromoxynil
N-phenylurea	Diuron, monuron(6), isoproturon, chlortoluron
s-Triazine	Atrazine(2), simazine
Triazinone	Metribuzin, metamitron(5)
Uracil	Lenacil, isocil

The identification of certain herbicide families, e.g. diphenylether, does not imply that all members of the family operate at this site. The numbers in parentheses refer to chemical structures in the text.

potential of B which makes electron transfer thermodynamically unfavourable (Pfister and Arntzen, 1979). Renger (1976) has demonstrated that mild trypsin treatment of isolated chloroplast membranes abolished the sensitivity to diuron by alteration of the 32 K dalton protein to a 16 K dalton species (Steinbeck *et al.*, 1982). The apparent susceptibility of this polypeptide site to herbicidal action raises certain questions: firstly, is this a natural regulatory site, and secondly do all herbicide classes which are electron transport inhibitors link to the same specific site of action?

It has been known for many years that CO_2 depletion limits electron transfer (Govindjee and Van Rensen, 1978). Recent work by Khanna *et al.* (1981) has indicated that the CO_2 site of action is located between Q and B and appears to be at the same polypeptide as that which binds to the herbicide atrazine (**2**). The exact significance of CO_2 regulation remains to be established. Evidence for common or partially common sites of action came initially from the work of Tischer and Strotmann (1977). By using binding constant studies these workers showed that the phenylurea herbicides interact with a similar site to that of the triazines, triazinones and pyridazinones and that binding is also quantitatively related to the degree of inhibition with a maximum of 1 molecule per 300–500 chlorophyll molecules. This is lower than that originally suggested by Izawa and Good (1965). In recent years a number of plants have developed

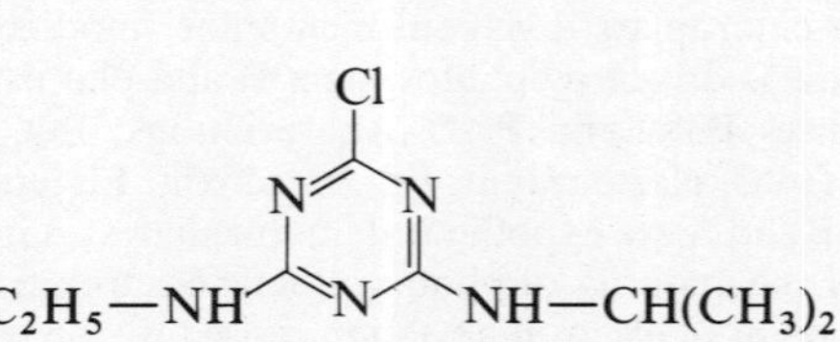

(2)

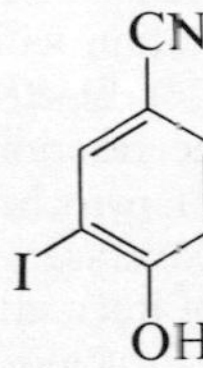

(3)

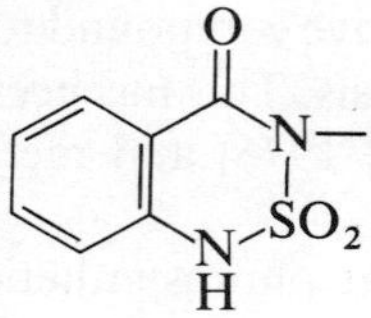

(4)

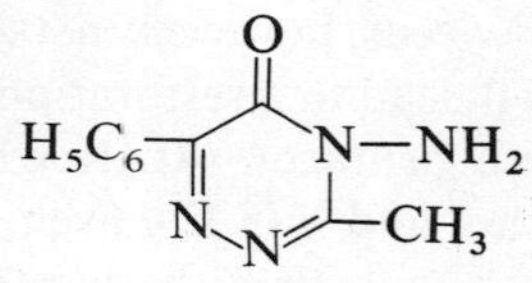

(5)

resistance to atrazine (**2**), These include *Amaranthus retroflexus*, *Chenopodium album* and *Senecio vulgaris* (Pfister and Arntzen, 1979). In general the strong resistance to triazine inhibition is shared to a lesser extent by the triazinones, pyridazinones and uracils. They are partially resistant to ureas and to an even lesser extent to bentazon (**4**). In some instances a difference in the protein components of the resistant biotype chloroplasts has been shown by gel electrophoresis (Pfister and Arntzen, 1979, Steinbeck *et al.*, 1982). This work indicates that the protein binding site could consist of an essential area common to all herbicides and a further area which had unique group binding properties. Interaction with both areas would be essential for inhibition (see Figure 2).

In spite of the apparent uniformity of action site, it is clear that some herbicides have different or additional sites of action as has been shown for the hydroxybenzonitrile herbicide ioxynil (**3**) (Pallett and Dodge, 1979a) together with certain other phenolic inhibitors which might interact with a 41 K dalton protein located beneath the 32 K dalton component (Oettmeier *et al.*, 1982).

The impedance of photosynthetic electron flow initiated by these herbicides will lead directly to an inhibition of CO_2 incorporation. With certain herbicides

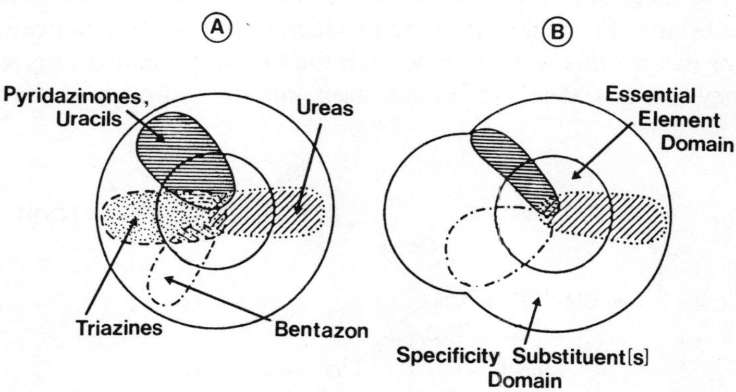

Figure 2 Schematic representation of possible binding sites for photosynthetic electron transport inhibitors (after Pfister and Arntzen, 1979). The areas of interaction consist of a central 'essential element domain', which links with all herbicides and an outer 'specificity substituent domain', which is responsible for the unique binding properties of each family of herbicides. **A** represents a normal binding site system; **B** represents a modified system from an atrazine resistance biotype

and tolerant weeds, however, metabolism of the active compound to an inactive derivative will lead to a restoration of photosynthesis. This has been shown for example with bentazon (**4**) (Retzlaff and Hamm, 1976) and metamitron (**5**) (Van Oorschot and Van Leeuwen, 1979).

Early work with these herbicides suggested that photosynthetic inhibition would lead to death by starvation (Gast, 1958). Subsequent experiments have indicated that phytotoxic symptoms were more apparent in light and were directly related to an increase in light intensity (Minshall, 1957; Van Oorschot and Van Leeuwen, 1974; Pallett and Dodge, 1979b). Experiments with isolated chloroplasts (Stanger and Appleby, 1972; Ridley, 1977) and whole leaves (Pallett and Dodge, 1980) have indicated that excitation energy, absorbed by the photosynthetic pigments and prevented from driving electron flow, overloads the natural carotenoid protective system and this will lead to cellular damage and death. The role of carotenoids in protection will be discussed below.

In summary, the work of Pallett and Dodge, 1980, and Dodge 1982a; 1982b indicates that the following events might occur (summarized in Figure 3). Excitation energy absorbed by the chloroplast chlorophyll pigments leads to the initial excitation to the short-lived singlet state (lifetime $\sim 10^{-8}$ s). Normally, in the absence of inhibitors, energy transfer to adjacent molecules and thence to the reaction centres P680 and P700 will follow (Figure 1). If this does not occur, intersystem crossing leads to the generation of a longer-lived state, triplet chlorophyll (lifetime $\sim 10^{-3}$ s). If this is unquenched by the carotenoids for example, deleterious reactions occur in two possible ways (Figure 3). Firstly and possibly to a lesser extent, by type 1 reactions in which electron or hydrogen abstraction from molecules such as membrane unsaturated fatty acids occurs; this will lead to the production of lipid free radicals. In the presence of oxygen this will combine with the lipid free radical to give rise to lipid peroxy radicals. The radical can itself initiate hydrogen abstraction in

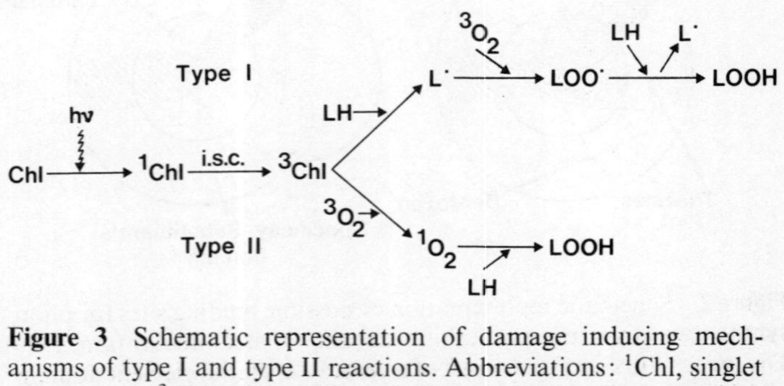

Figure 3 Schematic representation of damage inducing mechanisms of type I and type II reactions. Abbreviations: ^{1}Chl, singlet chlorophyll; ^{3}Chl, triplet chlorophyll; i.s.c., intersystem crossing; $^3O_2$, triplet or ground state oxygen; $^1O_2$, singlet oxygen; LH, unsaturated fatty acid; L˙, lipid free radical; LOO˙, peroxidized lipid radical; LOOH, lipid peroxide

type 1 reactions. Of more importance, as oxygen will be freely available, are type 2 reactions in which triplet chlorophyll interacts with oxygen to give singlet oxygen ($^1O_2$). Although singlet oxygen can be quenched by carotenoids and free radical scavengers, it is possible that these systems are overloaded and, among other possible deleterious reactions, this species will initiate the formation of lipid peroxides from unsaturated membrane fatty acids. Experiments with both monuron (**6**) and ioxynil (**3**) (Pallett and Dodge, 1980) have shown that carotenoid breakdown preceded that of chlorophyll, indicating the preferential destruction of the carotenoid protective system. Furthermore, the increasing formation of ethane, a breakdown product of lipid peroxides (Riely *et al.*, 1974), indicated that lipid destruction was occurring. The acyl lipids of chloroplast thylakoid membranes consist of a high amount of unsaturated fatty acids, especially of linolenic acid, and are therefore particularly susceptible to peroxidative attack. The disorganization of chloroplast and cellular membranes as shown by electron microscopy for example with the triazine herbicide atrazine (**2**) (Ashton *et al.*, 1963) and for monuron (**6**) and ioxynil (**3**) (Pallett and Dodge, 1980) correlated well with such deleterious reactions and the appearance of chlorotic leaves. The role of singlet oxygen in herbicide action indicates that specific singlet oxygen generators could be used as herbicides (Percival and Dodge, 1983) whether this species is generated by visible or ultraviolet radiation (Zweig and Nachtigall, 1975). They have incidentally been used as effective experimental insecticides (Callaham *et al.*, 1975).

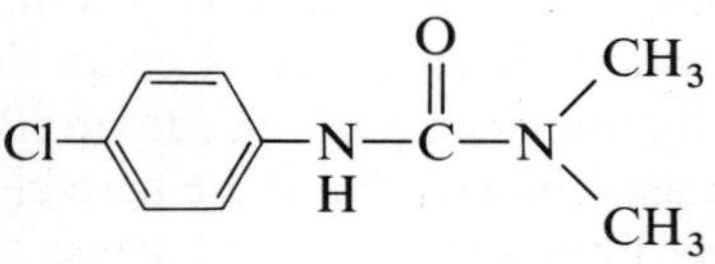

(**6**)

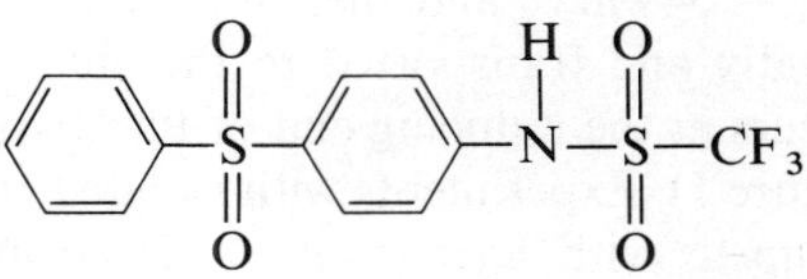

(**7**)

Inhibition of photosynthetic phosphorylation

Linked to electron transport and the generation of a proton gradient across the thylakoid membrane (Figure 1) is the synthesis of ATP from ADP and Pi according to the tenets of the chemiosmotic hypothesis (Mitchell, 1966). Phosphorylation will not occur if electron transport and hence the proton gradient is inhibited, but more specifically some herbicides or potential herbicides (Wright *et al.*, 1980) uncouple by penetrating the thylakoid membrane and dissipating the proton gradient. Certain amines are well known as uncouplers (Good, 1960) and the herbicide perfluidone (**7**) is thought to operate in a similar manner (Alsop and Moreland, 1975). Among the various actions ascribed to the diphenylether herbicides, certain molecules such as nitrofen (**8**) have a greater propensity towards the inhibition of photosynthetic phosphorylation (Sandmann *et al.*, 1981). With this group of herbicides a higher degree of energy transfer inhibition (i.e. phosphorylation) was shown when the diphenyl ethers were substituted with one or two chlorines (Sandmann *et al.*, 1981). Although the exact action of the new herbicide fluazifop-butyl (**9**) has not been ascertained, it too is thought to affect phosphorylation (Plowman *et al.*, 1980).

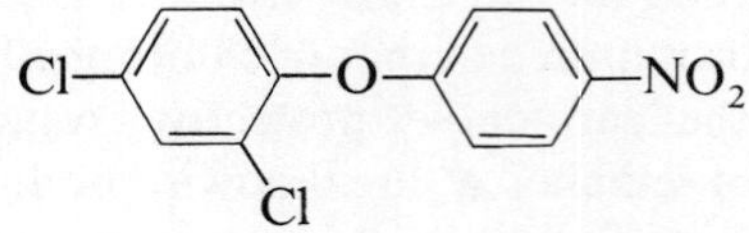

(8)

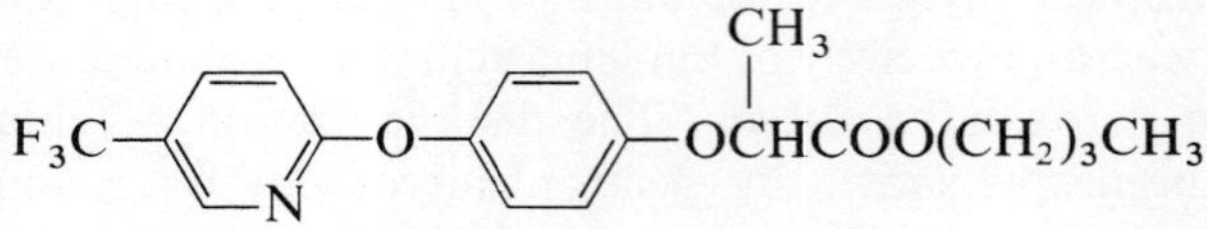

(9)

Divertors of photosynthetic electron transport

In this category are the well known bipyridyl herbicides paraquat (**10**) and diquat (**11**) (where the reoxidation of the reduced molecule by molecular oxygen is the primary facet of herbicidal action), and certain diphenylether herbicides.

The negative redox potentials of paraquat (E'_0 – 446 mV) and diquat (E'_0 – 349 mV) and their high water solubility enables them to be taken up rapidly and transported to the chloroplast where diversion of electron flow occurs at the reducing end of Photosystem I in the vicinity of ferredoxin (P in Figure 1). Experiments with isolated chloroplasts showed that these herbicides compete with ferredoxin for electrons from Photosystem I and cause the inhibition of $NADP^+$ reduction (Zweig *et al.*, 1965) and consequently of CO_2 incorporation in whole leaves (Harris and Dodge, 1972b) and algae (Turner *et al.*, 1970). Early experiments with these herbicides showed that, in addition to electron transport, oxygen was essential for action (Mees, 1960). The light-induced reduction of paraquat (and diquat) by one electron transfer to a free radical is followed by the immediate reoxidation by oxygen to yield superoxide. Pulse radiolysis studies demonstrated the feasibility of superoxide formation in paraquat treated leaves (Farrington *et al.*, 1973) and superoxide formation was also promoted by paraquat in experiments with isolated chloroplasts (Youngman *et al.*, 1980). A superoxide scavenger has also been shown to retard

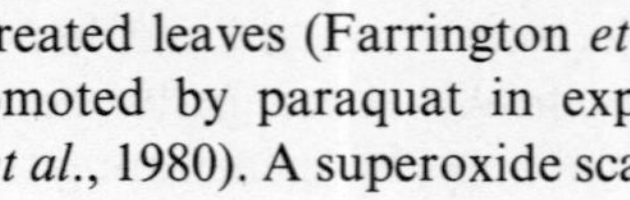

(10)

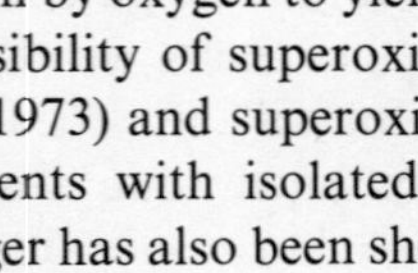

(11)

paraquat action (Youngman *et al.*, 1979; Youngman and Dodge, 1979). It is now generally accepted that the superoxide anion ($O_2^{\cdot -}$) is not capable of inducing deleterious reactions itself. However its toxicity is due to the range of more toxic species which are generated from this free radical. *In vivo* it is possible that hydroxyl radicals ($OH^{\cdot}$) might be generated from superoxide in a Haber–Weiss type reaction (McCord, 1979). Alternatively or in addition, hydroxyl radicals may be formed by the interaction of hydrogen peroxide, generated from superoxide, and the paraquat free radical (Winterbourne, 1981). There is also some evidence for a paraquat-hydroxyl linked radical (Youngman and Elstner, 1981). These ideas are summarized in Figure 4.

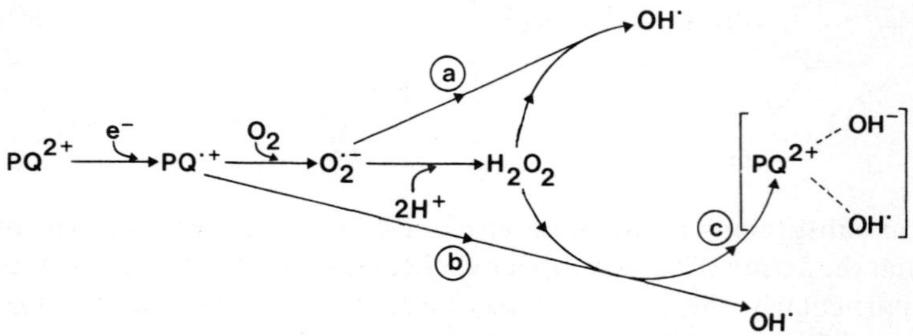

Figure 4 Schematic representation of possible interaction between reduced paraquat $PQ^{\cdot+}$ and oxygen; **a**, a Haber–Weiss type reaction (see McCord, 1979); **b**, the reaction between reduced paraquat and hydrogen peroxide (Winterbourne, 1981); **c**, the formation of a crypto-hydroxyl radical (Youngman and Elstner, 1981)

The actual effects of the bipyridyl herbicides on green plants commence therefore with an inhibition of carbon dioxide uptake and then the appearance of rapid chlorotic symptoms (Dodge, 1971). During this time a disorganization of cellular compartmentalization and membranes occurs (Harris and Dodge, 1972a) and it is assumed that this is induced by the hydroxyl free radicals generated as described above, which induce lipid peroxidation and hence membrane destruction (Harris and Dodge, 1972b, Van Rensen, 1975; Boehler-Kohler *et al.*, 1981). Ethane generation, a conspicuous feature of lipid peroxidation is produced in parallel to a decrease in pigment content (Youngman *et al.*, 1979).

In spite of the efficiency of the bipyridyl herbicides to kill all vegetation on contact, certain plants have been found which have considerable tolerance to paraquat. This has been shown in certain lines of *Lolium perenne* (Harvey *et al.*, 1978; Harper and Harvey, 1978; Harvey and Fraser, 1980) and certain biotypes of *Conyza linefolia* (Youngman and Dodge, 1981). Although in both instances an increase in the activity of the superoxide scavenging enzyme, superoxide dismutase, was evident, this was clearly only part of the mechanism of resistance. It could also include limited penetration to the chloroplast (Harvey and Fraser,

1980; Youngman and Dodge, 1981). Of considerable interest is the fact that paraquat tolerant biotypes of *Conzya linefolia* are susceptible to diquat (Dodge and Meacham 1982 unpublished). The mechanism of this difference has yet to be explained.

There are strong similarities between the factors which affect the action of the bipyridyl herbicides and those which affect some of the diphenylethers. The multi-site action of these herbicides as inhibitors of electron transport (e.g. fluorodifen (**12**)) and energy transfer in the chloroplast (e.g. nitrofen (**8**)) and as bleaching herbicides (e.g. oxyflurofen (**13**)) is clearly dependent upon

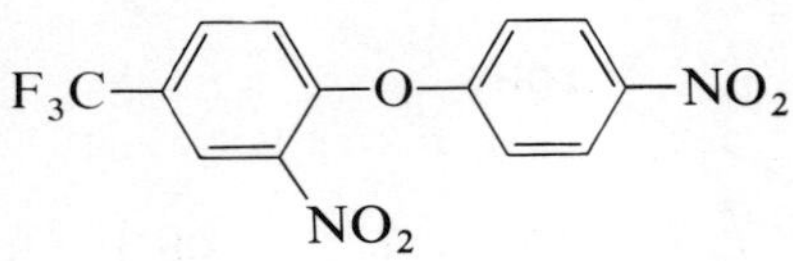

(**12**)

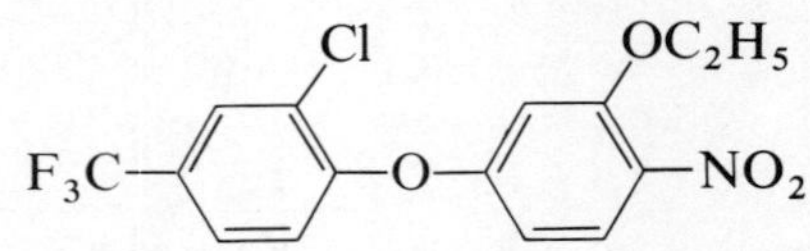

(**13**)

chemical substituents (Sandmann *et al.*, 1981). It has been known for many years that the action of certain diphenylethers is promoted by light (Matsunaka, 1969) particularly the action of oxyflurofen (Vanstone and Stobbe, 1979; Kunert and Böger, 1981). In the experiments of Vanstone and Stobbe (1979) with buckwheat seedlings, membrane damage was apparent before an effect on pigment breakdown. By contrast, Kunert and Böger (1981) working with the algae *Scenedesmus* showed a rapid bleaching effect with a concurrent increase in ethane, indicating membrane lipid peroxidation. Furthermore, the action of the herbicides was inhibited by a photosynthetic electron transport inhibitor (Kunert and Böger, 1981). Although these workers failed to demonstrate an increased oxygen uptake, as one would expect for a compound identical to the bipyridyls, their work indicated that the herbicide oxyflurofen was reduced by electron transport to a radical anion. This was the first stage of reduction to an amino derivative. The radical anion, it is suggested, reacts with oxygen to generate superoxide, and this would lead to the generation of other toxic species (as mentioned above) and induce membrane damage. In the experiments of Vanstone and Stobbe (1979) the failure to observe bleaching was probably a consequence of the short duration of the experiments. Experiments with paraquat clearly showed that membrane damage preceded by a considerable time the induction of chlorophyll breakdown (Dodge, 1971).

A somewhat different mechanism of energy diversion has been proposed for the diphenyl ether herbicide acifluorofen-methyl (**14**). Orr and Hess (1981; 1982a, b) have proposed that the molecule is light activated, but in this instance by some form of interaction with the carotenoid pigments of the chloroplast. The light activated aciflurofen-methyl molecule so generated is proposed to induce membrane damage by lipid peroxidation, as mentioned above for other groups of herbicides, and lead to cellular disorganization and death.

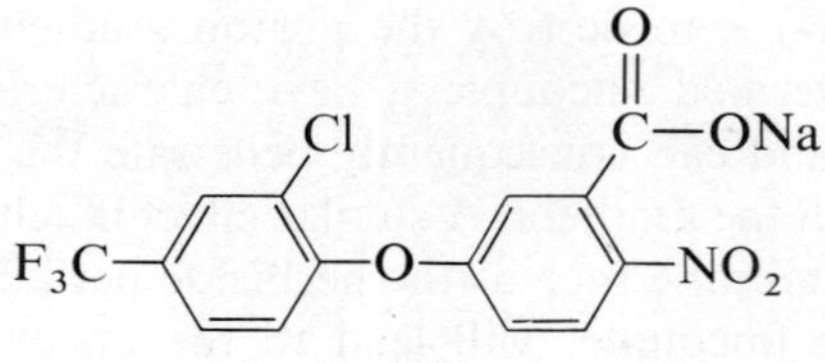

(14)

RESPIRATORY PROCESSES

In comparison with photosynthesis, respiration is of considerably less importance as a primary site of herbicide action, although historically the first organic herbicides, the nitrophenols, operate on this process. In principle, respiratory metabolism and mitochondria in particular are not good sites for herbicide action for conservative biochemistry has retained almost identical systems in autotrophic and heterotrophic organisms. There are, however, certain unique features of plant mitochondria (Palmer, 1979) but these have not been used as herbicide target sites.

The physical organization of plant mitochondria indicates that the components of the electron transport chain are organized in the inner membrane. The re-oxidation of $NADH^+$ or in case of succinate dehydrogenase $FADH_2$, via this chain to oxygen via cytochrome oxidase, occurs with the concurrent release of protons into the intermembrane space. As with chloroplasts, this generates a proton motive force which is linked to phosphorylation (Mitchell, 1966). The importance of ATP in the energy economy of the cell is too well known to need restatement, however the action of dinoseb (**15**) DNOC (**16**), of the hydroxylation product of dichlobenil (**17**) (Moreland *et al.*, 1974), and to a lesser extent of the hydroxybenzonitriles, ioxynil (**3**) and bromoxynil

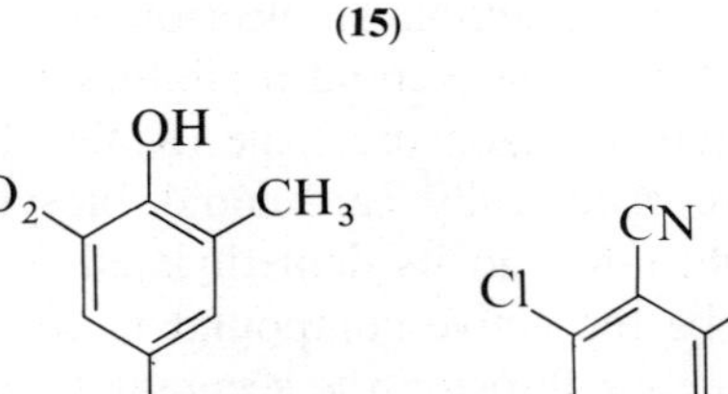

(15)

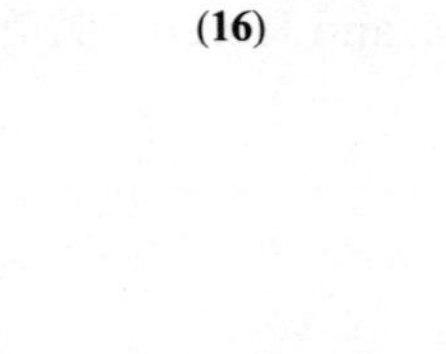

(16)

(17)

(Kerr and Wain, 1964) is to destroy the proton gradient. Such compounds, which are generally termed uncouplers, have characteristics of lipid-soluble weak acids or bases and can consequently penetrate the inner mitochondrial membrane and abolish the gradient. A similar effect is achieved by compounds which perturb the membrane such as the herbicide perfluidone (**7**) (Moreland, 1981). In practice an uncoupler will lead to the uncontrolled oxidation of substrate and hence not only to a disruption of the energy economy of the cell, but also to a total loss of food reserves. This can be demonstrated as a considerable increase in CO_2 evolution (Beevers, 1953). Some experiments have shown that light had a retarding effect on the action of uncoupler herbicides (Mellor and Salisbury, 1965; Wojtaszek, 1966) indicating perhaps a transient replacement of food reserves, but this could not possibly overcome a fundamental breakdown of cellular energy control.

INHIBITORS OF BIOSYNTHESIS

Carotenoids

Chloroplast carotenoid pigments have a dual role. On the one hand they act as light absorbers for photosynthesis and on the other as photoprotectors (Krinsky, 1979). As photoprotectors they act either to reduce the excitation of triplet chlorophyll or of singlet oxygen possibly generated by energy transfer from triplet chlorophyll. Although chloroplasts contain a number of carotenoid pigments, both carotenes (e.g. α-carotene and β-carotene) and xanthophylls (e.g. lutein, zeaxanthin and violaxanthin) their precise individual function has not been fully elucidated. Since the initial work of Griffiths *et al.* (1955) followed by that of Anderson and Robertson (1960) with maize mutants, it has been known that carotenoid-less or limited plants are increasingly susceptible to photo-oxidative attack. During the last few years an increasing number of herbicides have been developed which primarily act by inhibiting carotenoid biosynthesis therefore rendering plants more susceptible to damage. The biosynthetic pathway of carotenoid pigments is outlined in Figure 5 and shows that phytoene is converted by a series of desaturation reactions into lycopene, from which initially the carotene pigments arise by cyclization. The pyridazinone herbicides have multi-sites of action as electron transport inhibitors, fatty acid biosynthesis inhibitors or carotenoid inhibitors according to the specific substituents on the parent pyrazon molecule (Ridley 1982; St. John, 1982). Those which are more specifically carotenoid biosynthesis inhibitors are metflurazon (SAN 6706) (**18**) and its demethylated form norflurazon (SAN 9789) (**19**) which may be the active compound *in vivo* (Strang and Rogers, 1974). With the herbicide metflurazon the accumulation of phytoene and to a lesser extent phytofluene has been shown in dark-grown wheat seedlings (Bartels and McCullough, 1972; Ben-Aziz and Koren, 1974), radish and barley

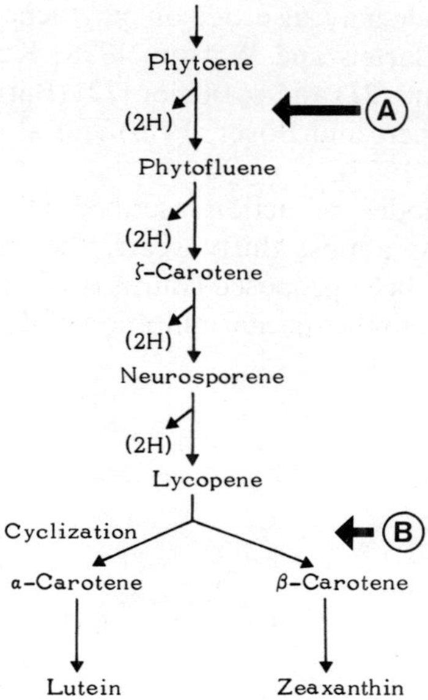

Figure 5 Biosynthetic pathway of carotenoid formation in plants. **A** represents the possible site of interaction of desaturase reaction inhibitors; **B** represents the point of interaction of cyclization inhibitors

(Lichtenthaler and Kleudgen, 1977) and algae including *chlorella* (Kümmel and Grimme, 1975). Likewise the related herbicide norflurazon induced a similar response (Devlin *et al.*, 1976; Bartels and Watson, 1978; Kunert and Böger, 1979). In general, these investigations support the idea that the specific point of action is at phytoene 'desaturase'. At present there is no evidence available for enzyme inhibition at this site, although it is possible that it could be an indirect effect mediated by a deficiency in chloroplast ribosomes (Feierabend and Schubert, 1978). Sandmann *et al.* (1980) have suggested that in the fungus *Phycomyces blakesleeanus*, the action of norflurazon is before phytoene and specifically on the enzyme phytoene synthetase.

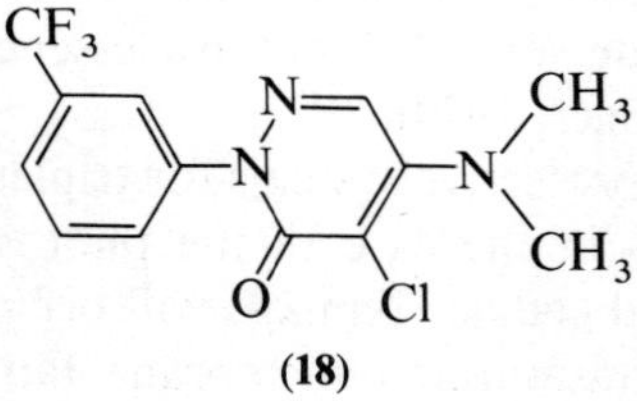

(18)

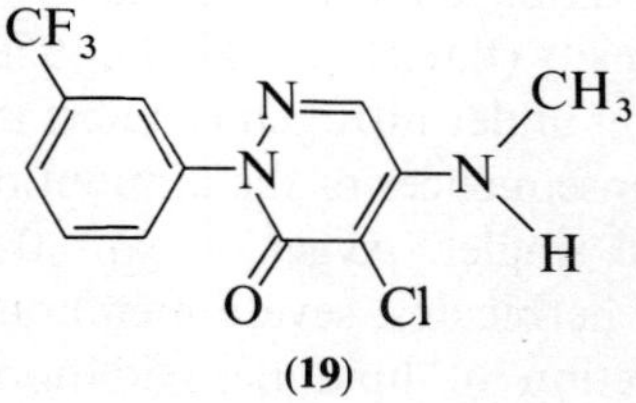

(19)

Other herbicides which may also act on phytoene 'desaturase' reactions include fluridone (**20**) (Bartels and Watson, 1978; Kunert and Böger, 1979; Ridley, 1982), dichlormate (**21**) and pyrichlor (**22**) (Burns *et al.*, 1971), difunon EMD-IT 5914 (**23**) (Kunert and Böger, 1979) and also possibly fluometuron (**24**) (Wergin and Potter, 1975).

Among the many modes of action ascribed to aminotriazole (**25**), an established herbicide for almost thirty years, the inhibition of carotenoid biosynthesis has recently been proposed (Burns *et al.*, 1971). Grumbach (1981) has shown that this is due to the specific inhibition of the cyclization of lycopene to α- and β-carotene.

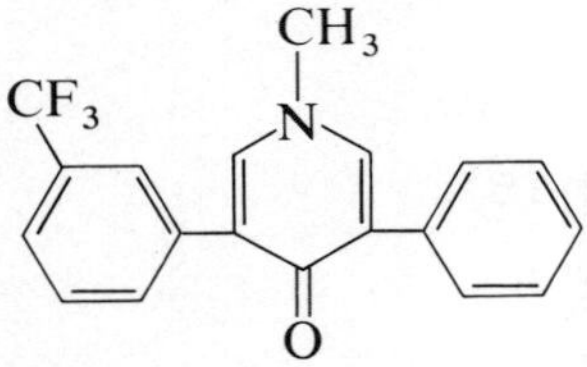

(20)

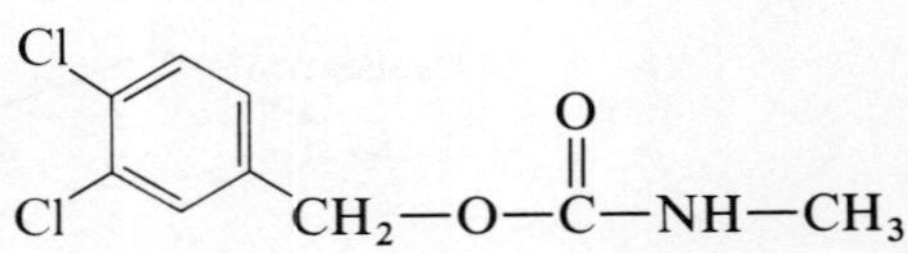

(21)

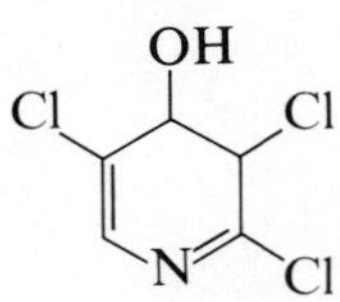

(22)

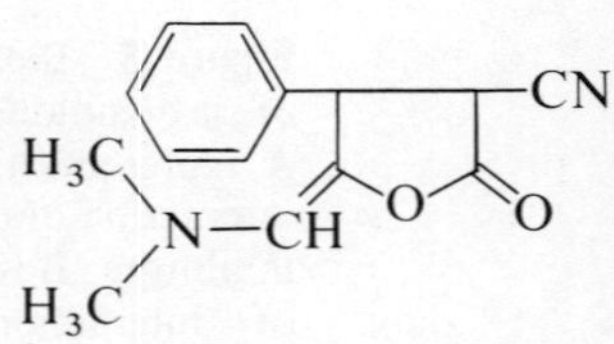

(23)

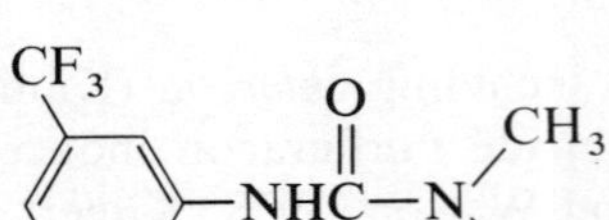

(24)

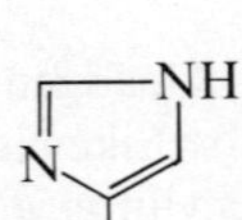

(25)

Experiments with dark grown black pine seedlings (*Pinus nigra*) (Bogdanovic *et al.*, 1981) and *Picea abies* (Thompson and Dodge, 1982 unpublished) which form chlorophylls and carotenoids in darkness, have shown that the action of these herbicides is quite specific for carotenoids. However, with light grown material, damage to both pigments was shown to be accelerated by increasing light intensity (Bartels and Hyde, 1970; Anderson, 1981) but to a lesser extent by incubation under nitrogen (Kunert and Böger, 1979).

The consequences of the diminution of protective systems for triplet chlorophyll and singlet oxygen, means that, as in the case of the photosynthetic inhibitor herbicides, severe membrane and cellular damage will occur due to the induction of lipid peroxidation of unsaturated membrane fatty acids.

Several investigators have shown this type of damage by electron microscope studies, for example for metflurazon (Bartels and Hyde, 1970), norflurazon (Devlin *et al.*, 1976; Frosch *et al.*, 1979), fluridone (Bartels and Watson, 1978), fluometuron (Wergin and Potter, 1975) and for the experimental herbicide haloxydine (Dodge and Lawes, 1972).

Lipid biosynthesis

The external surface of plants is protected by a waxy cuticle, the constituents of which include long chain esters (formed by the esterification of primary alcohols with very long chain fatty acids, i.e. up to C_{32}) and long chain alkanes of C_{21} to C_{31}. It is possible that the alkanes are produced by decarboxylation of the long chain fatty acids. Any interference, therefore, with the fatty acid elongation system would interfere with cuticular development and render plants more susceptible to pathogen attack, water loss and herbicide penetration.

Work by Gentner (1966) with the thiocarbamate EPTC (**26**) demonstrated that this herbicide reduced the cuticular wax deposit on cabbage leaves, and this was confirmed by Wilkinson and Hardcastle (1970) with sicklepod leaves as a conspicuous reduction in cuticle thickness. Work with a number of herbicides including diallate (**27**) and EPTC by Kolattukudy and Brown (1974)

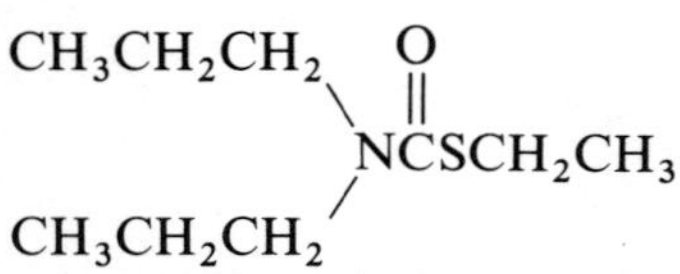

(**26**)

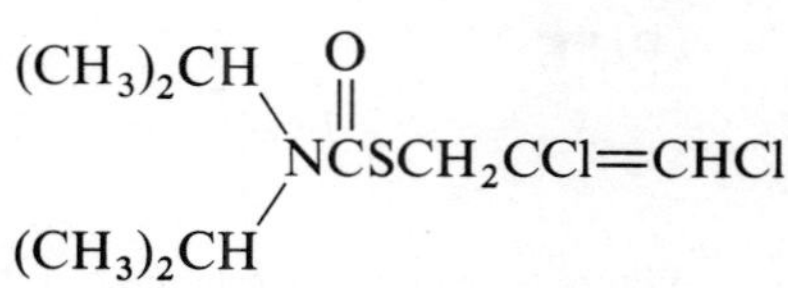

(**27**)

showed that they reduced the incorporation of acetate into the cuticular very long chain (C_{31}) alkanes and secondary alcohols, but there was an accumulation of short chain cuticular lipids. This work pointed to the specific inhibition of various chain-elongating enzymes systems and was confirmed by the experiments of Bolton and Harwood (1976) with EPTC, diallate and triallate (**28**) which inhibited the incorporation of acetate into very long chain fatty acids of the suberin of potato tuber slices. It has been known for many years that trichloroacetic reduced cuticular waxes (Juniper, 1959) and this is thought similarly to operate on the elongating system (Kolattukudy *et al.*, 1974).

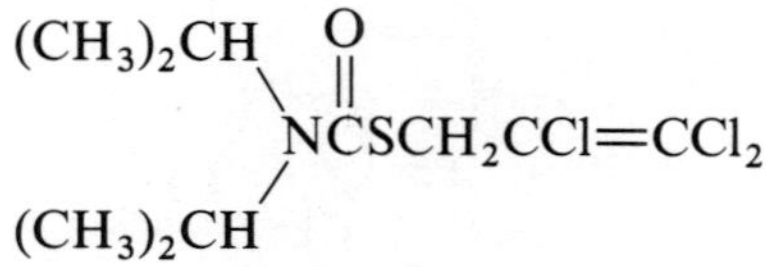

(**28**)

Although fatty acid elongation reactions appear to be a primary site for the action of thiocarbamate herbicides, work by Wilkinson and Smith (1975) and Wilkinson *et al.* (1977) indicated a reduction in the level of chloroplast and leaf unsaturated fatty acids and in particular of 18:3 linolenic acid a major component of the galactolipids of chloroplast membranes (Figure 6).

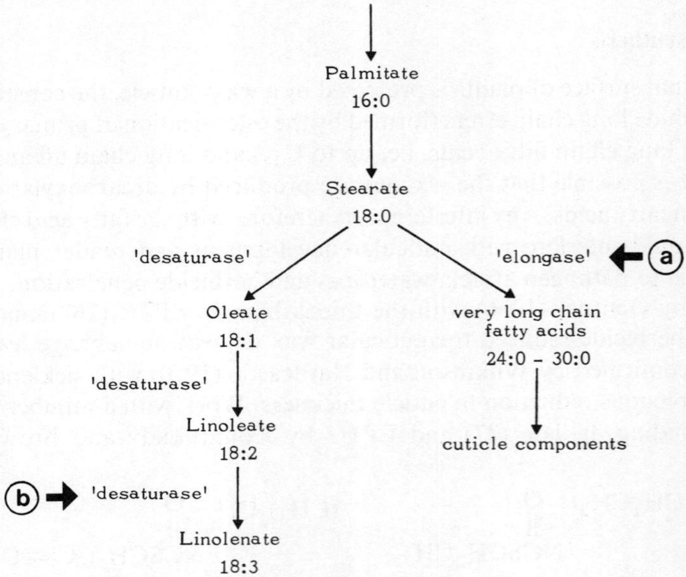

Figure 6 Summary scheme of fatty acid biosynthesis in plants with point of interaction of **a**, elongase inhibitors and **b** desaturase inhibitors. This integrated scheme does not intend to imply that these reactions occur in the same portion of the cell

The effect of certain pyridazinone herbicides on carotenoid biosynthesis has already been mentioned. The removal of the CF_3 group substituent of metflurazon produces SAN 9785 (BAS 13,338) (**29**) and this has been shown to be an effective inhibitor of unsaturated fatty acid formation. In particular this herbicide decreased the relative level of the 18 : 3 fatty acid linolenic acid to the 18:2 linoleic acid in the galactolipids of chloroplast membranes (St. John, 1976). Further work by Khan *et al.* (1979) also confirmed this site of action on

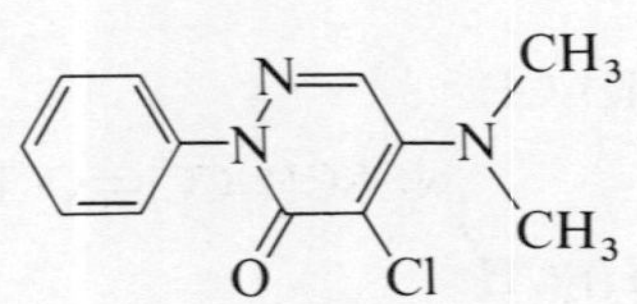

(**29**)

the fatty acid (linoleic) desaturase enzyme system. One interesting side effect of the herbicide SAN 9785, is that it reduced the chilling resistance of seedlings (St. John and Christiansen, 1976). This was possibly due to the prevention of linolenic acid formation, which is a possible prerequisite to frost resistance (Willemot, 1977). It is clear, therefore, that any modification of membrane fatty acid constituents will lead to abnormal membrane formation and hence cellular dysfunction.

Aromatic amino acid biosynthesis

The shikimic acid pathway, which occurs only in plants and micro-organisms, leads to the formation of certain essential amino acids and numerous other secondary products such as flavonoids and anthocyanims (Figure 7). The herbicidal possibilities of chemicals which inhibit this pathway were investigated and in particular of the enzyme shikimate dehydrogenase which catalyses the reaction of dehydroshikimate to shikimate (Baille *et al.*, 1972; Dowsett *et al.*, 1972). A number of compounds related to dehydroshikimate, for example 1,6-dihydroxy-2-oxoisonictinic acid, were effective inhibitors when the enzyme was assayed in the normally reverse direction, shikimate → dehydroshikimate, yet were ineffective in reverse or as herbicides.

The herbicidal properties of the post-emergence herbicide glyphosate (**30**) were announced in 1971 (Baird *et al.*, 1971) and in the following year studies by Jaworski (1972) indicated a possible inhibitory effect on the formation of

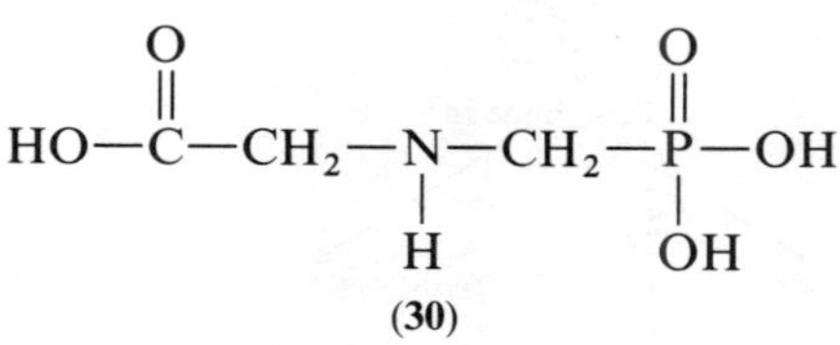

(30)

aromatic amino acids. In his experiments the growth of *Lemna gibba* and *Rhizobium japonicum* was inhibited by glyphosate, but relieved by the addition of L-phenylalanine, L-tyrosine and certain other intermediates of the shikimate pathway. On the basis of this work, Jaworski (1972) tentatively suggested that glyphosate affected the enzymes chorismate mutase and prephenate dehydratase by inhibition or by the modification of regulation. The identification of this particular biosynthetic pathway as a possible site of action, led to the instigation of numerous investigations with a variety of plant systems. Reversal of glyphosate-induced growth inhibition by aromatic amino acids was demonstrated by Haderlie *et al.* (1977) with cultures of carrot and tobacco cells. Tymonko and Foy (1978) also showed a similar reversal with soybean leaf cells by phenylalanine, tyrosine and tryptophan. In general the work was confirmed by the experiments of Roisch and Lingens (1974) and the subsequent

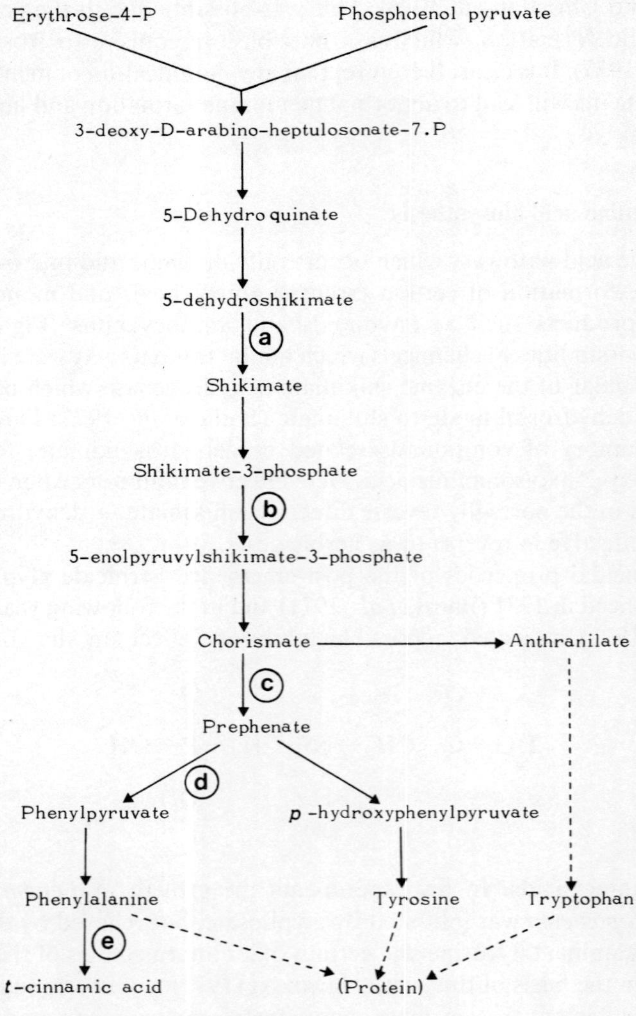

Figure 7 The shikimic acid pathway for the biosynthesis of aromatic amino acids. In addition to the generation of phenylalanine, tyrosine and tryptophan for protein biosynthesis, these acids are the starting point for secondary metabolism leading to the formation of many compounds including flavones, lignin and tannins. Enzymes specifically mentioned in the text, **a**, shikimate dehydrogenase; **b**, 5-enolpyruvylshikimate-3-phosphate synthase; **c**, chorismate mutase; **d**, prephenate dehydratase; **e**, phenylalanine ammonia lyase

work of Gresshoff (1979). On the other hand experiments by Brecke (1976) and Duke and Hoagland (1978) failed to repeat the aromatic amino acid reversal effect. More recent work by Amrhein and co-workers has shown that the formation of flavonoid compounds in buckweat seedling was inhibited by glyphosate (Holländer and Amrhein, 1980) but partially overcome by the addition of L-phenylalanine. Furthermore, glyphosate inhibited the incorporation of [^{14}C]shikimate into phenylalanine, tyrosine and trytophan, and also led to the accumulation of shikimate in treated plant tissue (Steinrücken and Amrhein, 1980). In attempting to define the precise site of inhibition between shikimate and phenylalanine, enzymes studies showed that glyphosate specifically inhibited 5-enolpyruvyl shikimate-3-phosphate synthase which catalyses the reaction between shikimate-3-phosphate and 5-enolpyruvyl shikimate-3 phosphate (Figure 8). This has been demonstrated with enzyme systems from *Aerobacter aerogenes* (Amrhein *et al.*, 1980) and generally confirmed for *E. coli* and mung beans (Steinrücken and Amrhein, 1980).

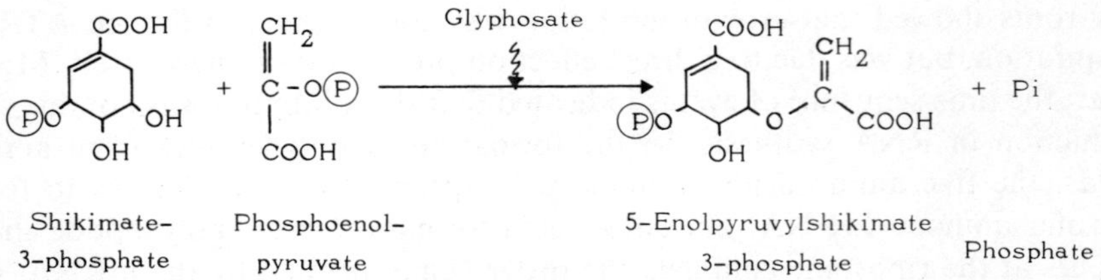

Figure 8 A specific site of action of glyphosate at the enzyme 5-enolpyruvylshikimate-3-phosphate synthase (see Amrhein, *et al.*, 1980)

Extensive work has indicated that a consequence of glyphosate treatment of plant tissue is an increase in the activity of the enzyme phenylalanine ammonia lysase (PAL) which catalyses the reaction between phenylalanine and *trans*-cinnamic acid. This has been demonstrated in a number of plant tissues including maize roots (Duke and Hoagland, 1978) soybean seedlings (Duke *et al.*, 1979; Hoagland *et al.*, 1979; Hoagland 1980) and in rhizome buds of *Agropyron repens* (Cole *et al.*, 1980). Rather than being a direct result of glyphosate action it appears more likely that a decrease in the level of phenylalanine leads to a derepression of PAL. It was thought possible that increased PAL activity would promote the formation of toxic phenols (Hoagland *et al.*, 1978) however, the failure of the PAL inhibitor L-α-aminoxy-β-phenylpropionic acid to alleviate toxicity (Duke *et al.*, 1980; Cole *et al.*, 1980) supports the contention that the increase in PAL activity is an effect, rather than a direct cause, of toxicity.

The consequences of an effect on the shikimate pathway is therefore not only the failure to form an extensive range of secondary products, but more importantly, the limited biosynthesis of certain key amino acids. Experiments by

Cole *et al.* (1980) showed that glyphosate also inhibited the incorporation of $[^{14}C]$leucine into *Agropyron repens* nodes, indicating that a reduction in the phenylalanine protein precursor pool had an indirect effect on the incorporation of other amino acids into protein. How far these effects represent the total picture of the action of glyphosate remains to be ascertained (Hoagland and Duke, 1982).

Protein synthesis

It is evident that protein synthesis will be inhibited indirectly by many means including the action of respiratory uncoupler herbicides (Mann *et al.*, 1965; Moreland *et al.*, 1969), or by glyphosate which prevents the formation of certain essential amino acids. However, certain compounds appear to have a more specific action on the formation of polypeptides. These include the chloroacetamide herbicides, alachlor (**31**), metolachlor (**32**), propachlor (**33**) and CDAA (**34**). The experiments by Duke *et al.* (1975) with propachlor and cucumber roots showed that growth inhibition was not an effect of limited ATP or respiration, but was due to a direct effect on protein biosynthesis itself. Moreover, the time sequence of events indicated that the inhibition was not due to a reduction in RNA synthesis. In the formation of polypeptides from amino acids, the free amino acids combine with specific tRNA molecules to form specific aminoacyl tRNA. The subsequent formation of the polypeptide chain occurs at the ribosome complex, the order being dictated by the mRNA. On

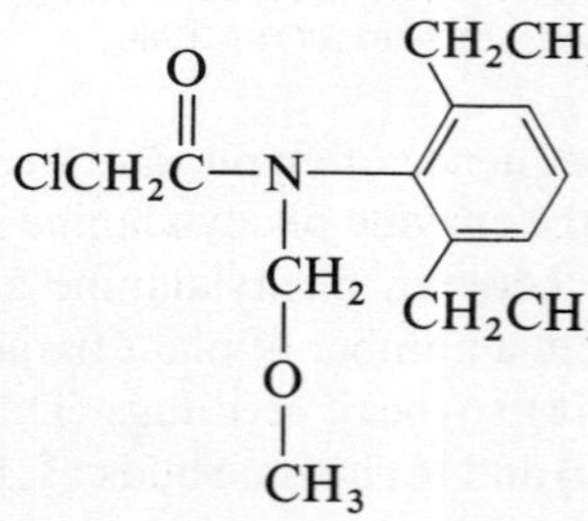

(31)

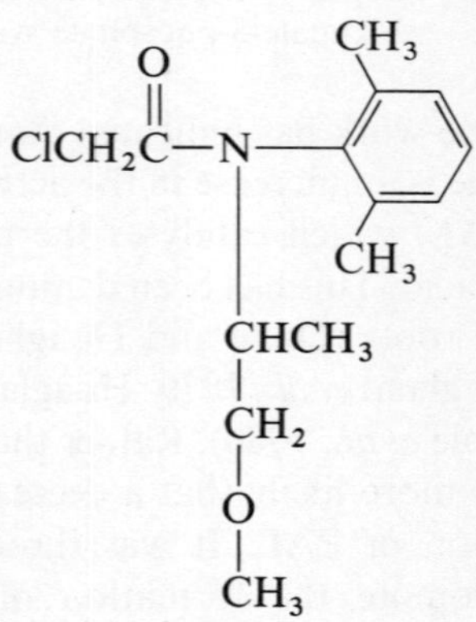

(32)

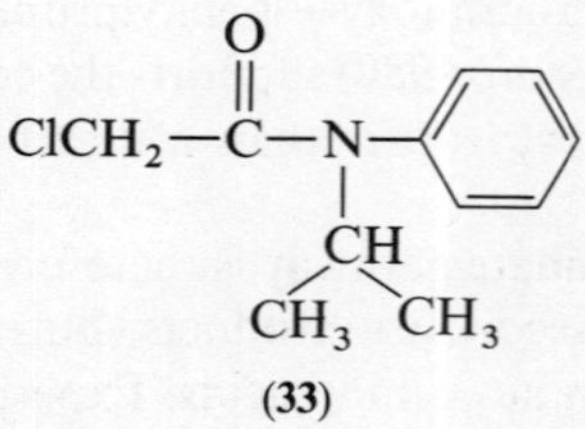

(33)

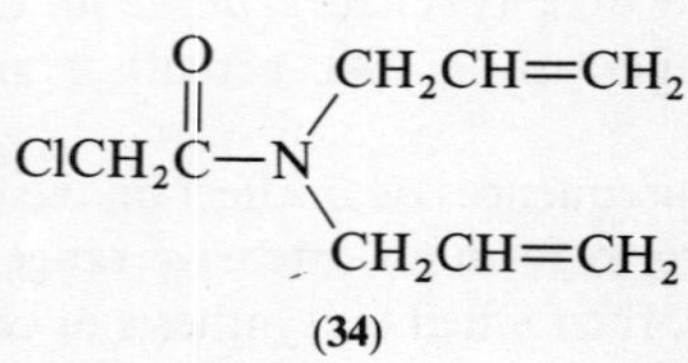

(34)

the basis of further experiments with propachlor, Duke *et al.* (1975) suggested that the action of this herbicide could be due to an inhibition of amino acid activation, thus preventing aminoacyl-tRNA formation, or a failure to form the polypeptide chains from these amino acids. This had been speculated in earlier work with the chloroacetamide herbicides (Jaworski, 1969). Subsequent experiments by Pillai *et al.* (1979) with the related herbicide metolachlor failed to confirm this hypothesis and suggested that the inhibition of protein synthesis was due to an inhibition of leucine uptake in the cucumber root system under investigation. Similarly experiments by Deal *et al.* (1980) with all four chloroacetamide herbicides failed to show an inhibition of polypeptide elongation or of translocation. These workers indicated that the failure to observe *in vivo* incorporation of amino acids could be due to a metabolic disruption not directly related to protein synthesis itself.

In spite of the conflicting nature of these experiments, with the chloroacetamides, elegant experiments by Baxter *et al.* (1973) and Weeks and Baxter (1972) with the experimental dinitroaniline herbicide MDMP (**35**), have indicated a very specific action for this compound. The growth of wheat seed embryos was inhibited and the more detailed biochemical experiments suggested that protein synthesis was inhibited by the prevention of initiation on the polyribosome complex.

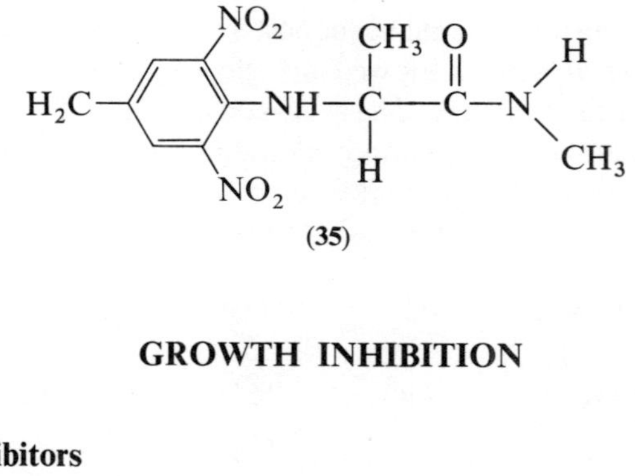

(**35**)

GROWTH INHIBITION

Cell division inhibitors

The complexities of the cell cycle and the interrelationship of component stages has been discussed by Rost and Bayer (1976) together with particular problems of identifying sites of herbicide action. For example cell division can be prevented by mechanisms as diverse as the failure of DNA synthesis in the S-phase to the failure of microtubule formation. The potential confusion was identified by Rost and Bayer (1976) who showed that propham at 10^{-5} M in experiments with pea root-tip cells, produced abnormal mitotic figures, but at a higher concentration of 10^{-3} M it also inhibited DNA, RNA and protein synthesis.

In the cell cycle the organization of the mitotic spindle is a prelude to the separation of daughter chromosomes at anaphase. The construction of the microtubules of the spindle is thought to involve a microtubule organizing centre, and the individual components are globular proteins (α and β-tubulin) of around 6.7×10^4 m.wt. which are organized into a helix. Experiments by Collis and Weeks (1978) with *Chlamydomonas reinhardii* suggested that amiprophosmethyl (APM) (**36**) prevented tubulin formation by a depletion of tubulin

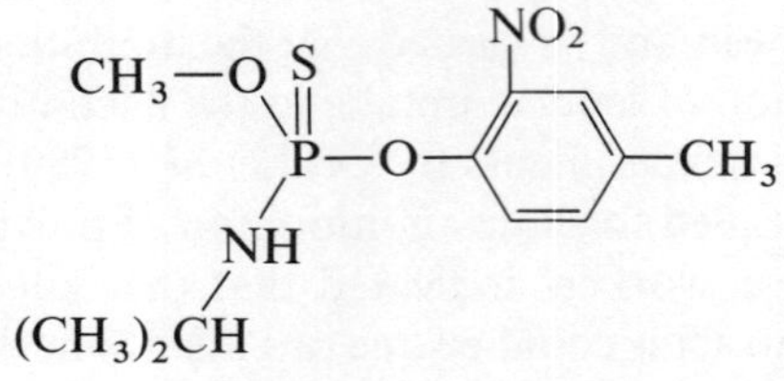

(36)

RNA. It is known that the polymerization of microtubules is affected by calcium (Ca^{2+}) (Marcum *et al.*, 1978), thus if the concentration of Ca^{2+} increases, depolymerization of the microtubules will occur. Plant mitochondrial experiments by Hertel *et al.* (1980) with oryzalin (**37**), trifluralin (**38**), propham (**39**), chlorpropham (**40**) and APM showed that these herbicides diminished Ca^{2+} uptake. Thus if the level of cytoplasmic Ca^{2+} is raised the organization of microtubules could be inhibited. In addition to Ca^{2+} accumulation by mitochondria, it is possible that it could be sequestered by the endoplasmic reticulum. However, Hertel *et al.* (1980) showed no effect of these herbicides on Ca^{2+} uptake by isolated microsomes. The well-known mitotic inhibitor colchicine was without effect on both mitochondrial and microsomal systems. The action of this particular chemical is thought to be by complexing with tubulin dimers (Margolis and Wilson, 1977) and leads to the disruption of the microtubules.

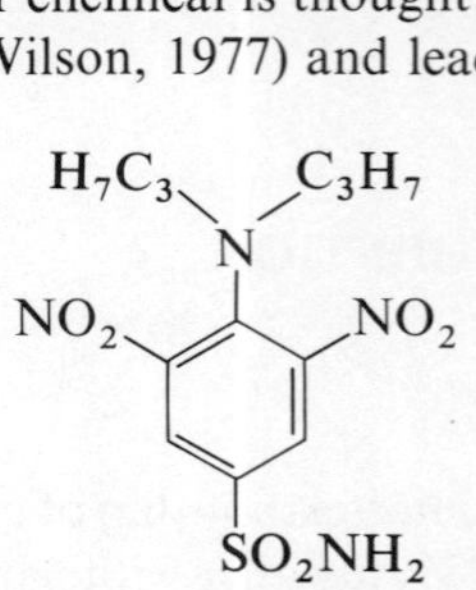

(37)

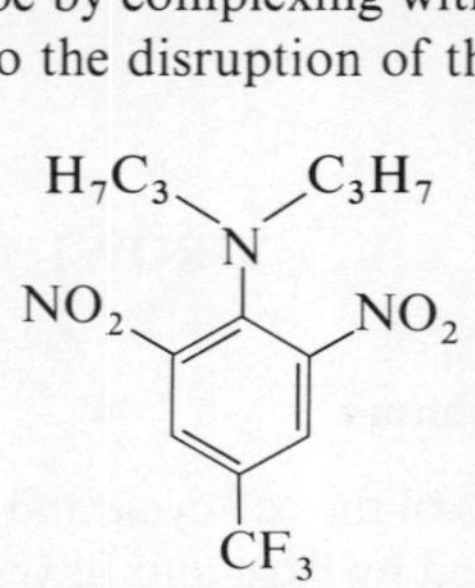

(38)

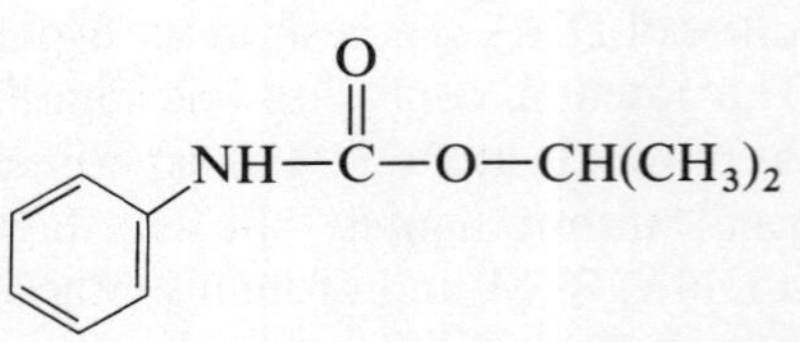

(39)

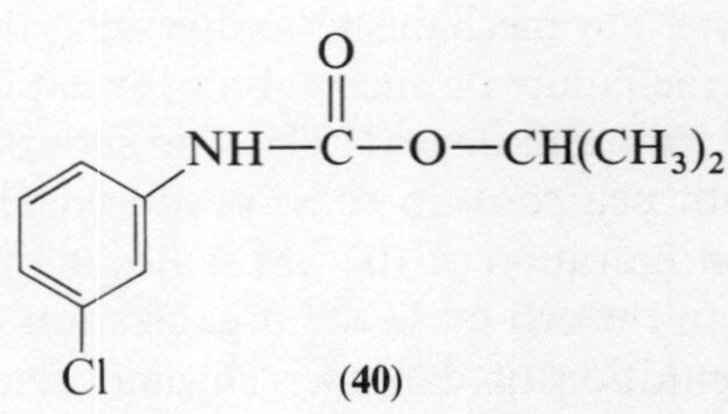

(40)

The differences between this inhibitor, and the herbicide induced effects which prevent microtubule formation was shown by previous ultrastructural studies (Bartels and Hilton, 1973; Hepler and Jackson, 1969). Other microscopic evidence of herbicide action has shown that *Allium* root-tips treated with APM contain giant nuclei due to the failure of cell division (Draber and Fedtke, 1979). Furthermore, asulam (**41**) which induced arrested metaphases and anaphases and chromosome bridges in onion root-tip cells (Sterrett and Fretz, 1975) and the related herbicide barban (**42**) are thought to operate in a similar manner.

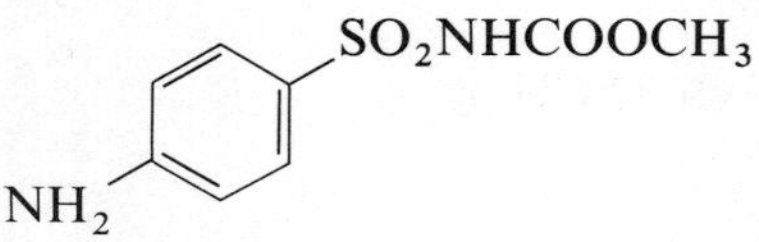

(**41**)

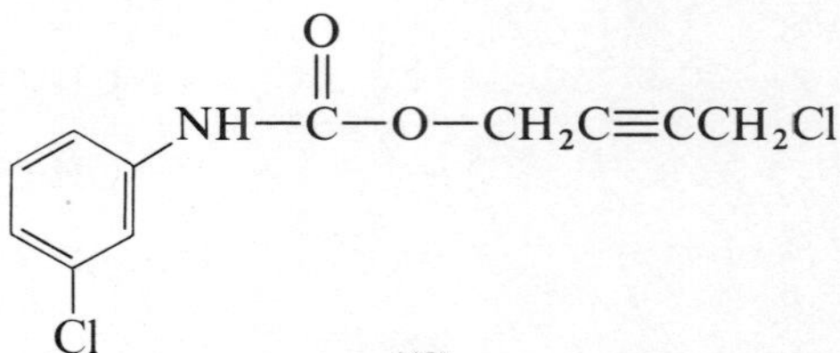

(**42**)

The recently introduced herbicide chlorsulfuron (**1**) (DPX-4189) used for the control of weeds in cereals is an effective inhibitor of cell division and plant growth. Experiments by Ray (1980) showed that in treated *Vicia faba* roots the number of mitotic figures was greatly reduced although there was no percentage difference in the frequency of various mitotic phases. Further experiments with thymidine uptake and incorporation into DNA indicated that the former process was not affected whereas incorporation was reduced to 20% after 6 h treatment. Although care should be taken in the interpretation of this data (McLoad and Thompson, 1981) it could indicate a disruption in DNA synthesis during the S phase. An inhibition of DNA synthesis could in part account for the action of the wild oat herbicide difenzoquat (**43**) (Pallett and Caseley, 1980).

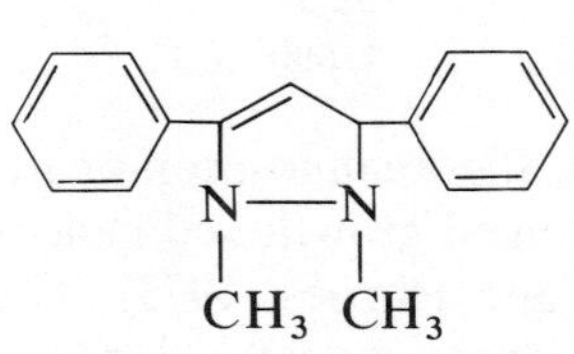

(**43**)

Cell elongation inhibitors

In addition to growth inhibition by cell division, a number of herbicides act by apparently inhibiting cell elongation and expansion. These include a number of herbicides which are particularly useful for the control of wild oats in cereals, such as the analinopropionates (benzoylprop-ethyl (**44**), flamprop-methyl (**45**) and flamprop-isopropyl (**45**)) together with chlorfenprop-methyl (**47**) and

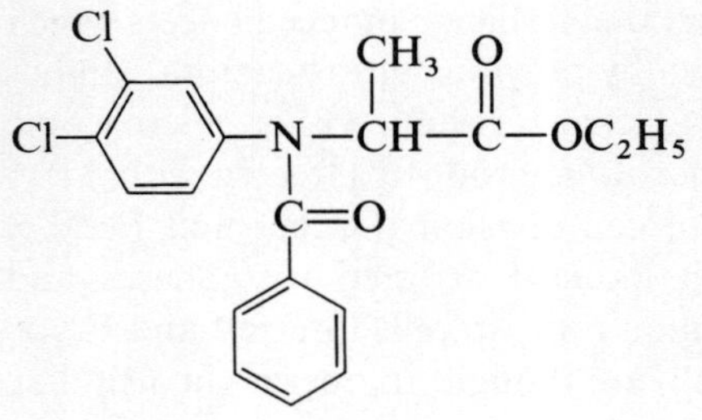

(44)

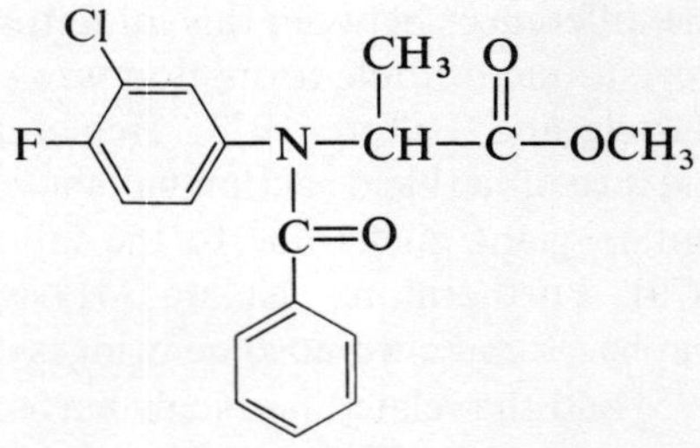

(45)

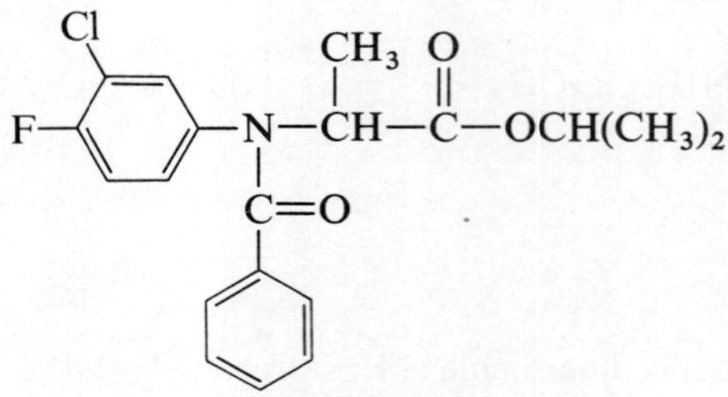

(46)

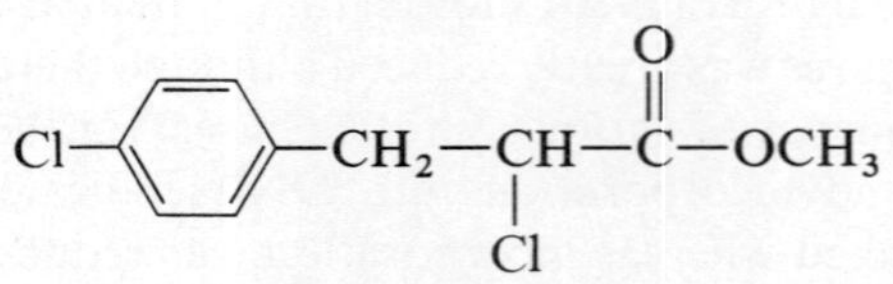

(47)

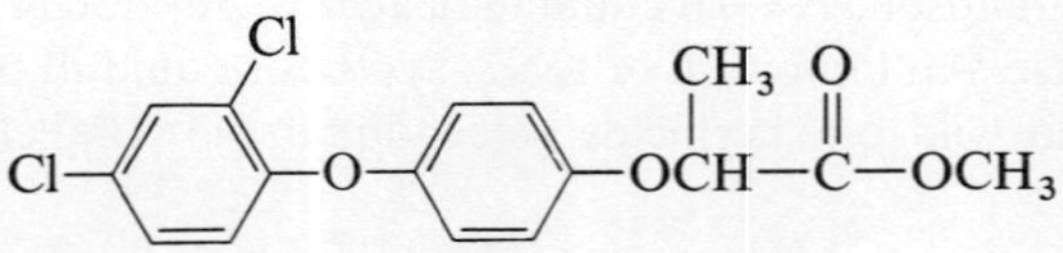

(48)

diclofop methyl (**48**). With the analino-propionate herbicides, action at the stem apical cells leads to stunted growth and hence unfavourable competition with crop plants (Jeffcoat and Harries, 1973). The selective action of these herbicides is dependent upon the greater degree of de-esterification of the applied compound, which is then translocated more readily via the phloem from the leaves to the stem apices. The free acid is more susceptible to removal by the formation of inactive conjugates, but in practice the balance *in vivo* between de-esterification and conjugation produces sufficiently toxic concentrations of free acid in the susceptible plants (Jeffcoat and Harries, 1973; Jeffcoat and Harries, 1975; Jeffcoat *et al.*, 1977). An esterase enzyme has been partially purified from wild oats (Hill *et al.*, 1978) and appeared to be a relatively non-specific carboxylesterase which also hydrolysed dichlofopmethyl to diclofop. More recent work by Venis (1982) has shown the existence of two esterase enzymes in oat tissue. One partially purified enzyme was virtually stereospecific for (−) isomers of the 'benzoylprop' and 'flamprop' esters. Although the esterase for (+) esters was not amenable to isolation, experiments indicated that it would only act upon ethyl and isopropyl esters. There was thus clear enantiomeric discrimination between the methyl esters.

Chlorfenprop-methyl is also rapidly hydrolysed in wild oat plants (Fedtke and Schmidt, 1977) by a methyl esterase to chlorfenprop which is probably the herbicidally active compound. In contrast to the analino-propionate herbicides, the degree of esterification appears to be similar in both susceptible and tolerant cultivars of wild oat (Fedtke and Schmidt, 1977) although the (−) enantiomer is the more active compound. Andreev and Amrhein (1976) have suggested that the action of this herbicide might be to compete with auxin at a specific membrane site or possibly to alter membrane integrity in such a way that auxin action is limited.

With the herbicide diclofop methyl, the ester and acid appear to be almost equally phytotoxic, although dichlofop had a greater effect on root growth (Shimabukuro *et al.*, 1978). Selectivity in this instance may be due to a greater degree of ring hydroxylation followed by phenolic conjugation in resistant plants (Dusky *et al.*, 1980). As with chlorfenprop methyl, the diclofop-methyl ester was a little more effective at reversing IAA-stimulated elongation of oat coloeptiles than its free acid (Shimabukuro *et al.*, 1978). Whether auxin antagonism is the primary mechanism of all cell elongation inhibitor herbicides remains to be established in more detail.

Auxin-type herbicides

Despite the historical and economic importance of the phenoxyacetic acid and related herbicides, a fundamental knowledge and understanding of their mode of action still remains to be elucidated. The most important chemicals to be considered in this group include 2,4-D (**49**), 2,4,5-T (**50**), 2,3,6-TBA (**51**), MCPA (**52**) and its butyric analogue MCPB. The behaviour and mode of action of these compounds has recently been extensively reviewed in this series

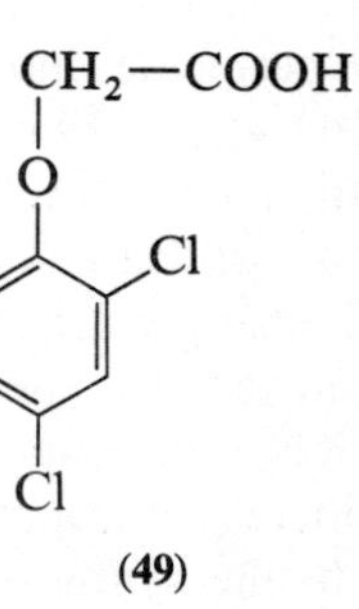

(49)

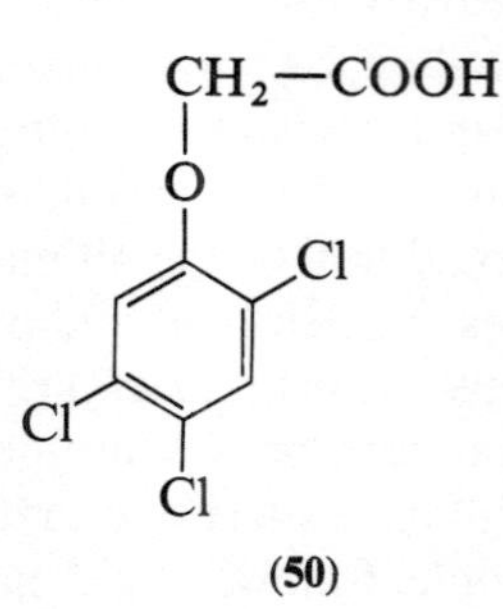

(50)

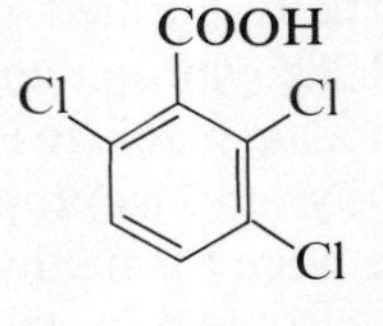

(51)

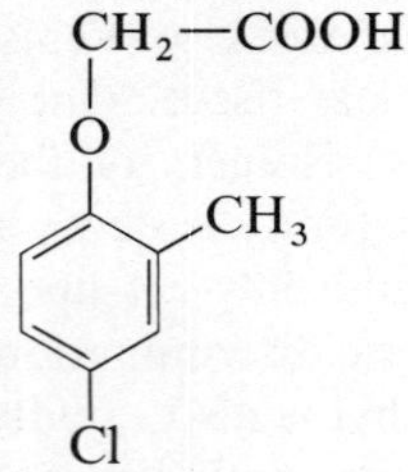

(52)

(Pillmoor and Gaunt, 1981) and the chemical requirements of active molecules has been reviewed by (Garraway and Wain, 1976).

It is generally assumed that these compounds imitate the action of the natural auxin IAA; however the plant has been calculated to receive an overdose of approximately 1000-fold (Pillmoor and Gaunt, 1981). In addition, they are less readily inactivated by IAA oxidase or by metabolic hydroxylation and conjugation, although these facets may play an important part in herbicide selectivity (Garraway and Wain, 1976).

The morphological aberrations which these herbicides induce such as stem enlargement, secondary root and callus growth and the stunting of leaves indicate a severe disruption of normal physiological processes (Cardenas *et al.*, 1968). Although it is assumed that plants 'grow themselves to death', there is no strict correlation between growth effects and toxicity (Eliasson, 1963).

In spite of much work over the last thirty or so years, we cannot say with any degree of certainty how these herbicides instigate growth aberrations and plant death. It is possible that an initial response is to bind with specific membrane binding proteins (Venis, 1979), although all binding proteins should not necessarily be regarded as natural receptors (Murphy, 1979). Extensive experiments into the mechanism of action of auxin have clearly identified short and longer term responses (Vanderhoef, 1981). The short term responses which can occur within the first hour possibly involve the activation of membrane bound ATPase enzymes which lead to acidification of the cell wall and thus extension (Hager *et al.*, 1971; Ray, 1974). The major growth aberrations discussed are clearly a result of a fundamental change in regulated metabolism. Many workers have shown that auxin type herbicides induce major changes in RNA, protein and ribosome number (Key *et al.*, 1966; Hanson and Slife, 1969; Zurfluh and Guilfoyle, 1980). A totally modified cellular enzyme content thus obtained could well be the ultimate cause of death (Coble and Slife, 1971). Exactly how these nucleic acid and protein changes are induced is still largely unknown, however Hardin *et al.* (1972) showed that the interaction of auxin with extranuclear membranes promoted RNA polymerase activity. Teissere *et al.* (1975; 1976) have suggested that the increased transcription of messenger RNA which would follow, could promote the synthesis of specific protein 'transcription factors' which themselves could stimulate nucleolar RNA polymerase. This would lead to increased synthesis of ribosomal RNA and thus an activation of protein synthesis.

CONCLUSION

An attempt has been made in this review to identify primary sites of herbicide action, and to describe the events which lead to plant death. It is evident that a major number of herbicides disturb the basic energetics of the cell either by interfering with photosynthesis or respiration. Protective devices are overtaxed or, in the case of cuticular protection, limited in formation. Cellular dysfunction is promoted by the aberration of membranes and enzyme proteins. Growth is modified by the modification of cell division, or extension, or by severe growth aberration.

With the photosynthetic electron transport inhibitor herbicides we are possibly nearer than any other group to describing specific sites of molecular interaction and the subsequent events which lead to plant death. Information of this type with all herbicides could be of considerable advantage in understanding herbicide resistance and in the production of herbicide antidotes.

The design of herbicides on rational grounds has proved to be of negligible value, although many new compounds have been produced following the leads given by screening programmes. It is to be hoped that a greater understanding of plant biochemical processes and specific enzyme sites might lead to a degree of success in this area in the future.

Mode of action studies will continue to be a constant challenge to students and scientists. It is hoped that more multidisciplinary effort will be expended on herbicide development and action with the result that chemicals used throughout the world are not only more efficient, but environmentally less toxic. Perhaps a trend to the use of naturally occurring molecules or related molecules, as in the case of the insecticidal synthetic pyrethroids, will be an area of advance in future decades.

REFERENCES

Alsop, W. R., and Moreland, D. E. (1975). 'Effects of herbicides on the light activated, magnesium-dependent ATPase of isolated spinach chloroplasts', *Pestic. Biochem. Physiol.*, **5**, 163–170.

Amrhein, N., Schab, J., and Steinrücken, H. C. (1980). 'The mode of action of the herbicide glyphosate', *Naturwiss.*, **67**, 356–357.

Anderson, I. C., and Robertson, D. S. (1960). 'Role of carotenoids in protecting chlorophyll from photodestruction', *Pl. Physiol.*, **35**, 531–534.

Anderson, L. W. J. (1981). 'Effect of light on the phytotoxicity of fluridone in American pondweed and Sago pondweed', *Weed Sci.*, **29**, 723–728.

Andreev, G. K., and Amrhein, N. (1976). 'Mechanism of action of the herbicide 2-chloro-3(4-chlorophenyl) propionate and its methyl ester: interaction with cell reponses mediated by auxin', *Physiol. Plant.*, **37**, 175–182.

Ashton, F. M., Gifford, E. M., and Bisalputra, T. (1963). 'Structural changes in *Phaseolus vulgaris* induced by atrazine. II. Effect on fine structure of chloroplasts', *Bot. Gaz.*, **124**, 336–343.

Baillie, A. C., Corbett, J. R., Dowsett, J. R., and McCloskey, P. (1972). 'Inhibitors of shikimate dehydrogenase as potential herbicides', *Pestic. Sci.*, **3**, 113–130.

Baird, D. D., Upchurch, R. P., Homesley, W. B., and Franz, J. E. (1971). 'Introduction of a new broad spectrum postemergence herbicide class with utility for herbaceous weed control', *Proc. N. Cent. Weed Contr. Conf.*, **26**, 64–68.

Bartels, P. G., and Hilton, J. L. (1973). 'Comparison of trifluralin, pronamide, propham and colchicine treatment on microtubules', *Pestic. Biochem. Physiol.*, **3**, 462–472.

Bartels, P. G., and Hyde, A. (1970). 'Chloroplast development in Sandoz 6706 treated wheat seedlings', *Pl. Physiol.*, **45**, 807–810.

Bartels, P. G., and McCullough, C. (1972). 'A new inhibitor of carotenoid synthesis in higher plants: 4-chloro-5-(dimethylamino)-2-α,α,α, (trifluoro-*m*-tolyl)-3(2H)-pyridazinone (Sandoz 6706)', *Biochem. Biophys. Res. Commun.*, **48**, 16–22.

Bartels, P. G., and Watson, C. W. (1978). 'Inhibition of carotenoid synthesis by fluridone and norflurazon', *Weed Sci.*, **26**, 198–203.

Ben-Aziz, A., and Koren, E. (1974). 'Interference in carotenogenesis as a mechanism of action of the pyridazinone herbicide Sandoz 6706', *Pl. Physiol.*, **54**, 916–920.

Baxter, R., Knell, V. C., Somerville, H. J., Swain, H. M., and Weeks, D. P. (1973). 'Effect of MDMP on protein synthesis in wheat and bacteria', *Nature* (Lond.), **243**, 139–142.

Beevers, H. (1953). '2,4-dinitrophenol and plant respiration', *Am. J. Bot.*, **40**, 91–96.

Boehler-Kohler, B. A., Schopf, M., and Böger, P. (1981). 'The effect of paraquat on membrane lipids', in *Photosynthesis*, VI (ed. G. Akoyunoglou), pp. 531–535. Balaban International Science Services, Philadelphia.

Bogdanovic, M., Brkić, P., Tamindžija, M., and Plesnićar, M. (1981). 'The effect of herbicide SAN 9789 on chlorophyll synthesis and photochemical activities in Black pine cotyledons', in *Photosynthesis*, VI (ed. G. Akoyunoglou), pp. 545–553. Balaban International Science Services, Philadelphia.

Bolton, P. and Harwood, J. L. (1976). 'Effect of thiocarbamate herbicides on fatty acid synthesis by potato', *Phytochem.*, **15**, 1507–1509.

Brecke, B. J. (1976). 'Studies into the mechanism of action of *N*-(phosphonomethyl) glycine', Ph.D. thesis, Cornell University.

Burns, E. R., Buchaman, G. A., and Carter, M. C. (1971). 'Inhibition of carotenoid synthesis as a mechanism of action of amitrole, dichlormate, and pyriclor', *Pl. Physiol.*, **47**, 144–148.

Callaham, M. F., Broome, J. R., Lindig, O. H., and Heitz, J. R. (1975). 'Dye-sensitized photo-oxidation reactions in the boll weevil, *Anthonomus grandis*', *Environ. Entomol.*, **4**, 837–84.

Cardenas, J., Slife, F. W., Hanson, J. B., and Butler, H. (1968). 'Physiological changes accompanying the death of cocklebur plants treated with 2,4-D', *Weed Sci.*, **16**, 96–100.

Coble, H. D., and Slife, F. W. (1971). 'Root disfunction in honeyvine milkweed caused by 2,4-D', *Weed Sci.*, **19**, 1–3.

Cole, D. J., Dodge, A. D., and Caseley, J. C. (1980). 'Some biochemical effects of glyphosate on plant meristems', *J. Exp. Bot.*, **31**, 1665–1674.

Collis, P. S., and Weeks, D. (1978). 'Selective inhibition of tubulin synthesis by amino-prophosmethyl during flagellar regeneration in *Chlamydomonas reinhardii*', *Science*, **202**, 440–442.

Deal, L. M., Reeves, J. T., Larkins, B. A., and Hess, F. D. (1980). 'Use of an in vitro protein synthesizing system to test the mode of action of chloroacetamides', *Weed Sci.*, **28**, 334–340.

Devlin, R. M., Kisiel, M. J., and Karczmarczyk, S. J. (1976). Chlorophyll production and chloroplast development in norflurazon-treated plants', *Weed Res.*, **16**, 125–129.

Dodge, A. D. (1971). 'The mode of action of the bipyridylium herbicides paraquat and diquat', *Endeavour*, **30**, 130–135.

Dodge, A. D. (1982a). 'The role of light and oxygen in the action of photosynthetic inhibitor herbicides', in *Biochemical Responses Induced by Herbicides*' (eds D. E. Moreland, J. B. St. John, and F. D. Hess). A.C.S. Symposia Series No. 181, pp. 57–77.

Dodge, A. D. (1982b). 'Oxygen radicals and herbicide action', *Biochem. Soc. Trans.*, **10**, 73–75.

Dodge, J. D., and Lawes, G. B. (1972). 'Some effects of the herbicide haloxydine on the structure and development of chloroplasts', *Ann. Bot.*, **36**, 315–323.

Dowsett, J. R., Middleton, B., Corbett, J. R., and Tubbs, P. K. (1972). 'The anomalous inhibition of shikimate dehydrogenase by analogues of dehydroshikimate', *Biochim. Biophys. Acta.*, **276**, 344–349.

Draber, W., and Fedtke, C. (1979). 'Herbicide interaction with plant biochemical systems', *Adv. Pestic. Sci.* (ed. H. Geissbühler), pp. 475–486, Pergamon, Oxford.

Duke, S. O., and Hoagland, R. E. (1978). 'Effects of glyphosate on metabolism of phenolic compounds. I. Induction of phenylalanine ammonia-lyase activity in dark grown maize roots', *Pl. Sci. Lett.*, **11**, 185–190.

Duke, S. O., Hoagland, R. E., and Elmore, C. D. (1979). 'Effects of glyphosate on metabolism of phenolic compounds. IV. Phenylalanine ammonia-lyase activity, free amino acids and soluble hydroxyphenolic compounds in axes of light-grown soybeans', *Physiol. Plant*, **46**, 307–317.

Duke, S. O., Hoagland, R. E., and Elmore, C. D. (1980). 'Effect of glyphosate on metabolism of phenolic compounds. V. L-α-ammonia-lyase in soybean seedlings', *Pl. Physiol.*, **65**, 17–21.

Duke, W. B., Slife, F. W., Hanson, J. B., and Butler, H. S. (1975). 'An investigation on the mechanism of action of propachlor', *Weed Sci.*, **23**, 142–147.

Dusky, J. A., Davis, D. G., and Shimabukuro, R. H. (1980). 'Metabolism of diclofop methyl (methyl-2-[4-(2′,4′-dichlorophenoxy)phenoxy]propanoate) in cell suspensions of diploid wheat', *Physiol. Plant*, **49**, 151–156.

Eliasson, L. (1963). 'The toxic effects of chlorinated phenoxyacetic acids on aspen', *Physiol. Plant*, **16**, 255–268.

Farrington, J. A., Ebert, M., Land, E. J., and Fletcher, K. (1973). 'Bipyridylium quaternary salts and related compounds V. Pulse radiolysis studies on the reaction of paraquat radical with oxygen. Implications for the mode of action of bipyridyl herbicides', *Biochim. Biophys. Acta.*, **314**, 372–381.

Fedtke, C., and Schmidt, R. R. (1977). 'Chlorfenprop-methyl: its hydrolysis *in vivo* and *in vitro* and a new principle for selective herbicidal action', *Weed Res.*, **17**, 233–239.

Feierabend, J., and Schubert, B. (1978). 'Comparative investigation of the action of several chlorosis-inducing herbicides on the biogenesis of chloroplasts and leaf microbodies', *Pl. Physiol.*, **61**, 1017–1022.

Frosch, S., Jabben, M., Bergfeld, R., Kleinig, H., and Mohr, H. (1979). 'Inhibition of carotenoid biosynthesis by the herbicide SAN 9789 and its consequences for the action of phytochrome on plastogenesis', *Planta*, **145**, 497–505.

Garraway, J. L., and Wain, R. L. (1976). 'The design of auxin-type herbicides', *Drug Design*, **7**, 113–164.

Gast, G. (1958). 'Uber pflanzenwachstumsregulatoren', *Experientia*, **14**, 134–136.

Gentner, W. A. (1966). 'The influence of EPTC on external foliage wax deposition', *Weeds*, **14**, 27–31.

Good, N. E., (1960). 'Activation of the Hill reaction by amines', *Biochim. Biophys. Acta.*, **40**, 502–517.

Govindjee, and van Rensen, J. J. S. (1978). 'Bicarbonate effects on the electron flow in isolated broken chloroplasts', *Biochim. Biophys. Acta.*, **505**, 183–213.

Graham-Bryce, I. J. (1981). 'The current status and future potential of chemical approaches to crop protection', *Phil. Trans. Roy. Soc. B.*, **295**, 5–16.

Gresshoff, P. M. (1979). 'Growth inhibition by glyphosate and reversal of its action by phenylalanine and tyrosine', *Aust. J. Pl. Physiol.*, **6**, 177–185.

Griffiths, M., Sistrom, W. R., Cohen-Bazine, G., and Stainer, R. Y. (1955). 'Function of carotenoids in photosynthesis', *Nature*, **176**, 1211–1214.

Grumbach, K. H. (1981). 'Effect of SAN 6706, amitrole, and photosystem II herbicides on chloroplast development', in *Photosynthesis*, VI (ed. G. Akoyunoglou), pp. 625–636. Balaban International Science Service, Philadelphia.

Haderlie, L. C., Widholm, J. M., and Slife, F. W. (1977). 'Effect of glyphosate on carrot and tobacco cells', *Pl. Physiol.*, **60**, 40–43.

Hager, A., Menzel, H., and Krauss, A. (1971). 'Experiments and hypothesis concerning the primary action of auxin in elongation growth', *Planta*, **100**, 47–75.

Hanson, J. B., and Slife, F. W. (1969). 'Role of RNA metabolism in the action of auxin-herbicides', *Residue Rev.*, **25**, 59–67.

Hardin, J. W., Cherry, J. H., Moore, D. J., and Lembi, C. A. (1972). 'Enhancement of RNA polymerase activity by a factor released by auxin from plasmamembrane', *Proc. Natl. Acad. Sci. USA*, **69**, 3146–3150.

Harper, D. B., and Harvey, B. M. R. (1978). 'Mechanism of paraquat tolerance in perennial rye grass. II. Role of superoxide dismutase, catalase and peroxidase', *Plant Cell Environ.*, **I**, 211–215.

Harris, N., and Dodge, A. D. (1972a). 'The effect of paraquat on flax cotyledons: Changes in fine structure', *Planta*, **104**, 201–209.

Harris, N., and Dodge, A. D. (1972b). 'The effect of paraquat on flax cotyledons: physiological and biochemical changes', *Planta*, **104**, 210–219.

Harvey, B. M. R., and Fraser, T. W. (1980). 'Paraquat tolerant and susceptible perennial ryegrass: effects of paraquat treatment on carbon dioxide uptake and ultrastructure of photosynthetic cells', *Plant Cell Environ.*, **3**, 107–117.

Harvey, B. M. R., Muldoon, J., and Harper, D. B. (1978). 'Mechanism of paraquat tolerance in perennial ryegrass. I. Uptake, metabolism and translocation of paraquat', *Plant, Cell Environ.*, **1**, 203–209.

Hepler, P. K., and Jackson, W. T. (1969). 'Isopropyl-*N*-phenyl carbonate affects spindle microtubule orientation in dividing endosperm cells of *Haemanthus katherinae*', *J. Cell Sci.*, **5**, 727–743.

Herstel, C., Quader, H., Robinson, D. G., and Marmé, D. (1980). 'Anti-microtubular herbicides and fungicides affect Ca^{2+} transport in plant mitochondria', *Planta*, **149**, 336–340.

Hill, B. D., Stobbe, E. H., and Jones, B. L. (1978). 'Hydrolysis of the herbicide benzoylprop ethyl by wild oat esterase', *Weed Res.*, **18**, 149–154.

Hoagland, R. E. (1980). 'Effect of glyphosate on metabolism of phenolic compounds VI. Effects of glyphosine and glyphosate metabolites on phenylalanine ammonia-lyase activity, growth, and protein, chlorophyll and anthocyanin levels in soybean seedlings', *Weed Sci.*, **28**, 393–400.

Hoagland, R. E., Duke, S. O., and Elmore, D. (1978). 'Effects of glyphosate on metabolism of phenolic compounds. II. Influence on soluble hydroxyphenolic compound, free amino acid and soluble protein levels in dark-grown maize roots', *Pl. Sci. Lett.*, **13**, 291–299.

Hoagland, R. E., Duke, S. O., and Elmore, C. D. (1979). 'Effects of glyphosate on metabolism of phenolic compounds. III. Phenylalanine ammonia-lyase activity, free amino acids, soluble protein and hydroxyphenolic compound in axes of soybeans', *Physiol. Plant.*, **46**, 357–366.

Hoagland, R. E., and Duke, S. O. (1982). 'Biochemical effects of glyphosate', in *Biochemical Responses Induced by Herbicides* (eds D. E. Moreland, J. B. St. John, and F. D. Hess). A.C.S. Symposia Series, No. 181, pp. 175–205.

Holländer, H., and Amrhein, N. (1980). 'The site of the inhibition of the shikimate pathway by glyphosate. I. Inhibition by glyphosate of phenylpropanoid synthesis in buckwheat', *Pl. Physiol.*, **66**, 823–829.

Izawa, S., and Good, N. E. (1965). 'The number of sites sensitive to DCMU, CMU and Atrazine', *Biochim. Biophys. Acta.*, **102**, 20–38.

Jaworski, E. G. (1969). 'Analysis of the mode of action of herbicidal α-chloroacetamides', *J. Agr. Food Chem.*, **17**, 165–170.

Jaworski, E. G. (1972). 'Mode of action of *N*-phosphonomethyl glycine: inhibition of aromatic amino acid biosynthesis', *J. Agr. Food Chem.*, **20**, 1195–1198.

Jeffcoat, B., and Harries, W. N. (1973). 'Selectivity and mode of action of ethyl-2-(*N*-benzoyl-3,4-dichloroanilino) propionate in the control of *Avena fatua* in cereals', *Pestic. Sci.*, **4**, 891–899.

Jeffcoat, B., and Harries, W. N. (1975). 'Selectivity and mode of action of flamprop-isopropyl in the control of *Avena fatua* in barley', *Pestic. Sci.*, **6**, 283–296.

Jeffcoat, B., Harries, W. N., and Thomas, D. B. (1977). 'Factors favouring the choice of flamprop-methyl for the control of *Avena* species in wheat', *Pestic. Sci.*, **8**, 1–12.

Juniper, B. E. (1959). 'The effect of pre-emergent treatment of peas with trichloroacetic acid in the sub-microscopic structure of the leaf surface', *New Phytol.*, **58**, 1–4.

Kerr, M. W., and Wain, R. L. (1964). 'The uncoupling of oxidation and phosphorylation in pea shoot mitochondria by ioxynil', *Ann. App. Biol.*, **54**, 441–446.

Key, J. L., Lin, C. Y., Gifford, E. M., and Dengler, R. (1966). 'Relation of 2,4-D-induced growth aberrations to changes in nucleic acid metabolism in soybean seedlings', *Bot. Gaz.*, **127**, 87–94.

Khan, M-U, Lem, N. W., Chandorkar, K. R., and Williams, J. P. (1979). 'Effects of substituted pyridazinones (San 6706, San 9774, San 9785) on glycerolipids and their associated fatty acids in the leaves of *Vicia faba* and *Hordeum vulgare*', *Pl. Physiol.*, **64**, 300–305.

Khanna, R., Pfister, K., Keresztes, A., Van Rensen, J. J. S., and Govindjee. (1981). 'Evidence for a close spatial location of the binding sites for CO_2 and for photosystem II inhibitors', *Biochim. Biophys. Acta.*, **634**, 105–116.

Kolattukudy, P., and Brown, L. (1974). 'Inhibition of cuticular lipid biosynthesis in *Pisum sativum* by thiocarbamates', *Pl. Physiol.*, **53**, 903–906.

Kolattukudy, P. E., Croteau, R., and Brown, L. (1974). 'Structure and biosynthesis of cuticular lipids', *Pl. Physiol.*, **54**, 670–677.

Krinsky, N. I. (1979). 'Carotenoid protection against oxidation', *Pure and Appl. Chem.*, **51**, 649–660.

Kümmel, H. W., and Grimme, L. H. (1975). 'The inhibition of carotenoid biosynthesis in green algae by SANDOZ 6706: Accumulation of phytoene and phytofluene in *Chlorella fusca*', *Z. Naturforsch.*, **30c**, 333–336.

Kunert, K.-J., and Böger, P. (1979). 'Influence of bleaching herbicides on chlorophyll and carotenoids', *Z. Naturforsch.*, **34c**, 1047–1051.

Kunert, K. J., and Böger, P. (1981). 'The bleaching effect of the diphenyl ether oxyfluorfen', *Weed Sci.*, **29**, 169–173.

Lichtenthaler, H. K., and Kleudgen, H. K. (1977). 'Effect of the herbicide San 6706 on biosynthesis of photosynthetic pigments and prenylquinones in *Raphanus* and in *Hordeum* seedlings', *Z. Naturforsch.*, **32c**, 236–240.

Mann, J. D., Jordan, L. S., and Day, B. E. (1965). 'A survey of herbicides for their effect upon protein synthesis', *Pl. Physiol.*, **40**, 840–843.

Marcum, J. M., Dedman, J. R., Brinkley, B. R., and Means, A. R. (1978). 'Control of microtubule assembly-disassembly by calcium-dependent regulator protein', *Proc. Natl. Acad. Sci. USA*, **75**, 3771–3775.

Margolis, R. L., and Wilson, L. (1977). 'Addition of colchicine-tubulin complex to microtubule ends: the mechanism of substoichiometric colchicine poisoning', *Proc. Natl. Acad. Sci. USA*, **74**, 3466–3470.

Matsunaka, S. (1969). 'Acceptor of light energy in photoactivation of diphenylether herbicides', *J. Agr. Food Chem.*, **17**, 171–175.

McCord, J. M. (1979). 'Superoxide, superoxide dismutase and oxygen toxicity', in *Reviews in Biochemical Toxicology* (ed. E. Hodgson, J. R. Bend, and R. M. Philpot), Vol. 1, pp. 109–124, Elsevier/North Holland, New York.

McLoad, R. D., and Thompson, A. (1981). 'Cell proliferation during the development of lateral root primordia', in *Structure and Function of Plant Roots* (eds R. Brouner, O. Gasparikova, J. Kolek, and B. C. Loughman), pp. 35–42. Martinus Nijhoff, The Hague.

Mees, G. C. (1960). 'Experiments on the herbicidal action of 1,1′-ethylene-2,2′-dipyridylium dibromide', *Ann. Appl. Biol.*, **48**, 601–612.

Mellor, R. S., and Salisbury, F. B. (1965). 'Influence of light regime on the toxicity and physiological activity of herbicides', *Pl. Physiol.*, **40**, 506–512.

Minshall, W. H. (1957). 'Influence of light on the effect of 3-*p*-(chlorophenyl)-1-1-dimethyl urea on plants', *Weeds*, **5**, 29–33.

Mitchell, P. (1966). 'Chemiosmotic coupling in oxidative and photosynthetic phosphorylation', *Biol. Rev.*, **41**, 445–502.

Moreland, D. E. (1981). 'Interaction of perfluidone with mitochondrial, thylakoid, and liposome membranes', *Pestic. Biochem. Physiol.*, **15**, 21–31.

Moreland, D. E., Hussey, G. G., and Farmer, F. S. (1974). 'Comparative effects of dichlobenil and its phenolic alteration products on photo- and oxidative phosphorylation', *Pestic. Biochem. Physiol.*, **4**, 356–364.

Moreland, D. E., Malhotra, S. S., Gruenhagen, R. D., and Shokraii, E. H. (1969). 'Effects of herbicides on RNA and protein synthesis', *Weed Sci.*, **17**, 556–563.

Murphy, G. T. P. (1979). 'Plant hormone receptors: comparison of NAA binding by maize extracts and by a non-plant protein', *Pl. Sci. Lett.*, **15**, 183–191.

Oettmeier, W., Masson, K., and Johnningmeir, U. (1982). 'Evidence for two different herbicide-binding proteins at the reducing side of photosystem II', *Biochim. Biophys. Acta.*, **679**, 376–383.

Orr, G. L., and Hess, F. D. (1981). 'Characterization of herbicidal injury by acifluorfen-methyl in excised cucumber (*Cucumis sativus*) cotyledons', *Pestic. Biochem. Physiol.*, **16**, 171–178.

Orr, G. L., and Hess, F. D. (1982a). 'Mechanism of action of the diphenyl ether herbicide acifluorfen-methyl in excised cucumber (*Cucumis sativus*) cotyledons', *Pl. Physiol.*, **69**, 502–507.

Orr, G. L., and Hess, F. D. (1982b). 'Proposed site(s) of action of new diphenyl ether herbicides', in *Biochemical Responses Induced by Herbicides* (eds D. E. Moreland, J. B. St. John, and F. D. Hess). A.C.S. Symposia Series. No. 181, pp. 131–152.

Pallett, K. E., and Caseley, J. C. (1980). 'Differential inhibition of DNA synthesis in difenzoquat tolerant and susceptible United Kingdom Spring wheat cultivars', *Pestic. Biochem. Physiol.*, **14**, 144–152.

Pallett, K. E., and Dodge, A. D. (1979a). 'Sites of action of photosynthetic inhibitor herbicides: experiments with trypsinated chloroplasts', *Pestic. Sci.*, **10**, 216–220.

Pallett, K. E., and Dodge, A. D. (1979b). 'The role of light and oxygen in the action of the photosynthetic inhibitor herbicide monuron', *Z. Naturforsch.*, **34c**, 1058–1061.

Pallett, K. E., and Dodge, A. D. (1980). 'Studies into the action of some photosynthetic inhibitor herbicides', *J. Exp. Bot.*, **31**, 1051–1066.

Palmer, J. M. (1979). 'The uniqueness of plant mitochondria', *Biochem. Soc. Trans.*, **7**, 246–252.

Percival, M., and Dodge, A. D. (1983). 'Photodynamic effects of Rose Bengal on senescent flax cotyledons', *J. Exp. Bot.* **34**, 47–54.

Pfister, K., and Arntzen, C. J. (1979). 'The mode of action of photosystem II specific inhibitors in herbicide-resistant weed biotypes', *Z. Naturforsch.*, **34c**, 996–1009.

Pfister, K., Steinbeck, K. E., Gardner, G., and Arntzen, C. J. (1981). 'Photoaffinity labelling of an herbicide receptor protein in chloroplast membranes', *Proc. Natl. Acad. Sci. USA*, **78**, 981–985.

Pillai, P., Davis, D. E., and Truelove, B. (1979). 'Effects of metolachlor on germination, growth, leucine uptake and protein synthesis', *Weed Sci.*, **27**, 634–637.

Pillmoor, J. B., and Gaunt, J. K. (1981). 'The behaviour and mode of action of the phenoxy-acetic acids in plants', *Prog. Pestic. Biochem.*, **1**, 147–218.

Plowman, R. E., Stonebridge, W. C., and Hawtree, J. N. (1980). 'Fluazifop butyl—a new selective herbicide for the control of annual and perennial grass weeds', *Proc. Brit. Crop Prot. Conf. Weeds*, 29–37.

Ray, P. M. (1974). 'The biochemistry of the action of indoleacetic acid on plant growth', *Recent Adv. Phytochem.*, **7**, 93–123.

Ray. T. B. (1980). 'Studies on the mode of action of DPX-4189', *Proc. Brit. Crop. Prot. Conf. Weeds*, 7–14.

Ray, T. B. (1982). 'The mode of action of chlorsulfuron: a new herbicide for cereals', *Pestic. Biochem. Physiol.*, **17**, 10–17.

Renger, G. (1976). 'Studies on the structural and functional organisation of system II of photosynthesis. The use of trypsin as a structurally selective inhibitor at the outer surface of the thylakoid membrane', *Biochim. Biophys. Acta.*, **440**, 287–300.

Retzlaff, G., and Hamm, R. (1976). 'The relationship between CO_2 accumulation and the metabolism of bentazone in wheat plants', *Weed Res.*, **16**, 263–266.

Ridley, S. M. (1977). 'Interaction of chloroplasts with inhibitors: induction of chlorosis by diuron during prolonged illumination *in vitro*', *Pl. Physiol.*, **59**, 724–732.

Ridley, S. M. (1982). 'Carotenoids and herbicide action', in 'Carotenoid Chemistry and Biochemistry' (eds. G. Britton and T. W. Goodwin) pp. 353–369. Pergamon Press, Oxford.

Riely, C. A., Cohen, G., and Lieberman, M. (1974). 'Ethane evolution: a new index of lipid peroxidation', *Science*, **183**, 208–210.

Roisch, H. U., and Lingens, F. (1974). 'Effect of the herbicide *N*-(phosphonomethyl) glycine on the biosynthesis of aromatic amino acids', *Angew. Chem.* (Internat. Edit.), **13**, 400.

Rost, T. L., and Bayer, D. E. (1976). 'Cell cycle population kinetics of pea root tip meristems treated with propham', *Weed Sci.*, **24**, 81–87.

Sandmann, G., Bramley, P. M., and Böger, P. (1980). 'The inhibitory mode of action of the pyridazinone herbicide, norflurazon on a cell-free carotenogenic enzyme system', *Pestic. Biochem. Physiol.*, **14**, 185–191.

Sandemann, G., Lambert, R., and Böger, P. (1981). 'Multifunctional mode of action of substituted nitrodiphenyl-ethers in *Scenedesmus* cells', *Z. Naturforsch.*, **36c**, 633–637.

Shimabukuro, M. A., Shimabukuro, R. H., Nord, W. S., and Hoerauf, R. A. (1978). 'Physiological effects of methyl 2-[4(2,4-dichlorophenoxy) phenoxy] propanoate on oat, wild, oat, and wheat', *Pestic. Biochem. Physiol.*, **8**, 199–207.

Shipman, L. L. (1981). 'Theoretical study of the binding site and mode of action for photosystem II herbicides', *J. theor. Biol.*, **20**, 123–148.

Shipman, L. L. (1982). 'Binding sites associated with inhibition of photosystem II', in *Biochemical Responses Induced by Herbicides* (eds D. E. Moreland, J. B. St. John, and F. D. Hess). A.C.S. Symposia Series No. 181, pp. 23–35.

Smith, A. E., and Secoy, D. M. (1976). 'Early chemical control of weeds in Europe', *Weed Sci.*, **24**, 594–597.

Stanger, C. E., and Appelby, A. P. (1972). 'A proposed mechanism for diuron-induced phytotoxicity', *Weed Sci.*, **20**, 357–363.

Steinbeck, K. E., Pfister, K., and Arntzen, C. J. (1982). 'Identification of the receptor site for triazine herbicides in chloroplast thylakoid membranes', in *Biochemical Responses Induced by Herbicides* (eds D. E. Moreland, J. B. St. John, and F. D. Hess). A.C.S. Symposia Series, No. 181, pp. 37–55.

Steinrücken, H. C., and Amrhein, N. (1980). 'The herbicide glyphosate is a potent inhibitor of 5-enol pyruvyl-shikimic acid-3 phosphate synthase', *Biochem. Biophys. Res. Commun.*, **94**, 1207–1212.

Sterrett, R. B., and Fretz, T. A. (1975). 'Asulam-induced mitotic irregularities in onion root-tips', *Hort. Sci.*, **10**, 161–162.

St. John, J. B. (1976). 'Manipulation of galactolipid fatty acid composition with substituted pyridazinones', *Pl. Physiol.*, **57**, 38–40.

St. John, J. B. (1982). 'Effects of herbicides on the lipid composition of plant membranes', in *Biochemical Responses Induced by Herbicides* (eds D. E. Moreland, J. B. St. John, and F. D. Hess). A.C.S. Symposia Series No. 181, pp. 97–109.

St. John, J. B., and Christiansen, M. N. (1976). 'Inhibition of linolenic acid synthesis and modification of chilling resistance in cotton seedlings', *Pl. Physiol.*, **57**, 257–259.

Strang, R. H., and Rogers, R. L. (1974). 'Behaviour and fate of two penylpyridazinone herbicides in cotton, corn and soybean', *J. Agr. Food Chem.*, **22**, 1119–1125.

Teissere, M., Penon, P., Van Huystee, R. B., Azou, Y., and Ricard, J. (1975). 'Hormonal control of transcription in higher plants', *Biochim. Biophys. Acta*, **402**, 391–402.

Teissere, M., Penon, P., Azou, Y., and Ricard, J. (1976). 'On the mode of action of transcription factors in higher plants', *Pl. Sci. Lett.*, **6**, 49–55.

Tischer, W., and Strotmann, H. (1977). 'Relationship between inhibitor binding by chloroplasts and inhibition of photosynthetic electron transport', *Biochim. Biophys. Acta.*, **460**, 113–125.

Trebst, A. (1974). 'Energy conservation in photosynthetic electron transport of chloroplasts', *Ann. Rev. Pl. Physiol.*, **25**, 423–458.

Trebst, A., and Draber, W. (1979). 'Structure activity correlations of recent herbicides in photosynthetic reactions', *Adv. Pestic. Sci.* (ed. H. Geissbühler), pp. 223–234, Pergamon, Oxford.

Turner, J. S., Stokes, D. M., and Gilmore, L. B. (1970). 'The effects of the dipyridyl diquat on the metabolism of *chlorella vulgaris*. I. Gas exchange in the light', *Aust. J. biol. Sci.*, **23**, 43–61.

Tymonko, J. M., and Foy, C. C. (1978). 'Inhibition of protein synthesis by glyphosate', *Pl. Physiol.*, **61**, Supp. p. 41.

Van Assche, C. J., and Carles, P. M. (1982). 'Photosystem II inhibiting chemicals', in *Biochemical Responses Induced by Herbicides* (eds D. E. Moreland, J. B. St. John, and F. D. Hess). A.C.S. Symposia Series No. 181, pp. 1–21.

Vanderhoef, L. (1981). 'Auxin-regulated elongation', *What's New in Pl. Physiol.*, **12**, 25–28.

Van Oorschot, J. L. P., and Van Leeuwen, P. H. (1974). 'Photosynthetic inhibition in *Phaseolus vulgaris* II. The influence of light on herbicidal action', *Weeds Res.*, **14**, 81–86.

Van Oorschot, J. L. P., and Van Leeuwen, P. H. (1979). 'Recovery from inhibition of photosynthesis by metamitron in various plant species', *Weed Res.* **19**, 63–67.

Van Rensen, J. J. S. (1975). 'Lipid peroxidation and chlorophyll destruction caused by diquat during photosynthesis in *Scenedesmus*', *Physiol. Plant.*, **33**, 42–46.

Vanstone, D. E., and Stobbe, E. M. (1979). 'Light requirement of the diphenylether herbicide oxyfluoren', *Weed Sci.*, **27**, 88–90.

Venis, M. A. (1979). 'Receptors for plant growth regulators', *Adv. Pestic. Sci.* (ed. H. Geissbühler), pp. 487–493, Pergamon, Oxford.

Venis, M. A. (1982). '(L-flamprop-isopropyl). The rational design of an optically active wild oat herbicide' *Pestic. Sci.* **13**, 309–317.

Weeks, D. P., and Baxter, R. (1972). 'Specific inhibition of peptide-chain initiation by 2-(4-methyl-2,6-dinitroanilino)-*N*-methylpropionamide', *Biochemistry*, **11**, 3060–3064.

Wergin, W. P., and Potter, J. R. (1975). 'The effects of fluometuron on the ultrastructural development, chlorophyll accumulation and photosynthetic competence in developing velvet leaf seedlings', *Pestic. Biochem. Physiol.*, **5**, 265–279.

Wessels, J. S. C., and Van der Veen, R. (1956). 'The action of some derivatives of phenylurethan and of 3-phenyl-1,1-dimethylurea on the Hill reaction', *Biochim. Biophys. Acta.*, **19**, 548–549.

Wilkinson, R. E., and Hardcastle, W. S. (1970). 'EPTC effects on total leaflet fatty acids and hydrocarbons', *Weed Sci.*, **18**, 125–128.

Wilkinson, R. E., and Smith, A. E. (1975). 'Thiocarbamate inhibition of fatty acid biosynthesis in isolated spinach chloroplasts', *Weed Sci.*, **23**, 100–104.

Wilkinson, R. E., Smith, A. E., and Michel, B. (1977). 'Alteration of soybean complex lipid biosynthesis by 5-ethyl dipropylthiocarbamate', *Pl. Physiol.*, **60**, 86–88.

Willemot, C. (1977). 'Simultaneous inhibition of linolenic acid synthesis in winter wheat roots and frost hardening by BASF 13-338, a derivative of pyridazinone'. *Pl. Physiol.*, **60**, 1–4.

Winterbourne, C. C. (1981). 'Production of hydroxyl radicals from paraquat radicals and H_2O_2', *FEBS Lett.*, **128**, 339–342.

Wojtaszek, T. (1966). 'Relationship between susceptibility of plants to DNBP and their capacity for ATP generation', *Weeds*, **14**, 125–129.

Wright, B. J., Baillie, A. C., Wright, K., Dowsett, J. R., and Sharpe, T. M. (1980). 'Synthesis of potential herbicides designed to uncouple photophosphorylation', *Phytochem.*, **19**, 61–65.

Youngman, R. J., and Dodge, A. D. (1979). 'Mechanism of paraquat action: inhibition of the herbicidal effect by a copper chelate with superoxide dismutating activity', *Z. Naturforsch.*, **34c**, 1032–1035.

Youngman, R. J., and Dodge, A. D. (1981). 'On the mechanism of paraquat resistance in *Conyza* sp.', in *Photosynthesis* VI (ed. G. Akoyunoglou), pp. 537–543. Balaban International Science Services, Philadelphia.

Youngman, R. J., Dodge, A. D., Lengfelder, E., and Elstner, E. F. (1979). 'Inhibition of paraquat phytoxicity by a novel copper chelate with superoxide dismutating activity', *Experientia*, **35**, 1295–1296.

Youngman, R. J., and Elstner, E. F. (1981). 'Oxygen species in paraquat toxicity: The Crypto-OH radical', *FEBS Lett.*, **129**, 265–268.

Youngman, R. J., Pallett, K. E., and Dodge, A. D. (1980). 'Active oxygen species in herbicide action', in *Chemical and Biochemical Aspects of Superoxide and Superoxide Dismutase* (ed. J. V. Bannister and H. A. O. Hill), pp. 402–411, Elsevier/North-Holland, New York.

Zurfluh, L. L., and Guilfoyle, T. J. (1980). 'Auxin-induced changes in the patterns of protein synthesis in soybean hypocotyl', *Proc. Natl. Acad. Sci. USA*, **77**, 357–361.

Zweig, A., and Nachtigall, G. W. (1975). 'Photosensitized herbicidal action', *Photochem. Photobiol.*, **22**, 257–259.

Zweig, G., Shavit, N., and Avron, M. (1965). 'Diquat in photoreactions of isolated chloroplasts', *Biochim. Biophys. Acta.*, **109**, 332–346.

Progress in Pesticide Biochemistry and Toxicology, Volume 3
Edited by D. H. Hutson and T. R. Roberts

CHAPTER 4

Oxidation of xenobiotics in plants

D. Cole

INTRODUCTION

A knowledge of pesticide metabolism in plants is desirable from several viewpoints. Of primary importance is the problem of persistence and residues in crops, but metabolism may also provide a mechanism for selectivity of herbicides and the activation/deactivation of systemic pesticides. From a purely biological point of view, there is much speculation on the relationship of xenobiotic metabolism to the normal metabolism of the plant, particularly with regard to enzymology. Such studies give an insight into the way in which plants may deal with their own toxic waste products.

The capacity of plants to metabolize foreign compounds appears to be more limited than that of other organisms, the evolution of such abilities in animals having possibly resulted from the encounter of secondary plant products in the diet. The stability of aromatic rings is in marked contrast to the degradative mechanisms available to micro-organisms, though other types of degradative oxidation can be of great importance, e.g. in the activation of the phenoxyalkanoic acids. Oxygenation is a common process, shared with other organisms and is an important initial mode of attack when organisms encounter what are often highly lipophilic compounds; it serves to increase water solubility, with the production of phenols and alcohols and provides an opportunity for conjugation via glycosidic bond formation. In animals, such mechanisms are important in facilitating the elimination of xenobiotics from the body; in the case of plants, residues are more persistent and effective removal from the cell may include sequestering via secretion into the vacuole or association with cell wall material. Little is known about these processes.

Structural modification and possible subsequent conjugation of xenobiotics will invariably alter their biological properties; this contrasts with the linkage of the intact compound with natural plant constituents, e.g. sugars and amino acids, which result in conjugates forming a pool from which the parent may be released (Frear *et al.*, 1978). Comprehensive studies on oxidative transformations for any given xenobiotic in plants are confined to relatively few examples, which is somewhat surprising in view of the fact that ingestion of pesticide residues in food is a major pathway of exposure to foreign compounds in man. The accumulating knowledge does, however, point to considerable similarities with those of other organisms. The various mechanisms of oxidation provide the major means of xenobiotic transformation in plants, though other forms of attack, particularly hydrolysis, can be important. Although the subject is approached here by consecutive consideration of major types of oxidation, it is recognized that attack on a xenobiotic will often involve a combination of mechanisms. In some cases, this results in a rather complex pattern of metabolites, although one pathway is usually dominant. Photoconversion can also be important in determining residue patterns, such as for those of the synthetic pyrethroid and cyclodiene insecticides.

Table 1 Metabolism of xenobiotics in plants: a summary of reviews

Class	References
Pesticides	Casida and Lykken, 1969; Klein, 1972; Matsunaka, 1972; Crosby, 1973; Menzer, 1973; Menn and Still, 1977.
Herbicides	Frear *et al.*, 1972; Kearney and Kaufman, 1975; Naylor, 1976; Schütte, 1977.
Insecticides	
general	O'Brien, 1967; Fukuto and Metcalf, 1969; Menn, 1972.
organochlorine	Brooks, 1974.
carbamate	Schlagbauer and Schlagbauer, 1972; Kuhr and Dorough, 1976.
organophosphorus	Frehse, 1971; Miyamoto, 1972; Bull, 1972a; b; Menn and McBain, 1974; Eto, 1974.
Xenobiotics (including pesticides)	Sandermann *et al.*, 1977; Baldwin, 1977; Lamoureux and Frear, 1979.

The function of this review is to consider the most important classes of oxidation together with the compounds most commonly attacked. An attempt to exhaustively catalogue transformations is not made. Thus more recent examples are discussed at the expense of earlier ones; references to these may be found in previous reviews. Where relevant, mention is made of the importance of oxidation in detoxication and activation, together with what is currently known about the enzymology of these transformations. Both the general metabolism of pesticides and xenobiotics in plants and that of specific classes has been extensively reviewed (Table 1).

β-OXIDATION

The selectivity of herbicides is usually a function of several independent factors among which differential metabolism may contribute to varying degrees. An outstanding example of selectivity based primarily upon differential metabolism is the selective capacity of plants to degrade initially inert substituted phenoxyalkanoic acids to the herbicidal acetic acid analogues. This utilizes the well-known *β*-oxidase system whereby organisms degrade fatty acids by the successive removal of two-carbon fragments. The basis of selectivity is that in some plants, substitution of the phenoxy ring can attenuate the capacity to degrade the alkanoic chain, thus conferring tolerance, as opposed to susceptible plants which carry out oxidation. Thus differences among the *β*-oxidase systems of plants led to the first example of herbicide selectivity by design. Alkanoic acid degradation has been the subject of several reviews (Robertson and Kirkwood, 1970; Loos, 1975; Wain and Smith, 1976); thus it is intended here only to cover some essential features.

The susceptibility of plants to some higher alkanoic derivatives of auxin analogues was initially demonstrated by Grace (1939) and Synerholm and Zimmerman (1947), who found that activity alternated with increase in length of the alkanoic chain, molecules having an odd number of methylene units being active, and those with an even number being inactive. These observations were extensively developed by Wain's group who showed that the metabolic basis of alternation was the ability of plants to *β*-oxidize odd-numbered chains to the active acetic acids whereas even numbered chains gave rise to the inactive phenol (Fawcett *et al.*, 1954; 1958) (Figure 1). The influence of ring substitution on side chain degradation was discovered by Wain and Wightman (1954). In auxin assays with wheat (*Triticum aestivum* L.), pea (*Pisum sativum* L.) and tomato (*Lycopersicon esculentum* Mill.), higher odd-numbered analogues of 4-chlorophenoxyacetic acid were active in all species, whereas those of 2,4,5-trichlorophenoxyacetic acid were active only in wheat, hence the additional ring substitution appeared to be inhibitory to the *β*-oxidase system of pea and tomato, but not to wheat. These studies were extended to 2,4-D homologues by Fawcett *et al.* (1958). The commercial legacy of this principle is the use of the butyric

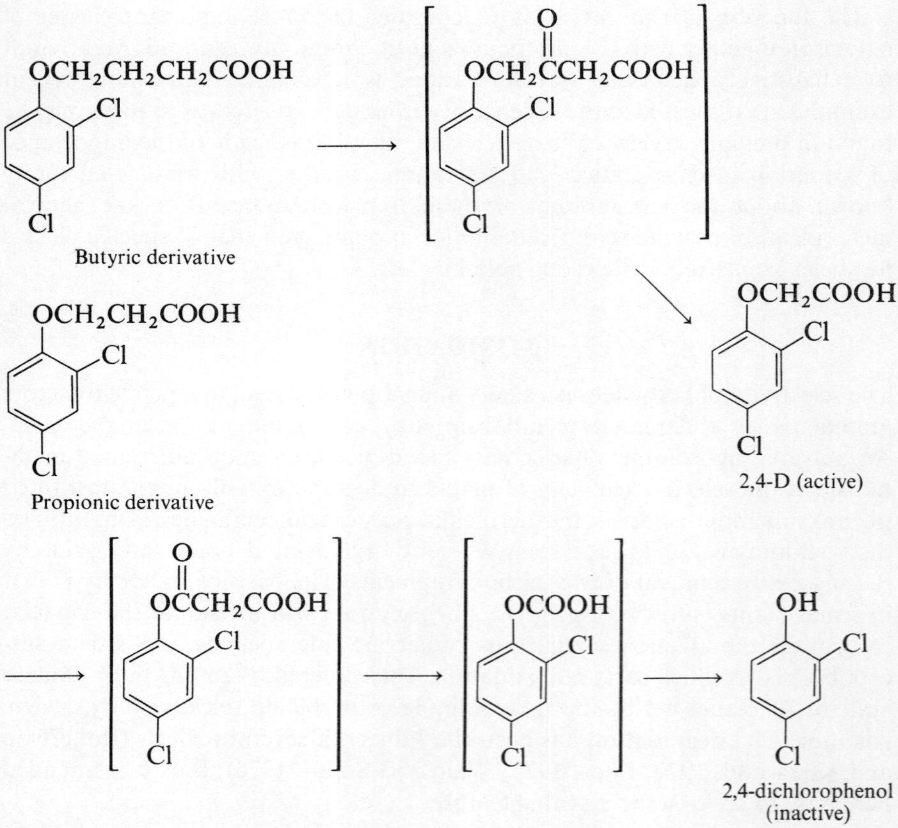

Figure 1 β-Oxidation of higher alkanoic homologues of 2,4-D. (From Wain and Smith, 1976; reproduced by permission of Academic Press)

acid analogues 2,4-DB and MCPB to which some weeds are susceptible whilst some crops, notably legumes, are tolerant (Wain, 1955a; b).

Subsequent evidence, however, suggests that β-oxidation may not be the sole mechanism of selectivity. Selectivity in the resistant legumes alfalfa (*Medicago sativa* L.) and red clover (*Trifolium pratense* L.) and susceptible *Amaranthus retroflexus* L. and *Xanthium pensylvanicum* Wallr. could be ascribed largely to differential uptake of 2,4-DB. Also, both tolerant and susceptible species degraded this with equal facility, but the tolerant legumes metabolized the resultant 2,4-D more rapidly (Hawf and Behrens, 1974). Linscott *et al.* (1968) had previously shown that alfalfa could metabolize 2,4-DB by alkanoic extension, in addition to β-oxidation. Cell suspension cultures of white clover (*Trifolium repens* L.) were able to metabolize both 2,4-DB and 2,4,5-TB rapidly (Smith and Oswald, 1979), but the major product of 2,4-DB was not 2,4-D, but a conjugate which upon hydrolysis yielded an aglycone slightly more polar than 2,4-DB and postulated to be a ring hydroxylated product of this (Smith, 1979). Amino acid conjugates of 2,4-DB also occurred to a lesser extent. Tolerance in white clover may thus be conferred by an ability to rapidly detoxify 2,4-DB via hydroxylation, rather than by an innate inability to carry out β-oxidation. Further oxidative degradation of the acetic acid side chain can occur which is in many plants a minor pathway but in some can be a major degradatory mechanism. The tolerance of red currant (*Ribes sativum* Syme) to 2,4-D compared with the susceptible black currant (*Ribes nigrum* L.) was associated with an ability of leaves to carry out side chain degradation, which was also observed in leaves of the tolerant Cox apple (*Malus sylvestris* Mill.) as opposed to that of the susceptible Bramley apple (Luckwill and Lloyd-Jones, 1960a; b). Radioactivity from side chain labelled 2,4-D liberated in red currant appeared largely as glyoxylic acid and it was suggested that oxidation resulted in the removal of a C_2 unit (Fleeker, 1973). Side chain degradation can be prevented by methylation of the α-carbon, e.g. to give mecoprop from MCPA which is herbicidal to some weeds tolerant to MCPA, e.g. *Stellaria media* (L) Vill. and *Galium aparine* L. (Wain and Smith, 1976).

RING HYDROXYLATION

Phenoxyacetic acids

The metabolism of the phenoxyacetic acids, notably that of 2,4-D, has received wide attention. It exhibits a diverse pattern across a variety of species in which ester, amide and ether conjugates are produced. Metabolites vary in phytoxicity, but generally, hydroxylation with the subsequent formation of phenolic glucosides may be regarded as a mechanism of detoxication, whereas ester and amide conjugates may hydrolyse, after transport to other sites, yielding the parent and thus retain auxin activity. For example an auxin-requiring soybean (*Glycine max.* (L.) Merr.) root callus converted 2,4-D principally into amino acid conjugates, but upon depletion of 2,4-D in the medium the concentration of conjugates declined with a concomitant rise in the pool of free 2,4-D (Davidonis *et al.*, 1980). Differential metabolism is not, however, thought to be of primary importance in the selectivity of the phenoxyacetic acids. It is intended here to specifically cover ring hydroxylation of phenoxyacetic acids; metabolism in general has been reviewed by Loos (1975) and certain aspects by Pillmoor and Gaunt (1981).

Metabolism of the phenoxyacetic acids is strongly influenced by ring substitution and the non-herbicidal parent, phenoxyacetic acid (POA) is most readily attacked. Hydroxylation at the 4-position is the favoured modification for all derivatives and for POA is a common step in wheat (Fawcett *et al.*, 1959), pea (Fawcett *et al.*, 1959; Hutber *et al.*, 1978), barley (*Hordeum vulgare* L.) and maize

(*Zea mays* L.) (Wilcox *et al.*, 1963), oat (*Avena sativa* L.) (Thomas *et al.*, 1963; Hutber *et al.*, 1978) and potato (*Solanum tuberosum* L.) tuber (Cole and Loughman, 1982). Fawcett *et al.* (1959) showed that pea and wheat 4-hydroxylated higher unsubstituted phenoxyalkanoic acids, whereas in oat, barley and maize roots, odd methylene numbered phenoxyalkanoic acids yielded 4-hydroxyphenoxyacetic acid (4-OH-POA) as the only hydroxylated product (Wilcox *et al.*, 1963). The major end product of POA metabolism is the phenolic β-glucoside of 4-OH-POA (Thomas, 1964; Hutber *et al.*, 1978; Cole and Loughman, 1982). Other products are chromatographically immobile material and a minor unknown. Two other minor compounds are produced in potato tuber slices which are more polar than the glucoside. Kinetic evidence suggests that in potato slices the formation of immobile material, the three unknowns and glucoside are all competing reactions for 4-OH-POA. Levels of immobile material and unknowns reach a maximum level at 4 h, thereafter declining. When labelled 4-OH-POA itself was fed to slices, all products observed from POA were found but immobile material became the major sink for radioactivity and the glucoside remained a minor metabolite. Immobile material was also the major product when 4-OH-POA was fed to excised oat coleoptiles and pea buds (Hutber *et al.*, 1978). It is thus likely that an enzyme-bound form of 4-OH-POA is a better substrate for glucosylation than free 4-OH-POA. Since xenobiotics which become hydroxylated are normally rapidly glycosylated it is reasonable to assume that the two enzymes may be in intimate association.

The immobilization of a fraction of POA metabolites, possibly with cell wall material suggests that POA may be metabolized by enzymes normally associated with the synthesis of lignin. Observations on the hydroxylation of POA in this laboratory go some way to supporting this, one of which is that the rate of hydroxylation in potato slices may be enhanced by ageing in water. This is to some extent analogous to the increase in the amount of extractable cinnamic acid hydroxylase found in aged potato tissues (Camm and Towers, 1973a; Rich and Lamb, 1977), although since an increase in activity is evident at 4 h, with a doubling of activity at 8 h, it is unlikely that new enzyme synthesis is induced; an alternative explanation may be the leaching out of an inhibitor into the medium. Potato tuber tissues also vary in their ability to hydroxylate POA. Slices cut from the narrow vascular band are more active than those from the outer cortex and inner parenchyma which it delineates. The central pith is least active. Enzymes involved in lignin synthesis have often been found to be particularly associated with lignification (Camm and Towers, 1973b; Rhodes and Wooltorton, 1978). POA hydroxylation was not enhanced by light, unlike cinnamic acid hydroxylase (Benveniste *et al.*, 1978), although subsequent metabolism was (Hutber *et al.*, 1978). Recent work in this laboratory has shown that cell cultures of bean (*Phaseolus vulgaris* L.) produce similar metabolites from POA to those seen in tissue from the intact plant, although important quantitative differences are evident. The alga *Chlorella protothecoides* Krüger failed completely to metabolize POA, which appeared to be due to an inability to absorb the compound.

2,4-D undergoes hydroxylation in many species, but carboxylic acid conjugations are also often important. Hydroxylation occurs normally at the 4-position; thus the chlorine at this position becomes displaced and may either be eliminated or may undergo migration to the adjacent *ortho* positions via the 'NIH' shift mechanism for aryl hydroxylation (Guroff *et al.*, 1967) (Figure 2). The latter mechanism accounts for the major products of 2,4-D hydroxylation, namely 4-OH-2,5-D and to a lesser extent 4-OH-2,3-D. Exceptions to this generality were maize plants (Feung *et al.*, 1978) and callus (Feung *et al.*, 1975) and rice (*Oryza sativa* L.) callus (Feung *et al.*, 1976), in which these proportions were reversed. Other derivatives which have been identified are 3-OH-2,4-D (Feung *et al.*, 1975), 5-OH-2,4-D (Feung *et al.*, 1978), 6-OH-2,4-D (Bristol *et al.*, 1977), 4-OH-2-chlorophenoxyacetic acid (4-OH-2CPA) (Feung *et al.*, 1975) and 2-OH-4-CPA (Bristol *et al.*, 1977) (Figure 3). The production of 4-OH-2,5-D and 4-OH-2,3-D may be regarded as a mechanism of detoxication, since these lack auxin activity (Feung *et al.*, 1971; Hamilton *et al.*, 1971), however 3-OH-2,4-D retains auxin activity (Feung *et al.*, 1971).

Metabolic hydroxylation of 2,4-D was first suspected by Holley (1952) who found that bean converted 2,4-D into a water-soluble compound which yielded an ether-soluble acid upon hydrolysis. Similar results were obtained by Crosby (1964) and the bean metabolite was identified by Thomas *et al.* (1964a) as the *O*-*β*-glucoside of 4-OH-2,5-D, the glucoside of 4-OH-2,3-D being present to a lesser extent. Independent identification of 4-OH-2,5-D had also been made at the same time as a minor metabolite in *Aspergillus niger* (Faulkner and Woodcock, 1964). Hydroxylation in bean was also demonstrated by Montgomery *et al.* (1971), Hamilton *et al.* (1971) and Chkanikov *et al.* (1971) and also in a number of weed species (Fleeker and Steen, 1971).

Marked species differences exist towards the metabolism of 2,4-D. In common with bean, hydroxylation was the major mode of metabolism in cell suspension cultures of diploid wheat (*Triticum monococcum* L.) (Bristol *et al.*, 1977) and hexaploid wheat (Scheel and Sandermann, 1981a). In contrast, rice callus was a poor hydroxylating tissue, with the accumulation of a glycosyl ester of 2,4-D as the major metabolite (Feung *et al.*, 1976). Amino acid conjugates were the major metabolites in soybean cotyledon callus culture (Feung *et al.*, 1971; 1973) although the proportion of phenolic glucosides increased with time. Amino acid conjugates were also major metabolites in whole plants of soybean (Chkanikov *et al.*, 1972) and soybean cell suspension culture (Scheel and Sandermann, 1981a). Although soybean callus preferentially produced amino acid conjugates, notably 2,4-D-glutamic acid, callus fed directly with this substrate produced comparatively more phenolic glucosides than callus fed with 2,4-D, suggesting that the amide conjugate, or the 2,4-D released from it was a more direct

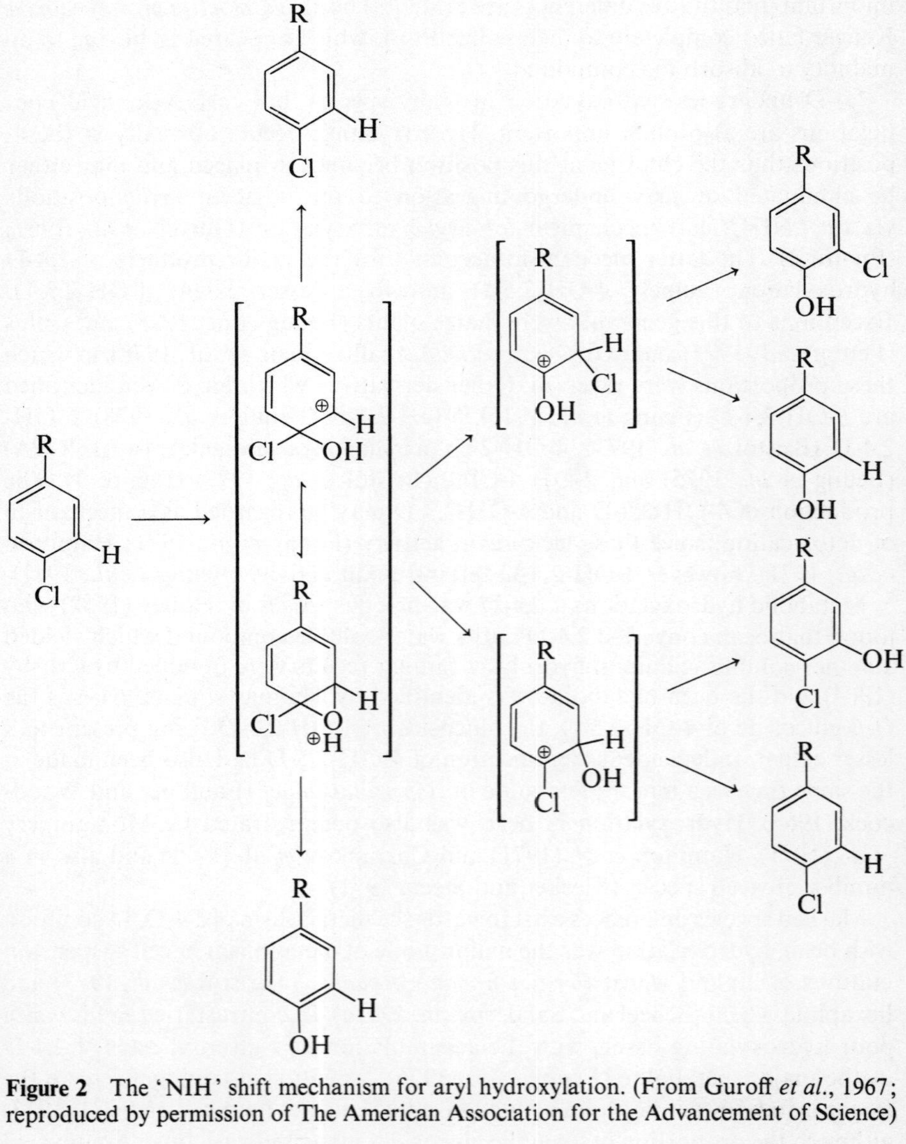

Figure 2 The 'NIH' shift mechanism for aryl hydroxylation. (From Guroff *et al.*, 1967; reproduced by permission of The American Association for the Advancement of Science)

substrate for hydroxylation (Feung *et al.*, 1973). Although no direct hydroxylation of the conjugate was found, it was later observed that hydrolysis of ether-soluble metabolites yielded both amino acids and hydroxylated derivatives (Feung *et al.*, 1978). Amino acid conjugates of phenolic glucosides were identified as major metabolites in cucumber (*Cucumis sativa* L.) plants (Chkanikov *et al.*, 1977).

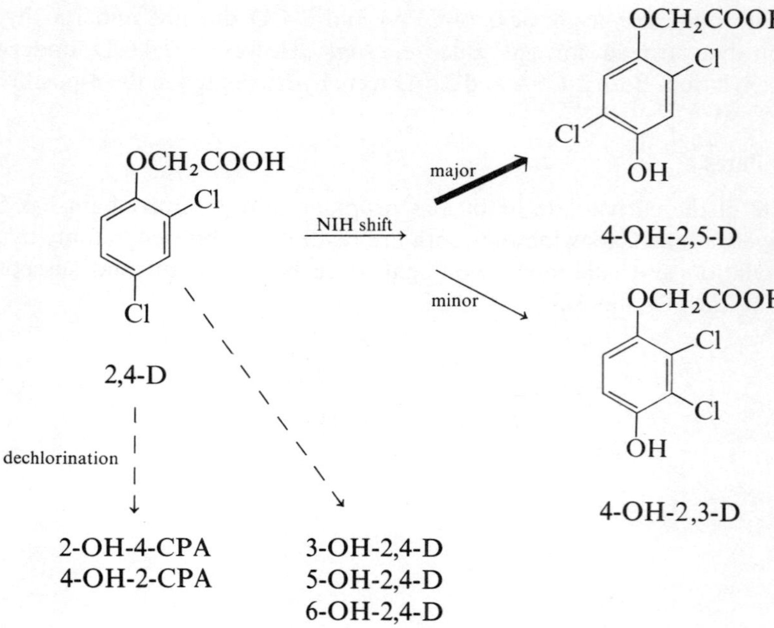

Figure 3 Products formed from the hydroxylation of 2,4-D

Quantitative differences may also exist in the metabolism of 2,4-D by plants and their derivative callus culture. Soybean callus metabolized 2,4-D more rapidly than plants, but plants produced relatively higher proportions of hydroxylated derivatives (Feung *et al.*, 1978). Variation in proportions of metabolites may also occur with time. In bean and maize, hydroxylated derivatives were initially dominant metabolites, but the production of 2,4-D conjugates increased with time (Montgomery *et al.*, 1971). Conversely, in soybean callus tissue, the level of phenolic glucosides increased with time relative to the level of amino acid conjugates (Feung *et al.*, 1972). In cucumber, a plant which metabolizes 2,4-D rapidly, 2,4-D-glucoside was the initial major metabolite, which declined after 3 h and was concurrent with an increase in the level of hydroxylated derivatives (Chkanikov *et al.*, 1977). It thus appears from these last two cases that the hydroxylated 2,4-D may be provided in part directly from amide and ester conjugates.

In contrast to 2,4-D, hydroxylation of 2,4,5-T is of minor importance, although only limited studies have been undertaken. Hamilton *et al.* (1971) identified 4-OH-2,5-D as a minor product in bean, and the major metabolites in soybean callus were amino acid conjugates (Arjmand *et al.*, 1978). Thomas *et al.* (1964b) investigated the ability of oat to hydroxylate a number of other chlorinated derivatives and found that compounds substituted at the 4-position, i.e. 4-chlorophenoxyacetic acid (4-CPA) and 2,4-D did not undergo hydroxylation but instead formed glucose esters. However, 2,4,6-D underwent 3-hydroxylation. Both 2-CPA and 2,6-D were hydroxylated in the 4-position.

Carbanilates

The fate of the carbanilate herbicides propham and chlorpropham has been extensively studied, showing that both are readily metabolized, mainly by ring hydroxylation and subsequent conjugation in both tolerant and susceptible species (Figures 4 and 5).

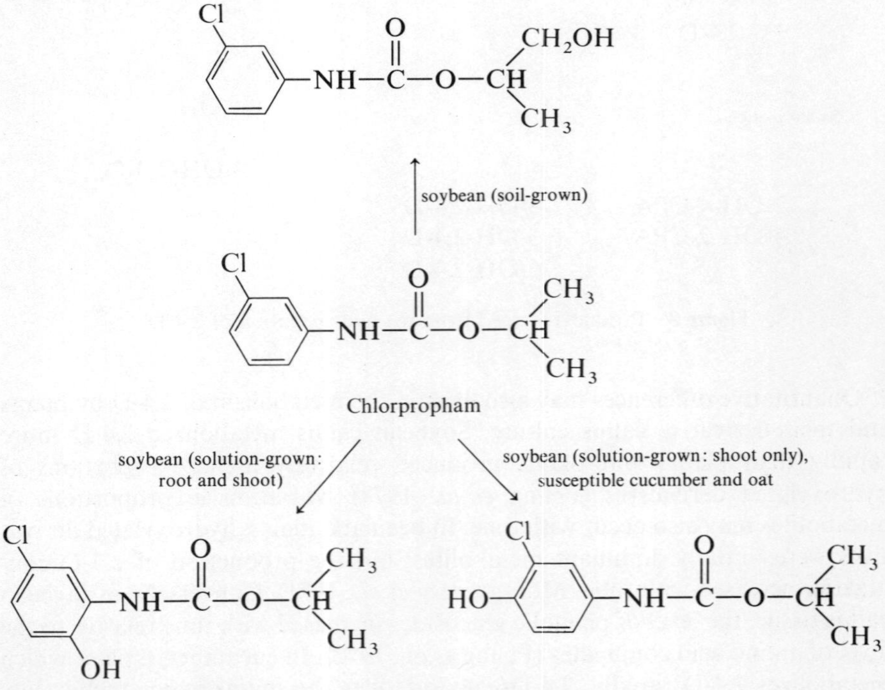

Figure 4 Hydroxylation of chlorpropham

Initial studies by James and Prendeville (1969) indicated that leaf-applied chlorpropham was metabolized to water-soluble compounds, the aglycones released upon β-glucosidase treatment being thought to be alkyl hydroxylated derivatives. Metabolism was more rapid in the susceptible *Polygonum lapathifolium* L., *Amaranthus retroflexus* L. and tomato than in the resistant parsnip (*Pastanica sativa* L.). Much of the subsequent work on chlorpropham has utilized hydroponically grown plants in which the herbicide is absorbed by the root. In soybean, polar metabolites were found in both root and shoot (Still

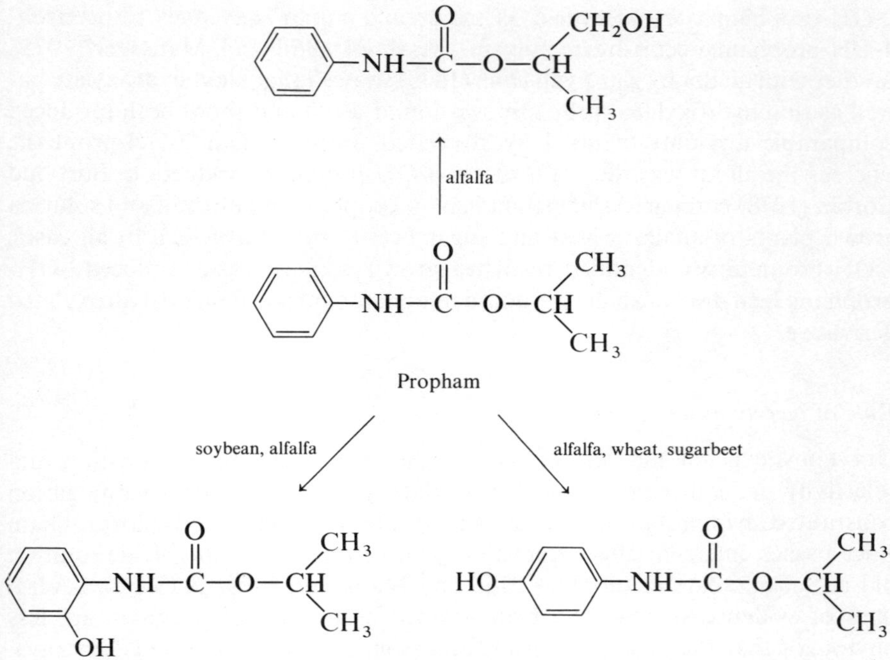

Figure 5 Hydroxylation of propham

and Mansager, 1971) and the aglycones identified as 2-OH-chlorpropham in both root (Still and Mansager, 1972) and shoot (Still and Mansager, 1973a), plus 4-OH-chlorpropham in the shoot only (Still and Mansager, 1973a). The two hydroxy derivatives were found in the shoot in equal amounts. A third metabolite in soybean was identified as the isopropyl hydroxylated derivative which occurred as a minor component in solution-grown plants, but was the major aglycone in soil-grown plants (Wiedmann *et al.* 1976). Thus the growth environment may play a part in directing the metabolism of chlorpropham and results such as these illustrate the importance of experimental design with respect to residue studies. In contrast to the pattern of chlorpropham metabolism in soybean, a tolerant plant, the susceptible species cucumber and oat produced only 4-OH-chlorpropham (Still and Mansager, 1973b; Still and Rusness, 1977). Metabolism has also been investigated in cell suspension cultures (Davis *et al.*, 1977). In cultures of soybean and carrot (*Daucus carota* L.) chlorpropham was poorly metabolized over a period of 48 h, but when 2-OH-chlorpropham and 4-OH-chlorporpham were fed, these were converted to water-soluble compounds.

Propham has been shown to undergo ring hydroxylation in several species. The only aglycone identified in soybean was 2-OH-propham, produced by both root and shoot (Still and Mansager, 1973c). In alfalfa 4-OH-propham and

2-OH-propham were identified as major and minor aglycones respectively, 4-OH-propham occurring mainly in the shoot (Still and Mansager, 1975). Studies with alfalfa by Zurqiyah *et al.* (1976) showed that alkyl hydroxylated as well as ring hydroxylated propham was found. Root and shoot both produced comparable amounts of alkyl hydroxylated propham and 2-OH-propham, whereas the shoot was the main site of 4-OH-propham production. Burt and Corbin (1978) compared the metabolism of propham in both soil and solution grown plants of alfalfa, wheat and sugar beet (*Beta vulgaris* L.). In all cases, 4-OH-propham was identified, but wheat grown in solution also produced 2-OH-propham. Leaf discs of alfalfa produced minor amounts of the alkyl hydroxylated derivative.

Role of metabolism in selectivity

Our knowledge of the role of metabolism in carbanilate detoxication and selectivity is equivocal. Ring hydroxylation and subsequent conjugation constituted an immobilization mechanism for both propham and chlorpropham since tissues emergent after removal of plants from the ^{14}C-herbicide solution did not accumulate metabolites (Still and Mansager, 1971; 1973b; c). Several lines of evidence suggest that hydroxylated derivatives of propham are less phytotoxic than the parent (Shaw and Swanson, 1953; Baskakov and Zemskaya, 1959; Moreland and Hill, 1959; Still and Mansager, 1975). The carbanilates are thought to exert phytotoxicity via the inhibition of mitosis (Corbett, 1974) and chlorpropham and its metabolites also interfere with mitochondrial reactions (Rusness and Still, 1974). Davis *et al.* (1977) have examined the effects of chlorpropham and its metabolites upon cell division in soybean and carrot, both tolerant to the herbicide. Chlorpropham at 10^{-5} M inhibited the growth of soybean cell suspension cultures, whereas both 2-OH-chlorpropham and 4-OH-chlorpropham were without effect. However, 2-OH-chlorpropham was more effective in arresting mitosis in soybean roots than the parent, although 4-OH-chlorpropham was less effective. In contrast to soybean cells, suspension cultures of carrot were susceptible to these metabolites. When 2,4-D was included in the medium, metabolites were less inhibitory than the parent, with 4-OH-chlorpropham being least effective, but in the presence of 2,4-D, 2-OH-chlorpropham was more toxic than the parent, whilst 4-OH-chlorpropham was again less inhibitory. The biological activity of chlorpropham metabolites was thus dependent upon both species and experimental design. Rusness and Still (1974) showed that 2-OH-chlorpropham inhibited oxidative phosphorylation in isolated mitochondria whereas the parent was not itself inhibitory. The inhibitory properties of 2-OH-chlorpropham were lost upon glucosylation. Conversely, 4-OH-chlorpropham did not inhibit oxidative phosphorylation and unlike the parent, did not inhibit the oxidation of NADH in phosphorylating mitochondria. As has been mentioned previously, the resistant soybean metabolizes chlorpropham to both 2-OH-chlorpropham and 4-OH-chlorpropham whereas 4-OH-chlorpropham was the only product of the susceptible cucumber and oat. Thus resistance might be correlated with the ability to 2-hydroxylate chlorpropham. However, the evidence outlined suggests that 2-hydroxylation can enhance the herbicidal properties of chlorpropham, in contrast to 4-hydroxylation which appears to detoxify the herbicide. Thus the selectivity of chlorpropham appears not to be directly related to qualitative differences in metabolism.

Miscellaneous compounds

One of the most elegant examples correlating herbicide selectivity with differential metabolism is provided by the herbicide diclofop-methyl, a phenoxy-phenoxy-propionic acid ester used for the control of wild oat (*Avena fatua* L.) in cereals (Figure 6). Gorbach *et al.* (1977) showed that the major product in wheat following hydrolysis to the carboxylic acid diclofop was the 5-OH-dichlorophenyl derivative, together with minor amounts of the 3-OH and 6-OH isomers, accumulating as conjugates. Subsequent reports confirmed

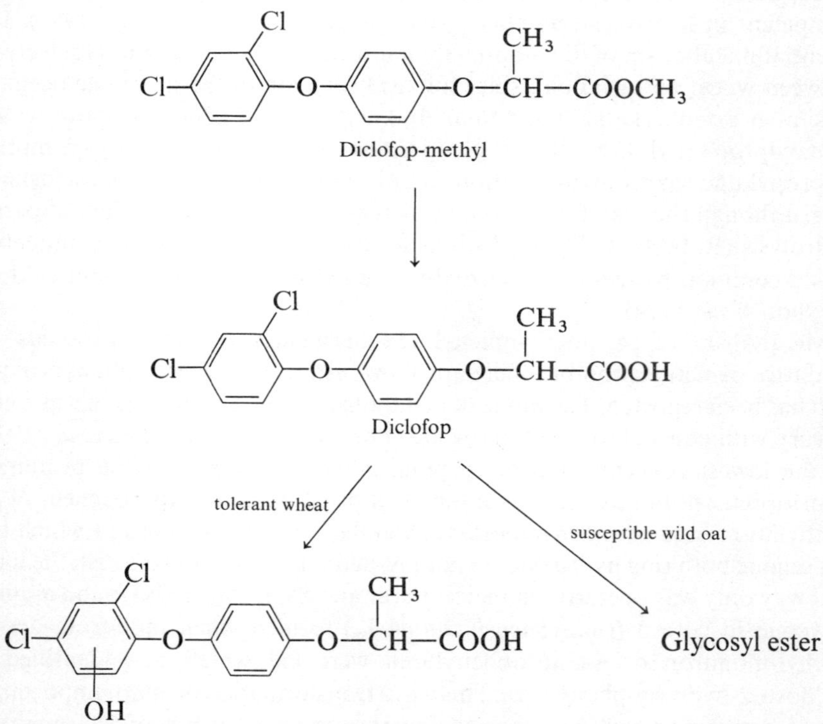

Figure 6 Differential metabolism of diclofop-methyl in tolerant and susceptible plants

dichlorophenyl hydroxylation as the major route of metabolism in wheat (Shimabukuro *et al.*, 1979; Donald and Shimabukuro, 1980; Gorecka *et al.*, 1981), although in these cases the exact position of hydroxylation was not established. Diclofop-methyl was similarly metabolized in cell suspension cultures of diploid wheat (*Triticum monococcum* L.). Hydrolysis to diclofop occurred rapidly, with subsequent hydroxylation and conjugation, metabolism of diclofop being effectively complete after 8 h (Dusky *et al.*, 1980). In contrast to wheat, the susceptible wild oat produced predominantly a glycosyl ester of diclofop, which was a minor reaction in wheat (Shimabukuro *et al.*, 1979; Donald and Shimabukuro, 1980). Analogous results were obtained by Boldt and Putnam (1981) who found polar products in five species, which on hydrolysis yielded diclofop in the susceptible *Echinochloa crus-galli* (L.) Beauv and *Panicum miliaceum* L., but unspecified ring hydroxylated derivatives in the case of the tolerant soybean, cucumber and *Cenchrus longspinus* (Hack.) Fern. In a study of the metabolism of diclofop-methyl and clofop-isobutyl in excised coleoptiles or shoots of three species, differential metabolism again correlated with selectivity. Hence in maize, which was susceptible to both, esterification was prominent whereas wheat hydroxylated both acids. Oat is susceptible to diclofop-methyl but tolerant to clofop-isobutyl and was more competent in hydroxylating clofop than diclofop (Gorecka *et al.*, 1981). Differential metabolism of diclofop-methyl appears likely to explain its selectivity between wheat and wild oat, since foliar absorption of the herbicide occurred to similar extents (Donald and Shimabukuro, 1980). Flamprop-isopropyl and benzoylprop-ethyl are related compounds but in barley, only flamprop-isopropyl underwent hydroxylation, at either the 3 or 4 position of the benzoyl ring, although the major metabolite was the carboxylic acid product of parent hydrolysis (Roberts, 1977). Hydrolysis followed by glycosyl ester conjugation was a common pathway for benzoylprop-ethyl in wheat, barley and wild oat (Beynon *et al.*, 1974).

Metabolism of the methylphenylurea herbicides normally proceeds via oxidative demethylation but there are a few instances in which ring hydroxylation has been reported. The mode of metabolism of monuron in bean was found to vary with concentration of herbicide (Lee and Fang, 1973; Lee *et al.*, 1973). At the lowest concentration of 16 ppm, 2-hydroxylation was the prominent transformation, but as the maximum level of 54 ppm was approached, *N*-demethylation became more important. Metabolites were also found which had undergone both ring hydroxylation and *N*-demethylation. In contrast, the latter pathway only was operative in maize. Frear and Swanson (1974) found a minor aglycone in cotton (*Gossypium hirsutum* L.) formed when monuron, des-*N*-methyl-monuron or 4-chlorophenylurea were fed which was identified as 4-chloro-2-hydroxy-phenylurea. This was a transformation of minor importance, which, it was suggested, was operating on the terminal product of *N*-demethylation. The urea derivative cisanilide has been shown to readily undergo hydroxylation on the phenyl and pyrrolidine rings in excised leaves of carrot and cotton; in contrast, the respective cell suspension cultures were able only to form the alcohol, which was directly excreted into the medium whereas the leaf metabolites became conjugated (Frear and Swanson, 1975) (Figure 7). The tolerance of plants to cisanilide appeared to correlate with their ability to convert it to unidentified products (Holm and Stallard, 1974).

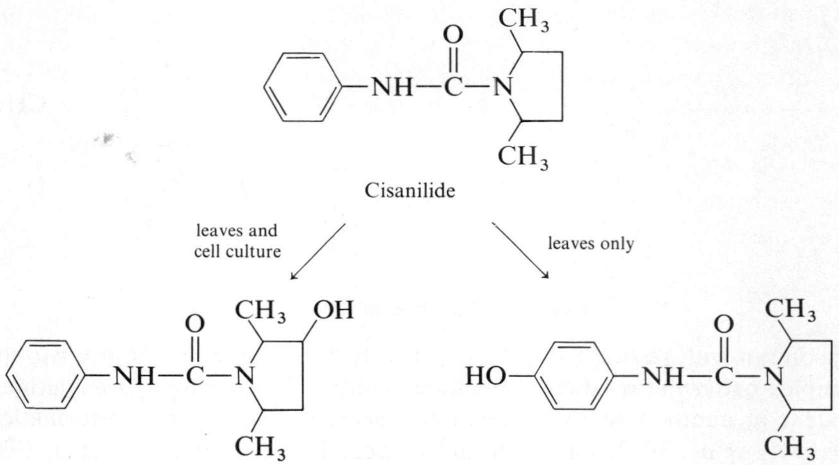

Figure 7 Aryl and alkyl hydroxylation of cisanilide in carrot and cotton: leaves versus cell culture

The carbamate insecticide carbaryl underwent ring hydroxylation in bean (Kuhr and Casida, 1967; Dorough and Wiggins, 1969; Mumma *et al.*, 1971), pea, pepper (*Capsicum frutescens* L.) and maize (Mumma *et al.*, 1971) to produce 4-OH-carbaryl and 5-OH-carbaryl. Dihydroxycarbaryl was also identified in bean by Kuhr and Casida (1967). Ring hydroxylation occurred in addition to *N*-methyl hydroxylation (Dorough and Wiggins 1969; Mumma *et al.*, 1971), metabolites accumulating as conjugates. The systemic insecticide carbofuran is susceptible to oxidation at the furan ring (Figure 8) giving 3-OH-carbofuran which was further oxidized to 3-keto-carbofuran in cotton, maize (Metcalf *et al.*, 1968; Turner and Caro, 1973), tobacco (*Nicotina tabacum* L.) (Ashworth and Sheets, 1972) and oat (Fuhremann and Lichenstein, 1980). The butylamino sulphenyl and morpholino sulphenyl derivatives of carbofuran underwent hydrolysis to carbofuran followed by either furan oxidation, or to a lesser extent *N*-methyl hydroxylation or both (Umetsu *et al.*, 1979). Oxidized metabolites of carbofuran may hydrolyse, but the corresponding phenols are of minor importance. Oxidized metabolites of carbofuran are less effective anticholinesterases (Dorough, 1968), hence the plant effectively detoxified

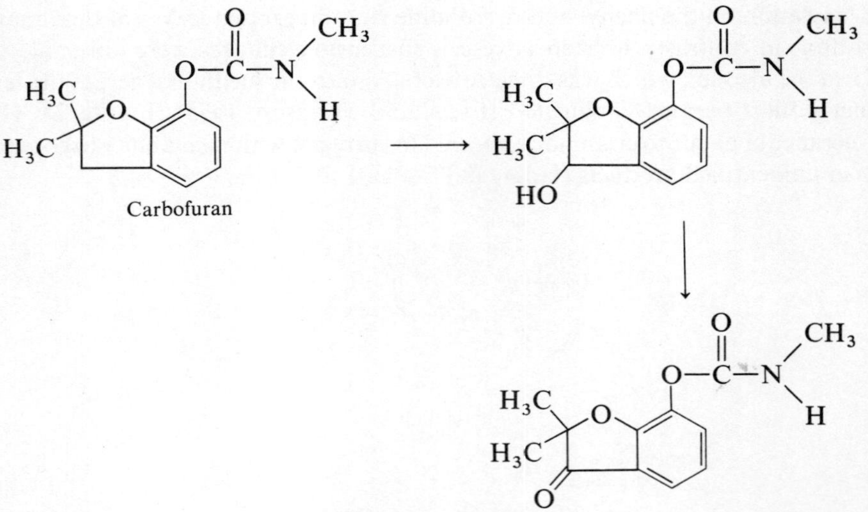

Figure 8 Oxidation of carbofuran

carbofuran with respect to the insect. The synthetic pyrethroids give rise to a complex pattern of residues in plants in which phenoxy ring hydroxylation is evident in addition to the major ester cleavage and other transformations (Okhawa *et al.*, 1977; Gaughan and Casida, 1978; Ruzo and Casida, 1979; Okhawa *et al.*, 1980). Ring hydroxylation of a number of miscellaneous herbicides has been noted. Inactivation of bentazone via hydroxylation was shown to be the major determinant of selectivity between the tolerant rice and susceptible *Cyperus serotinus* Rottb. (Mine *et al.*, 1975). Additional evidence for hydroxylation as a detoxication mechanism in wheat was provided by Retzlaff and Hamm (1976) who were able to correlate recovery from initial inhibition of CO_2 uptake with metabolism of bentazone. Hydroxylated derivatives were themselves less toxic. Napropamide (Figure 9) became extensively hydroxylated in tomato, although *N*-de-ethylation also occurred to some extent (Murphy *et al.*, 1973). Thiobencarb was metabolized in the tolerant rice and susceptible *Echinochloa utilis* L. by common pathways which include ring hydroxylation (Nakamura *et al.*, 1977). Alfalfa hydroxylated buthidazole on the imidazolidine ring; this occurred to a lesser extent in maize (Yu *et al.*, 1980). Diphenamid underwent ring hydroxylation in pepper, although this was secondary to *N*-methyl hydroxylation as the major mode of attack (Hodgson and Hoffer, 1977). Keller *et al.* (1979) investigated the ability of plant cell suspension cultures to metabolize chloridazon. Those of sugar beet and poppy (*Papaver bracteatum* L.) possessed a limited ability to ring hydroxylate chloridazon whilst other species were inactive. These cultures were also able to both ring- and methyl-hydroxylate the drug antipyrine. Other herbicides susceptible to ring hydroxylation are dicamba and dichlobenil (Frear *et al.*, 1972).

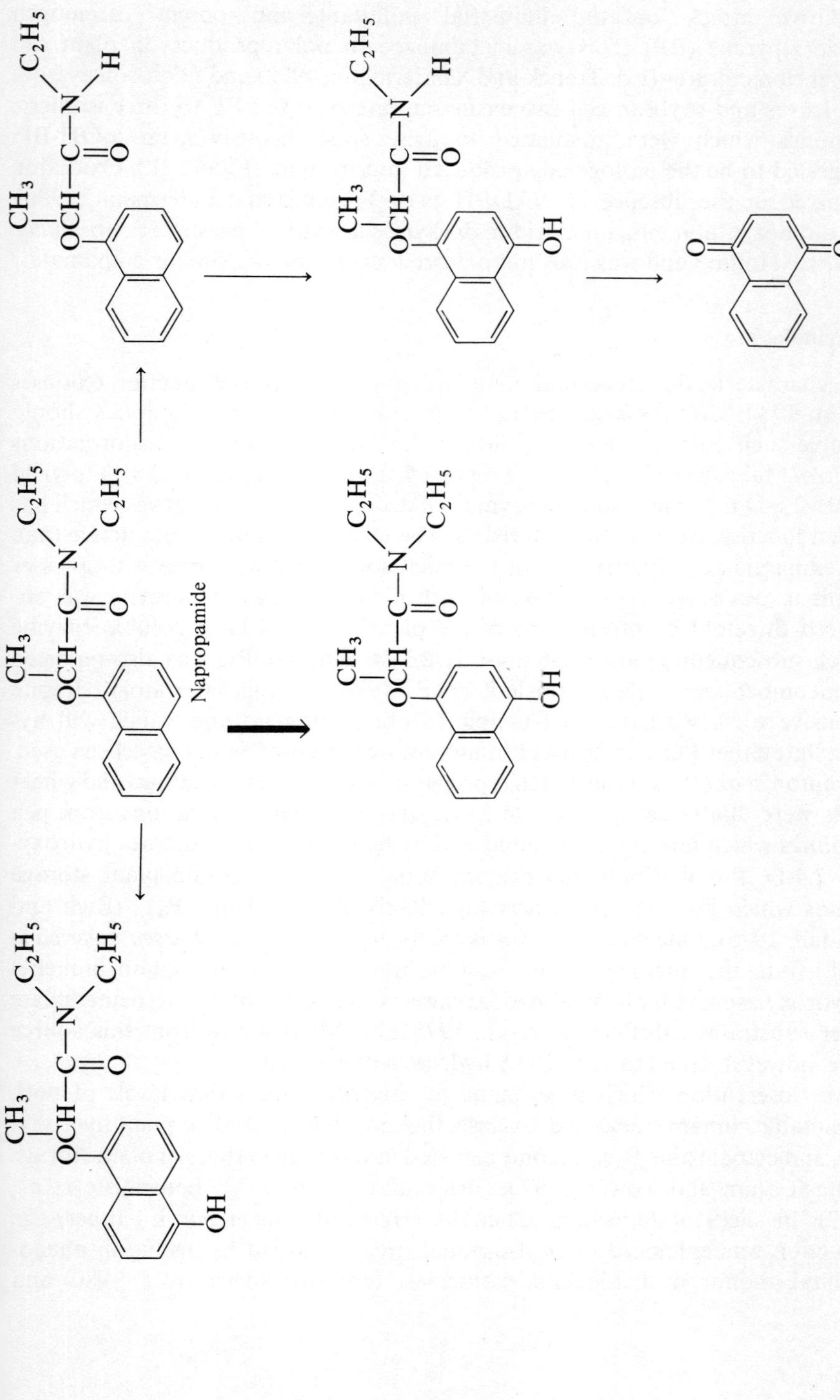

Figure 9 Metabolism of napropamide in tomato

Quinone formation rather than hydroxylation has been shown to result from oxidative attack on the industrial pollutant and potent carcinogen benzo(α)pyrene (BP). This was metabolized to polar products in plant cell suspension cultures (v.d. Trenck and Sandermann, 1978) and microsomes from pea leaves and soybean cell suspension culture oxidized BP to three isomeric quinones which were postulated to arise spontaneously from 6-OH-BP, suggested to be the biologically produced intermediate (Figure 10). Oxidation occurred in the absence of NADPH (v.d. Trenck and Sandermann, 1980). BP did not inhibit cinnamic acid hydroxylase activity of parsley (*Petroselinum hortense* Hoffm.) and was thus not oxidized at the binding site for cinnamate.

Enzymology

The characterization of several plant hydroxylases as mixed function oxidases (West, 1980) strongly suggests that ring hydroxylation of xenobiotics should involve such enzymes, but very little is known about these transformations *in vitro*. Makeev *et al.* (1977) have reported the hydroxylation of 2,4-D to yield 4-OH-2,5-D by a microsomal enzyme isolated from cucumber leaves which has mixed function oxidase characteristics. The enzyme was much less active than the cinnamic acid hydroxylase of the microsomes and was present to a lesser extent in pea leaves. In the case of both plants, extractable activity was enhanced threefold by prespraying of the plants with 2,4-D. A soluble enzyme which subsequently glucosylated 4-OH-2,5-D using UDPG was also reported in cucumber leaves (Makoveichuk *et al.*, 1978). In this laboratory, despite extensive work we have been unable to obtain preparations which will hydroxylate either POA or 2,4-D, although a wide variety of tissues have been used. Hamilton *et al.* (1971) reported that preparations from bean seedlings and wheat roots were unable to hydroxylate 2,4-D and also that preparations from pea seedlings which possessed cinnamic acid hydroxylase activity did not hydroxylate 2,4-D. Possible potential enzyme sources may be certain plant storage organs which possess particularly high levels of cytochrome P_{450} (Rich and Bendall, 1975). One such material is mesocarp of avocado (*Persea americana* Mill.) fruit, the microsomes and soluble fraction of which contain biphenyl 2-hydroxylase and biphenyl 4-hydroxylase as well as the ability to demethylate other substrates (McPherson *et al.*, 1975a; b). Microsomes from this source have, however, failed to yield POA hydroxylating activity.

An observation which is germane to this problem is that levels of both extractable cinnamic acid hydroxylase, the most widely studied plant hydroxylase, and cytochrome P_{450} become elevated in water-aged slices of plant storage tissue (Camm and Towers, 1973a; Rich and Lamb, 1977; Benveniste *et al.*, 1977). In slices of Jerusalem artichoke (*Helianthus tuberosus* L.) tuber, age induction was enhanced or post-optimal decay retarded by ageing in phenobarbital, monuron, dichlobenil, manganese ions (Reichhart *et al.*, 1980) and

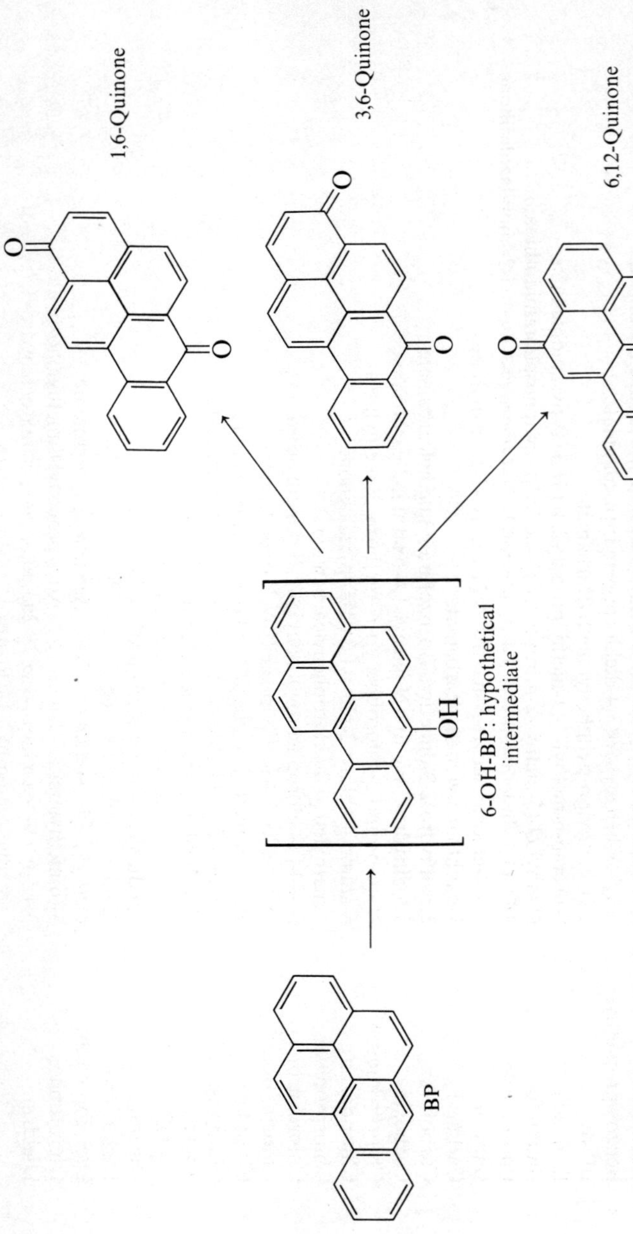

Figure 10 Microsomal oxidation of benzo(α)pyrene

Table 2 Systematic names of chemicals mentioned in the text

Common name	Systematic name
Aldicarb	2-methyl-2-(methylthio)propanal *O*-[(methylamino)carbonyl]oxime
Aldrin	1,2,3,4,10,10-hexachloro-1α,4α,4aβ,5α,8α,8aβ-hexahydro-1,4:5,8-dimethano naphthalene
Antipyrine	1,2-dihydro-1,5-dimethyl-2-phenyl-3 H-pyrazol-3-one
Bentazone	3-(1-methylethyl)-(1 H)-2,1,3-benzothiadiazin-4(3 H)-one 2,2-dioxide
Benzoylprop-ethyl	ethyl-*N*-benzoyl-*N*-(3,4-dichlorophenyl)-DL-alaninate
BPMC	2-(1-methylpropyl)phenyl methylcarbamate
Bromacil	5-bromo-6-methyl-3-(1-methylpropyl)-2,4(1 H,3 H)-pyrimidinedione
Butamifos	*O*-ethyl *O*-(5-methyl-2-nitrophenyl)1-methylpropylphosphoramidothioate
Buthidazole	3-[5-(1,1-dimethylethyl)1,3,4-thiadiazol-2-yl]-4-hydroxy-1-methyl-2-immidazolidinone
Buturon	*N'*-(4-chlorophenyl)-*N*-methyl-*N*-(1-methyl-2-propynyl)urea
Carbaryl	1-naphthalenyl methylcarbamate
Carbofuran	2,3-dihydro-2,2-dimethyl-7-benzofuranylmethylcarbamate
Carboxin	5,6-dihydro-2-methyl-*N*-phenyl-1,4-oxathiin-3-carboxamide
Chlorbromuron	*N'*-(4-bromo-3-chlorophenyl)-*N*-methoxy-*N*-methylurea
Chloridazon	5-amino-4-chloro-2-phenyl-3 (2 H)-pyridazinone
Chlorpropham	1-methylethyl 3-chlorophenylcarbamate
Chloroxuron	*N'*-[4-(4-chlorophenoxy)phenyl]-*N*,*N*-dimethylurea
Chlortoluron	*N'*-(3-chloro-4-methylphenyl)-*N*,*N*-dimethylurea
Cisanilide	*cis*-2,5-dimethyl-*N*-phenyl-1-pyrrolidinecarboxamide
Clofop-isobutyl	isobutyl 2-[4-(4-chlorophenoxy)phenoxy]propanoate
2,4-D	(2,4-dichlorophenoxy)acetic acid
2,4-DB	4(2,4-dichlorophenoxy)butanoic acid
Dicamba	3,6-dichloro-2-methoxybenzoic acid
Dichlobenil	2,6-dichlorobenzonitrile
Diclofop-methyl	methyl-2-[4(2,4-dichlorophenoxy)phenoxy]propanoate
Dicrotophos	3-(dimethylamino)-1-methyl-3-oxo-1-propenyl dimethylphosphate
Dieldrin	3,4,5,6,9,9-hexachloro-1aα,2β,2aα,3β,6β,6aα,7β,7aα-octahydro-2,7:3,6-dimethanonaphth[2,3-b]oxirene
Dimethoate	*O*,*O*-dimethyl *S*-[2-(methylamino)-2-oxoethyl]phosphorodithioate
Diphenamid	*N*,*N*-dimethyl-phenylbenzeneacetamide
Disulfoton	*O*,*O*-diethyl-*S*-2-ethylthioethylphosphorodithioate

Endrin	3,4,5,6,9,9-hexachloro-1aα,2β,2aβ,3a,6α,6aβ,7β,7aα-octahydro-2,7:3,6-di-
Carbaryl	methanonaphth [2,3-b]oxirene
EPTC	*S*-ethyl-dipropylcarbamothioate
Ethiofencarb	2-(ethylthiomethyl)phenylmethylcarbamate
Flamprop-isopropyl	1-methylethyl *N*-benzoyl-*N*-(3-chloro-4-fluorophenyl)-DL-alaninate
Fluometuron	*N*,*N*-dimethyl-*N*′-[3-(trifluoromethyl)phenyl]urea
Flurenol-butyl	butyl ester of 9-hydroxy-9H-fluorene-9-carboxylic acid
Fonofos	(±)-*O*-ethyl-*S*-phenyl ethylphosphonodithioate
Heptachlor	1,4,5,6,7,8,8-heptachloro-3a,4,7,7a-tetrahydro-4,7-methano-1H-indene
Isodrin	1,2,3,4,10,10-hexachloro-1α,4α,4aβ,5β,8β,8aβ-hexahydro-1,4:5,8-dimethanonaphthalene
Leptophos	*O*-4-bromo-2,5-dichlorophenyl *O*-methyl phenylphosphonothioate
Linuron	*N*′-(3,4-dichlorophenyl)-*N*-methoxy-*N*-methylurea
MCPA	(4-chloro-2-methylphenoxy)acetic acid
MCPB	4-(4-chloro-2-methylphenoxy)butanoic acid
Mecoprop	2-(4-chloro-2-methylphenoxy)propanoic acid
Mephosfolan	diethyl(4-methyl-1,3-dithiolan-2-ylidene)phosphoramidate
Metflurazon	4-chloro-5-(dimethylamino)-2-[(3-trifluoromethyl)phenyl]-3(2H)-pyridazinone
Methabenzthiazuron	*N*-2-benzothiazolyl-*N*,*N*′-dimethylurea
Methazole	2-(3,4-dichlorophenyl)-4-methyl-1,2,4-oxadiazolidine-2,5-dione
MIPC	2-(1-methylethyl)phenyl methylcarbamate
Monocrotophos	dimethyl(E)-1-methyl-3-(methylamino)-3-oxo-1-propenylphosphate
Monolinuron	*N*′-(4-chlorophenyl)-*N*-methoxy-*N*-methylurea
Monuron	*N*′-(4-chlorophenyl)-*N*,*N*-dimethylurea
Napropamide	*N*,*N*-diethyl-2-(1-naphthalenyloxy)propanamide
Parathion	*O*,*O*-diethyl *O*-4-nitrophenyl phosphorothioate
Phorate	*O*,*O*-diethyl *S*-ethylthiomethyl phosphorodithioate
Piperophos	*S*-[2-(2-methyl-1-piperidinyl)-2-oxoethyl]*O*,*O*-dipropyl phosphorodithioate
POA	phenoxyacetic acid
Propaphos	4-(methylthio)phenyl dipropyl phosphate
Propham	1-methylethyl phenylcarbamate
2,4,5-T	(2,4,5-trichlorophenoxy)acetic acid
2,4,5-TB	4(2,4,5-trichlorophenoxy) butanoic acid
Terbacil	5-chloro-3-(1,1-dimethylethyl)-6-methyl-2,4(1H, 3H)-pyrimidinedione
Thiobencarb	*S*-[(4-chlorophenyl)methyl]diethylcarbamothioate
Tiocarbazil	*S*-phenylmethyl bis(1-methylpropyl)carbamothioate

2,4-D (Adele *et al.*, 1981). This work favours the concept that plant detoxication mechanisms may to some extent be inducible, thus it may be possible that active hydroxylating enzyme may be obtained from tissue aged in phenoxyacetic acids. Such an approach yielded a limited amount of an unidentified polar product in Jerusalem artichoke, the appearance of which was dependent upon NADPH; but this did not represent a satisfactory enzyme preparation (Adele *et al.*, 1981). A similar effect was also elicited by ageing in $MnCl_2$, but in our laboratory, neither of these approaches has yielded *in vitro* hydroxylating activity from potato slices. We have also found that these treatments had no effect on the rate of hydroxylation *in vitro* of POA in potato slices; however, subsequent glucosylation of 4-OH-POA was promoted by ageing in 2,4-D, glucoside occurring at the expense of immobile material and the unknown metabolites mentioned previously. Enhancement by 2,4-D exhibited a broad optimum concentration of 10^{-4}–10^{-3} M, and a similar effect was elicited by MCPA and IAA but not by POA. Enhancement of glucoside formation thus appears to be an auxin mediated effect.

An alternative hydroxylating system dependent upon hydroperoxide has been found in microsomes from imbibed pea seeds and is associated with a CO-binding haemoprotein which is distinct from cytochrome P_{450} (Ishimaru and Yamazaki, 1977a; b). This system hydroxylated indole, phenol, 1-naphthol and aniline and was supported by linoleic acid hydroperoxide and to a lesser extent H_2O_2, *t*-butyl hydroperoxide and cumene hydroperoxide. Unlike the other agents, linoleic acid hydroperoxide was an allosteric effector of hydroxylation and since it can be produced by lipoxygenase it has been suggested that this may be the natural cofactor. ^{18}O-Labelling showed that oxygen incorporated into indole was derived directly from linoleic acid hydroperoxide (Figure 11). Thus the reaction was only indirectly dependent upon molecular oxygen. The use of *p*-chloromercuribenzoate, which converts P_{450} into P_{420}, did not inhibit hydroperoxide-dependent hydroxylation.

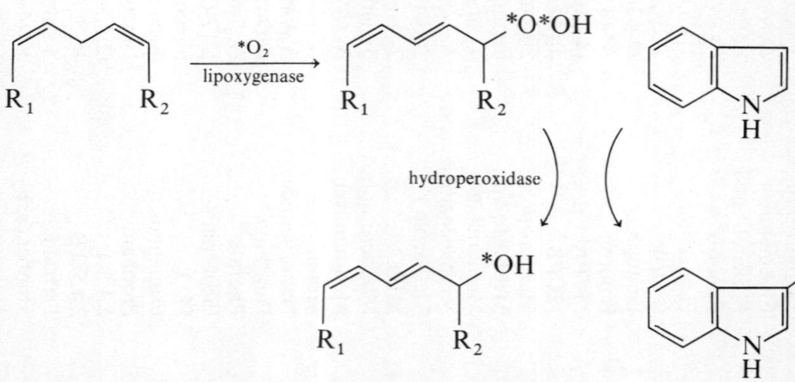

Figure 11 Hydroperoxide-dependent hydroxylation mechanism of pea seed microsomes

ALKYL HYDROXYLATION

Alkyl hydroxylation, usually that of methyl groups, is a common mode of metabolism of pesticides in plants, and occurs in a wide variety of chemical classes. *N*-Methylol derivatives have varying degrees of stability, normally decaying with subsequent demethylation although relatively stable methylols form *O*-glucosides. Oxidative demethylation is particularly important in detoxication of the methylphenylurea herbicides (Geissbühler *et al.*, 1975). The *O*-dealkylation of some organophosphorus pesticides is also probably an oxidative process (Menn and Still, 1977; Kulkarni and Hodgson, 1980).

Hydroxylation of methyl groups as aromatic ring substituents occurs in the case of the herbicides MCPA and chlortoluron. MCPA, unlike the other phenoxyacetic acids has not been reported to undergo ring hydroxylation, but rather becomes methyl hydroxylated (Collins and Gaunt, 1970; 1971). OH-Methyl-MCPA was a major metabolite in pea and *Silene dioica* (L.) Clairv., both tolerant species, whereas this was a lesser metabolite in rape (*Brassica napus* L.), the major product being MCPA aspartic acid conjugate. This was present in minor amounts in the other two species and retained auxin activity whereas OH-methyl-MCPA was less active and its glucoside completely inactive. However, since translocation of MCPA was more rapid in rape, tolerance could not be explained purely in terms of differential metabolism. In this laboratory, we have examined the metabolism of MCPA in a variety of species by vacuum infiltration into excised tissue (Table 3). In several cases, MCPA was effectively completely metabolized in 48 h. In all cases, two products were observed, one of which co-chromatographed with OH-methyl-MCPA, the other being an ether soluble acid and postulated to be an amino acid conjugate. Generally, plants could be separated into those that were primarily hydroxylators or those which produced mainly the amino acid conjugate. In addition,

Table 3 Metabolism of MCPA in excised plant tissues

	Major mode of metabolism	
Species	Methyl hydroxylation	Amino acid conjugation
Maize coleoptiles	✓	
Potato tuber slices	✓	
Carrot root slices		✓
Rice coleoptiles	✓	
Barley coleoptiles	✓	
Oat coleoptiles	✓	✓
Beet root slices		✓
Soybean apical bud		✓
Jerusalem artichoke tuber slices		✓
Swede root slices		✓

up to four minor aglycones were observed which accounted for between 1–5% of the total recovered radioactivity; these are postulated to represent ring-hydroxylated derivatives of MCPA. Slices of potato aged in 10^{-4} M MCPA had an increased capacity to produce OH-methyl-MCPA, although the major effect was to induce the formation of a glycosyl ester of MCPA. In contrast, slices of carrot were not sensitive to this pretreatment.

Chlortoluron has been shown to be metabolized in wheat via two routes, the conventional *N*-demethylation, in common with other methyl-phenylureas and ring hydroxylation, the dominant pathway (Figure 12). Conjugates of ring-methyl hydroxylated chlortoluron predominated in young solution grown plants but in the field at harvest, a further oxidized product, the carboxyphenyl derivative and also its demethylated analogues were found (Gross *et al.*, 1979). It was concluded that the early conjugation was thus reversed but since the two experiments differed it cannot be assumed that chlortoluron was metabolized in the same way in young plants in the field as in the laboratory solution grown plants. Ring methyl hydroxylation occurs in the case of the insecticides 3,4,5- and 2,3,5-tri-methylphenyl methylcarbamate (Slade and Casida, 1970) and *m*-tolyl-*N*-methylcarbamate and 3,4-xylyl-*N*-methylcarbamate (Ohkawa *et al.*, 1974).

Hydroxylation of higher alkyl moieties also occurs. That of the herbicides propham (Zurqiyah *et al.*, 1976) and chlorpropham (Wiedmann *et al.*, 1976) has been mentioned previously. Other examples are the plant growth regulator flurenol-butyl (Wotschokowsky, 1972) and the substituted uracil herbicide bromacil. In orange (*Citrus sinensis* L.) and mandarin (*Citrus reticulata*), bromacil underwent both butyl and methyl hydroxylation (Figure 13). The leaves of the more tolerant mandarin were able to metabolize bromacil more rapidly than those of orange (Jordan and Clerx, 1981). The related compound terbacil, however, only underwent methyl hydroxylation in alfalfa (Rhodes, 1977). The carbamate insecticides BPMC and MIPC became alkyl hydroxylated in rice (Ogawa *et al.*, 1976; 1977) and in cotton the systemic insecticide mephosfolan underwent hydroxylation of the methyl group as a substituent of the dithiolane ring (Zulalian and Blinn, 1977).

N-Methyl hydroxylation and demethylation

The identification of methylol derivates as intermediates in oxidative *N*-demethylation may be established by direct observation of a relatively stable *N*-methylol or its glucoside, or by the observation of a methoxy derivative after extraction in methanol. The free *N*-methylol of the systemic vinyl phosphate insecticide dicrotophos was observed together with its demethylated product (monocrotophos) in cotton (Bull and Lindquist, 1964) and bean (Menzer and Casida, 1965). Further demethylation did not occur but if monocrotophos itself was administered, its *N*-methylol and unsubstituted amide products were

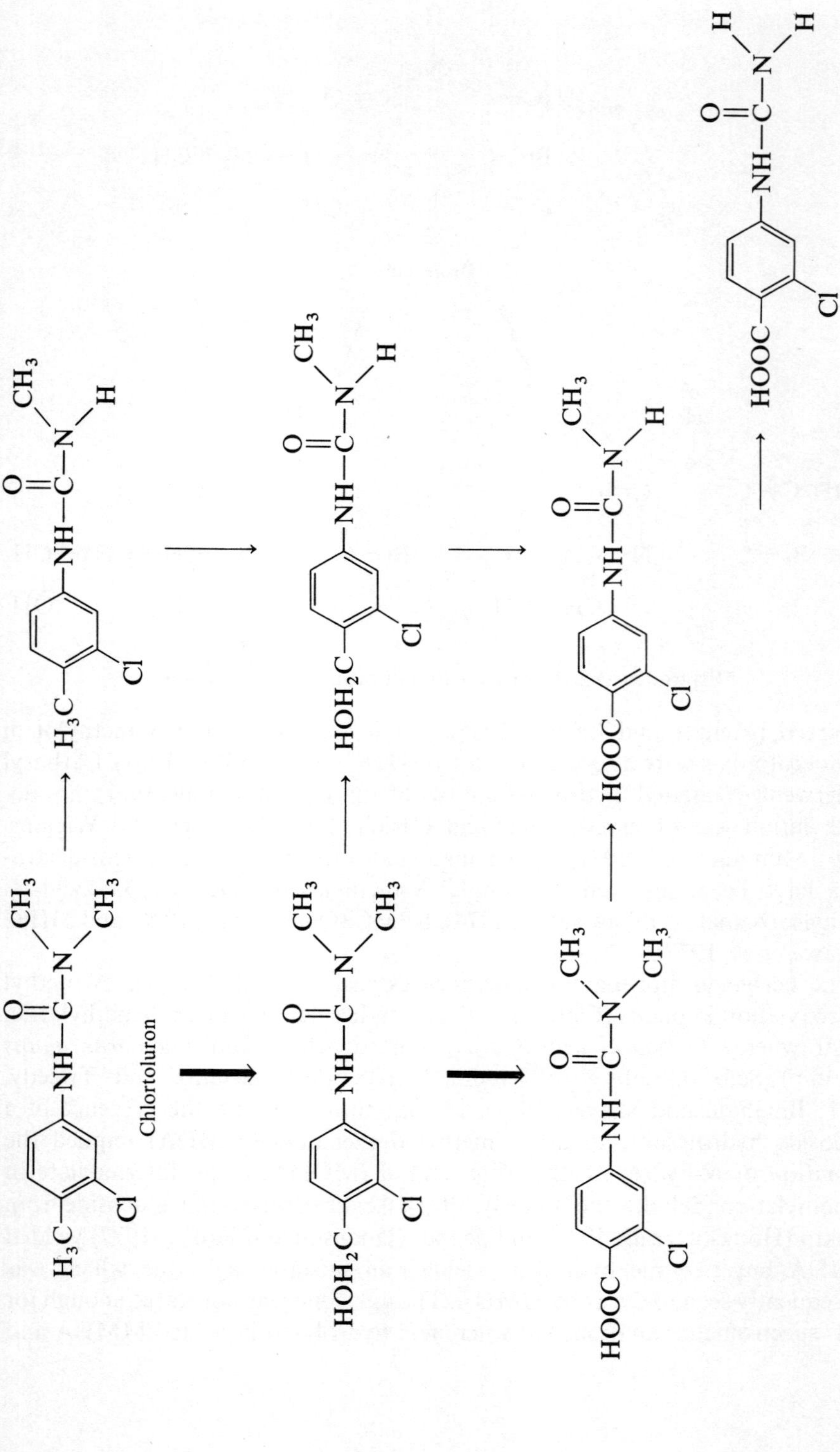

Figure 12 Metabolism of chlortoluron in wheat

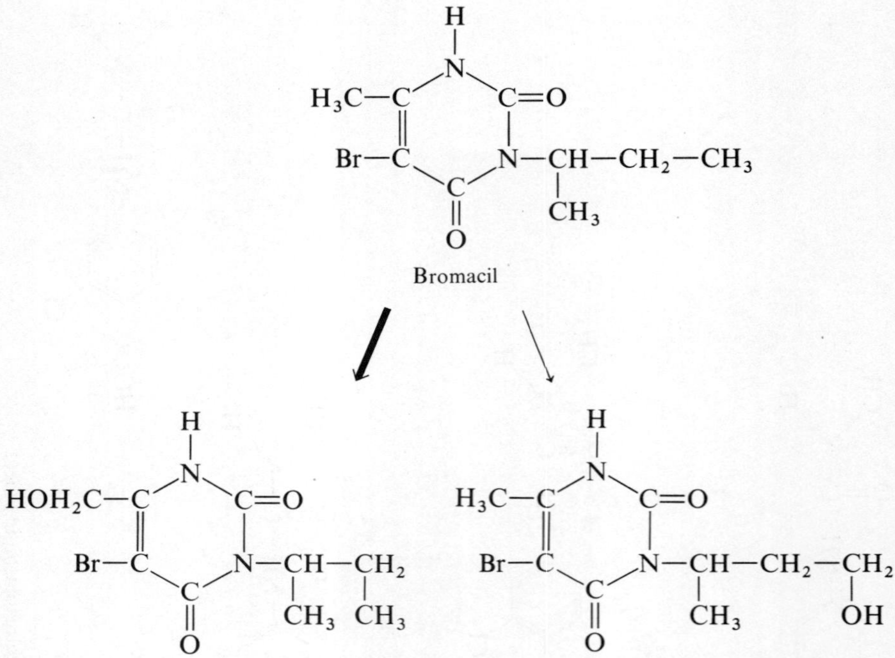

Figure 13 Alkyl hydroxylation of bromacil in *Citrus* spp.

observed (Menzer and Casida, 1965). Small amounts of the *N*-methylol of monocrotophos were also found in cotton (Lindquist and Bull, 1967). Carbaryl underwent *N*-methyl hydroxylation with subsequent conjugation; no demethylation was observed (Kuhr and Casida, 1967; Dorough and Wiggins, 1969; Mumma *et al.*, 1971). Other insecticides for which *N*-methylol derivatives have been observed are *m*-tolyl-*N*-methylcarbamate and 3,4-xylyl-*N*-methylcarbamate (Ohkawa *et al.*, 1974), BPMC (Ogawa *et al.*, 1976) and MIPC (Ogawa *et al.*, 1977).

The herbicide diphenamid undergoes extensive metabolism via *N*-methyl hydroxylation in plants (Figure 14). Early studies demonstrated demethylation in strawberry (*Fragaria grandiflora*), tomato, wheat and *Euonymus alatus* (Thunb.) Seib (Golab *et al.*, 1966; Lemin, 1966; Schultz and Tweedy, 1971; Bingham and Shaver, 1971). In tomato and wheat, the presence of a glucoside hydrolysable to des-*N*-methyl diphenamid (MMDA) implied the formation of *N*-hydroxymethyl diphenamid (MODA) as an intermediate in demethylation (Schultz and Tweedy, 1971). Acid hydrolysis of a glucoside from tomato (Hodgson *et al.*, 1973) and pepper (Hodgson and Hoffer, 1977) yielded MMDA but enzymic hydrolysis yielded an unstable aglycone which was subsequently seen to decay to MMDA. The aglycone was not stable enough for mass spectrometric analysis, but when acid hydrolysed it yielded MMDA and

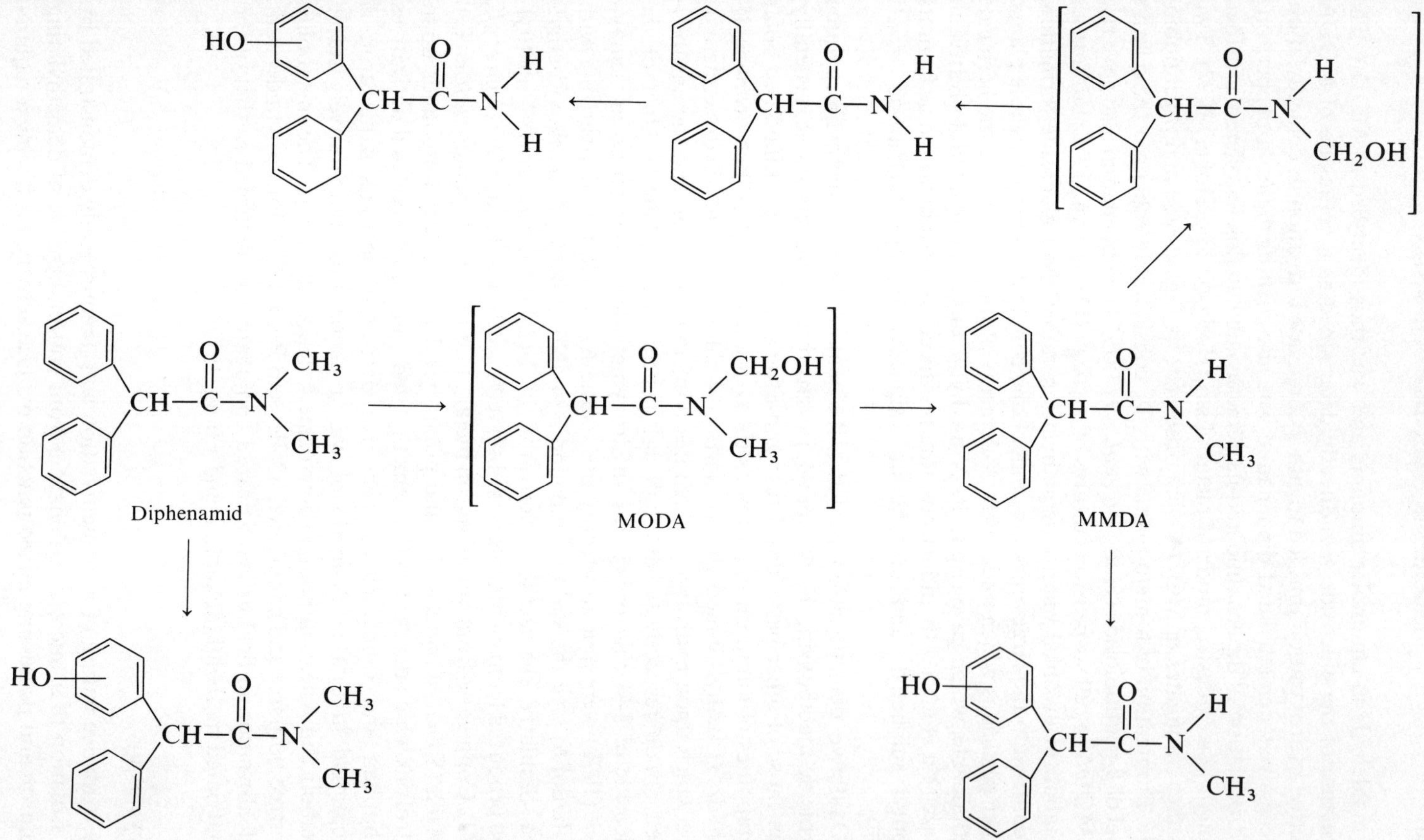

Figure 14 Oxidation of diphenamid in pepper

formaldehyde in equimolar amounts and was thus identified as MODA. The structure of the glucoside was directly determined as a conjugate of MODA. Methyl hydroxylation thus led to either conjugate formation or demethylation. Successively demethylated diphenamid was also identified and in pepper, ring hydroxylation of diphenamid and its demethylated analogues offered a further, though less extensive, mode of metabolism (Hodgson and Hoffer, 1977). Two distinct conjugates of MODA were observed: its β-glucoside and β-gentiobioside. Fumigation of tomato plants with O_3 considerably enhanced the relative level of β-gentiobioside (Hodgson *et al.*, 1973; 1974). Diphenamid also gave rise to MODA in cell suspension cultures of soybean (Davis *et al.*, 1978) but further metabolism involved mainly demethylation rather than glucoside formation, in contrast to the large amounts of alcoholic conjugates found in whole plants. Other pesticides which underwent *N*-demethylation in plants are the herbicide buthidazole, which gave rise to its *N*-methylol and demethylated derivative in maize (Yu *et al.*, 1980) and carbofuran, produced in cotton and maize from its dibutyl sulphenyl- and morpholino sulphenyl-derivatives (Umetsu *et al.*, 1979).

Oxidative demethylation provides the major route of methylphenylurea herbicide metabolism, the *N*-methylol derivatives being less stable, generally, than those of other pesticides. Frear and Swanson (1972) identified two polar metabolites of monuron in cotton as the *O*-β-glucoside of 3-(4-chlorophenyl)-*N*-hydroxymethyl-*N*-methylurea and 3-(4-chlorophenyl)-*N*-hydroxymethylurea which upon hydrolysis yielded des-*N*-methyl monuron and the unsubstituted amide respectively although a transient intermediate product of the glucoside of 3-(4-chlorophenyl)-*N*-hydroxymethylurea was observed. Tanaka *et al.* (1972) were able to identify this as the *N*-methylol by derivatizing with methanol to give the stable compound 3-(4-chlorophenyl)-*N*-methoxymethylurea. Similarly, the methoxy derivative of 1-(3,4-dichlorophenyl)-*N*-methylurea produced from methazole was observed in cotton and bean (Dorough *et al.*, 1973). Conjugates from monolinuron metabolites in cress (*Lepidium sativum* L.), spinach (*Spinacea oleracea* L.) and potato yielded unstable compounds upon hydrolysis which decayed to the demethylated and demethoxylated derivatives (Schuphan and Ebing, 1978). The urea derivative, methabenzthiazuron is exceptional in that its *N*-methylol was a major and comparatively stable metabolite in wheat, although some demethylation did occur. The metabolite occurred both free and conjugated (Collet and Pont, 1974). Buturon underwent *N*-dealkylation in field grown wheat; a conjugate was identified as that of the debutylated methylol (Schuphan and Ebing, 1977).

Enzymology

The oxidative nature of *N*-demethylation was unequivocally demonstrated by the isolation of a microsomal enzyme from cotton, capable of demethylating monuron and possessing mixed function oxidase activity, i.e. having a requirement for O_2 and NAD(P)H, susceptible to inhibition by CO and inhibited by *p*-chloromercuribenzoate (Frear, 1968; Frear *et al.*, 1969). The reaction proceeded via the equimolar liberation of [^{14}C]formaldehyde from [^{14}C-methyl]-monuron. (Figure 15) The enzyme was not capable of successively demethylating monuron, but if des-*N*-methyl monuron was used as a substrate, this could be further demethylated to *p*-chlorophenylurea. Diuron and fluometuron also served as substrates but diphenamid did not. Enzyme activity was inhibited competitively by *N*-methylcarbamates suggesting that these may possibly be demethylated by the same system. Cotton is particularly resistant to monuron and of a range of species assayed for *in vitro* activity, was the most active source. Those species least resistant to monuron yielded microsomes with the lowest *N*-demethylating activities. One problem with *in vitro* studies is the presence of low molecular weight inhibitors. Thus microsomes which were dialysed or subjected to gel filtration showed increased activity. Microsomes from carrot and okra (*Hibiscus esculentus* L.) were reported to be completely inactive unless subject to gel filtration (Frear *et al.*, 1972). The *N*-demethylase system is also subject to end-product inhibition. The demethylated products of monuron and diuron were inhibitory to fluometuron *N*-demethylase of cotton microsomes (Frear *et al.*, 1972), the completely demethylated products being most effective. An *N*-demethylase has also been isolated from castor bean (*Ricinus communis* L.) endosperm. Microsomes which possessed cinnamic acid hydroxylase

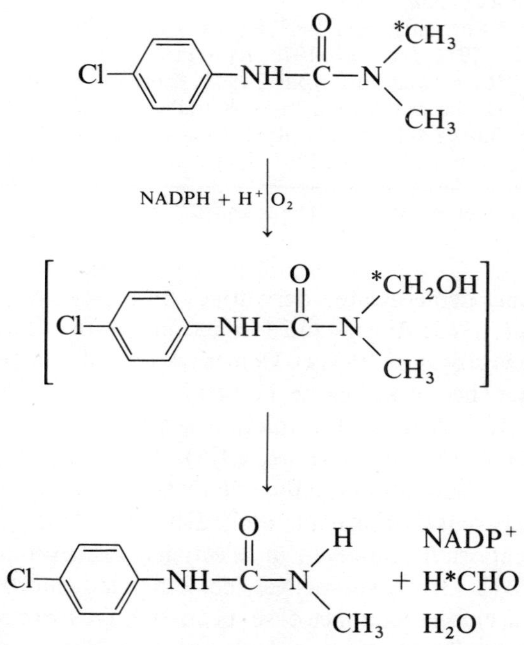

Figure 15 Microsomal oxidative demethylation of monuron

activity were also able to *N*-demethylate the non-pesticidal xenobiotic *p*-chloro-*N*-methylaniline, and, to a lesser extent, *N*-methylaniline and *N*,*N*-dimethylaniline with mixed function oxidase characteristics (Young and Beevers, 1976). The substrate liberated formaldehyde and the requirement for NADPH could be fulfilled by cumene hydroperoxide. In the case of both these *N*-demethylases, activity was located principally in the endoplasmic reticulum. Microsomes and the soluble fraction from avocado mesocarp were also able to demethylate *p*-chloro-*N*-methyl-aniline and to carry out the *O*-demethylation of *p*-nitro anisole (McPherson *et al*., 1975b).

Role of metabolism in selectivity of methylphenylurea herbicides

A large body of evidence exists which suggests that the *N*-demethylation of urea herbicides is an important factor in selectivity (Table 4). Generally, des-*N*-methyl derivatives are significantly less phytotoxic than the parent compounds

Table 4 Contribution of metabolism to selectivity of methylphenylurea herbicides

Herbicide	References
Chlorbromuron	Nashed *et al*., 1970
Chloroxuron	Geissbühler *et al*., 1963
Chlortoluron	Ryan *et al*., 1981
Diuron	Smith and Sheets, 1967; Swanson and Swanson, 1968a; Osgood *et al*., 1972; Liu *et al*., 1978; Ryan *et al*., 1981
Fluometuron	Rogers and Funderburk, 1968; Eshel and Rubin, 1971
Linuron	Hogue and Warren, 1968; Kuratle *et al*., 1969
Methazole[a]	Butts and Foy, 1974; Keeley and Thullen, 1979; Verity *et al*., 1981a,b
Monuron	Smith and Sheets, 1967; Swanson and Swanson, 1968a

[a] Activated to the phytotoxic *N'*-(3,4-dichlorophenyl)-*N*-methylurea which can subsequently be demethylated.

and the completely demethylated derivatives possess very low phytotoxicity (Geissbühler *et al*., 1963; Rogers and Funderburk, 1968; Goren, 1969; Rubin and Eshel, 1971; Verity *et al*., 1981a). Demethylation of other compounds may decrease toxicity. The pyridazinone herbicide metflurazon was successively demethylated more rapidly in the tolerant soybean than in the susceptible *Cassia obtusifolia* L. (Motooka *et al*., 1977). The completely demethylated product was much less inhibitory in the Hill reaction and to chlorophyll formation than was the parent (Hilton *et al*., 1969). Differential rates of demethylation are not always important, however, in selectivity. For example, soybean and *Ipomoea purpurea* (L.) Roth. successively demethylated chloroxuron to similar extents despite the greater tolerance of soybean. In this case selectivity appeared to be due to differential translocation (Feeny *et al*., 1974).

SULPHOXIDATION

A major metabolic pathway of organophosphorus and carbamate pesticides containing oxidizable sulphur is progressive oxidation to the sulphoxide and sulphone analogues (Schlagbauer and Schlagbauer, 1972; Bull, 1972a; b; Miyamoto, 1972; Matsunaka, 1972; Menzer, 1973; Baldwin, 1977; Menn and Still, 1977). As a general rule, based upon the study of the metabolism of numerous such compounds, initial sulphoxidation is a rapid process, producing a sulphoxide, often of enhanced insecticidal or herbicidal properties. Subsequent conversion to the sulphone is less rapid and often decreases toxicity. An example of this type of metabolism may be afforded by the well studied systemic carbamate insecticide aldicarb (Figure 16). In cotton, aldicarb sulphoxide was rapidly produced with either subsequent oxidation to the sulphone or hydrolysis to the oxime sulphoxide and further degradation, significant variation in quantitative metabolism being apparent between reports. In greenhouse grown plants, uptake by excised leaves or by stem injection into the whole plant gave rise to the oxime sulphoxide as the major residue (Metcalf *et al.*, 1966), whilst in a study by Coppedge *et al.*, (1967) the sulphoxide was more stable, with lesser amounts of the oxime sulphoxide and nitrile sulphoxide. Root uptake in greenhouse plants gave rise to the sulphoxide which was readily converted into the oxime and nitrile analogues, followed by further degradation (Bartley *et al.*, 1970). However, in field conditions, the sulphoxide underwent limited further conversion (Bull, 1968). In potato, experimental design had a marked effect on the fate of aldicarb (Andrawes *et al.*, 1971). In stem injected greenhouse plants the sulphoxide was readily hydrolyzed, but in the field, the sulphone became the major residue. In a long term study of aldicarb metabolism in a sugar beet crop, sulphoxide and sulphone analogues of the parent were minor metabolites at harvest, other metabolites being degradation products of these, mainly of the sulphone (Rouchard *et al.*, 1980). From these results it appears that there may be a tendency for field conditions to favour the further oxidation of aldicarb sulphoxide, which is a minor pathway in greenhouse plants. Although aldicarb sulphoxide was shown to be considerably more potent as an anticholinesterase than the parent, LD_{50} values for insects were similar, whilst aldicarb sulphone was less toxic (Payne *et al.*, 1966; Bull *et al.*, 1967). The toxicity of systemic insecticide metabolites to the plants themselves is not a subject which is given great consideration. Spurr and Sousa (1974) investigated the toxicity of oxidized aldicarb metabolites to several crops. At the high concentrations used (56 kg/ha) metabolites were generally innocuous, though aldicarb sulphoxide was toxic in pre-emergence application to rye grass (*Lolium perenne* L.) and millet (*Setaria italica* L.) and aldicarb nitrile sulphone was inhibitory to *Amaranthus retroflexus* L. Recent examples of insecticide sulphoxidation in plants are provided by propaphos in rice (Fujii *et al.*, 1980) and *O*-[4-[(4-chlorophenyl)thio]phenyl]

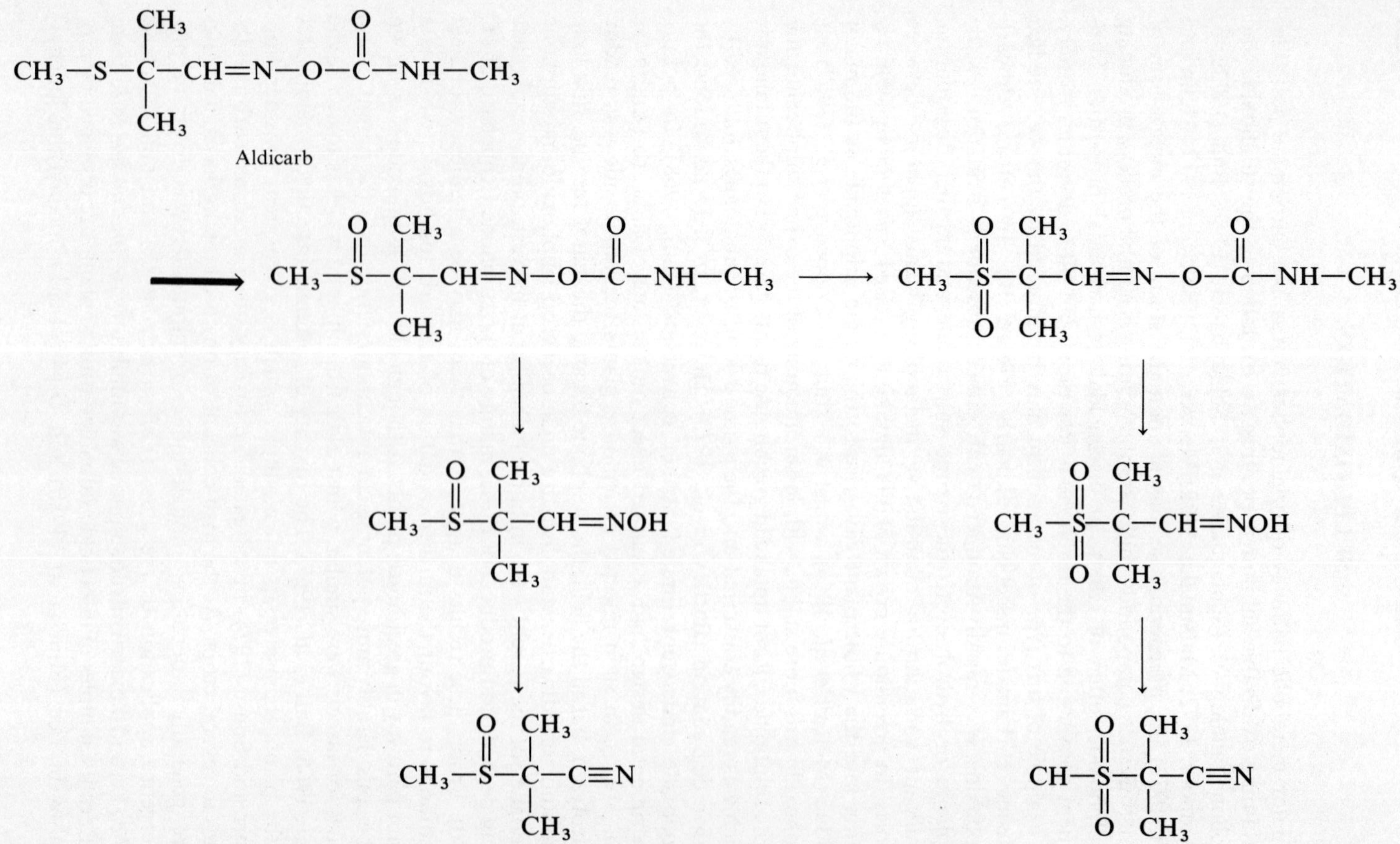

Figure 16 Sulphoxidation of aldicarb and subsequent metabolites

O-ethyl-*S*-propyl phosphorothioate in cotton (Bull and Ivie, 1981). In both cases, these compounds were metabolized by progressive oxidation, followed by hydrolysis to give phenolic products. The systemic fungicide carboxin also undergoes oxidation in plants. Carboxin sulphoxide was the major metabolite in wheat, barley (Chin *et al.*, 1970) and bean (Snel and Edgington, 1970) and possesses very low fungitoxicity (Mathre, 1971). Interaction within the plant may thus attenuate the efficacy of carboxin.

Some organophosphorus compounds may not themselves be susceptible to sulphoxidation, but liberate hydrolysis products which are. The non-systemic insecticide fonofos underwent thioester cleavage in potato yielding a thiophenyl moiety which underwent *S*-methylation followed by progressive sulphoxidation. These transformations decreased animal toxicity (McBain *et al.*, 1970). A similar pathway for the thioester cleavage product of piperophos (or its oxygen analogue) was observed in paddy rice (Figure 17). Although piperophos was taken up from paddy soil, in which similar metabolites were found, plant oxidation was demonstrated by the formation of sulphoxidized products from the directly injected *S*-methylated cleavage product (Mayer *et al.*, 1981).

Enzymology

Studies on the sulphoxidation of insecticides *in vitro* have been limited to the organophosphorus compound phorate (Krueger, 1975) and aldicarb (Krueger, 1977). In both cases an enzyme was isolated from soybean roots which formed the sulphoxides as the sole products. Two enzymes are involved, since phorate sulphoxidase activity was associated predominantly with the 25 000 × g pellet, whereas the major part of aldicarb sulphoxidase activity was retained in the corresponding supernatant. Both enzymes could be inhibited by a number of sulphur-containing pesticides, suggesting that fairly broad substrate specificities may exist. However, phorate inhibited aldicarb sulphoxidase but aldicarb did not inhibit phorate sulphoxidase. Both enzymes possess a pH optimum of 5.5 and were more active in roots than other parts of the plant. Soybean and bean roots were more active enzyme sources than those of several other species. Activity was not dependent upon NAD(P)H. The plant enzymes appear to differ from the microsomal mixed function oxidases known to catalyse sulphoxidation of pesticides in animals (Hajjar and Hodgson, 1980).

Role of metabolism in phytotoxicity of thiocarbamate herbicides

The sulphoxides of the thiocarbamate herbicides are considerably more phytotoxic than the parent compounds and Casida's group have proposed that toxicity depends upon this transformation taking place in the plant. The

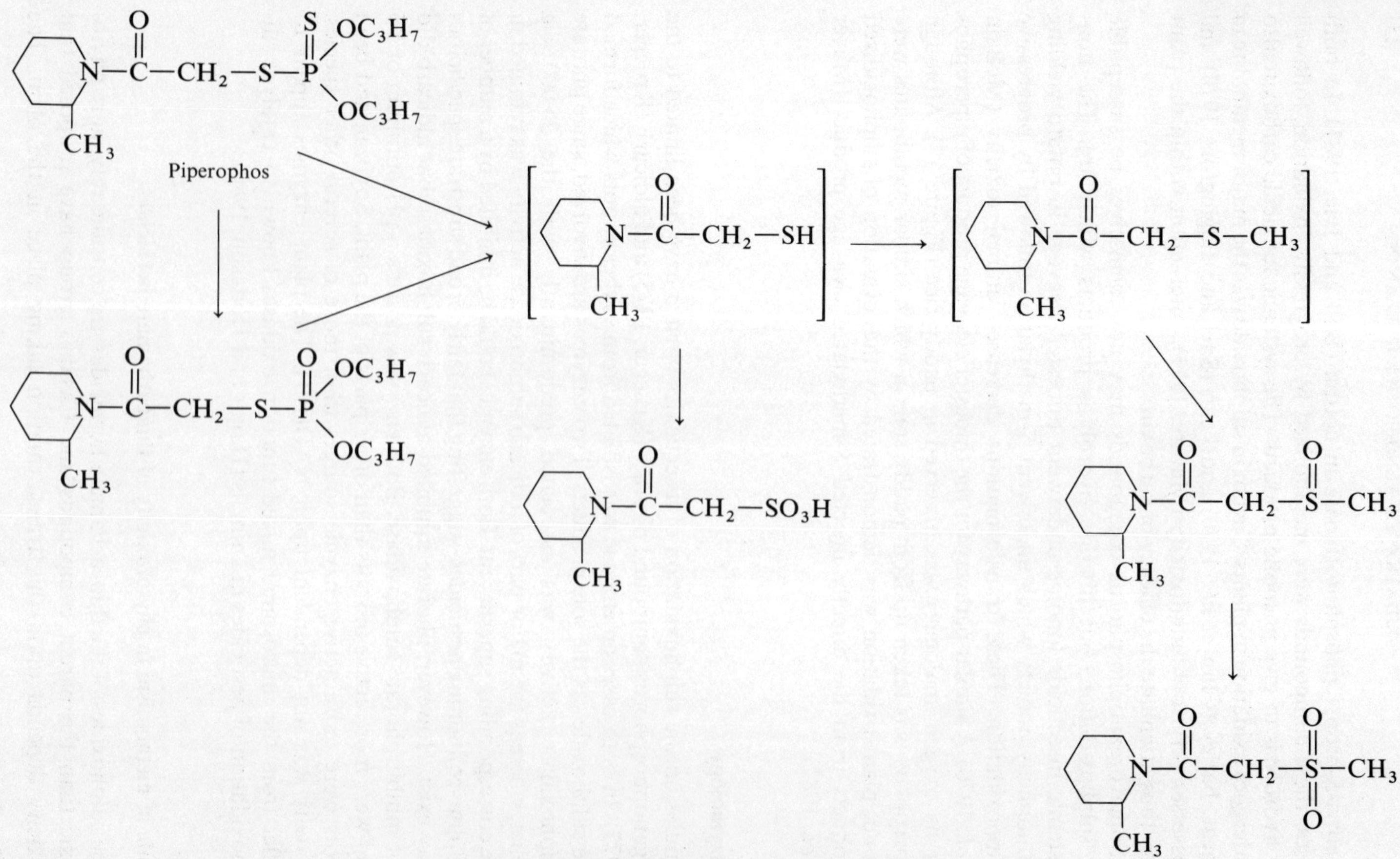

Figure 17 Sulphoxidation products from the cleavage of piperophos in paddy rice

sulphones are much less phytotoxic (Casida *et al.*, 1974). In maize and oat, EPTC was converted into the sulphoxide which was subsequently rendered inactive by carbamoylating glutathione (Figure 18) (Lay *et al.*, 1975; Lay and Casida, 1976; Hubbell and Casida, 1977; Carringer *et al.*, 1978). Such conjugation occurs non-enzymically, though there is evidence that a glutathione-*S*-transferase may also be involved. The proposal of the sulphoxide as the phytotoxic principle is supported by the observation that oat, which is more susceptible to EPTC than maize, had lower levels of glutathione and glutathione-*S*-transferase, and that in maize protection from EPTC was conferred by the use of dichloroacetamide antidotes which enhanced levels of both glutathione and transferase enzyme (Lay *et al.*, 1975; Lay and Casida, 1976). Antidote also enhanced the amount of carbamoylation when EPTC sulphoxide itself was administered to maize (Lay and Casida, 1976; Hubbell and Casida, 1977).

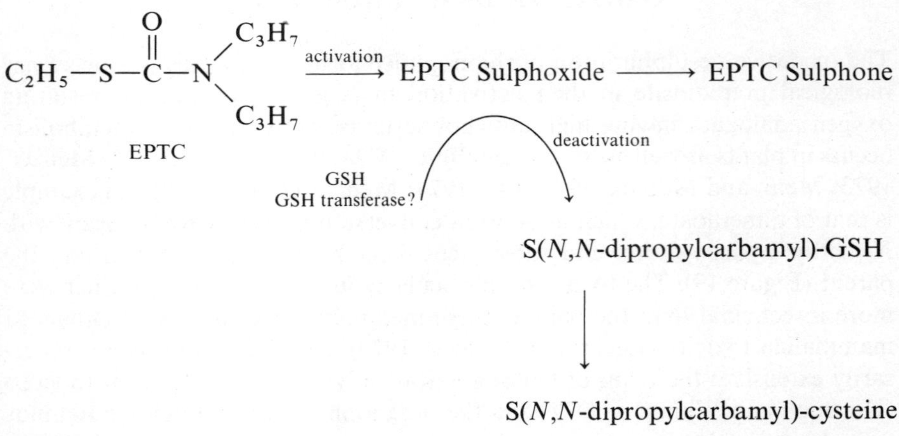

Figure 18 Proposed mechanism for the activation and deactivation of EPTC

Carringer *et al.* (1978) found that conjugation of EPTC sulphoxide and glutathione could be non-enzymic as it occurred *in vitro* in boiled extract and also in buffer alone. In contradiction to previous findings, no glutathione-*S*-transferase activity was found in maize root extracts using either EPTC sulphoxide or atrazine as substrate. Non-enzymic conjugation was also demonstrated by Leavitt and Penner (1979). Later findings, by Horváth and Pulay (1980) were further contrary in that it was proposed that the sulphone of EPTC is the true carbamoylating agent, rather than the sulphoxide. In young maize seedlings, these workers identified the glutathione conjugate as the main metabolite. However, EPTC sulphoxide was reported not to carbamoylate glutathione in aqueous solution, but EPTC sulphone did. Moreover, when EPTC and its oxidized analogues were added to plant homogenates, the glutathione conjugate was produced only with EPTC sulphone. EPTC was oxidized partly to sulphoxide, and EPTC sulphoxide itself was not metabolized. Of all these *in vitro* transformations, carbamoylation of glutathione by EPTC sulphone was the most rapid, and by extrapolating to the intact plant it was concluded that this would be the rate-limiting reaction, thus explaining why free EPTC sulphone is not observed as a major metabolite *in vivo*.

Although ability to conjugate an EPTC metabolite may explain the differing susceptibilities of maize and oat to EPTC, the differential toxicity of tiocarbazil appeared to be unrelated to this. The tolerant rice and susceptible *Echinochloa* spp. both metabolized tiocarbazil to its sulphoxide, and the sulphoxide itself was more toxic to rice. Resistance in rice appeared to be due to slower uptake of the herbicide (Santi and Gozzo, 1976).

OXIDATIVE DESULPHURATION

The oxidative desulphuration of thionophosphorus insecticides is a necessary biological prerequisite in their activation in target organisms, the resulting oxygen analogues having high anticholinesterase activities. Such metabolism occurs in plants as well as in animals (Bull, 1972a; b; Miyamoto, 1972; Menzer, 1973; Menn and McBain, 1974; Eto, 1974; Menn and Still, 1977). An example is that of dimethoate, which underwent conversion into the oxon in bean, with *N*-methyl hydroxylation and subsequent demethylation of both this and the parent (Figure 19). The oxon and particularly its demethylated product were more insecticidal than the parent; these metabolites also possessed enhanced mammalian toxicity (Lucier and Menzer, 1970). Desulphuration is not necessarily extensive; the oxon of leptophos was only slowly produced in tobacco (Dorough and Whitacre, 1977) and the organophosphate herbicide butamifos was also only slowly oxidized in bean, rice and carrot (Mihara *et al*., 1976). Many organophosphates can, of course, give rise to residues formed by either desulphuration or sulphoxidation or both. Such is the case of disulfoton where desulphuration of its sulphoxide and sulphone derivatives was observed (Metcalf *et al*., 1957). Oxidized metabolites were considerably more potent anticholinesterases than the parent. The oxygen analogue of piperophos was only a minor metabolite in rice, the major residues being sulphur-oxidized metabolites of the thioester cleavage product. It was, however, unclear whether the parent underwent cleavage or if desulphuration was a prerequisite (Mayer *et al*., 1981).

Oxidative desulphuration in animals and insects by mixed function oxidases has been well established (Eto, 1974; Kulkarni and Hodgson, 1980) but little is known about the corresponding enzymology in plants though Knaak *et al*. (1962) showed that plant peroxidases could catalyse the oxidation of parathion to paraoxon.

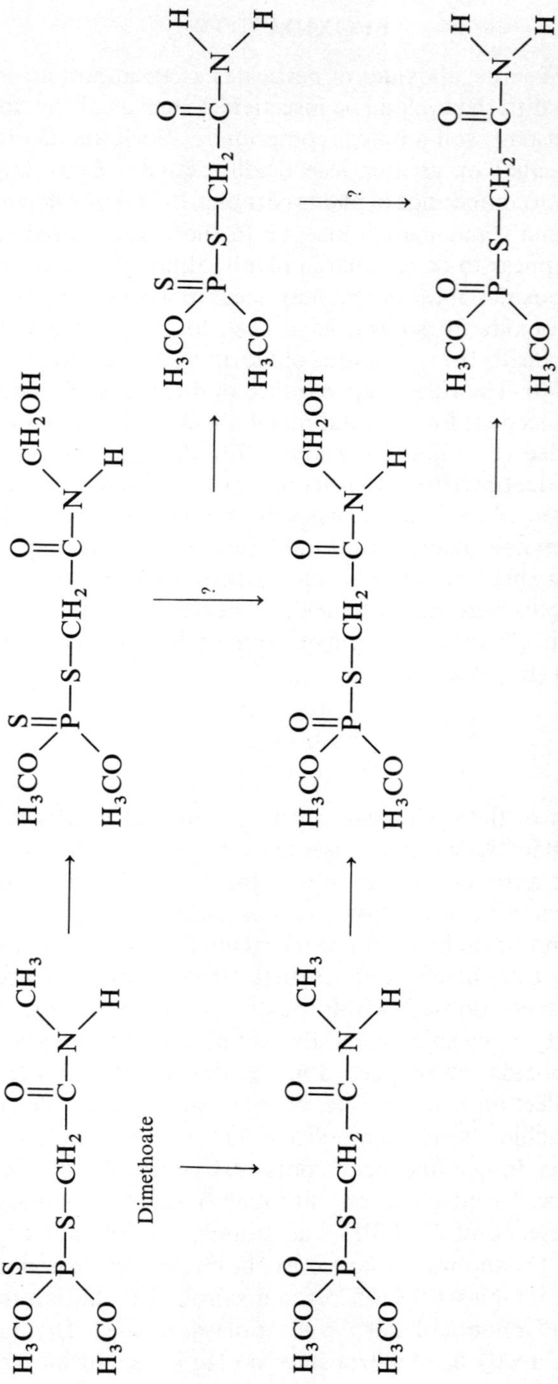

Figure 19 Oxidation of dimethoate

EPOXIDATION

The formation of stable epoxides of pesticides as an important mode of metabolism is confined to the cyclodiene insecticides such as aldrin and heptachlor, which are persistent soil-applied compounds. Such metabolism produces compounds of equal or greater insecticidal activity. Epoxidation is a well known fate of such compounds in plants (Brooks, 1974) but whereas the epoxides formed in mammals and insects may be further metabolized to more polar products these appear to be residual in plants, although the formation of other products from epoxides or the parent may occur to some extent. Several products of endrin, the epoxide of isodrin, have been found in plants (Bayless *et al.*, 1970; Nash and Beall, 1971) and also of aldrin and its epoxide dieldrin (Weisgerber *et al.*, 1970). The rate of appearance of dieldrin from aldrin in pea was not sufficient to account for the total metabolism of aldrin, but other products were not identified (Earl and Kennedy, 1973) although these would be likely to include those identified from *in vitro* preparations (McKinney and Mehendale, 1973). Metabolism of aldrin in cell suspension cultures of bean and potato was similar to that in whole plants (Brain and Lines, 1977), i.e. mainly by epoxidation to dieldrin, with small amounts of aldrin *trans*-diol. No diol formation was observed when cells were fed with dieldrin suggesting that the diol was formed directly from aldrin rather than by hydration of dieldrin, which is the favoured route in animals (Brooks, 1974).

Enzymology

The epoxidation of the cyclodienes in mammals and insects is known to be carried out by mixed function oxidases (Brooks, 1974; Kulkarni and Hodgson, 1980) and some authors have proposed that a similar mechanism exists in plants. The evidence for this, is however, insubstantial. The *in vitro* epoxidation of aldrin has been studied by several workers but the reported enzyme properties vary somewhat. Lichenstein and Corbett (1969) initially demonstrated the presence of aldrin epoxidase in cell-free extracts from pea roots, with the major fraction of activity being soluble and the rest distributed between low and high speed pellets. Epoxidation proceeded in the absence of NADPH and addition of this had no effect on reaction rate. Neither isodrin, the *endo-endo* isomer of aldrin nor heptachlor, were metabolized. Yu *et al.* (1971) found that aldrin epoxidase activity in pea and bean roots was pelletable to a large extent by low and high speed centrifugation, although a significant amount remained soluble. The presence of NADPH was stimulatory but not obligatory. The enzyme required O_2 and was inhibited by the characteristic mammalian mixed function oxidase inhibitors SKF 525A and safrole. In addition to epoxidation to dieldrin, a small amount of aldrin *trans*-diol was formed. This again appeared to be produced directly from aldrin since no lag phase for diol formation was observed, also when dieldrin itself was used as substrate, no formation of aldrin diol took place. Isodrin was oxidized to endrin ketone but heptachlor was not metabolized. Bean roots were more active sources of aldrin epoxidase than pea, and maize contained very low activity. Mehendale *et al.* (1972) reported that aldrin epoxidase of bean roots was recovered from both low and high speed pellets, little activity being detected in the high speed supernatant. Again, activity was essentially independent of NADPH, although this did to a limited extent enhance activity. Sources in order of activity were: bean > pea > broad bean (*Vicia faba* L.) > maize and significant variations were found between pea varieties. None of the sources epoxidized heptachlor. Subsequently it was reported that aldrin epoxidase could be mainly recovered in the microsomal fraction of bean and pea roots (Mehendale, 1973). Microsomes of pea and lupin (*Lupinus luteus* L.) roots contained aldrin epoxidase which was stimulated by NADPH and Mg^{2+} (Earl and Kennedy, 1973). The degrees of stimulation were not stated, nor was the percentage yield of epoxidase activity in the microsomes given. These workers observed that pretreatment of pea plants with aldrin precipitated a rise in the rate of subsequent aldrin metabolism, but the amount of dieldrin produced was no greater than that in the controls. Hence there appeared to be an induction of conversion into unknown metabolites. Extracts of plants treated with aldrin and dieldrin contained an enhanced capacity to convert aldrin *in vitro* into unknown metabolites. In a subsequent paper, aldrin epoxidase of pea was assayed in the low speed supernatant; activity was associated to a lesser extent with various low speed pellets (Earl and Kennedy, 1975). NADPH was stimulatory but not essential. Activity was enhanced by a variety of electron acceptors and by chelators and solubilizers which inhibit mammalian cytochrome P_{450}-dependent oxidations. It was suggested that aldrin may be epoxidized by a terminal oxidase competing with P_{450} for electrons. In addition to these studies, Oloffs (1970) reported the epoxidation of aldrin by an extracellular activity associated with pea roots. The enzyme was inhibited by heptachlor but did not metabolize it and had a pH optimum of 6.5, similar to those reported for the intracellular enzymes.

Metabolites of aldrin in pea and bean roots *in vitro* other than dieldrin were identified as the *cis* and *trans* aldrin diols, aldrin ketone and the *exo*- and *endo*-alcohols of aldrin, although dieldrin was by far the major metabolite (McKinney and Mehendale, 1973). The failure of preparations to interconvert any of these metabolites suggested that all metabolites, with the exception of the alcohols, may arise from a common oxygenated intermediate, or that metabolites may arise directly from aldrin, but by separate reactions (Figure 20).

The studies carried out on aldrin epoxidase indicate that it is microsomal to a certain extent, but not exclusively so. The observations of Yu *et al.* (1971) have suggested some mixed function oxidase characteristics, but more rigorous studies need to be undertaken. An external reducing agent appears to be unnecessary but this may be supplied by endogenous NADPH in the tissue

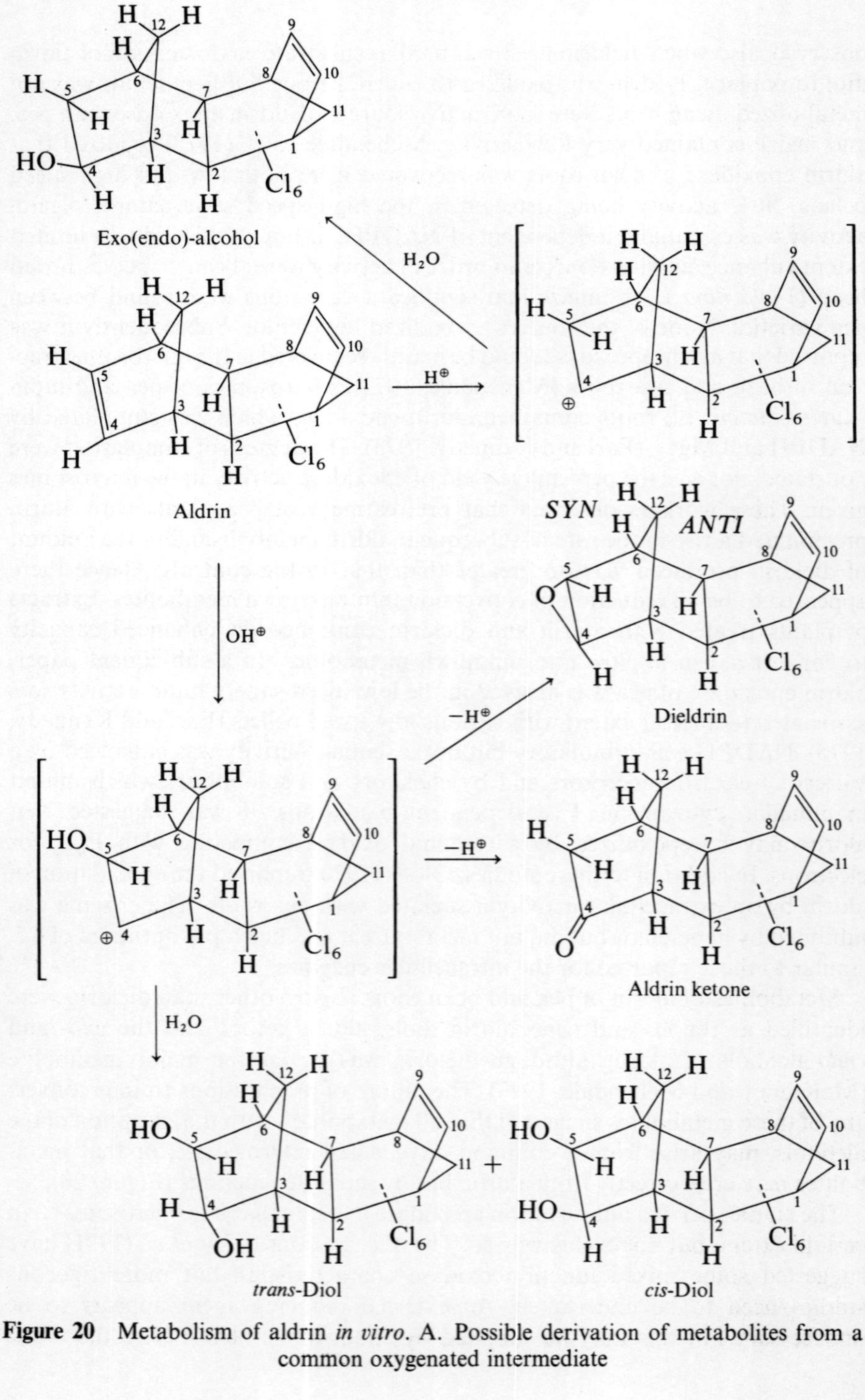

Figure 20 Metabolism of aldrin *in vitro*. A. Possible derivation of metabolites from a common oxygenated intermediate

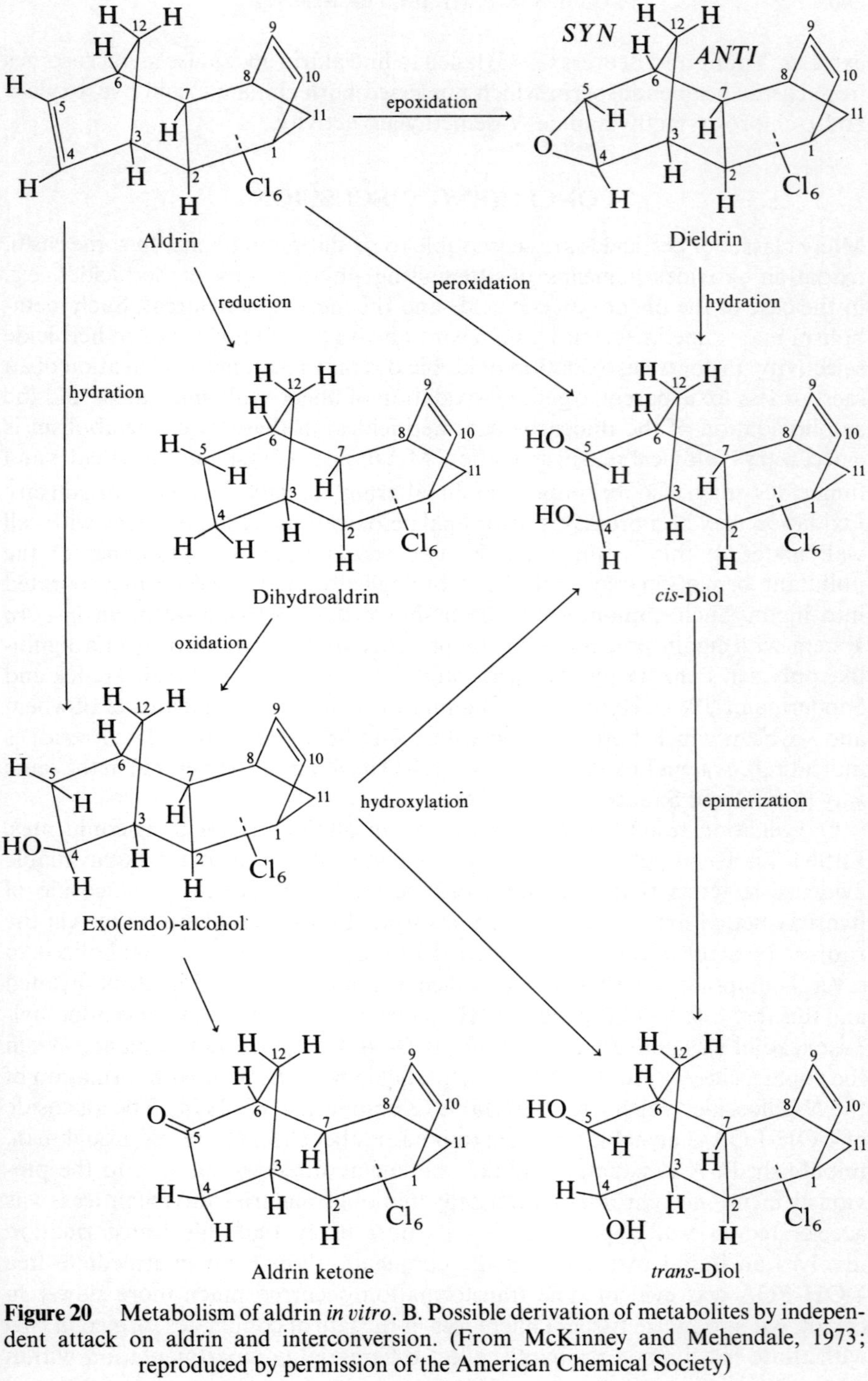

Figure 20 Metabolism of aldrin *in vitro*. B. Possible derivation of metabolites by independent attack on aldrin and interconversion. (From McKinney and Mehendale, 1973; reproduced by permission of the American Chemical Society)

extracts. Young and Beevers (1976) failed to find aldrin epoxidase in microsomes from castor bean endosperm which possessed both cinnamic acid hydroxylase and *p*-chloro-*N*-methylaniline *N*-demethylase activity.

CONCLUDING DISCUSSION

Many classes of pesticides are susceptible to oxidation in plants. In some cases, oxidation provides a means of attenuating phytotoxicity of herbicides, e.g. in the case of the phenoxyacetic acids and the methylphenylureas. Such metabolism may sometimes, but by no means always be a major factor in herbicide selectivity. In contrast, oxidation of herbicides may result in the activation of an inert or less toxic parent, e.g. the β-oxidation of phenoxyalkanoic acids and the sulphoxidation of the thiocarbamate herbicides; in these cases metabolism is a necessary biological step in activation. Metabolism of systemic insecticides and fungicides may also be important in altering toxicity to target organisms. Oxidation may be a prerequisite for final detoxication via sequestering with cell wall material; this is illustrated by two recent examples. Quinones of the pollutant benzo(α)pyrene, produced biologically, may become incorporated into lignin. Such quinones have been shown to co-polymerize in an *in vitro* system with lignin precursors in the presence of peroxidase to form a lignin-like polymer. Benzo(α)pyrene itself could not be incorporated (v.d. Trenck and Sandermann, 1981). Hydrolysis of lignin from cell suspension cultures of wheat and soybean which were supplied with 2,4-D liberated hydroxylated residues indicating covalent linkage with lignin. 2,4-D itself was also liberated in the same way (Scheel and Sandermann, 1981b).

Oxygenation results in residues which can often accumulate as conjugates. Little is known about the stability of such conjugates in the plant, but available evidence suggests that conjugates can be readily cleaved. The glucoside of hydroxymethyl monolinuron, when absorbed by spinach plants either via the root or by stem injection accumulated in the leaves and was metabolized to several compounds, tentatively identified as the aglycone and its demethylated and unsubstituted amide products (Haque *et al.*, 1978). The fungicide 5-methyl-3-isoxazolol was converted into both its *O*- and *N*-glucosides in plants. When the *O*-glucoside was itself fed, the free pesticide was liberated with formation of the *N*-glucoside, which was itself stable (Kamimura *et al.*, 1974). The glucoside of 4-OH-POA is unstable when fed to potato tuber slices (Cole and Loughman, unpublished). A substantial amount was metabolized within 48 h to the previously mentioned chromatographically immobile material and the process was accelerated in water-aged slices. It is most likely that this transformation involves an initial hydrolysis of the conjugate, though no intermediate free 4-OH-POA was evident. The transformation occurred much more slowly in carrot, oat and maize tissue. Differences in the fate of conjugates formed *in vivo* with those fed directly probably reflect differential compartmentation within the cell. Vacuolar secretion may conceivably account for the relative stability of conjugates formed *in vivo* although little is known about the possible importance of such a process. Conjugates in crops of course, represent a pool of potential aglycone release if ingested by animals. Labelled conjugates of carbaryl and ethiofencarb residues isolated from bean plants and fed to rats resulted in radioactivity being excreted via the urine, indicating absorption from the gut. It was not clear, however, if the original conjugates were excreted intact or were hydrolysed with subsequent reconjugation in the body. Studies on the metabolism of 1-naphthol conjugates in the rat indicate that the latter possibility can occur (Dorough, 1976).

The physiological activity of systemic insecticide/fungicide metabolites towards the plant is an area given little attention. The phenolic hydrolysis products of carbofuran and its furan-oxidized metabolites promote plant growth which appears to be due to a competitive inhibition of IAA oxidase (Lee, 1977; Franco and Durán, 1981). Competition for sites of oxidation may underlie some of the synergistic properties of certain pesticide mixtures. Two carbamates, carbaryl and 4-benzothiophane *N*-methylcarbamate, specifically inhibited the second demethylation step of monuron in cotton leaf discs (Swanson and Swanson, 1968b), whilst carbaryl also inhibited the detoxication of linuron in carrot leaves (Del Rosario and Putnam, 1973). The inhibition of methylphenylurea demethylation *in vitro* by *N*-methyl-carbamates has been mentioned previously.

Much more remains to be discovered about mechanisms of oxidation. In several instances mixed function oxidases have been implicated, similar to those established in mammals and insects, although in only two cases for xenobiotics has an association with the presence of cytochrome P_{450} been suggested (Frear *et al.*, 1969; Young and Beevers, 1976). The variety of endogenous substrates known to be oxidized by such a system (West, 1980) implies that these enzymes should be important in xenobiotic metabolism, but the known list of substrates is a fraction of that known for mammals and insects. Peroxidases may also be important (Lamoureux and Frear, 1979). In contrast to cases of similarity, several instances are evident where plants produce the same oxidation products as animals from a given xenobiotic, but by enzymes with substantially different properties.

An important point in the study of pesticide metabolism in plants is the ability to assess the likely nature of residues in crops. Most studies make use of experimental designs from which results cannot be extrapolated to the field. Plants grown hydroponically are widely favoured, where the pesticide is introduced via the root bathing solution and callus and particularly cell suspension cultures are gaining increasing popularity. The experimental advantages of such systems are obvious, allowing controlled conditions in which external factors may be eliminated. Suspension cultures offer a sterile system and other problems such as photoconversion can be eliminated. Rapidity of results, reproducibility and ease of manipulation are other attractive facets. However, notable examples of such 'unnatural" systems have been pointed out as producing correspondingly varied metabolite profiles, and the patterns of metabolism even in the whole plant can be greatly influenced by environment and mode of pesticide introduction to the plant. Hence, whilst such systems may be invaluable in probing the biochemistry and physiology of plants, as far as human toxicology is concerned, there appears to be no substitute for approximating field conditions as closely as possible, where environment clearly plays a part in determining the pattern of terminal residues.

ACKNOWLEDGEMENTS

The support of the Agricultural Research Council and the Cecil Pilkington Charitable Trust is gratefully acknowledged.

REFERENCES

Adele, P., Reichhart, D., Salaün, J.-P., Benveniste, I., and Durst, F. (1981). 'Induction of cytochrome P_{450} and monooxygenase activity by 2,4-dichlorophenoxyacetic acid in higher plant tissue', *Pl. Sci. Lett.*, **22**, 39–46.

Andrawes, N. R., Bagley, W. P., and Herrett, R. A. (1971). 'Metabolism of 2-methyl-2-(methylthio) propionaldehyde-*O*-(methylcarbamoyl)oxime (Temik aldicarb pesticide) in potato plants', *J. Agr. Food Chem.*, **19**, 731–737.

Arjmand, M., Hamilton, R. H., and Mumma, R. O. (1978). 'Metabolism of 2,4,5-trichlorophenoxyacetic acid. Evidence for amino acid conjugates in soybean callus tissue', *J. Agr. Food Chem.*, **26**, 1125–1128.

Ashworth, R. J., and Sheets, T. J. (1972). 'Metabolism of carbofuran in tobacco', *J. Agr. Food Chem.*, **20**, 407–412.

Baldwin, B. C. (1977). 'Xenobiotic metabolism in plants', in *Drug Metabolism: From Microbe to Man* (eds D. V. Parke, and R. L. Smith), pp. 191–217. Taylor and Francis Ltd., London.

Bartley, W. J., Andrawes, N. R., Chancey, E. L., Bagley, W. P., and Spurr, H. W. (1970). 'The metabolism of Temik aldicarb pesticide [2-methyl-2-(methylthio)propionaldehyde *O*-(methylcarbamoyl)oxime] in the cotton plant', *J. Agr. Food Chem.*, **18**, 446–453.

Baskakov, Yu. A., and Zemskaya, V. A. (1959). 'The possibility of transformation of carbanilic esters in plants', *Sov. Pl. Physiol.*, **6**, 63–68.

Bayless, A., Weisgerber, I., Klein, W., and Korte, F. (1970). 'Contributions to ecological chemistry XXV. Transformation and residues of ^{14}C-endrin in cotton', *Tetrahedron*, **26**, 775–778.

Benveniste, I., Salaün, J.-P., and Durst, F. (1977). 'Wounding-induced cinnamic acid hydroxylase in Jerusalem artichoke tuber', *Phytochem*, **16**, 69–73.

Benveniste, I., Salaün, J.-P., and Durst, F. (1978). 'Phytochrome-mediated regulation of a monooxygenase hydroxylating cinnamic acid in etiolated pea seedlings', *Phytochem.*, **17**, 359–363.

Beynon, K. I., Roberts, T. R., and Wright, A. N. (1974). 'The degradation of the herbicide benzoylprop-ethyl on the foliage of cereal seedlings', *Pestic. Biochem. Physiol.*, **4**, 98–107.

Bingham, S. W., and Shaver, R. (1971). 'Uptake, translocation and degradation of diphenamid in plants', *Weed Sci.*, **19**, 639–643.

Boldt, P. F., and Putnam, A. R. (1981). 'Selectivity mechanism for foliar applications of diclofop-methyl II. Metabolism', *Weed Sci.*, **29**, 237–241.

Brain, K. R., and Lines, D. S. (1977). 'Uptake and metabolism of aldrin in plant tissue cultures', in *Plant Tissue Culture and its Bio-Technological Applications. Proceedings of the 1st International Congress on Medicinal Plant Research Section B* (eds W. Barz, E. Reinhard, and M. H. Zenk), pp. 197–203. Springer-Verlag, Berlin.

Bristol, D. W., Ghanuni, A. M., and Oleson, A. E. (1977). 'Metabolism of 2,4-Dichlorophenoxyacetic acid by wheat cell suspension cultures', *J. Agr. Food Chem.*, **25**, 1309–1314.

Brooks, G. T. (1974). *Chlorinated Insecticides, vol. II. Biological and Environmental Aspects.* CRC Press, Cleveland, Ohio.

Bull, D. L. (1968). 'Metabolism of UC-21149[2-methyl-2-(methylthio)-propionaldehyde *O*-(methylcarbamoyl)] oxime in cotton plants and soil in the field', *J. Econ. Entomol.*, **61**, 1598–1602.

Bull, D. L. (1972a). 'Metabolism of organophosphorus insecticides in animals and plants', *Res. Rev.*, **43**, 1–23.

Bull, D. L. (1972b). 'Metabolism of organophosphorus insecticides in plants', in *Radiotracer Studies of Chemical Residues in Food and Agriculture*. pp. 25–33, International Atomic Energy Agency, Vienna.

Bull, D. L., and Ivie, G. W. (1981). 'Fate of *O*-[4-[(4-chlorophenyl)thio]phenyl]*O*-ethyl *S*-propyl phosphorothioate (RH-0994) in cotton', *J. Agr. Food Chem.*, **29**, 121–125.

Bull, D. L. and Lindquist, D. A. (1964). 'Metabolism of 3-hydroxy-*N*,*N*-dimethyl-crotonamide dimethyl phosphate by cotton plants, insects and rats', *J. Agr. Food Chem.*, **12**, 310–317.

Bull, D. L., Lindquist, D. A., and Coppedge, J. R. (1967). 'Metabolism of 2-methyl-2(methylthio)propionaldehyde *O*-(methylcarbamoyl)oxime (Temik, UC-21149) in insects', *J. Agr. Food Chem.*, **15**, 610–616.

Burt, M. E., and Corbin, F. T. (1978). 'Uptake, translocation and metabolism of propham by wheat (*Triticum aestivum*), sugar beet (*Beta vulgaris*) and alfalfa (*Medicago sativa*),' *Weed Sci.*, **26**, 296–303.

Butts, E. R., and Foy, C. L. (1974). 'Comparative uptake and metabolism of methazole in prickly sida and cotton', *Pestic. Biochem. Physiol.*, **4**, 44–55.

Camm, E. L., and Towers, G. H. N. (1973a). 'Is cinnamic acid 4-hydroxylase of potato a light-induced enzyme?' *Can. J. Bot.*, **51**, 824–825.

Camm, E. L., and Towers, G. H. N. (1973b). 'Phenylalanine ammonia lyase', *Phytochem.*, **12**, 961–973.

Carringer, R. D., Rieck, C. E., and Bush, L. P. (1978). 'Metabolism of EPTC in corn (*Zea mays*)', *Weed Sci.*, **26**, 157–160.

Casida, J. E., Gray, R. A., and Tilles, H. (1974). 'Thiocarbamate sulfoxides: Potent, selective and biodegradable herbicides', *Science*, **184**, 573–574.

Casida, J. E., and Lykken, C. (1969). 'Metabolism of organic pesticide chemicals in higher plants', *Ann. Rev. Pl. Physiol.*, **20**, 607–636.

Chin, W. T., Stone, G. M., and Smith, A. E. (1970). 'Metabolism of carboxin by barley and wheat plants', *J. Agr. Food Chem.*, **18**, 709–712.

Chkanikov, D. I., Makeev, A. M., Pavlova, N. N., and Dubovoi, V. P. (1971). 'Water soluble 2,4-D metabolites in green maize and bean plants', *Sov. Pl. Physiol.* **18**, 88–94.

Chkanikov, D. I., Makeev, A. M., Pavlova, N. N., and Dubovoi, V. P. (1972). '*N*-(2,4-dichlorophenoxyacetyl)-L-glutamic acid: A new metabolite of 2,4-D', *Sov. Pl. Physiol.*, **19**, 364–369.

Chkanikov, D. I., Pavlova, N. N., Makeev, A. M., Nazarova, T. A., and Makoveichuk, A. Yu. (1977). 'Paths of detoxication and immobilization of 2,4-D in cucumber plants', *Sov. Pl. Physiol.*, **24**, 457–463.

Cole, D. J., and Loughman, B. C. (1982). 'Metabolism of phenoxyacetic acid in potato (*Solanum tuberosum* L.) tuber slices', *Pl. Sci. Lett.*, **27**, 289–298.

Collet, G. F., and Pont, V. (1974). 'Distribution and metabolism of methabenzthiazuron in plant species', *Weed Res.*, **14**, 151–165.

Collins, D. J., and Gaunt, J. K. (1970). 'The metabolic fate of 4-chloro-2-methylphenoxyacetic acid in peas', *Biochem. J.*, **118**, 54P.

Collins, D. J., and Gaunt, J. K. (1971). 'The metabolism of 4-chloro-2-methylphenoxyacetic acid in plants', *Biochem. J.*, **124**, 9P.

Coppedge, J. R., Lindquist, D. A., Bull, D. L., and Dorough, H. W. (1967). 'Fate of 2-methyl-2-(methylthio)propionaldehyde *O*-(methylcarbamoyl)oxime (Temik) in cotton plants and soil', *J. Agr. Food Chem.*, **15**, 902–910.

Corbett, J. R. (1974). *The Biochemical Mode of Action of Pesticides.* Academic Press, London, New York.

Crosby, D. G. (1964). 'Metabolites of 2,4-Dichlorophenoxyacetic acid (2,4-D) in bean plants', *J. Agr. Food Chem.*, **12**, 3–6.

Crosby, D. G. (1973). 'The fate of pesticides in the environment', *Ann. Rev. Pl. Physiol.*, **24**, 467–492.

Davidonis, G. H., Hamilton, R. H., and Mumma, R. O. (1980). 'Metabolism of 2,4-dichlorophenoxyacetic acid (2,4-D) in soybean root callus. Evidence for the conversion of 2,4-D amino acid conjugates to free 2,4-D', *Pl. Physiol.*, **66**, 537–540.

Davis, D. G., Hodgson, R. H., Dusbabek, K. E., and Hoffer, B. L. (1978). 'The metabolism of the herbicide diphenamid (*N-N*-dimethyl-2,2-diphenyl-acetamide) in cell suspensions of soybean (*Glycine max*)', *Physiol. Plant.*, **44**, 87–91.

Davis, D. G., Hoerauf, R. A., Dusbabek, K. E., and Dougall, D. K. (1977). 'Isopropyl-*m*-chlorocarbanilate and its hydroxylated metabolites: their effects on cell suspensions and cell division in soybean and carrot', *Physiol. Plant.*, **40**, 15–20.

Del Rosario, D. A., and Putnam, A. R. (1973). 'Enhancement of foliar activity of linuron with carbaryl', *Weed Sci.*, **21**, 465–468.

Donald, W. W., and Shimabukuro, R. H. (1980). 'Selectivity of diclofop-methyl between wheat and wild oat', *Physiol. Plant.*, **49**, 459–464.

Dorough, H. W. (1968). 'Metabolism of Furadan (NIA-10242) in rats and houseflies', *J. Agr. Food Chem.*, **16**, 319–325.

Dorough, H. W. (1976). 'Biological activity of pesticide conjugates', in *Bound and Conjugated Pesticide Residues* (eds D. D. Kaufman, G. G. Still, G. D. Paulson, and S. K. Bandal), pp. 11–34, American Chemical Society Symposium Series Vol. 29, Washington DC.

Dorough, H. W., and Whitacre, D. M. (1977). 'Persistence and metabolism of ^{14}C-Leptophos in Burley tobacco', *J. Econ. Entomol.*, **70**, 153–155.

Dorough, H. W., Whitacre, D. M., and Cardona, R. A. (1973). 'Metabolism of the herbicide methazole in cotton and beans, and fate of certain of its polar metabolites in rats', *J. Agr. Food Chem.*, **21**, 797–803.

Dorough, H. W. and Wiggins, O. G. (1969). 'Nature of the water soluble metabolites of Carbaryl in bean plants and their fate in rats', *J. Econ. Entomol.*, **62**, 49–53.

Dusky, J. A., Davis, D. G., and Shimabukuro, R. H. (1980). 'Metabolism of diclofop-methyl(methyl-2-[2′,4-dichlorophenoxyphenoxy]propanoate) in cell suspensions of diploid wheat (*Triticum monococcum* L.)', *Physiol. Plant.*, **49**, 151–156.

Earl, J. W., and Kennedy, I. R. (1973). 'Pretreatment effects on the rates of aldrin metabolism in pea plants (*Pisum sativum*)', *Aust. J. Biol. Sci.*, **26**, 341–347.

Earl, J. W., and Kennedy, I. R. (1975). 'Aldrin epoxidase from pea roots', *Phytochem.*, **14**, 1507–1512.

Eshel, Y., and Rubin, B. (1971). 'Metabolism of fluometuron in cotton and weeds as a basis for selective action', in *Herbicides, Fungicides, Formulation Chemistry. Proceedings of the 2nd International IUPAC Congress of Pesticide Chemistry* (ed. A. S. Tahori), pp. 113–123. Gordon and Breach, New York, London.

Eto, M. (1974). *Organophosphorus Pesticides: Organic and Biological Chemistry*. CRC Press, Cleveland, Ohio.

Faulkner, J. K., and Woodcock, D. (1964). 'Metabolism of 2,4-Dichlorophenoxyacetic acid ('2,4-D') by *Aspergillus niger* van Tiegh', *Nature* (London), **203**, 865.

Fawcett, C. H., Ingram, J. M. A., and Wain, R. L. (1954). 'The β-oxidation of ω-phenoxyalkylcarboxylic acids in the flax plant in relation to their plant-growth regulating ability', *Proc. Roy. Soc., Lond. B*, **142**, 60–72.

Fawcett, C. H., Pascal, R. M., Pybus, M. B., Taylor, H. F., Wain, R. L., and Wightman, F. (1959). 'Plant growth regulating activity in homologous series of ω-phenoxyalkanecarboxylic acids and the influence of ring substitution on their breakdown by β-oxidation within plant tissues', *Proc. Roy. Soc., Lond. B*, **150**, 95–119.

Fawcett, C. H., Taylor, H. F., Wain, R. L., and Wightman, F. (1958). 'The metabolism of certain acids, amides and nitriles within plant tissues', *Proc. Roy. Soc. Lond. B*, **148**, 543–570.

Feeny, R. W., Parochetti, J. V., and Colby, S. R. (1974). 'Selective action of chloroxuron on Soybean and tall morning glory', *Weed Sci.*, **22**, 143–150.

Feung, C-S., Hamilton, R. H., and Mumma, R. O. (1973). 'Metabolism of 2,4-dichlorophenoxyacetic acid V. Identification of metabolites in soybean callus tissue cultures', *J. Agr. Food Chem.*, **21**, 637–640.

Feung, C-S., Hamilton, R. H., and Mumma, R. O. (1975). 'Metabolism of 2,4-dichlorophenoxyacetic acid VII. Comparison of metabolites from five species of plant callus tissue cultures', *J. Agr. Food Chem.*, **23**, 373–376.

Feung, C-S., Hamilton, R. H., and Mumma, R. O. (1976). 'Metabolism of 2,4-dichlorophenoxyacetic acid X. Identification of metabolites in root callus tissues', *J. Agr. Food Chem.*, **24**, 1013–1015.

Feung, C-S., Hamilton, R. H., and Witham, F. H. (1971). 'Metabolism of 2,4-dichlorophenoxyacetic acid by soybean cotyledon callus culture tissues', *J. Agr. Food Chem.*, **19**, 475–479.

Feung, C-S., Hamilton, R. H., Witham, F. H., and Mumma, R. O. (1972). 'The relative amounts and identification of some 2,4-dichlorophenoxyacetic acid metabolites isolated from soybean cotyledon callus cultures', *Pl. Physiol.*, **50**, 80–86.

Feung, C-S., Loerch, S. L., Hamilton, R. H., and Mumma, R. O. (1978). 'Comparative metabolic fate of 2,4-dichlorophenoxyacetic acid in plants and plant tissue culture', *J. Agr. Food Chem.*, **26**, 1064–1067.

Fleeker, J. R. (1973). 'Removal of the acetate moiety of 2,4-dichlorophenoxyacetic acid in *Ribes sativum*', *Phytochem.*, **12**, 757–762.

Fleeker, J., and Steen, R. (1971). 'Hydroxylation of 2,4-D in several weed species', *Weed Sci.*, **19**, 507–510.

Franco, C., and Duran, N. (1981). 'Metabolites of carbofuran: effect on indole-3-acetic acid degradation', *Pestic. Biochem. Physiol.*, **16**, 136–140.

Frear, D. S. (1968). 'Microsomal *N*-demethylation by a cotton leaf oxidase system, of 3-(4-chlorophenyl)-1,1-dimethylurea (monuron)', *Science*, **162**, 674–675.

Frear, D. S., Hodgson, R. H., Shimabukuro, R. H., and Still, G. G. (1972). 'Behaviour of herbicides in plants', *Adv. Agron.*, **24**, 327–378.

Frear, D. S., and Swanson, H. R. (1972). 'New metabolites of monuron in excised cotton leaves', *Phytochem.*, **11**, 1919–1929.

Frear, D. S., and Swanson, H. R. (1974). 'Monuron metabolism in excised *Gossypium hirsutum* leaves: aryl hydroxylation and conjugation of 4-chlorophenylurea', *Phytochem.*, **13**, 357–360.

Frear, D. S., and Swanson, H. R. (1975). 'Metabolism of Cisanilide (*cis*-2,5-dimethyl-1-pyrrolidine carboxanilide) by excised leaves and cell suspension cultures of carrot and cotton', *Pestic. Biochem. Physiol.*, **5**, 73–80.

Frear, D. S., Swanson, H. R., Mansager, E. R., and Wien, R. G. (1978). 'Chloramben metabolism in plants: isolation and identification of glucose ester', *J. Agr. Food Chem.*, **26**, 1347–1351.

Frear, D. S., Swanson, H. R., and Tanaka, F. S. (1969). '*N*-demethylation of substituted 3-(phenyl)-1-methylureas: isolation and characterization of a microsomal mixed function oxidase from cotton', *Phytochem.*, **8**, 2157–2169.

Frear, D. S., Swanson, H. R., and Tanaka, F. S. (1972). 'Herbicide metabolism in plants', in *Recent Advances in Phytochemistry, Vol. 5* (eds V. C. Runeckles and T. C. Tso), pp. 225–246. Academic Press, London, New York.

Frehse, H. (1971). 'Terminal residues of organophosphorus insecticides in plants', in *Pesticide Terminal Residues. International Symposium on Pesticide Terminal Residues. International Union of Pure and Applied Chemistry* (ed. A. S. Tahori), pp. 9–32. Butterworths, London.

Fuhremann, T. W., and Lichenstein, E. P. (1980). 'A comparative study of the persistence, movement and metabolism of six ^{14}C-insecticides in soil and plants', *J. Agr. Food Chem.*, **28**, 446–452.

Fujii, Y., Asaka, S., and Misato, T. (1980). 'Fate of propaphos (Kayaphos) in rice plants', *J. Pestic. Sci.*, **5**, 55–62.

Fukuto, T. R., and Metcalf, R. L. (1969). 'Metabolism of insecticides in plants and animals, *Ann. N.Y. Acad. Sci.*, **160**, 97–111.

Gaughan, L. C., and Casida, J. E. (1978). 'Degradation of *trans*- and *cis*-permethrin on cotton and bean plants', *J. Agr. Food Chem.*, **26**, 525–528.

Geissbühler, H., Haselbach, C., Aebi, H., and Ebner, L. (1963). 'The fate of *N*-(4-chlorophenoxy)-phenyl-*N*,*N*-dimethyl urea (C-1983) in soils and plants. III. Breakdown in soils and plants', *Weed Res.*, **3**, 277–297.

Geissbühler, H., Martin, H., and Voss, G. (1975). 'The substituted ureas', in *Herbicides—Chemistry, Degradation and Mode of Action* (eds P. C. Kearney and D. D. Kaufman), 2nd edition, vol. 1, pp. 209–291. Marcel Dekker, New York.

Golab, T., Herberg, R. J., Parka, S. J., and Tepe, J. B. (1966). 'The metabolism of carbon-14 diphenamid in strawberry plants', *J. Agr. Food Chem.*, **14**, 592–596.

Gorbach, S. G., Kuenzler, K., and Asshauer, J. (1977). 'On the metabolism of HOE 234080H in wheat', *J. Agr. Food Chem.*, **25**, 507–511.

Gorecka, K., Shimabukuro, R. H., and Walsh, W. C. (1981). 'Aryl hydroxylation: a selective mechanism for the herbicides diclofop-methyl and clofop-isobutyl in gramineous species.' *Physiol. Plant.*, **53**, 55–63.

Goren, R. (1969). 'The effect of fluometuron on the behaviour of *Citrus* leaves', *Weed Res.*, **9**, 121–135.

Grace, N. H. (1939). 'Physiological activity of a series of naphthyl acids', *Can. J. Res.* (*C*), **17**, 247–255.

Gross, D., Laanio, T., Dupuis, G., and Esser, H. O. (1979). 'The metabolic behaviour of chlortoluron in wheat and soil', *Pestic. Biochem. Physiol.*, **10**, 49–59.

Guroff, G., Daly, J. W., Jerina, D. M., Renson, J., Witkop, B., and Udenfriend, S. (1967). 'Hydroxylation-induced migration: The NIH shift', *Science*, **157**, 1524–1530.

Hajjar, N. P., and Hodgson, E. (1980). 'Flavin adenine dinucleotide-dependent monooxygenase: its role in the sulfoxidation of pesticides in mammals', *Science*, **209**, 1134–1136.

Hamilton, R. H., Hurter, J., Hall, J. K., and Ercegovich, C. D. (1971). 'Metabolism of phenoxyacetic acids: metabolism of 2,4-D and 2,4,5-T by bean plants', *J. Agr. Food Chem.*, **19**, 480–483.

Haque, A., Schuphan, I., and Ebing, W. (1978). 'On the metabolism of phenylurea herbicides X. Movement and behaviour of a glucoside conjugate in plant and soil', *Chemosphere*, **7**, 675–680.

Hawf, L. R., and Behrens, R. (1974). 'Selectivity factors in the response of plants to 2,4-DB', *Weed Sci.*, **22**, 245–249.

Hilton, J. L., Schlaren, A. L., St. John, J. B., Moreland, D. E., and Norris, K. H. (1969). 'Modes of action of pyridazinone herbicides', *Weed Sci.*, **17**, 541–547.

Hodgson, R. H., Dusbabek, K. E., and Hoffer, B. L. (1974). 'Diphenamid metabolism in tomato: time course of an ozone fumigation effect', *Weed Sci.*, **22**, 205–210.

Hodgson, R. H., Frear, D. S., Swanson, H. R., and Regan, L. A. (1973). 'Alteration of diphenamid metabolism in tomato by ozone', *Weed Sci.*, **21**, 542–548.

Hodgson, R. H., and Hoffer, B. L. (1977). 'Diphenamid metabolism in pepper and an ozone effect II. Herbicide metabolite characterisation', *Weed Sci.*, **25**, 331–337.

Hogue, E. J., and Warren, G. F. (1968). 'Selectivity of linuron on tomato and parsnip', *Weed Sci.*, **16**, 51–54.

Holley, R. W. (1952). 'Studies on the fate of radioactive 2,4-dichlorophenoxyacetic acid in bean plants. II. A water-soluble transformation product of 2,4-D', *Arch. Biochem. Biophys.*, **35**, 171–175.

Holm, R. E., and Stallard, D. E. (1974). 'Metabolism of a pyrrolidine herbicide in corn and weeds', *Weed Sci.*, **22**, 10–14.

Horváth, L., and Pulay, A. (1980). 'Metabolism of EPTC in germinating corn: sulfone as the true carbamoylating agent', *Pestic. Biochem. Physiol.*, **14**, 265–270.

Hubbell, J. P., and Casida, J. E. (1977). 'Metabolic fate of the *N*,*N*-dialkylcarbamoyl moiety of thiocarbamate herbicides in rats and corn', *J. Agr. Food Chem.*, **25**, 404–413.

Hutber, G. N., Lord, E. I., and Loughman, B. C. (1978). 'The metabolic fate of phenoxyacetic acids in higher plants', *J. Exp. Bot.*, **29**, 619–629.

Ishimaru, A., and Yamazaki, I. (1977a). 'The carbon monoxide-binding hemoprotein reducible by hydrogen peroxide in microsomal fractions of pea seeds', *J. Biol. Chem.*, **252**, 199–204.

Ishimaru, A., and Yamazaki, I. (1977b). 'Hydroperoxide-dependent hydroxylation involving "H_2O_2-reducible hemoprotein" in microsomes of pea seeds', *J. Biol. Chem.*, **252**, 6118–6124.

James, C. S., and Prendeville, G. N. (1969). 'Metabolism of chlorpropham (isopropyl *m*-chlorocarbanilate) in various plant species', *J. Agr. Food Chem.*, **17**, 1257–1260.

Jordan, L. S., and Clerx, W. A. (1981). 'Accumulation and metabolism of bromacil in pineapple sweet orange (*Citrus sinensis* L.) and Cleopatra mandarin (*Citrus reticulata*)', *Weed Sci.*, **29**, 1–14.

Kamimura, S., Nishikawa, M., Saeki, H., and Takahi, Y. (1974). 'Absorption and metabolism of 3-hydroxy-5-methylisoxazole in plants and biological activity of its metabolites', *Phytopathology*, **64**, 1273–1281.

Kearney, P. C., and Kaufman, D. D. (1975). *Herbicides—Chemistry, Degradation and Mode of Action*, 2nd Edition, vols. 1 and 2, Marcel Dekker, New York.

Keeley, P. E., and Thullen, R. J. (1979). 'Metabolic fate of methazole in purple and yellow nutsedge', *Pestic. Biochem. Physiol.*, **10**, 275–286.

Keller, E., Eberspacher, J., and Lingens, F. (1979). 'Metabolism of chloridazon and antipyrine in plant cell suspension cultures', *Z. Naturforsch. Sect. C*, **34**, 914–922.

Klein, W. (1972). 'Metabolism of pesticides in higher plants', in *Environmental Quality and Safety. Chemistry, Toxicology and Technology Vol. I Global Aspects of Chemistry, Toxicology and Technology as Applied to the Environment* (eds F. Coulston and F. Korte), pp. 164–179, Georg Thieme, Stuttgart, Academic Press, New York.

Knaak, J. B., Stahmann, M. A., and Casida, J. E. (1962). 'Peroxidase and ethylenediaminetetraacetic acid-ferrous iron catalysed oxidation and hydrolysis of parathion to paroxon', *J. Agr. Food Chem.*, **10**, 154–158.

Krueger, H. R. (1975). 'Phorate sulfoxidation by plant root extracts', *Pestic. Biochem. Physiol.*, **5**, 396–401.

Krueger, H. R. (1977). 'Aldicarb sulfoxidation by plant root extracts', *Pestic. Biochem. Physiol.*, **7**, 154–160.

Kuhr, R. J., and Casida, J. E. (1967). 'Persistent glycosides of metabolites of methylcarbamate insecticide chemicals formed by hydroxylation in bean plants', *J. Agr. Food Chem.*, **15**, 814–824.

Kuhr, R. J., and Dorough, H. W. (1976). *Carbamate Insecticides: Chemistry, Biochemistry and Toxicology*. CRC Press, Cleveland, Ohio.

Kulkarni, A. P., and Hodgson, E. (1980). 'Metabolism of insecticides by mixed function oxidase systems', *Pharmacol. Ther.*, **8**, 379–475.

Kuratle, H., Rahn, E. M., and Woodmansee, C. W. (1969). 'Basis for selectivity of linuron on carrot and common ragweed', *Weed Sci.*, **17**, 216–219.

Lamoureux, G. L., and Frear, D. S. (1979). 'Pesticide metabolism in higher plants: *in vitro* enzyme studies', in *Xenobiotic Metabolism. In vitro Methods* (eds G. D. Paulson, D. S. Frear and E. P. Marks), pp. 72–128. American Chemical Society Symposium Series, Vol. 97, Washington DC.

Lay, M. M., and Casida, J. E. (1976). 'Dichloroacetamide antidotes enhance thiocarbamate sulfoxide detoxification by elevating corn root glutathione content and glutathione-*S*-transferase activity', *Pestic. Biochem. Physiol.*, **6**, 442–456.

Lay, M. M., Hubell, J. P., and Casida, J. E. (1975). 'Dichloroacetamide antidotes for thiocarbamate herbicides: mode of action', *Science*, **189**, 287–288.

Leavitt, J. R. C., and Penner, D. (1979). '*In vitro* conjugation of glutathione and other thiols with acetanilide herbicides and EPTC sulfoxide and the action of the herbicide antidote R-25788', *J. Agr. Food Chem.*, **27**, 533–536.

Lee, S. S., and Fang, S. C. (1973). 'Metabolism of monuron in excised leaves of corn and bean plants', *Weed Res.*, **13**, 59–66.

Lee, S. S., Griffin, D. A., and Fang, S. C. (1973). 'Identification of β-D-glucosides of ring-hydroxylated ureas in monuron-treated bean leaves', *Weed Res.*, **13**, 234–235.

Lee, T. T. (1977). 'Promotion of plant growth and inhibition of enzymic degradation of indole-3-acetic acid by metabolites of carbofuran, a carbamate insecticide', *Can. J. Bot.*, **55**, 574–579.

Lemin, A. J. (1966). 'Absorption, translocation and metabolism of diphenamid-1-C^{14} by tomato seedlings', *J. Agr. Food Chem.*, **14**, 109–111.

Lichenstein, E. P., and Corbett, J. R. (1969). 'Enzymatic conversion of aldrin to dieldrin with subcellular components of pea plants, *J. Agr. Food Chem.*, **17**, 589–594.

Lindquist, D. A., and Bull, D. L. (1967). 'Fate of 3-hydroxymethyl *cis*-crotonamide dimethylphosphate in cotton plants', *J. Agr. Food Chem.*, **15**, 267–272.

Linscott, D. L., Hagin, R. D., and Dawson, J. E. (1968). 'Conversion of 4(2,4-dichlorophenoxy) butyric acid to homologues by alfalfa. Mechanism of resistance to this herbicide', *J. Agr. Food Chem.*, **16**, 844–848.

Liu, L. C., Shimabukuro, R. H., and Nalewaja, J. D. (1978). 'Diuron metabolism in two sugar cane (*Saccharum officinarum*) cultivars', *Weed Sci.*, **26**, 642–646.

Loos, M. A. (1975). 'Phenoxyalkanoic acids', in *Herbicides—Chemistry, Degradation and Mode of Action* (eds P. C. Kearney and D. D. Kaufman), Vol. 1, 2nd edition, pp. 1–128, Marcel Dekker, New York.

Lucier, G. W., and Menzer, R. E. (1970). 'Nature of oxidative metabolites of dimethoate formed in rats, liver microsomes and bean plants', *J. Agr. Food Chem.*, **18**, 698–704.

Luckwill, L. C., and Lloyd-Jones, C. P. (1960a). 'Metabolism of plant growth regulators. I. 2,4-Dichlorophenoxyacetic acid in leaves of red and blackcurrant', *Ann. appl. Biol.*, **48**, 613–625.

Luckwill, L. C., and Lloyd-Jones, C. P. (1960b). 'Metabolism of plant growth regulators II. Decarboxylation of 2,4-Dichlorophenoxyacetic acid in leaves of apple and strawberry', *Ann. appl. Biol.*, **48**, 626–636.

Makeev, A. M., Makoveichuk, A. Yu, and Chkanikov, D. I. (1977). 'Mirosomal hydroxylation of 2,4-D in plants', *Doklady. Bot. Sci.*, **233**, 36–38.

Makoveichuk, A. Yu, Makeev, A. M., and Chkanikov, D. I. (1978). 'Compartmentation of certain metabolites of 2,4-D in cells of cucumber leaves', *Sov. Pl. Physiol.*, **25**, 320–322.

Mathre, D. E. (1971). 'Mode of action of oxathiin systemic fungicides Structure-activity relationships', *J. Agr. Food Chem.*, **19**, 872–874.

Matsunaka, S. (1972). 'Metabolism of pesticides in higher plants', in *Environmental Toxicology of Pesticides* (eds F. Matsumura, G. M. Boush, and T. Misato), pp. 341–364. Academic Press, New York, London.

Mayer, P., Kriemler, H. P., Hambock, H., and Laanio, T. L. (1981). 'Metabolism of *O*,*O*-dipropyl *S*-[2-(2′-methyl-1′-piperidinyl)-2-oxoethyl] phosphorodithioate (C19490) in paddy rice', *Agr. Biol. Chem.*, **45**, 355–360.

McBain, J. B., Hoffman, L. J., and Menn, J. J. (1970). 'Metabolic degradation of *O*-ethyl *S*-phenyl ethylphosphonodithioate (Dyfonate) in potato plants', *J. Agr. Food Chem.*, **18**, 1139–1144.

McKinney, J. D., and Mehendale, H. M. (1973). 'Formation of polar metabolites from aldrin by pea and bean root preparations', *J. Agr. Food Chem.*, **21**, 1079–1084.

McPherson, F. J., Markham, A., Bridges, J. W., Hartman, G. C., and Parke, D. V. (1975a). 'A comparison of the properties *in vitro* of biphenyl 2- and 4-hydroxylase in the mesocarp from Avocado Pear (*Persea americana*) and Syrian hamster hepatic tissue', *Biochem. Soc. Trans.*, **3**, 281–283.

McPherson, F. J., Markham, A., Bridges, J. W., Hartman, G. C., and Parke, D. V. (1975b). 'Effects of pre-incubation *in vitro* with 3,4-benzopyrene and phenobarbitone on the drug metabolising systems present in the microsomal and soluble fractions of the Avocado Pear (*Persea americana*)', *Biochem. Soc. Trans.*, **3**, 283–285.

Mehendale, H. M. (1973). 'Aldrin epoxidation by plant root extracts', *Phytochem.*, **12**, 1591–1594.

Mehendale, H. M., Skrentny, R. F., and Dorough, H. W. (1972). 'Oxidative metabolism of aldrin by subcellular root fractions of several plant species', *J. Agr. Food Chem.*, **20**, 398–407.

Menn, J. J. (1972). 'Absorption and metabolism of insecticide chemicals in plants', in *Degradation of Synthetic Organic Molecules in the Biosphere* National Academy of Science, Washington DC, pp. 206–243.

Menn, J. J., and McBain, J. B. (1974). 'New aspects of organophosphorus pesticides IV. Newer aspects of the metabolism of phosphonate insecticides', *Res. Rev.*, **53**, 35–51.

Menn, J. J., and Still, G. G. (1977). 'Metabolism of insecticides and herbicides in higher plants', *Crit. Rev.Toxicol.*, **5**, 1–21.

Menzer, R. E. (1973). 'Biological oxidation and conjugation of pesticidal chemicals', *Res. Rev.*, **48**, 79–116.

Menzer, R. E., and Casida, J. E. (1965). 'Nature of toxic metabolites formed in mammals, insects, and plants from 3(dimethoxyphosphinyloxy)-*N*,*N*-dimethyl-*cis*-crotonamide and its *N*-methyl analogue', *J. Agr. Food Chem.*, **13**, 102–112.

Metcalf, R. L., Fukuto, T. R., Collins, C., Borck, K., Abdel-Aziz, S., Munoz, R., and Cassil, C. C. (1968). Metabolism of 2,2-dimethyl-2,3-dihydrobenzofuranyl-7-*N*-methylcarbamate (Furadan) in plants, insects and mammals', *J. Agr. Food Chem.*, **16**, 300–311.

Metcalf, R. L., Fukuto, T. R., Collins, C., Borck, K., Burk, J., Reynolds, H. T., and Osman, M. F. (1966). 'Metabolism of 2-methyl-2-(methylthio)-propionamide *O*-(methylcarbamoyl)-oxime in plant and insect', *J. Agr. Food Chem.*, **14**, 579–584.

Metcalf, R. L., Fukuto, T. R., and March, R. B. (1957). 'Plant metabolism of Dithio-Systox and Thimet', *J. Econ. Entomol.*, **50**, 338–345.

Mihara, K., Nambu, K., Misaki, Y., and Miyamoto, J. (1976). 'Metabolism of ^{3}H-Cremart in rats and plants', *J. Pestic. Sci.*, **1**, 207–218.

Mine, A., Miyakado, M., and Matsunaka, S. (1975). 'The mechanism of bentazon selectivity', *Pestic. Biochem. Physiol.*, **5**, 566–574.

Miyamoto, J. (1972). 'Metabolism of organophosphorus pesticides in plants', in *Environmental Toxicology of Pesticides* (eds F. Matsumura, G. M. Boush and T. Misato), pp. 307–325, Academic Press, New York, London.

Montgomery, M. L., Chang, Y. L., and Freed, V. H. (1971). 'Comparative metabolism of 2,4-Dichlorophenoxyacetic acid by bean and corn plants', *J. Agr. Food Chem.*, **19**, 1219–1221.

Moreland, D. E., and Hill, K. L. (1959). 'The action of alkyl *N*-phenyl-carbamates on the photolytic activity of isolated chloroplasts', *J. Agr. Food Chem.*, **7**, 832–836.

Motooka, P. S., Corbin, F. T., and Worsham, A. D. (1977). 'Metabolism of Sandoz 6706 in soybean and sicklepod', *Weed Sci.*, **25**, 9–12.

Mumma, R. O., Khalifa, S., and Hamilton, R. H. (1971). 'Spectroscopic identification of metabolites of carbaryl in plants', *J. Agr. Food Chem.*, **19**, 445–451.

Murphy, J. J., Didriksen, J., and Gray, R. A. (1973). 'Metabolism of 2-(α-naphthoxy)-*N*-*N*-diethyl propionamide in tomato', *Weed Sci.*, **21**, 11–15.

Nakamura, Y., Ishikawa, K., and Kuwatsuka, S. (1977). 'Metabolic fate of benthiocarb herbicide in plants', *Agr. Biol. Chem.*, **41**, 1613–1620.

Nash, R. G., and Beall, M. L. (1971). 'Extraction and identification of endrin and heptachlor degradation products', *J. Assoc. Off. Anal. Chem.*, **54**, 959–963.

Nashed, R. B., Katz, S. E., and Ilnicki, R. D. (1970). 'The metabolism of ^{14}C-chlorbromuron in corn and cucumber', *Weed Sci.*, **18**, 122–125.

Naylor, A. W. (1976). 'Herbicide metabolism in plants', in *Herbicides Physiology, Biochemistry and Ecology* (ed. L. J. Audus), 2nd Edition, pp. 367–426. Academic Press, New York and London.

O'Brien, R. D. (1967). *Insecticides: Action and Metabolism.* Academic Press, New York and London.

Ogawa, K., Tsuda, M., Yamauchi, F., Yamaguchi, I., and Misato, T. (1976). 'Metabolism of 2-*sec*-butylphenyl *N*-methylcarbamate (Bassa, BPMC) in rice plants and its degradation in soils', *J. Pestic. Sci.*, **1**, 219–229.

Ogawa, K., Tsuda, M., Yamauchi, F., Yamaguchi, I., and Misato, T. (1977). 'Metabolism of 2-iso-propylphenyl *N*-methylcarbamate (Mipcin, MIPC) in rice plants and its degradation in soil', *J. Pestic. Sci.*, **2**, 51–57.

Ohkawa, H., Kaneko, H., and Miyamoto, J. (1977). 'Metabolism of permethrin in bean plants', *J. Pestic. Sci.*, **2**, 67–76.

Ohkawa, H., Nambu, K., and Miyamoto, J. (1980). 'Metabolic fate of fenvalerate (Sumicidin) in bean plants', *J. Pestic. Sci.*, **5**, 215–223.

Ohkawa, H., Yoshihara, R., Kohara, T., and Miyamoto, J. (1974). 'Metabolism of *m*-tolyl *N*-methylcarbamate in rats, houseflies and bean plants', *Agr. Biol. Chem.*, **38**, 1035–1044.

Oloffs, P. C. (1970). 'Epoxidation of aldrin by cell-free pea root preparations', *Pestic. Sci.*, **1**, 228–232.

Osgood, R. V., Romanowski, R. R., and Hilton, H. W. (1972). 'Differential tolerance of Hawaiian sugarcane cultivars to diuron', *Weed Sci.*, **20**, 537–539.

Payne, L. K., Stansbury, H. A., and Weiden, M. H. J. (1966). 'The synthesis and insecticidal properties of some cholinergic bisubstituted acetaldehyde *O*-(methylcarbamoyl)oximes', *J. Agr. Food Chem.*, **14**, 356–365.

Pillmoor, J. B., and Gaunt, J. K. (1981). 'The behaviour and mode of action of the phenoxyacetic acids in plants', in *Progress in Pesticide Biochemistry, Vol. 1* (eds D. H. Hutson and T. R. Roberts) pp. 147–218. John Wiley, Chichester.

Reichhart, D., Salaün, J-P., Benveniste, I., and Durst, F. (1980). 'Time course of induction of cytochrome P_{450} NADPH-cytochrome c reductase and cinnamic acid hydroxylase by phenobarbital, ethanol, herbicides and manganese in higher plant microsomes', *Pl. Physiol.*, **66**, 600–604.

Retzlaff, G., and Hamm, R. (1976). 'The relationship between CO_2 assimilation and the metabolism of bentazone in wheat plants', *Weed Res.*, **16**, 263–266.

Rhodes, R. C. (1977). 'Metabolism of 2-^{14}C-terbacil in alfalfa', *J. Agr. Food Chem.*, **25**, 1066–1068.

Rhodes, J. M., and Wooltorton, L. S. C. (1978). 'The biosynthesis of phenolic compounds in wounded plant storage tissues', in *Biochemistry of Wounded Plant Storage Tissues* (ed. G. Kahl), pp. 243–286, Walter de Gruyter, Berlin and New York.

Rich, P. R., and Bendall, D. S. (1975). 'Cytochrome components of plant microsomes', *Eur. J. Biochem.*, **55**, 333–341.

Rich, P. R., and Lamb, C. J. (1977). 'Biophysical and enzymological studies upon the interaction of *trans*-cinnamic acid with higher plant microsomal cytochrome P_{450}', *Eur. J. Biochem.*, **72**, 353–360.

Roberts, T. R. (1977). 'The metabolism of the herbicide flamprop-isopropyl barley', *Pestic. Biochem. Physiol.*, **7**, 378–390.

Robertson, M. M., and Kirkwood, R. C. (1970). 'The mode of action of foliage-applied translocated herbicides with particular reference to the phenoxy acid compounds. II. The mechanism and factors influencing translocation, metabolism and biochemical inhibition', *Weed Res.*, **10**, 94–120.

Rogers, R. L., and Funderburk, H. H. (1968). 'Physiological aspects of fluometuron in cotton and cucumber', *J. Agr. Food Chem.*, **16**, 434–440.

Rouchard, J., Moons, C., and Meyer, J. A. (1980). 'The metabolism of ^{14}C-aldicarb in the leaves of sugar beet plants', *Pestic. Sci.*, **11**, 483–492.

Rubin, B., and Eshel, Y. (1971). 'Phytotoxicity of fluometuron and its derivatives to cotton and weeds', *Weed Sci.*, **19**, 592–594.

Rusness, D. G., and Still, G. G. (1974). 'Effect of isopropyl-3-chlorocarbanilate and isopropyl-3-chloro-hydroxycarbanilate analogs upon oxidative phosphorylation in plant mitochondria', *Pestic. Biochem. Physiol.*, **4**, 24–30.

Ruzo, L. O., and Casida, J. E. (1979). 'Degradation of decamethrin on cotton plants', *J. Agr. Food Chem.*, **27**, 572–575.

Ryan, P. J., Gross, D., Owen, W. J., and Laanio, T. L. (1981). 'The metabolism of chlortoluron, diuron and CGA 43 057 in tolerant and susceptible plants', *Pestic. Biochem. Physiol.*, **16**, 213–221.

Sandermann, H., Diesperger, H., and Scheel, D. (1977). 'Metabolism of xenobiotics by plant cell cultures', in *Plant Tissue Culture and its Biotechnological Applications. Proceedings of the First International Congress on Medicinal Plant Research. Section B* (eds W. Barz, E. Reinhard, and M.-H. Zenk), pp. 178–196, Springer-Verlag, Berlin.

Santi, R., and Gozzo, F. (1976). 'Degradation and metabolism of drepamon in rice and barnyard grass', *J. Agr. Food Chem.*, **24**, 1229–1235.

Scheel, D., and Sandermann, H. (1981a). 'Metabolism of 2,4-dichlorophenoxyacetic acid in cell suspension cultures of soybean (*Glycine max.* L.) and wheat (*Triticum aestivum* L.) I. General results', *Planta*, **152**, 248–252.

Scheel, D., and Sandermann, H. (1981b). 'Metabolism of 2,4-dichlorophenoxyacetic acid in cell suspension cultures of soybean (*Glycine max* L.) and wheat (*Triticum aestivum* L.) II. Evidence for incorporation into lignin', *Planta*, **152**, 253–258.

Schlagbauer, B. G. L., and Schlagbauer, A. W. J. (1972). 'The metabolism of carbamate pesticides—a literature analysis—Part I', *Res. Rev.*, **42**, 1–84.

Schultz, D. P., and Tweedy, B. G. (1971). 'Uptake and metabolism of *N,N*-dimethyl-2,2-diphenylacetamide in resistant and susceptible plants', *J. Agr. Food Chem.*, **19**, 36–40.

Schuphan, I., and Ebing, W. (1977). 'Metabolism of phenylurea herbicides VII. Metabolism studies and balance of the fate of ^{14}C-buturon after application to wheat', *Weed Res.*, **17**, 181–189.

Schuphan, I., and Ebing, W. (1978). 'Metabolism and balance studies of ^{14}C-monolinuron after use in spinach followed by cress and potato cultures', *Pestic. Biochem. Physiol.*, **9**, 107–118.

Schütte, H. R. (1977). 'Metabolism of synthetic phytoeffectors in higher plants', in *Plant Growth Regulators; Proceedings of the 2nd International Symposium* (eds T. Kudrev, I. Ivanova, and E. Karanov), pp. 387–396. Bulgarian Academy of Sciences, Sofia.

Shaw, W. C., and Swanson, C. R. (1953). 'The relation of structural configuration to the herbicidal properties and phytoxicity of several carbamates and other chemicals', *Weeds*, **2**, 43–65.

Shimabukuro, R. H., Walsh, W. C., and Hoerauf, R. A. (1979). 'Metabolism and selectivity of diclofop-methyl in wild oat and wheat', *J. Agr. Food Chem.*, **27**, 615–623.

Slade, M., and Casida, J. E. (1970). 'Metabolic fate of 3,4,5- and 2,3,5-trimethylphenylmethylcarbamates, the major constituents in Landrin insecticide', *J. Agr. Food Chem.*, **18**, 467–474.

Smith, A. E. (1979). 'Metabolism of 2,4-DB by white clover (*Trifolium repens*) cell suspension cultures', *Weed Sci.*, **27**, 392–396.

Smith, A. E., and Oswald, T. W. (1979). 'Degradation of phenoxyalkylcarboxylic acids by white clover (*Trifolium repens*) cell suspensions', *Weed Sci.*, **27**, 389–392.

Smith, J. W., and Sheets, T. J. (1967). 'Uptake, distribution and metabolism of monuron and diuron by several plants', *J. Agr. Food Chem.*, **15**, 577–581.

Snel, M. and Edgington, L. V. (1970). 'Uptake, translocation and decomposition of systemic oxathiin fungicides in bean', *Phytopathology*, **60**, 1708–1716.

Spurr, H. W., and Sousa, A. A. (1974). 'Potential interactions of aldicarb and its metabolites on non-target organisms in the environment', *J. Environ. Qual.*, **3**, 130–133.

Still, G. G., and Mansager, E. R. (1971). 'Metabolism of isopropyl 3-chlorocarbanilate by soybean plants', *J. Agr. Food Chem.*, **19**, 879–884.

Still, G. G., and Mansager, E. R. (1972). 'Aryl hydroxylation of isopropyl-3-chlorocarbanilate by soybean plants', *Phytochem.*, **11**, 515–520.

Still, G. G., and Mansager, E. R. (1973a). 'Soybean shoot metabolism of isopropyl-3-chlorocarbanilate: *ortho* and *para* aryl hydroxylation', *Pestic. Biochem. Physiol.*, **3**, 87–95.

Still, G. G., and Mansager, E. R. (1973b). 'Metabolism of isopropyl-3-chlorocarbanilate by cucumber plants', *J. Agr. Food Chem.*, **21**, 787–791.

Still, G. G., and Mansager, E. R. (1973c). 'Metabolism of isopropyl carbanilate by soybean plants', *Pestic. Biochem. Physiol.*, **3**, 289–299.

Still, G. G., and Mansager, E. R. (1975). 'Alfalfa metabolism of propham', *Pestic. Biochem. Physiol.*, **5**, 515–522.

Still, G. G., and Rusness, D. G. (1977). '*S*-cysteinyl-hydroxychlorpropham: formation of the S-cysteinyl conjugate of isopropyl-3′-chloro-4′-hydroxy carbanilate in oat (*Avena sativa* L.)', *Pestic. Biochem. Physiol.*, **7**, 210–219.

Swanson, C. R., and Swanson, H. R. (1968a). 'Metabolic fate of monuron and diuron in isolated leaf discs', *Weed Sci.*, **16**, 137–143.

Swanson, C. R., and Swanson, H. R. (1968b). 'Inhibition of degradation of monuron in cotton leaf tissue by carbamate insecticides, *Weed Sci.*, **16**, 481–484.

Synerholm, M. E., and Zimmerman, P. W. (1947). 'Preparation of a series of ω-(2,4-dichlorophenoxy) aliphatic compounds and some related compounds with a consideration of their biochemical role as plant growth regulators', *Contrib. Boyce Thompson Inst. Pl. Res.*, **14**, 369–382.

Tanaka, F. S., Swanson, H. R., and Frear, D. S. (1972). 'An unstable hydroxymethyl intermediate formed in the metabolism of 3-(4-chlorophenyl)-1-methylurea in cotton', *Phytochem.*, **11**, 2701–2709.

Thomas, E. W. (1964). 'Aspects of the incorporation of substituted phenoxyacetic acids by stem tissues of higher plants', D. Phil. Thesis, University of Oxford.

Thomas, E. W., Loughman, B. C., and Powell, R. G. (1963). 'Hydroxylation of phenoxyacetic acid in stem tissue of *Avena sativa*', *Nature* (London), **199**, 73–74.

Thomas, E. W., Loughman, B. C., and Powell, R. G. (1964a). 'Metabolic fate of 2,4-Dichlorophenoxyacetic acid in the stem tissue of *Phaseolus vulgaris*', *Nature* (London), **204**, 884–885.

Thomas, E. W., Loughman, B. C., and Powell, R. G. (1964b). 'Metabolic fate of some chlorinated phenoxyacetic acids in the stem tissue of *Avena sativa*', *Nature* (London), **204**, 286.

Trenck, Th. v.d., and Sandermann, H. (1978). 'Metabolism of benzo (α)pyrene in cell suspension cultures of parsley (*Petroselinum hortense* (Hoffm.)) and soybean (*Glycine max* L.)', *Planta*, **141**, 245–251.

Trenck, Th. v.d., and Sandermann, H. (1980). 'Oxygenation of benzo(α) pyrene by plant microsomal fractions', *FEBS Lett.*, **119**, 227–231.

Trenck, Th. v.d., and Sandermann, H. (1981). 'Incorporation of benzo(α)pyrene quinones into lignin', *FEBS Lett.*, **125**, 72–76.

Turner, B. C., and Caro, J. H. (1973). 'Uptake and distribution of carbofuran and its metabolites in field-grown corn plants', *J. Environ. Qual.*, **2**, 245–247.

Umetsu, N., Fahmy, A. H., and Fukuto, T. R. (1979). 'Metabolism of 2,3 dihydro-2,2-dimethyl-7-benzofuranyl(di-*n*-butylaminosulfenyl) (methyl) carbamate and 2,3 dihydro-2,2-dimethyl-7 benzofuranyl (morpholinosulfenyl) (methyl) carbamate in cotton and corn plants', *Pestic. Biochem. Physiol.*, **10**, 104–119.

Verity, J., Walker, A., and Drennan, D. S. H. (1981a). 'Aspects of the selective phytotoxicity of methazole II. Behaviour in plants following root exposure', *Weed Res.*, **21**, 307–316.

Verity, J., Walker, A., and Drennan, D. S. H. (1981b). 'Aspects of the selective phytotoxicity of methazole III. Behaviour in plants following foliar treatment', *Weed Res.*, **21**, 317–324.

Wain, R. L. (1955a). 'Herbicidal selectivity through specific action of plants on compounds applied', *J. Agr. Food Chem.*, **3**, 128–130.

Wain, R. L. (1955b). 'A new approach to selective weed control', *Ann. Appl. Biol.*, **42**, 151–157.

Wain, R. L., and Smith, M. S. (1976). 'Selectivity in relation to metabolism', in *Herbicides, Physiology, Biochemistry and Ecology* (ed. L. J. Audus), Vol. 2, 2nd edition, pp. 279–302. Academic Press, London, New York.

Wain, R. L., and Wightman, F. (1954). 'The growth-regulating activity of certain ω-substituted alkylcarboxylic acids in relation to their β-oxidation within the plant', *Proc. Roy. Soc. Lond. B*, **142**, 525–536.

Weisgerber, I., Klein, W., and Korte, F. (1970). 'Contributions to ecological chemistry XXVI. Transformation and residues of ^{14}C-aldrin and ^{14}C-dieldrin in white cabbage, spinach and carrot', *Tetrahedron*, **26**, 779–789.

West, C. A. (1980). 'Hydroxylases, monooxygenases and cytochrome P_{450}', in *The Biochemistry of Plants: A Comprehensive Treatise* (eds-in-chief P. K. Stumpf and E. E. Conn), *Vol. 2, Metabolism and Respiration* (ed. D. D. Davies), pp. 317–364, Academic Press, New York, London.

Wiedmann, J. L., Ecke, G. G., and Still, G. G. (1976). 'Synthesis and isolation of 1-hydroxy-2-propyl-3-chlorocarbanilate from soybean plants treated with isopropyl 2-chlorocarbanilate', *J. Agr. Food Chem.*, **24**, 588–592.

Wilcox, M., Moreland, D. E., and Klingman, G. C. (1963). 'Aryl hydroxylation of phenoxyaliphatic acids by excised roots', *Physiol. Plant.*, **16**, 565–571.

Wotschokowsky, M. (1972). 'On the metabolism of flurenol-*n*-butyl ester in higher plants', *Weed Res.*, **12**, 80–89.

Young, O., and Beevers, H. (1976). 'Mixed function oxidases from germinating castor bean endosperm', *Phytochem.*, **15**, 379–385.

Yu, C. C., Atallah, Y. H., and Whitacre, D. M. (1980). 'Metabolism of the herbicide buthidazole in corn seedlings and alfalfa plants', *J. Agr. Food Chem.*, **28**, 1090–1095.

Yu, S. J., Kiigemagi, U., and Terriere, L. C. (1971). 'Oxidative metabolism of Aldrin and Isodrin by bean root fractions', *J. Agr. Food Chem.*, **19**, 5–9.

Zulalian, J., and Blinn, R. C. (1977). 'Metabolism of Cytrolane systemic insecticide (mephosfolan), propylene (diethoxyphosphinyl)dithioimidiocarbonate, in cotton plants', *J. Agr. Food Chem.*, **25**, 1033–1039.

Zurquiyah, A. A., Jordan, L. S., and Jolliffe, V. A. (1976). 'Metabolism of isopropyl carbanilate (propham) in alfalfa grown in nutrient solution', *Pestic. Biochem. Physiol.*, **6**, 35–45.

Progress in Pesticide Biochemistry and Toxicology, Volume 3
Edited by D. H. Hutson and T. R. Roberts

CHAPTER 5

Plant tissue culture and pesticide metabolism

R. O. Mumma and G. H. Davidonis

INTRODUCTION

The US produced approximately 6.43×10^8 kg of pesticides in 1978 and this production is expected to increase albeit not as rapidly as in the past (Fowler and Mahan, 1980). Use of pesticides in lesser developed countries will probably account for a larger proportion of the increase. Due to the concern of the public and regulatory agencies industrial companies are required to run a battery of tests for candidate compounds to ensure the public a reasonable degree of safety. With ever-rising inflation the costs of these tests currently run into millions of dollars. Metabolism information is needed for registration of new pesticides, for reregistration of old pesticides, for guidance of occupational exposure studies, for testing on animals other than humans and for pollution and ecological concerns. All of these tests are costly and industrial companies are always looking for ways to reduce cost. The target of most pesticides is a plant irrespective of whether they are fungicides, insecticides, herbicides, growth regulators, harvest aids, etc. Therefore, data on plant metabolism of a pesticide is of primary importance. The use of plant tissue cultures to investigate plant metabolism of a pesticide offers time saving methods and thus reduces cost. In addition to determining the metabolic fate of the pesticide, plant tissue cultures can be used to bioassay the metabolites, to screen for new compounds and to further our understanding of the mechanism of action of pesticides in order to assist in designing new, more selective and toxic pesticides.

The use of plant tissue cultures in metabolism studies has been previously reviewed (Sandermann *et al.*, 1977; Barz, 1977; Mumma and Hamilton, 1979). This chapter focuses on pesticide metabolism by plant tissue cultures by emphasizing the metabolism of several selected pesticides. In addition, the expanding literature dealing with studies on the mechanism of action of pesticides and the use of plant tissue cultures in bioassays and phytotoxicity investigations is reviewed.

What are plant tissue cultures? How can they help the pesticide chemist? How can they be used to reduce time and cost? These are some of the questions we will hope to answer in this review.

The nature of plant tissue cultures

Plant tissue cultures can include differentiated as well as undifferentiated plant tissue. Roots and meristems in culture are examples of differentiated tissue in culture. We will focus our attention for this review on undifferentiated tissue culture since almost all pesticide metabolism studies have been conducted with these cultures. Undifferentiated plant cultures can be grown as single cells or small clumps of cells in sterile medium or as larger masses of cells growing on the surface of a sterile solidified agar medium. The selection of the composition of the medium is critical for the species of plant to be cultured. Single cells or small clumps of cells in liquid culture are referred to as suspension cultures and behave like bacterial cultures in that they go through a lag, log and stationary phase during a period between several days and a week. Large masses of undifferentiated plant cells in culture on an agar medium are referred to as callus cultures. These cultures grow more slowly and require several weeks or months before becoming senescent. Plant tissue cultures can be derived from almost any part of the plant. Each culture expresses biochemical characteristics of the plant species and may also demonstrate characteristics of the source of the explant. For example, we have investigated the metabolism of 2,4-dichlorophenoxyacetic acid (2,4-D) in cotyledon, root and leaf cultures from one specific soybean variety (Davidonis *et al.*, 1978; 1980a). The preparation and culture of plant tissue has been adequately reviewed (Gamborg, 1975).

The pesticide chemist can use plant tissue cultures to determine the fate of pesticides in plant tissue. They can also be used to compare metabolism between species, to investigate mechanism of action, to determine levels of degradative enzymes, to bioassay for biological activity, to determine phytotoxicity and to mass-produce selected important metabolites.

Many factors affect metabolism of pesticides by plant tissue culture. These factors include the age of the tissue, duration of the incubation of the pesticide with the tissue, type and source of plant tissue, composition of culture medium, concentration of pesticide used, method of introduction of the pesticide and whether it is a suspension or callus culture. All these factors have to be taken into consideration in any metabolism study, but they can be easily controlled. On the other hand, by varying any of these factors the investigator may produce an intentionally desirable condition, e.g., the mass production of a minor metabolite.

PLANT TISSUE CULTURE AND PESTICIDE METABOLISM

Pesticides constitute a large group of chemicals exhibiting a diverse biological effect. Most are designed to control unwanted plants, insects or fungi but others are used as harvest-aids, desiccants, antidotes, fumigants, etc. Usually all of these chemicals are eventually applied to plant tissues, but only the metabolism of selected herbicides and insecticides have been investigated with plant tissue cultures. Most studies have involved the use of suspension cultures. Some of these studies were thorough investigations while others were superficial. The phenoxy herbicides seemed to have attracted most of the attention from a metabolic and mechanistic standpoint. We will summarize the pesticide metabolism by grouping the chemicals according to their structure somewhat irrespective of their biological activity.

Propanil and Diphenamid

The metabolism of two amide herbicides, propanil and diphenamid, has been examined with plant tissue cultures. Propanil (3,4-dichloropropionanilide) controls weeds in rice, which is resistant to this herbicide. Rice possesses a high level of an arylacyl amidase that hydrolyses propanil to 3,4-dichloroaniline and propionic acid. Older rice plants (four or more leaves) possess two to four times the enzyme levels of younger plants (less than four leaves). The metabolism of propanil has been investigated in rice plants and rice suspension cultures (Lieb *et al.*, 1973; Ray and Still, 1975) and rice suspension cultures hydrolysed propanil only after reaching the stationary phase of growth, 5.5 days. Thus, tissue cultures also demonstrated a similar age dependent hydrolytic activity.

The metabolism of diphenamid (*N*,*N*-dimethyl-2,2-diphenylacetamide), has been investigated with soybean root suspension cultures. No hydrolysis of the amide bond was detected at any stage of growth. The same metabolites were found in both suspension cultures and plants, these being the *N*-hydroxymethyl, *N*-dealkylated and glycoside metabolites (Figure 1) (Davis *et al.*, 1978a). The

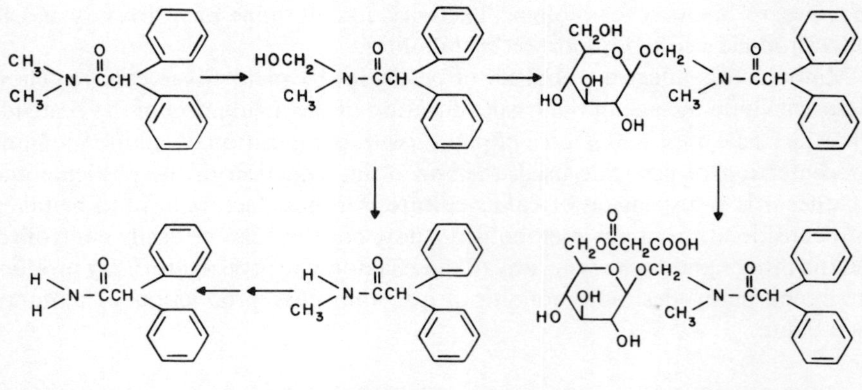

Figure 1 The metabolism of diphenamid by suspension cultures of soybean root

hydroxylated and dealkylated metabolites tended to accumulate in the media. Only 0.5–1.4% of the applied diphenamid was found as glycosides with soybean suspension cultures but glucosides were the major metabolites of soybean plants. The acidic glucoside was found only in the cell extract reflecting its poor membrane permeability. Evidently the hydroxylated metabolites (aglycons) leaked out of the cells prior to the glucosylation process or this particular soybean root suspension culture lacked the ability to form glycosides.

Carbaryl

The metabolism of the carbamate insecticide, carbaryl (1-naphthyl methylcarbamate) has been studied in tobacco suspension cultures over a period of 14 days (Locke *et al.*, 1976). Approximately 35% of the [^{14}C]carbaryl was metabolized and 77% of this [^{14}C]label (27%) was not identified but was primarily associated with the cell debris and water-soluble fractions. A summary of the metabolism of carbaryl is presented in Figure 2 and represents most of the identified metabolites (7.9%). The glycoside of α-naphthol represented ~74% of the characterized metabolites and was found entirely in the tobacco cells. Whole plants have shown similar hydrolytic pathways to α-naphthol (Mostafa *et al.*, 1966; Dorough and Wiggins, 1969) but in general it is not a major metabolite of plants. *N*-Hydroxymethyl carbaryl was a major metabolite (19%) and was found in the culture medium. Glycosides of *N*-hydroxymethyl carbaryl and of ring hydroxylated carbaryl were isolated from the tobacco cells in small amounts (2.5 and <1% respectively), however, these metabolites have been reported present in whole plants in higher levels (Kuhr and Casida, 1967; Mumma *et al.*, 1971). Perhaps the high percentage of uncharacterized metabolites of carbaryl is associated with the macromolecular fraction as has been shown with phenolic metabolites of other pesticides.

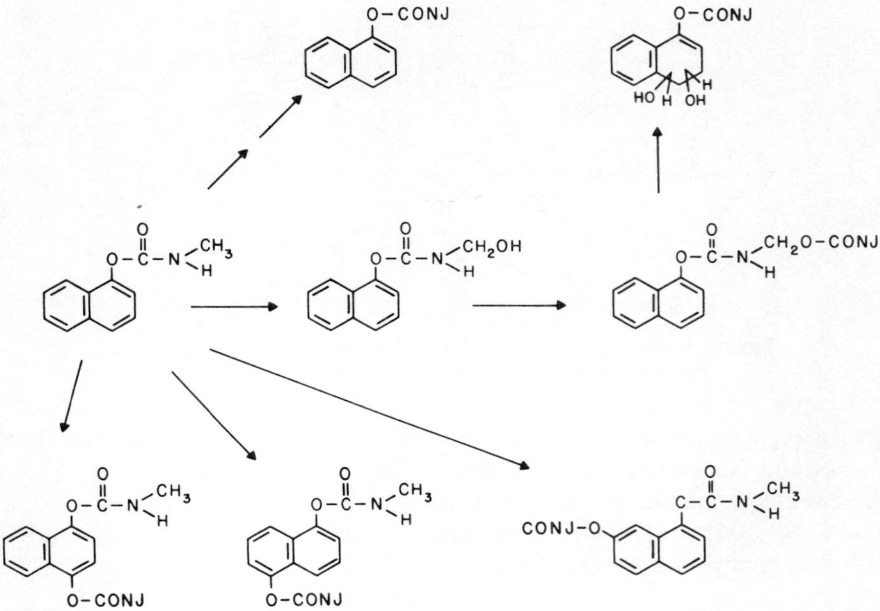

Figure 2 The matabolism of carbaryl by suspension cultures of tobacco

Cisanilide

The metabolism of one urea herbicide, cisanilide (*cis*-2,5-dimethyl-1-pyrolidine-carboxanilide), has been investigated in excised leaves and cell suspension cultures of carrot and cotton (Frear and Swanson, 1975). Hydrolytic products were not detected. The excised leaves accumulated glycosides of the 4-hydroxy-(phenyl) cisanilide (I) and of the 3-hydroxy-(pyrrolidine) cisanilide (II). In suspension cultures only aglycon-II was found and it was exclusively in the medium. Presumably the suspension cultures formed both aglycon-I and aglycon-II, but these diffused out of the cells. The phenolic aglycon-I was then incorporated into the extracellular insoluble residue leaving aglycon-II in the medium. The same metabolic processes seem to be operating in excised leaves as in suspension cultures, but wide quantitative differences exist.

EL-494

Another structurally related urea derivative is EL-494 (*N*-[[[5-(4-bromo-phenyl)-6-methyl-2-pyrazinyl]amino]carbonyl]-2,6-dichlorobenzamide) which is being considered as a possible insect growth regulator. AbdelMonem (1981) investigated its metabolism in soybean callus tissue cultures, in soybean plants

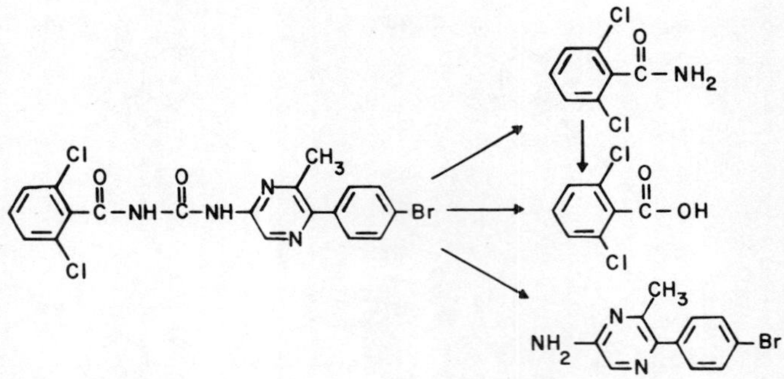

Figure 3 The metabolism of EL-494 by soybean callus tissue cultures, by soybean plants and by gypsy moth larvae

and in gypsy moth larvae. The metabolism proved to be the same in all three tissues and is summarized in Figure 3. The only metabolites detected were present in low concentration and were hydrolytic products: (1) dichlorobenzoic acid (0.54–2.04 %); (2) dichlorobenzamide (6.11–16.51 %); and (3) 2-amino-5-(4-bromophenyl-6-methylpyrazine).

Organochlorine compounds

The metabolism of several halogenated hydrocarbons has been examined with plant tissue cultures, these include the two cyclodiene insecticides aldrin and dieldrin as well as lindane, Kelthane and DDT. Epoxidation of aldrin to dieldrin is the primary metabolic route (Brain and Lines, 1977) with bean root and shoots and potato tuber suspension cultures. Aldrin-transdiol was only found in small amounts in the bean shoot and potato tuber cultures. Photodieldrin was formed from dieldrin in all cultures. No dicarboxylic acid derivative was detected even though it does occur in small quantities in whole plants.

The metabolism of ^{36}Cl-labelled lindane(α-hexachlorocyclohexane) was examined with twelve different plant suspension cultures (Stöckigt and Ries, 1977) and compared to similar studies with intact plants. Twelve to 28-day incubations metabolized 0 % to 6.8 % of the lindane with liberation of $^{36}Cl^-$. With intact plants very little $^{36}Cl^-$ was liberated. A glycoside of trichlorophenol was tentatively identified as the main metabolite in wild carrot suspension cultures. 1,2,4-Trichlorobenzene was identified in trace quantities in tobacco cultures, but was not detected in the other cultures. With incubations of DDT in parsley and soybean suspension cultures (44–48 h) 0.6 to 2.2 % of the applied DDT was converted to DDE (1,1-dichloro-2,2,-bis-[4-chlorophenyl]-ethylene) (Scheel and Sandermann, 1977). Small quantities of a polar (0.3 %) and non-polar (0.1 %) metabolite of Kelthane were detected in parsley suspension cultures by these same investigators.

2,4-Dichlorophenoxyacetic acid (2,4-D)

Feung *et al.* (1972) investigated the metabolism of [1-^{14}C]2,4-D in soybean cotyledon callus over a period of 32 days. Many metabolites were found and their relative composition varied greatly over the entire 32-day incubation. Two ether-soluble metabolites, the glutamic and aspartic acid conjugates of 2,4-D (2,4-D-Glu and 2,4-D-Asp), were shown to be present in large relative proportions. Subsequently, five additional amino acid conjugates have been isolated from soybean callus tissue: alanine, valine, leucine, phenylalanine and tryptophan (Feung *et al.*, 1973). The water-soluble metabolites increased with incubation time and primarily yielded upon β-glucosidase treatment, 4-hydroxy-2,3-dichlorophenoxyacetic acid (4-OH-2,3-D) and 4-hydroxy-2,5-dichlorophenoxyacetic acid (4-OH-2,5-D). Other ring hydroxylated 2,4-D derivatives were found in minor amounts in wheat suspension cultures (Bristol *et al.*, 1977). The water-soluble extract also gave about 1% free 2,4-D on β-glucosidase treatment which suggested that a carboxylic glucoside was present. Ojima and Gamborg (1968) had previously identified such a structure. Subsequently much higher levels of the carboxylic glucoside have been found with tissue cultures of other plant species (Feung *et al.*, 1975, 1976a).

To further understand species susceptibility and tolerance to 2,4-D, the metabolism of 2,4-D in callus tissue from five dicots and two monocot species were compared (Feung *et al.*, 1973; 1975; 1976a) and is presented in Table 1. The monocots, corn and rice, contained high levels of the carboxylic glycoside of 2,4-D and in rice it was the only significant metabolite. The dicots produced higher levels of the amino acid conjugates than the monocots. All produced hydroxylated products and these were the major metabolites of tobacco and corn. An ethyl ester of 2,4-D was isolated from corn but it was presumably an artefact of the isolation procedure.

Feung *et al.* (1978) investigated the comparative metabolism of 2,4-D in whole plants and in callus tissue of soybean and corn plants when 2,4-D was incubated with these tissues at herbicidal levels. In soybean callus 24.6% of the applied 2,4-D was found as the glucosides of 4-OH-2,3-D, 4-OH-2,5-D and 5-OH-2,4-D(5-hydroxy-2,4-dichlorophenoxyacetic acid), but at this level of treatment soybean plants only produced 4.7% of these metabolites. Corn plants also only produced 4.12% of the hydroxylated glycoside fraction. Both corn plants and callus tissue gave appreciable levels of free 2,4-D after treatment with β-glucosidase implying high levels of the carboxylic glucoside. The ether-soluble fraction contained numerous metabolites, chiefly the amino acid conjugates. These conjugates were found in lower amounts in whole plants than found in callus tissue. The ether-soluble fraction from the callus tissue also possessed small amounts of metabolites which were proposed to be amino acid conjugates of ring hydroxylated 2,4-D. These structures have been subsequently isolated from young wheat and barley plants (Nararova *et al.*, 1980). The levels

Table 1 Relative percentage of ether- and water-soluble [1-^{14}C]2,4-D metabolites isolated from seven species of plant callus cultures

	% Total tissue[a]						
Metabolites	Soybean	Carrot	Jackbean	Sunflower	Tobacco	Corn	Rice
Ether-Soluble							
Unknown	1.2					1.9	
Unknown						1.7	
2,4-D-Asp	3.7		1.2		0.8	1.6	
2,4-D-Glu	12.9	23.8	32.5	5.0	6.7	1.3	
Unknown	1.4	0.8	1.1	0.7	1.8	3.2	
Unknown	5.3[b]	5.8	5.2	5.7	5.5	0.8	
2,4-D	33.7	51.7	45.4	51.7	34.3	12.5	14.7
Unknown	4.0[c]	1.7		15.3	4.5	2.6	
Others							1.2
Total	62.2	83.8	85.4	78.4	53.6	25.6	15.9
Water-soluble[d]							
(4-OH-2,3-D, 4-OH-2,5-D)	26.3	6.9	6.9	8.3	17.4	31.7	0.4
Unknown	↑	3.2	2.6	4.1	8.3	19.4	
Unknown		0.7	0.4	0.8	2.0	0.5	
Unknown		0.4	0.3	0.5	1.9	1.2	
Unknown	6.3	0.1	0.3	0.8	2.3		
Unknown						1.6	
Unknown	↓	0.9	1.1	0.6	2.6		
2,4-D	0.8	1.0	1.2	3.2	7.7	10.2	14.9
Ethyl-2,4-D							12.9
Others							1.5
Total	33.4	13.2	12.8	18.3	42.2	64.6	29.7

[a] Feung *et al.*, 1973, 1975 and 1976a.
[b] In soybean this unknown has been identified as a mixture of alanine and valine conjugates of 2,4-D.
[c] In soybean this unknown has been identified as a mixture of leucine, phenylalanine and tryptophan conjugates of 2,4-D.
[d] Isolated as the alycones after β-glucosidase treatment of water-soluble metabolites.

of amino acid conjugates are dependent upon the levels of applied 2,4-D and will be discussed later.

Shvetsov and Gamburg (1980) also investigated the comparative metabolism of 2,4-D in suspension cultures of corn and soybean. Glucosides were the primary metabolites of corn and aspartic acid conjugates of soybeans. Corn suspension cultures decarboxylated 2,4-D more extensively than did soybean cultures. These data are in general agreement with results obtained in callus cultures (Feung *et al.*, 1978), however, it is difficult to compare Shvetsov and Gamburg's results with other studies since they did not report relative percentages nor do they report finding 2,4-D-Glu as has been found in all other investigations with soybean tissue in culture. The authors suggest that the ratio of free 2,4-D in the cells to that in the medium was constant after a few minutes with corn suspension cultures and after three days with soybean cells. These data are consistent with the conclusions of Davidonis *et al.* (1978) that the level of free 2,4-D found in the cells may be regulated by an age dependent process in soybean callus tissue.

Scheel and Sandermann (1981a) also examined the metabolism of 2,4-D in suspension cultures of soybean and wheat (Table 2). The distribution of each fraction is similar to that reported by Feung *et al.* (1973) and Bristol *et al.* (1977). They partially purified a low molecular weight protein fraction from both soybean and wheat which contained covalently bound ^{14}C-labelled 2,4-D or a derivative. They also isolated a lignin-like fraction from both soybean and wheat which yielded 2,4-D and hydroxylated 2,4-D on acid hydrolysis (Scheel and Sandermann, 1981b).

Montague *et al.* (1981a) have also compared the metabolism of 2,4-D in suspension cells of wild carrot and two varieties of soybean ('Dare' and 'Wayne'). After three days incubation of [1-^{14}C]2,4-D with wild carrot cells, 94% of the applied herbicide was found as free 2,4-D and 3% as 2,4-D-Glu. No detectable 2,4-D-Asp was found. Feung *et al.* (1975) reported high concentrations of free 2,4-D (51.7%) and of 2,4-D-Glu (23.8%) but no detectable 2,4-D-Asp. The 'Wayne' variety of soybean gave 52% 2,4-D-Glu, 28% 2,4-D-Asp and 13% free 2,4-D after three days. A higher proportion of 2,4-D-Asp (56%) was found with 'Dare'. These data emphasize the importance of varietal differences in regulating composition of metabolites. A summary of the metabolism of 2,4-D is shown in Figure 4.

Many factors affect metabolism studies with plant tissue cultures. The method of culturing the cells is important, e.g., whether suspension or callus tissue cultures are employed. Other factors such as age of the culture, concentration of pesticide, plant variety selection, and source of plant part for culture (roots, stems, cotyledon, leaves, etc.) also affect metabolism. These factors have recently been examined with the metabolism of 2,4-D and soybean callus tissue (Davidonis *et al.*, 1978; 1980a; b).

The amount of free 2,4-D found in 3-week-old soybean root callus tissue (*Glycine max* [L.] Merrill var. Amsoy) rapidly increased proportionally as the concentration of 2,4-D in the medium was increased from 10^{-6} to 10^{-5} M, whereas the level of glycosides and amino acid conjugates increased more slowly (Davidonis *et al.*, 1978). With 9-week-old tissue the amount of free 2,4-D remained at a low and constant level as the concentration of 2,4-D was increased in the medium. The glycoside fraction increased slightly but the amino acid conjugate fraction increased greatly. This older tissue regulated the level of free 2,4-D at 4 nmol/g fresh weight (4 μM) by shunting excess 2,4-D into the amino acid conjugates. In differentiated root cultures derived from the above callus tissue, the metabolism of 2,4-D resembled the metabolism in the older

Table 2 Relative percentage of selected [1-^{14}C]2,4-D metabolites from suspension cultures of various plant species

Metabolite	% Total in tissue								
	Wheat[a]	Wheat[b]	Soybean[c]	Soybean[d]	Soybean[e]	Soybean[f]	Corn[g]	Wild carrot[h]	Sycamore–Maple[i]
Amino acid conjugates	?	4	26						0
Glutamic acid				52	33			3	
Aspartic acid				28	56	major	minor		
Glucoside conjugates	60	19							
Carboxylic glucoside	24					minor	major		80
Ring hydroxylated glucoside	10								
Unchanged 2,4-D	16	17	32	13	5			94	20
Unidentified metabolites (Soluble and insoluble)		20	22						

[a] Bristol *et al.*, 1977.
[b] Scheel and Sandermann, 1981a, var. Koga II, 48 h incubation.
[c] Scheel and Sandermann, 1981a, var. Mandarin, 48 h incubation.
[d] Montague *et al.*, 1981a, var. Wayne, three days incubation.
[e] Montague *et al.*, 1981a, var. Dare, three days incubation.
[f] Shvetsov and Gamburg, 1980, cotyledon.
[g] Shvetsov and Gamburg, 1980.
[h] Montague *et al.*, 1981a, three days incubation
[i] Leguay and Guern, 1977, three days incubation.

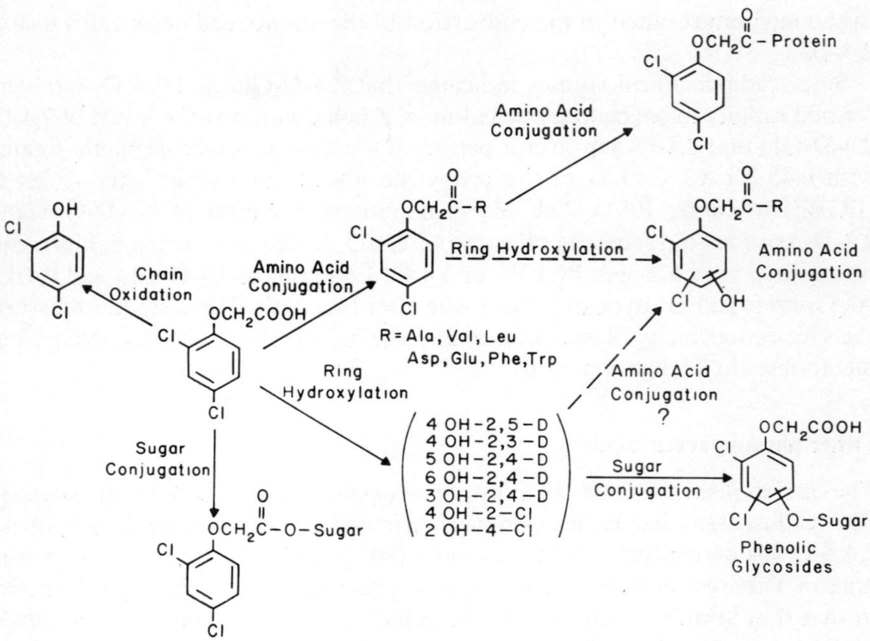

Figure 4 The metabolism of 2,4-D by plants

root callus tissue (the 9-week-old tissue); the level of free 2,4-D was regulated at 1–2 nmol/g (1–2 μM).

Leaf and cotyledon callus tissue derived from the same soybean variety (Amsoy) exhibited quantitative differences in metabolism of 2,4-D from that of root callus (Davidonis *et al.*, 1980a). The green leaf callus tissue gave a higher relative percentage of hydroxylated derivatives than did root or cotyledon cultures. Both leaf and cotyledon Amsoy callus cultures exhibited the same age dependent metabolism of 2,4-D as that of root callus. In 6–8-week-old tissue the level of 2,4-D was regulated at 1.5–3.5 μM, while the amino acid conjugates increased in concentration as the concentration of 2,4-D increased. In 3- or 6-week-old cotyledon callus from another variety of soybean (Acme) no age dependent metabolism was demonstrated. Rice suspension cultures demonstrated a similar age dependent metabolism of propanil (Ray and Still, 1975). Montague *et al.* (1981a) also showed varietal specificity in formation of the amino acid conjugates.

The metabolism of 2,4-D by root callus tissue was examined 9, 14, 24, 48 and 72 h after treatment with [1^{14}C]2,4-D (Davidonis *et al.*, 1980b). Maximum amino acid conjugate levels were observed at 24 h. In comparative 48 h chase experiments the amino acid pool was not affected by the presence or absence of free 2,4-D. However, longer incubation times in the absence of any free 2,4-D in the medium resulted in the conversion of the amino acid conjugates to free 2,4-D.

Since radiochemical studies indicated that 2,4-D-Glu and 2,4-D-Asp were formed rapidly in soybean tissue cultures, Zama quantified the levels of 2,4-D, 2,4-D-Glu and 2,4-D-Asp over a period of ten days in soybean plants treated with 0.45 Kg A.I./0.40 ha of the propylene glycol butyl ether ester of 2,4-D (PGBEE) (Zama, 1981). The plants contained 132 ppm of 2,4-D-Glu and 2,4-D-Asp four days after application of 2,4-D. In contrast, when callus tissue was treated with 0.2 ppm PGBEE of 2,4-D, 1.64 ppm of 2,4-D-Glu and 2,4-D-Asp were found in soybean callus tissue after two days. These studies illustrate the species specificity of metabolism and the need for further quantification of metabolites in field grown crops.

Other phenoxyacetic acids

The metabolism of 2,4,5-trichlorophenoxyacetic acid (2,4,5-T) by soybean root callus tissue has been examined (Arjmand *et al.*, 1978). Only 5% of the 2,4,5-T was converted into the glutamic (90%) and aspartic (10%) acid conjugates. These conjugates have subsequently been shown to be biologically active in that they stimulate cell growth and cell elongation (Davidonis *et al.*, 1982). No glucose ester of the carboxyl group was detected.

Smith and Oswald (1979) compared the tolerance and metabolism of 2,4-D, 2,4-DB [4-(2,4-dichlorophenoxy)butyric acid], 2,4,5-T, 2,4,5-TP [2-(2,4,5-trichlorophenoxy)propionic acid] and 2,4,5-TB [2-(2,4,5-trichlorophenoxy) butyric acid] in white clover cell suspension cultures. Legumes including white clover are recognized to have more tolerance for the longer chain homologues of 2,4-D. Herbicide tolerance increased in cell populations pretreated with 2,4-DB and 2,4,5-T. However, [^{14}C]-2,4-DB was rapidly degraded, 50% in 48 h, by the white clover cells and 81% by cells preincubated with 2,4-DB. Thus, the acquired tolerance of these cells by preincubations may have been due to changes in metabolism.

The metabolism of ^{14}C-ring-labelled 2,4-DB by white clover suspension cells was investigated over a 48 h period (Smith, 1979). No free 2,4-D was detected reflecting the lack of any appreciable β-oxidation of the chain. The major metabolite was a glycoside presumably of the ring hydroxylated 2,4-DB. Also, 34% of the radiolabel was incorporated into cell structural material. The parent herbicide also was conjugated to amino acids (8–10%). Under similar conditions 95.5% of 2,4-D treated clover suspension cells were recovered unchanged. Therefore, the enzyme(s) responsible for conjugation or hydroxylation have selectivity for 2,4-DB versus 2,4-D. These data suggest that white clover and perhaps legume tolerance to 2,4-DB is not due to the hitherto presumed rate of β-oxidation of the chain but due to enhanced hydroxylation and conjugation reactions selective for 2,4-DB.

Indoleacetic acid (IAA)

The metabolism of indoleacetic acid (IAA) by plant tissue cultures has been investigated by several authors (Troxler and Hamilton, 1965; Epstein and Lavee, 1975; Epstein *et al.*, 1975; Feung *et al.*, 1976b; Hamilton *et al.*, 1976 and others). Two major metabolic pathways are recognized: (1) the peroxidase–IAA oxidase oxidation of IAA and (2) the conjugation of IAA with amino acids or sugars.

Diclofop-methyl

Diclofop-methyl (methyl-2-[4-(2,4-dichlorophenoxy)phenoxy]-propanate) is a selective herbicide that controls many grass species in wheat. Its metabolism has been investigated in wheat, susceptible wild oat and in cell suspensions of diploid wheat (Dusky *et al.*, 1980; Donald and Shimabukuro, 1980). Ninety-nine percent of the diclofop-methyl was metabolized in 8 h with wheat suspension cultures. After 24 h 78.4% of the applied material was found present as glycoside conjugates. The first step in the metabolism is presumed to be the hydrolysis of the ester to diclofop, followed by hydroxylation of the 2,4-dichlorophenyl ring and conjugation with glucose (Figure 5). Ring hydroxylated diclofop is found only in the cells and a carboxylic ester conjugate is present only in the medium. Similar metabolism was found in resistant wheat plants (Shimabukuro *et al.*, 1979). However, in susceptible wild oat, diclofop-methyl was primarily conjugated via a neutral glycosyl ester.

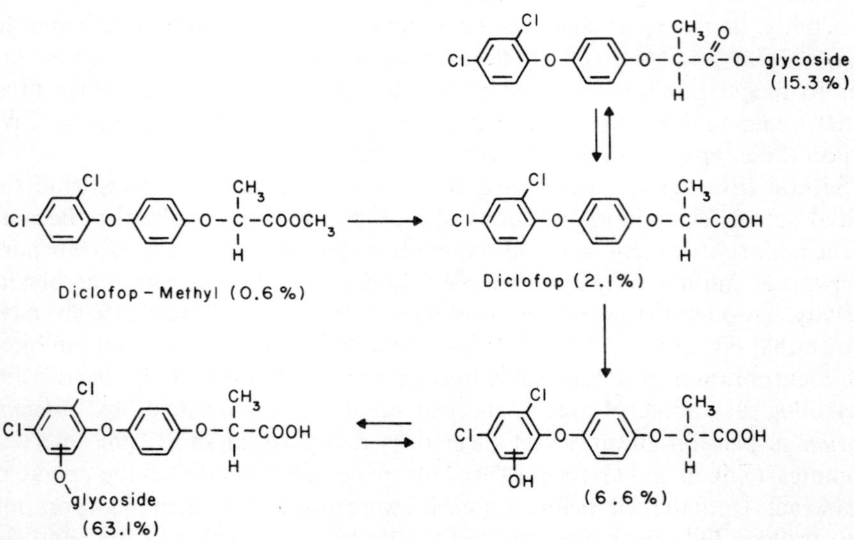

Figure 5 The metabolism of diclofop-methyl by wheat suspension cultures

Flurodifen

The metabolism of the phenoxy herbicide, flurodifen (*p*-nitrophenyl α,α,α-trifluror-2-nitro-*p*-tolyl ether), has been investigated in suspension cultures of tobacco cells (Locke and Baron, 1972). When [^{14}C]-nitrophenyl-flurodifen was used, the major ^{14}C-metabolites were 4-nitrophenol (7%), found only in the medium, and aqueous soluble conjugates of 4-nitrophenol (93%), primarily the β-D-glycoside of 4-nitrophenol. In similar studies with soybean and maize seedlings, the glucoside of 4-nitrophenol was the major degradation product.

Benzo[α]pyrene

Soluble metabolites of benzo[α]pyrene were found with suspension cultures of parsley (1–2.2%) and with soybean (10–28%) (v.d. Trenck and Sandermann, 1978). Although the metabolites were not identified they were chemically characterized in terms of their solubility in organic solvents, aqueous buffers and hot aqueous dodecylsulphate. In addition, a significant amount of the benzo[α]pyrene was associated with the lignin fraction (v.d. Trenck and Sandermann, 1981).

MODE OF ACTION AND MOLECULAR EFFECTS

Plant tissue cultures offer unusual opportunities to investigate the mode of action of pesticides and their effect at the molecular level. Because of the ease of standardized conditions and reduced dependency of transport on expression of activity, many experiments can be designed to ask questions concerning the physiological and biochemical effects of pesticides on plants. It is perhaps not fair to assign publications strictly to this section of the review since most articles cited in this section deal with both metabolism and mode of action. We expect these types of studies to increase rapidly in the future.

Several investigators have used plant cell suspensions to investigate the initial actions (kinetic information) of biologically active molecules. Because of the uniformity of cells, controlled conditions and short distances of transport, suspension cultures offer considerable promise in evaluating initial herbicidal activity. Through tissue culture experiments, Zilkah and Gressel (1978b) have shown that dikegulac (2,3:4,6-di-*O*-isopropylidine-2-keto-L-gulonate) inhibited the incorporation of amino acids into proteins and this was the basis of its physiological action. However, the cell uptake of amino acids by *Solanum nigrum* suspension cultures was affected by dikegulac in short time intervals (minutes) (Zilkah and Gressel, 1979). Dikegulac rapidly inhibited the uptake of amino acids into the cells before any effects were observed on their incorporation into protein. Dikegulac also induced the leakage of a preloaded dye and the leakage was less with cell suspensions in stationary phase of growth, paralleling the age-dependent results obtained with whole plants. These investigators concluded that the first observed effects of dikegulac were due to the effects on the permeability of the cell membrane and these data illustrate the importance of the use of plant tissue cultures in mode-of-action studies.

Leguay and Guern (1977) have investigated the metabolism of 2,4-D in suspension cultures of *Acer Pseudo-Platanus* and its influence on control of cell division. *Acer Pseudo-Platanus* metabolized 2,4-D to only the carboxylic ester glycoside. No amino acid conjugates or ring hydroxylated derivatives were detected. This simplified metabolism permitted these investigators to design experiments relating to the control of cell division as influenced by the extracellular and intracellular concentration of 2,4-D. The initial concentration of 2,4-D determined the ultimate number of cells produced. Since growth stopped with measurable amounts of 2,4-D inside and outside the cells there was a suggestion that a threshold concentration of auxin was needed for cell division to proceed. When the cells reached the stationary phase of growth the extracellular concentration of 2,4-D was a linear function of population density. At this phase of growth, the intracellular ratio of 2,4-D to the carboxylic glycoside of 2,4-D was 1:4. These investigators used the distribution of [2-^{14}C]-5,5-dimethyloxazolidine-2,4-dione to determine the intracellular and extracellular pH. The intracellular pH was lower as the population density at the stationary phase was increased. The intracellular pH seems to control the ratio between levels of intracellular and extracellular auxin. The intracellular auxin concentration is approximately 0.3 μM at onset of the stationary phase, and is independent of population density or of the concentration of 2,4-D-glycoside. This level was suggested as a threshold value of 2,4-D below which cell division could not occur.

In an interesting paper Montague *et al.* (1981b) have shown that cytokinins inhibit the formation of 2,4-D-amino acid conjugates in suspension cultures of soybean cells. Also, cytokinin-treated soybean cells released more of their [1-^{14}C]-2,4-D than did control cells when transferred to hormone-free medium.

The levels of several enzymes, RNase, DNase, phosphodiesterase and phosphomonoesterase were determined in callus cultures of wheat grown in the presence of optimal concentrations of α-naphthylacetic acid (NAA), 2,4,5-T and 3,6-dichloro-*o*-anisic acid (dicamba) (Udvardy *et al.*, 1976). With dicamba, all enzyme activities showed a negative correlation of growth. With NAA and 2,4,5-T a positive correlation with growth was observed with phosphodiesterase and phosphomonoesterase activity.

Davis and Shimabukuro (1980) have used callus-derived cell suspensions of peanut and soybean as well as enzymatically isolated peanut mesophyll cells to investigate the herbicide toxicity and mode of action of two bipyrilidium herbicides (diquat and paraquat) and of perfluidone. Mesophyll cells were very useful for bioassaying the photosynthetic inhibition of 1 or 10 μM bipyrilidiums.

A susceptible ('Coker 102') and a resistant ('Bragg') soybean cultivar to metribuzin (4-amino-6-*t*-butyl-3-[methylthio]-*cis*-triazin-5-[4H]-one), an unsymmetric triazine used as a preemergence herbicide on soybean, were examined in suspension cell culture. The cells exhibited the same susceptibility and resistance in culture as the whole plant (Oswald *et al.*, 1978). The authors concluded that detoxification of metribuzin did not occur in the susceptible cultivar because of the accumulation of a substance which inhibited the detoxifying enzyme.

USE OF PLANT TISSUE CULTURES TO EVALUATE BIOLOGICAL ACTIVITY OF TEST COMPOUNDS

Plant tissue cultures are very useful to evaluate the biological activity of test compounds. Examples of biological activities that have been examined are phytotoxicity, herbicidal activity, mutagenicity and growth regulator activities. The activity of the test compound can be measured with suspension or callus tissue cultures in a number of ways. Physical methods of measuring biological activity often require lengthy experiments and involve such parameters as gain and loss in dry weight, cell volume or cell number. More rapid biological assays expressing vitality or functionality of the cells have recently been evaluated (Zilkah and Gressel, 1978c). Metabolic activity of the cells can be expressed in terms of levels of selected enzymes or of the amount of the incorporation of small molecules such as amino acids or purines into macromolecules. The effect of test chemicals on the permeability of membranes to selected dyes (fluorescein diacetate assay) can also be employed. The cessation of cytoplasmic streaming has also been used to express vitality.

Susceptible varieties or species seem to possess these same susceptibility properties in tissue cultures. For this reason plant tissue cultures are being investigated as a means to rapidly select for resistance or other biochemical attributes with the ultimate goal of regeneration of a complete plant exhibiting the same resistance (Gressel and Segel, 1978). As previously mentioned, suspension cultures derived from susceptible and resistant soybean cultivars to metribuzin demonstrated the same susceptible and resistant characteristics as the whole plant (Oswald *et al.*, 1978). Diclofop-methyl is a selective herbicide used to control wild oat in wheat fields. Plant tissue cultures derived from a resistant wheat variety demonstrated the same qualitative and quantitative metabolism of diclofop-methyl as did resistant wheat plants (Hoerauf and Shimabukuro, 1979; Donald and Shimabukuro, 1980; Shimabukuro *et al.*, 1979; Dusky *et al.*, 1980). Since plant tissue cultures possess many biochemical characteristics of the intact plant it is appropriate to use plant tissue cultures for bioassay purposes with the expectation that the results can be extrapolated to whole plants.

Soybean callus tissue cultures have been utilized to evaluate the biological activity of a number of amino acid conjugates of growth regulator herbicides.

These include nineteen L-amino acid derivatives of 2,4-D (Feung *et al.*, 1974), six D-amino acid conjugates of 2,4-D (Davidonis *et al.*, 1982), fourteen L-amino acid conjugates of 2,4,5-T (Davidonis *et al.*, 1979) and fifteen amino acid conjugates of indoleacetic acid (IAA) (Feung *et al.*, 1977a). Various concentrations of these compounds were evaluated for their ability to stimulate the growth (increase in mass) of callus tissue cultures and elongation of *Avena* coleoptile sections. Field herbicidal activity was also determined with the L-amino acid conjugates of 2,4-D (Feung *et al.*, 1977b). A good correlation was obtained between results of the *Avena* elongation and growth of plant tissue cultures. Some striking differences occurred between the field herbicidal tests and the plant tissue culture test. The more polar amino acid conjugates such as 2,4-D-Asp and 2,4-D-Glu gave good stimulation of growth in the tissue culture test and elongation in the *Avena* test but were not as active on whole plants in the field. This difference presumably is due to the lack of permeability of these polar compounds through the cuticular waxes and cell surfaces in whole plants. Less polar amino acid conjugates, such as conjugates of valine, leucine and methionine exhibited strong herbicidal properties and presumably more readily penetrated the cell surface. These less polar amino acid conjugates also gave strong stimulation in plant tissue culture and *Avena* coleoptile bioassays. These data emphasize the importance of cautiously extrapolating results obtained with tissue culture experiments to field conditions.

The use of plant tissue cultures to evaluate phytotoxicity and herbicidal activity is an active area of research. Since this technique can be used to cheaply screen new synthetic compounds industrial companies are actively pursuing this approach. Unfortunately, these data do not appear in the published literature (Gressel, 1979). Gressel and associates have studied many factors involved in the use of plant tissue cultures to evaluate the phytotoxicity of test compounds. Various physical and chemical methods to estimate death of cells were compared with suspension cultures of five plants: *Chrysanthemum segetum*, *Circium arvense*, *Solanum nigrum*, *Rumex obtusifolius* and *Lycopersicon esculentum*, and fifteen herbicides (Zilkah *et al.*, 1977; Zilkah and Gressel, 1977a; b). These methods include packed volume of cells, changes in dry weight, cell numbers, permeability to fluorescein, reduction of triphenyltetrazolium chloride and incorporation of [^{14}C]-leucine into proteins. It was concluded that there were wide quantitative variations between these methods but there was good correlation when a ranking technique was used for comparison.

The phytotoxicity of twenty-two herbicides and growth regulators in seedlings and white calli have been examined (Zilkah *et al.*, 1977; Zilkah and Gressel, 1977b). These investigators concluded that photosynthetic inhibitors affected seedlings much more than white calli. On the other hand, a few compounds affected calli more than seedlings, probably because of reduced transport by the seedlings. For example, napropamide was much more toxic to calli than to tomato seedlings. It was also more toxic to achlorophyllous calli than to green calli (Zilkah *et al.*, 1978; Zilkah and Gressel, 1978a). Trifluralin and chloromequat affected *C. segetum* and *C. arvense* calli more than the seedlings. The authors concluded from these studies that calli have great potential in phytotoxicity screening.

Dikegulac (2,3:4,6-di-*O*-isopropylidene-2-keto-L-gulonate) is another example of a growth regulator being more toxic to calli than to the whole plant (Zilkah and Gressel, 1978b). Dikegulac is used commercially to break apical dominance and has very little effect on mature leaves. It is toxic to plant tissue cultures because it inhibits the incorporation of amino acids into proteins by inhibiting the uptake of the amino acids at the membranes. Dikegulac had very little affect on suspension cells in the stationary phase (undividing cells) but strongly affects dividing cells, paralleling the observed effects with whole plants (Zilkah and Gressel, 1978b). Herbicides such as simazine are generally thought to express their biological activity as photosynthetic inhibitors but they are also toxic to plant tissue cultures at higher concentrations than that used on intact plants. Thus, tissue cultures can be used to study selected biochemical mechanisms of activity that cannot be readily examined with intact plants. Zilkah and Gressel (1978a) concluded that green calli better mirror the effect of test chemicals on plant seedlings than does achlorophyllus calli, particularly photosynthetic inhibitors.

Jordan *et al.* (1966) examined the effect of five photosynthetic-inhibitors on achlorophyllous tobacco callus tissue cultures. Callus growth was inhibited by molar concentrations of 10^{-7} for atrazine and simazine, 10^{-8} for monuron, 10^{-4} for bromacil and 10^{-6} for paraquat and amitrole. At these concentrations the observed toxicity must be due to effects other than that of photosynthetic inhibitors. Tobacco callus tissue cultures were used to determine the effect of eleven 2,6-dinitroaniline herbicides at five different concentrations on growth (Huffman and Camper, 1978). The molar concentration required to inhibit the fresh weight by 50% (I_{50}) was determined from the regression equations. At 100 times the I_{50}, D-α-tocopherol, decreased the inhibition of growth caused by these dinitroaniline herbicides (Huffman and Camper, 1978).

In an interesting and important bioassay, Davis *et al.* (1978b) investigated the effects of small amounts (0–20%) of various organic solvents on the growth response and fine structure of suspension cultures of soybean, okra, wild carrot and wheat. The response varied with the plant species but the toxic effects followed the ranking: ethanol > isopropanol > acetone > methanol > dimethyl sulphoxide > chloroform. These results are significant since many laboratories are required to introduce small quantities of substances to both whole plants and tissue cultures via organic solvents. Ethanol and acetone are two commonly used solvents but both of these exhibited considerable toxicity. These data had significant effects on one of our own experiments. When amino acid conjugates of 2,4,5-T dissolved in 200 μl of ethanol were introduced to 50 ml of soybean callus tissue medium nearly complete inhibition of growth occurred and incorrect conclusions were initially made (Arjmand, 1978; Davidonis *et al.*, 1979). Even at concentrations of 0.1 % ethanol, some inhibition was noted.

CONCLUSIONS

Because of the limited number of compounds studied and the variety of conditions used in metabolism investigations with plant tissue cultures it is difficult to generalize these results. Most investigations have involved studies with suspension cultures but several laboratories, including our own, have conducted extensive metabolism investigations with plant callus tissue cultures. Several generalizations are evident. Plant tissue cultures offer a sterile, rapid, inexpensive procedure for the study of metabolism of pesticides in plants. The same metabolic pathways seem to be operating in plant tissue cultures as are operating in whole plants. Qualitatively, the metabolism is very similar but major quantitative differences may be observed. Using plant tissue culture, and in particular suspension cultures, more metabolites seem to leak out of the cells, before the plant tissue is able to conjugate these metabolites, than with whole plants. Thus a higher percentage of aglycons are found in the media. Also, with plant tissue culture, a greater proportion of the applied pesticide has often been reported in the insoluble residue fraction.

These quantitative differences may not be as great as first appears. Davidonis *et al.* (1978; 1980a) have shown that various factors such as type of culture, age of tissue, concentration of pesticide, plant variety and source of tissue all affect metabolism. With proper selection of these factors, conditions may be achieved to closely mimic the metabolism in whole plants, Also, conditions can be selected to mass produce desired products such as glycosides and aglycons which may be needed for structural characterization.

It is easier to isolate and characterize metabolites with plant tissue cultures than with whole plants because plant tissue cultures lack the large quantities of pigments and starches. Also, tissue culture studies can be performed with small amounts of ^{14}C-labelled compounds which permits the efficient use of an often expensive ^{14}C-labelled material.

The more recent metabolism studies with plant tissue cultures have generated information on the mechanisms of action of pesticides and metabolic understanding of species specificity. This kind of information may be of greatest benefit for future developments. Through a better understanding of the mechanism of action of pesticides, the synthetic chemist can more appropriately design newer chemicals to be more selective and take advantage of species specificity.

The literature is expanding rapidly on the use of plant tissue culture for bioassaying pesticides and their metabolites. It is a useful technique for rapidly and cheaply screening new compounds for herbicidal and growth regulatory activity. However, investigators must carefully interpret these results because of the variability of the methods used and the biological variability within each experiment. A need continues to exist for standardization of technique to better interpret and extrapolate results.

It is not foreseen that plant tissue culture studies will ever replace studies with whole plants but they will complement these investigations. Until we more thoroughly understand the various factors affecting metabolism studies with plant tissue cultures and learn to control these factors we cannot expect results with plant tissue culture to be extrapolated to whole plants.

REFERENCES

AbdelMonem, A. E. (1981). 'Toxicity, mode of action, analytical technique, and metabolism of a new molt inhibitor (EL-494)', Ph.D. Thesis, The Pennsylvania State University, University Park, Pennsylvania.

Arjmand, M. (1978). 'Identification, characterization, analytical techniques and biological activity of amino acid conjugates of chlorophenoxy herbicides', Ph.D. Thesis, The Pennsylvania State University, University Park, Pennsylvania.

Arimand, M., Hamilton, R. H., and Mumma, R. O. (1978). 'Metabolism of 2,4,5-trichlorophenoxyacetic acid: Evidence for amino acid conjugates', *J. Agric. Food Chem.*, **26**, 1125–1128.

Barz, W. (1977). 'Catabolism of endogenous and exogenous compounds by plant cell cultures', in *Plant Tissue Culture and Its Biotechnological Applications* (eds W. Barz, E. Reinhard, and M. H. Zenk), pp. 153–177, Springer-Verlag, Berlin, Heidelberg, New York.

Brain, K. R., and Lines, D. S. (1977). 'Uptake and metabolism of aldrin in plant tissue cultures' in *Plant Tissue Culture and Its Biotechnological Applications* (eds W. Barz, E. Reinhard, and M. H. Zenk), pp. 197–203, Springer-Verlag, Berlin, Heidelberg, New York.

Bristol, D. W., Ghanuni, A. M., and Oleson, A. E. (1977). 'Metabolism of 2,4-dichlorophenoxyactic acid by wheat cell suspension cultures', *J. Agric. Food Chem.*, **25**, 1308–1314.

Davidonis, G. H., Arjmand, M., Hamilton, R. H., and Mumma, R. O. (1979). 'Biological properties of amino acid conjugates of 2,4,5-trichlorophenoxyacetic acid', *J. Agric. Food Chem.*, **27**, 1086–1088.

Davidonis, G. H., Hamilton, R. H., and Mumma, R. O. (1978). 'Metabolism of 2,4-dichlorophenoxyacetic acid in soybean root callus and differentiated soybean root cultures as a function of concentration and tissue age', *Plant Physiol.*, **62**, 80–83.

Davidonis, G. H., Hamilton, R. H., and Mumma, R. O. (1980a). 'Comparative metabolism of 2,4-dichlorophenoxyacetic acid in cotyledon and leaf callus from two varieties of soybean', *Plant Physiol.*, **65**, 94–97.

Davidonis, G. H., Hamilton, R. H., and Mumma, R. O. (1980b). 'Metabolism of 2,4-D in soybean root callus: evidence for the conversion of 2,4-D amino acid conjugates to free 2,4-D', *Plant Physiol.*, **66**, 537–540.

Davidonis, G. H., Hamilton, R. H., Vallejo, R. P., Buly, R., and Mumma, R. O. (1982). 'Biological properties of D-amino acid conjugates of 2,4-D', *Plant Physiol.*, **70**, 357–360.

Davis, D. G., Hodgson, R. H., Dusbabek, K. E., and Hoffer, B. L. (1978a). 'The metabolism of the herbicide diphenamid (*N-N*-dimethyl-2,2-diphenylacetamide) in cell suspensions of soybean [*Glycine max* (L.) Merr.]', *Physiol. Plantarum*, **44**, 87–91.

Davis, D. G., and Shimabukuro, R. H. (1980). 'Studies of herbicide toxicity and mode of action using isolated mesophyll cells and callus-derived cell suspensions', *Can. J. Bot.*, **58**, 1482–1489.

Davis, D. G., Wergin, W. P., and Dusbabek, K. E. (1978b). 'Effects of organic solvents on growth and ultrastructure of plant cell suspensions', *Pestic. Biochem. Physiol.*, **8**, 84–97.

Donald, W. W., and Shimabukuro, R. H. (1980). 'Selectivity of diclofop-methyl between wheat and wild oat: growth and herbicide metabolism', *Physiol. Plantarum*, **49**, 459–464.

Dorough, H. W., and Wiggins, O. G. (1969). 'Nature of the water-soluble metabolites of carbaryl in bean plants and their fate in rats', *J. Econ. Entomol.*, **62**, 49–53.

Dusky, J. A., Davis, D. G., and Shimabukuro, R. H. (1980). 'Metabolism of diclofop-methyl (methyl-2-[4-(2′,4′-dichlorophenoxy)-phenoxy]-propanate) in cell suspensions of diploid cheat (*Triticum monococcum*)', *Physiol. Plantarum*, **49**, 151–156.

Epstein, E., Klein, I., and Lavee, S. (1975). 'Uptake and fate of IAA in apple callus tissues: metabolism of IAA-2-^{14}C', *Plant and Cell Physiol.*, **16**, 305–311.

Epstein, E., and Lavee, S. (1975). 'Uptake and fate of IAA in apple callus tissue using IAA-1-^{14}C', *Plant and Cell Physiol.*, **16**, 553–561.

Feung, C.-S., Hamilton, R. H., Witham, F. H., and Mumma, R. O. (1972). 'The relative amounts and identification of some 2,4-D metabolites isolated from soybean cotyledon tissue cultures', *Plant Physiol.*, **50**, 80–86.

Feung, C.-S., Hamilton, R. H., and Mumma, R. O. (1973). 'Metabolism of 2,4-dichlorophenoxyacetic acid. V. Identification of metabolites in soybean callus tissue cultures', *J. Agric. Food Chem.*, **21**, 637–640.

Feung, C.-S., Mumma, R. O., and Hamilton, R. H. (1974). 'Metabolism of 2,4-dichlorophenoxyacetic acid. VI. Biological properties of amino acid conjugates', *J. Agric. Food Chem.*, **22**, 307–309.

Feung, C.-S., Hamilton, R. H., and Mumma, R. O. (1975). 'Metabolism of 2,4-dichlorophenoxyacetic acid. VII. Comparison of metabolites from five species of plant callus tissue cultures', *J. Agric. Food Chem.*, **23**, 373–376.

Feung, C.-S., Hamilton, R. H., and Mumma, R. O. (1976a). 'Metabolism of 2,4-dichlorophenoxyacetic acid. 10. Identification of metabolites in rice root callus tissue cultures', *J. Agric. Food Chem.*, **24**, 1013–1015.

Feung, C.-S., Hamilton, R. H., and Mumma, R. O. (1976b). 'Metabolism of indole-3-acetic acid. III. Identification of metabolites isolated from crown gall callus tissue', *Plant Physiol.*, **58**, 666–669.

Feung, C.-S., Hamilton, R. H., and Mumma, R. O. (1977a). 'Metabolism of indole-3-acetic acid. IV. Biological properties of amino acid conjugates', *Plant Physiol.*, **59**, 91–93.

Feung, C.-S., Hamilton, R. H., and Mumma, R. O. (1977b). 'Metabolism of 2,4-dichlorophenoxyacetic acid. 11. Herbicidal activity of amino acid conjugates', *J. Agric. Food Chem.*, **25**, 898–900.

Feung, C.-S., Loerch, S. L., Hamilton, R. H., and Mumma, R. O. (1978). 'Metabolism of 2,4-dichlorophenoxyacetic acid. Comparative metabolic fate in plants and plant tissue cultures', *J. Agric. Food Chem.*, **26**, 1064–1067.

Fowler, D. D., and Mahan, J. N. (1980). *The Pesticide Review 1978*, US Dept. of Agriculture.

Frear, D. S., and Swanson, H. R. (1975). 'Metabolism of cisanalide (*cis*-2,5-dimethyl-1-pyrrolidinecarboxanilide) by excised leaves and cell suspension cultures of carrot and cotton', *Pestic. Biochem. Physiol.*, **5**, 73–80.

Gamborg, O. L. (1975). 'Callus and cell culture', in *Plant Tissue Culture Methods* (eds O. L. Gamborg, and L. R. Wetter), pp. 1–9, National Research Council of Canada, Prairie Regional Laboratory, Saskatoon, Saskatchewan.

Gressel, J. (1979). 'A review of the place of *In vitro* cell culture systems in studies of action, metabolism and resistance of biocides affecting photosynthesis', *Z. Naturforsch.*, **34c**, 905–913.

Gressel, J., and Segel, L. A. (1978). 'The paucity of plants evolving genetic resistance to herbicides: possible reasons and implications', *J. Theor. Biol.*, **75**, 349–372.

Hamilton, R. H., Meyer, H. E., Burke, R. E., Feung, C.-S., and Mumma, R. O. (1976). 'Metabolism of indole-3-acetic acid. II. Oxindole pathway in *Parthenocissus tricuspidata* crown-gall tissue cultures', *Plant Physiol.*, **58**, 77–81.

Hoerauf, R. A., and Shimabukuro, R. H. (1979). 'The response of resistant and susceptible plants to diclofop-methyl', *Weed Research*, **19**, 293–397.

Huffman, J. B., and Camper, N. D. (1978). 'Growth inhibition in tobacco (*Nicotiana tabacum*) callus by 2,6-dinitroaniline herbicides and protection by D-α tocopherol acetate', *Weed Sci.*, **26**, 527–530.

Jordan, L. S., Murashige, T., Mann, J. D., and Day, B. E. (1966). 'Effect of photosynthesis-inhibiting herbicides on non-photosynthetic tobacco callus tissue', *Weeds*, **14**, 134–136.

Kuhr, R. J., and Casida, J. E. (1967). 'Persistent glycosides of metabolites of methylcarbamate insecticide chemicals formed by hydroxylation in bean plants', *J. Agric. Food Chem.*, **15**, 814–824.

Leguay, J.-J., and Guern, J. (1977). 'Quantitative effects of 2,4-dichlorophenoxyacetic acid on growth of suspension-cultured *Acer pseudoplantanus* cells. II. Influence of 2,4-D metabolism and intracellular pH on the control of cell division by intracellular 2,4-D concentrations', *Plant Physiol.*, **60**, 265–270.

Lieb, H. G., Ray, T. B., and Still, C. C. (1973). 'Growth of rice root-derived callus tissue in suspension culture', *Plant Physiol.*, **51**, 1140–1141.

Locke, R. K., and Baron, R. L. (1972). 'Preforan metabolism by tobacco cells in suspension culture', *J. Agric. Food Chem.*, **20**, 861–867.

Locke, R. K., Chen, J. T., Damico, J. N., Dusold, L. R., and Sphon, J. A. (1976). 'Identification by physical means of organic moieties of conjugates produced from carbaryl by tobacco cells in suspension culture', *Archives of Environ. Contam. Toxicology*, **4**, 60–100.

Montague, M. J., Enns, R. K., Siegel, N. R., and Jaworski, E. G. (1981a). 'A comparison of 2,4-dichlorophenoxyacetic acid metabolism in cultured soybean cells and in embryogenic carrot cells', *Plant Physiol.*, **67**, 603–607.

Montague, M. J., Enns, R. K., Siegel, N. R., and Jaworski, E. G. (1981b). 'Inhibition of 2,4-dichlorophenoxyacetic acid conjugation to amino acids by treatment of cultured soybean cells with cytokinis', *Plant Physiol.*, **67**, 710–714.

Mostafa, I. Y., Hassan, A., and Zayed, S. M. A. D. (1966). 'Metabolism of carbamate drugs. III. Translocation and degradation of ^{14}C-labelled Sevin in cotton plant', *Z. Naturforsch.*, **21**, 1060–1062.

Mumma, R. O., and Hamilton, R. H. (1979). 'Xenobiotic metabolism in higher plants: in vitro tissue and cell culture techniques', in *Xenobiotic Metabolism: In vitro Methods* (eds G. D. Paulson, D. S. Frear, and E. P. Marks), pp. 35–72, ACS Symposium Series No. 97.

Mumma, R. O., Khalifa, S., and Hamilton, R. H. (1971). 'Spectroscopic identification of metabolites of carbaryl in plants', *J. Agric. Food Chem.*, **19**, 445–451.

Nazarova, T. A., Pavlova, N. N., Makeev, A. M., and Chkanikov, D. I. (1980). 'Amino acid conjugates of 2,4-D in cereals', *Fiziol. Rast.*, **27**, 740–745.

Ojima, K., and Gamborg, O. L. (1968). 'The metabolism of 2,4-dichlorophenoxyacetic acid by suspension cultures of soybean root', *Biochem. Physiol. Plant Growth Subst., Proc. Int. Conf. Plant Growth Subst.*, **6**, 857–865.

Oswald, T. H., Smith, A. E., and Phillips, D. V. (1978). 'Phytotoxicity and detoxification of metribuzin in dark-grown suspension cultures of soybean', *Pestic. Biochem. Physiol.*, **8**, 73–83.

Ray, T. B., and Still, C. C. (1975). 'Propanil metabolism in rice: a comparison of propanil amidase activities in rice plants and callus cultures', *Pestic. Biochem. Physiol.*, **5**, 171–177.

Sandermann, H., Diesperger, H., and Scheel, D. (1977). 'Metabolism of Xenobiotics by plant cell cultures', in *Plant Tissue Culture and Its Biotechnological Applications* (eds W. Barz, E. Reinhard, and M. H. Zenk), pp. 178–196, Springer-Verlag, Berlin, Heidelberg, New York.

Scheel, D., and Sandermann, H. (1977). 'Metabolism of DDT and Kelthane in cell suspension of parsley (*Petroselinum hortense*, Hoff.) and soybean (*Glycine max* L.)', *Planta*, **133**, 315–320.

Scheel, D., and Sandermann, H. (1981a). 'Metabolism of 2,4-dichlorophenoxyacetic acid in cell suspension cultures of soybean (*Glycine max* L.) and wheat (*Triticum aestiuum* L.). I. General Results', *Planta*, **152**, 248–252.

Scheel, D., and Sandermann, H. (1981b). 'Metabolism of 2,4-dichlorophenoxyacetic acid in cell suspension cultures of soybean (*Glycine max* L.) and wheat (*Triticum aestiuum* L.). II. Evidence for incorporation into lignin', *Planta*, **152**, 253–258.

Shimabukuro, R. H., Walsh, W. C., and Hoerauf, R. A. (1979). 'Metabolism and selectivity of diclofop-methyl in wild oat and wheat', *J. Agric. Food Chem.*, **27**, 615–623.

Shvetsov, S. G., and Gamburg, K. Z. (1980). 'Absorption and metabolism of 2,4-dichlorophenoxyacetic acid (2,4-D) in suspension cultures of corn and soybean cell', *Fiziol. Rast.*, **27**, 746–755.

Smith, A. E. (1979). 'Metabolism of 2,4-DB by white clover (*Trifolium repens*) cell suspension cultures', *Weed Sci.*, **27**, 392–396.

Smith, A. E., and Oswald, T. H. (1979). 'Degradation of phenoxyalkylcarboxylic acids by white clover (*Trifolium repens*) cell suspensions', *Weed Sci.*, **27**, 389–391.

Stöckigt, J., and Ries, B. (1977). 'Catabolism of γ-hexachlorocyclohexane (lindane) by plants and cell cultures, a comparison', in *Plant Tissue Culture and Its Biotechnological Applications*, (eds W. Barz, E. Reinhard, and M. H. Zenk), pp. 204–210, Springer-Verlag, Berlin, Heidelberg, New York.

v.d. Trenck, T., and Sandermann, H. (1978). 'Metabolism of benzo[α]-pyrene in cell suspension cultures of parsley (*Petroselinum hortense*, (Hoffm.) and soybean (*Glycine max* L.)', *Planta*, **141**, 245–251.

v.d. Trenck, T., and Sandermann, H. (1981). 'Incorporation of benzo[α]-pyrene quinones into lignin', *FEBS Lett.*, **125**, 72–76.

Troxler, R. F., and Hamilton, R. H. (1965). 'The metabolism of indole-3-acetic acid by geranium stem callus cultures', *Plant Physiol.*, **40**, 400–405.

Udvardy, J., Sivók, B., and Néet, G. (1976). 'Effect of naphtylacetic acid, 2,4,5-trichlorophenoxyacetic acid and 3,6-dichloro-*o*-anisic acid on nucleolytic enzymes in callus cultures from wheat root', *Z. Pflanzenphysiol.*, **87**, 33–40.

Zama, P. (1981). 'An improved method for the analysis of two amino acid conjugates of 2,4-dichlorophenoxyacetic acid in soybean plants and in tissue cultures by gas chromatography', Masters Thesis, The Pennsylvania State University, University Park, Pennsylvania.

Zilkah, S., Bocion, P. R., and Gressel, J. (1977). 'Cell cultures vs. whole plants for measuring phytotoxicity. II. Correlations between phytotoxicity in seedlings and calli', *Plant and Cell Physiol.*, **18**, 657–670.

Zilkah, S., Bocion, P. F., and Gressel, J. (1978). 'Target tissue for napropamide inhibition: effects on green and white callus cultures and seedlings', *Weed Sci.*, **26**, 711–713.

Zilkah, S., and Gressel, J. (1977a). 'Cell cultures vs. whole plants for measuring phytotoxicity. I. The establishment and growth of callus and suspension cultures; definition of factors affecting toxicity on calli', *Plant and Cell Physiol.*, **18**, 641–655.

Zilkah, S., and Gressel, J. (1977b). 'Cell cultures vs. whole plants for measuring phytotoxicity. III. Correlations between phytotoxicities in cell suspension cultures, calli and seedlings', *Plant and Cell Physiol.*, **18**, 815–820.

Zilkah, S., and Gressel, J. (1978a). 'Correlations in phytotoxicity between white and green calli of *Rumex obtusifolius*, *Nicotiana tabacum* and *Lycopersicon esculentum*', *Pestic. Biochem. Physiol.*, **9**, 334–339.

Zilkah, S., and Gressel, J. (1978b). 'Differential inhibition by dikegulac of dividing and stationary cells in vitro cultures', *Planta*, **142**, 281–286.

Zilkah, S., and Gressel, J. (1978c). 'The estimation of cell death in suspension cultures evoked by phytotoxic compounds: differences among techniques', *Plant Sci. Lett.*, **12**, 304–316.

Zilkah, S., and Gressel, J. (1979). 'Cell suspensions for kinetic studies of inhibitor action: the rapid initial action of dikegulac on membrane properties in situ', *Planta*, **145**, 272–278.

Progress in Pesticide Biochemistry and Toxicology, Volume 3
Edited by D. H. Hutson and T. R. Roberts

CHAPTER 6

Separation and purification of pesticide metabolites

W. Muecke

ABBREVIATIONS

c.c.d.	counter-current distribution
d.c.c.c.	droplet counter-current chromatography
g.c.	gas chromatography
g.l.c.	gas liquid chromatography
h.p.l.c.	high performance liquid chromatography
h.v.e.	high voltage electrophoresis
i.r.	infra-red (spectroscopy)
m.s.	mass spectrometry
n.m.r.	nuclear magnetic resonance (spectroscopy)
p.c.	paper chromatography
t.l.c.	thin-layer chromatography

Continued:

—gluc. (β-D-glucose)

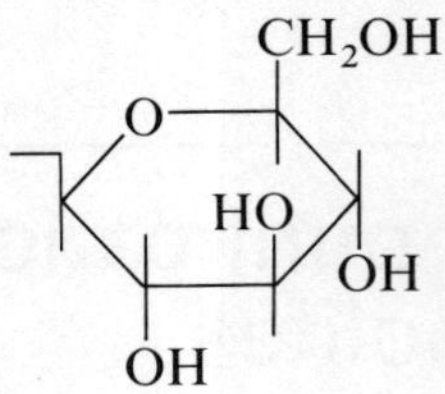

—gluc. ac. (β-D-glucuronic acid)

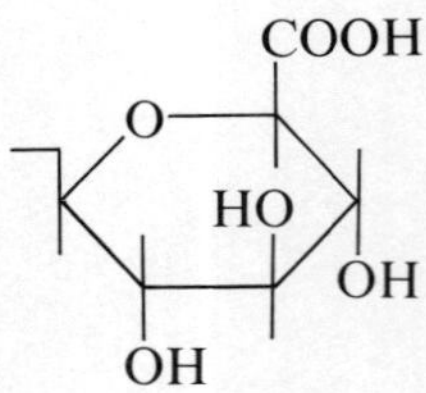

—gluc.-gluc. (β-D-glucose-β-D-glucose, gentiobiose)

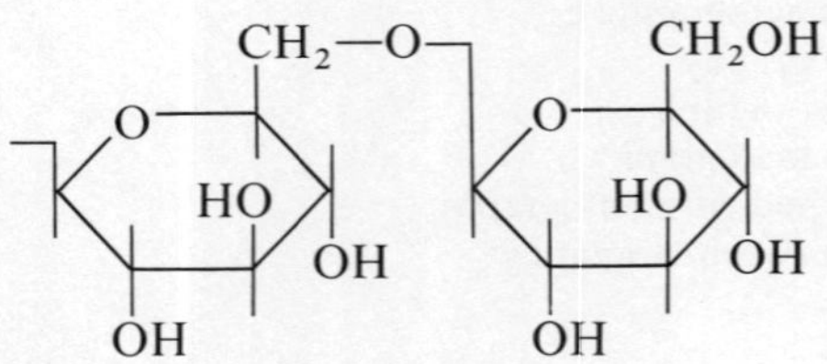

—gluc.-arab. (β-D-glucose-β-D-arabinose)

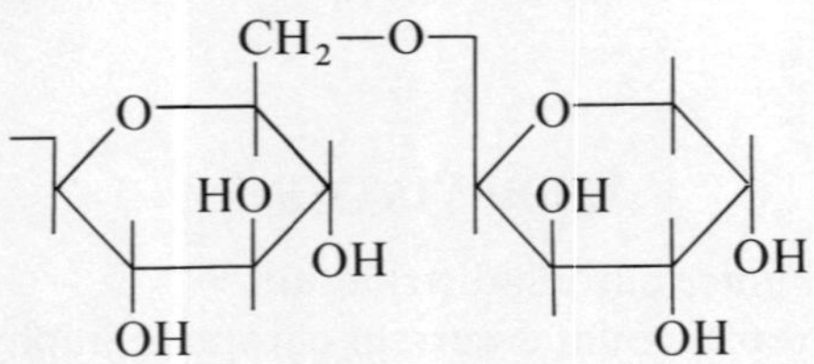

—gluc.-xyl. (β-D-glucose-β-D-xylose)

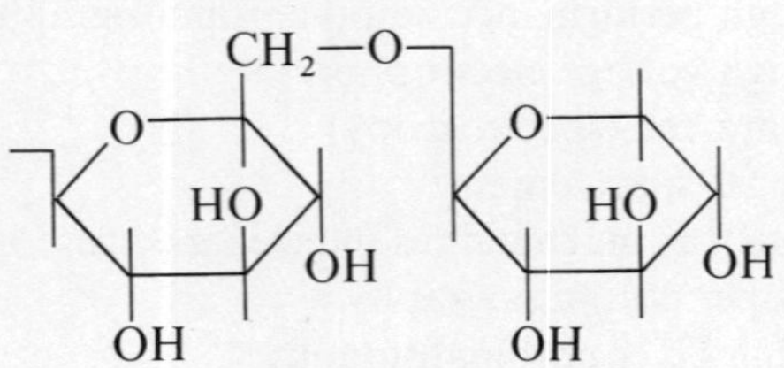

—gly. (glycine)

$—NH—CH_2—COOH$

—glutam. (glutamine)

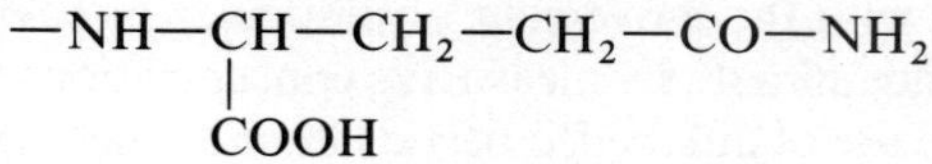

—cyst. (cysteine)

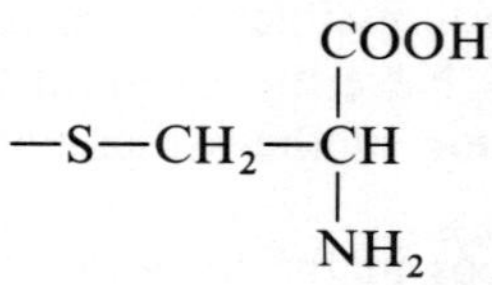

—*N*-acet. cyst. (*N*-acetyl cysteine, mercapturic acid)

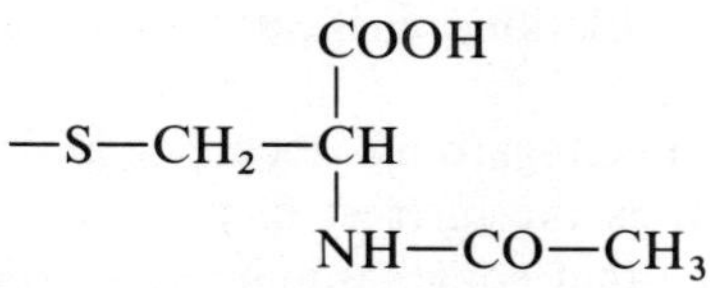

—*N*-mal. cyst. (*N*-malonyl cysteine)

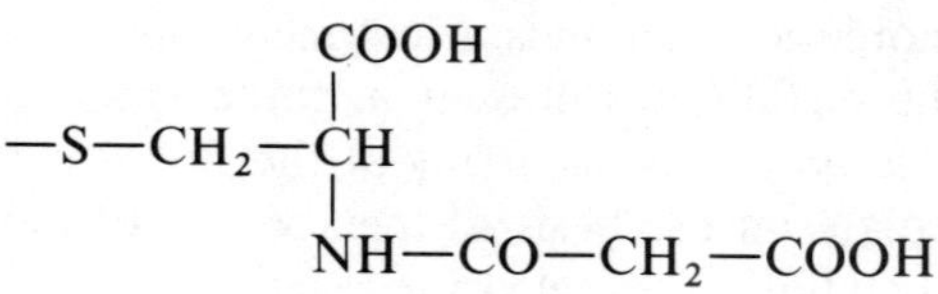

—glutath. (glutathione)

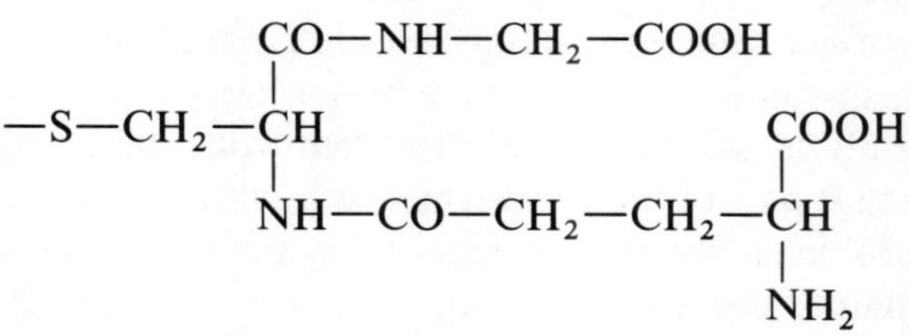

—sulph. ac. (sulphuric acid)

$—SO_2—OH$

—taur. (taurine)

$—NH—CH_2—CH_2—SO_2—OH$

INTRODUCTION

Metabolism studies represent together with other disciplines like toxicology and residue analysis a vital part of the approach to evaluate and assess the safety of pesticides to man and the environment. Early metabolism studies were mainly concerned with the movement of pesticides in environmental systems, including plants and animals, by measuring concentrations mostly of the parent pesticide only. The use of unlabelled pesticides and labelling with ^{32}P, common for metabolism studies of insecticides in the 1950s, was of limited value for the identification of degradation products in biological systems. When in the 1960s ^{14}C-labelled pesticides became readily accessible an important tool for chemical transformation studies was available. Nevertheless, until the early 1970s emphasis was placed on the identification of the rather non-polar metabolites obviously for two major reasons:

The knowledge of the primary steps in the metabolism pathways leading mostly to still non-polar products were assumed to provide the most important information for a safety evaluation. More importantly, the availability of methodological tools for the isolation, purification and identification of polar metabolites was rather limited.

Several recent developments regarding concepts and methodology have stimulated each other and thus the interest in pesticide metabolism studies. There is now ample evidence that some conjugates, in the past generally regarded as terminal detoxication products, must be considered as strong electrophilic reactants *in vivo*, which may play a significant role in the adverse effects of a particular xenobiotic in animals. Most metabolism studies in animals are concerned with the metabolites present in the excreta, which may not fully reflect the metabolic processes occurring at the site of concern. It is certainly important to speculate on the basis of metabolic pathways derived from the structures of the metabolites excreted for reactive intermediates and prove or disprove a corresponding hypothesis with properly designed experiments. However, there is an increasing awareness of the necessity to identify the nature of retained or accumulating metabolites in a particular tissue or organ.

Metabolism studies in plants have been often carried out with young plants and the metabolites isolated and identified from leaves, stems, roots, etc., but rarely with that part of the plant which is used for human or animal consumption. The problems involved in identifying the metabolites in the consumable parts are mainly due to the minute amounts of metabolites present and the limited availability of adequate isolation tools. The investigations of the transformation of pesticides during the various transfer processes within environmental chains, therefore, require new and improved methods.

High performance liquid chromatography (h.p.l.c.) is probably the most powerful tool introduced within the last twenty-five years for separating and purifying metabolites. The increasing number of different stationary phases available allow us to utilize various mechanisms for separation, thus covering a wide array of metabolites with respect to their physical-chemical properties. In contrast to preparative gas liquid chromatography (g.l.c.), variations in the mobile phase offer an additional variable to optimize the separation process. Interfacing a high performance liquid chromatograph with a mass spectrometer may add additional value to this method, when the problems associated with the large amounts of organic and aqueous solvent mixtures and buffers are solved (see Chapter 7).

The most informative spectroscopic methods used for the identification of pesticide metabolites, i.e. proton nuclear magnetic resonance spectroscopy (^{1}H n.m.r.) and mass spectrometry (m.s.) have progressed impressively during the 1970s. Applying high fields (>300 MHz) combined with Fourier transformation techniques and optimized electronic equipment have improved the sensitivity of ^{1}H n.m.r. by a factor of 1000 allowing the acquisition of interpretable spectra from sample amounts as low as about 10 μg.

The recently developed soft-ionization techniques in mass spectrometry, e.g. field desorption, plasma desorption and laser desorption form the ions from metabolites being investigated in the solid state. As here, in contrast to the traditional ionization techniques (electron-impact, chemical ionization) the sample does not have to be vaporized, spectra are obtained at lower temperature, yielding more intense molecular-ion peaks and reduced secondary fragmentation. This essentially means that often molecular ions are obtained where electron-impact ionization failed, that less volatile metabolites yield interpretable spectra and that metabolites may not have to be derivatized. These ionization techniques reveal more structural information and extend mass spectroscopy to metabolites with lower vapour pressure and higher molecular weight, e.g. polar conjugates (see Chapter 7).

In the framework outlined above, this Chapter will address the approaches and methods currently used for the isolation of pesticide metabolites, with special emphasis on conjugates. Furthermore, it will be restricted to methods leading finally to a substantial amount of a purified metabolite for multiple-method elucidation of the structure. Therefore, combined purification and identification methods, e.g. g.c.–m.s., g.c.–i.r., h.p.l.c.–m.s., reversed isotope dilution analysis, as well as simple comparative methods (with reference compounds) will not be covered.

ON STRATEGIES

'The situation [for pesticide metabolism chemists] is rather like looking for the proverbial "needle in the haystack" when one does not know the size, shape or composition of the needle, or even how many needles there are in the stack' (Harvey, 1980). Fortunately the 'needles' are tagged today, usually with ^{14}C and there is a considerable variety of methods available to establish their 'number, size, shape and composition'. Assuming the 'needles' were tagged in the proper position(s), the problem remains to select the most adequate method for each individual step and to arrange them in a meaningful sequence during the separation and purification procedure.

A strategy has to be developed in order to obtain the desired metabolites in sufficient amounts, with adequate purity and in their original form, or as a derivative thereof obtained by well understood chemical transformations for the elucidation of their genuine structures in an economical way, with respect to manpower, time and materials. There are very few papers in the pesticide literature demonstrating a systematic approach and stating the rationale for the application of a particular method. Most of the schemes followed are probably based on those methods of which the investigator has the most experience; these may not be the most suitable for a particular problem. Although this empirical approach has its justification, the increasing experience with a particular method leads essentially to its even more predetermined selection possibly at the expense of more suitable methods.

Another general problem involved in the optimization of a particular method is the fact that the metabolism chemist will proceed to the next step in the purification process when the procedure used fulfils his minimal requirement with respect to a desired purification or separation goal. In contrast to routine residue analysis, a certain separation problem might be considered unique and he may not be willing to make additional expenditure to proceed from the stage of minimal requirement to that of maximum economy. Although this hesitation might be justified regarding the project under study, it hampers the process of acquiring a rational base of experience.

Starting from a raw agricultural commodity, soil, animal excreta and tissues, etc. the process leading to purified metabolites might be subdivided into four interdependent steps. The metabolites present in solid matrices must be solubilized by solvent extraction or similar techniques (solubilization), and in turn enriched by methods like solvent partitioning or others (isolation). Thereafter, the more or less complex mixture of metabolites must be separated into individual compounds (separation) and the non-labelled impurities removed (purification).

This stepwise procedure is, of course, arbitrary as all solubilization, isolation and separation techniques and methods will influence the purity of metabolites in their mixtures. However, in discussing this subject, it is convenient to distinguish three major steps.

To design and apply successfully an isolation and purification strategy we require:

1. A sound understanding of the technology and the basic principles and mechanisms of the methods available.
2. Knowledge of the amount and nature of the matrix material concomitantly extracted or isolated.

3. Knowledge of the number, distribution and properties of the metabolites to be isolated and purified.

Regarding the first requirement, which will not be addressed in this chapter, the excellent treatise of Karger, Snyder and Horvath (1973) is highly recommended. Reference to comprehensive reviews on individual methods will be given in the corresponding sections.

The amount and the nature of co-extractables or co-isolates are determined by the type of matrix and by the selectivity of the isolation procedure used. In most cases soil does not pose a serious problem, as the solubility of the matrix in aqueous or organic solvents is rather limited and thus usually relatively pure extracts are obtained.

The amount and the nature of the ingredients present in urine are well-known, so that the removal of the major ones from the metabolites can be readily integrated in the design of a dedicated strategy. The same holds basically true for the isolation of metabolites from milk, eggs and animal tissues, at least with respect to the major constituents. But for these matrices the isolation is often complicated by the rather small amounts of metabolites present. The purification of metabolites extracted from faeces is a very tedious task, as these extracts contain almost a continuum of ingredients with respect to their polarity, i.e. ranging from very polar to extremely non-polar. This is probably why the isolation of faecal metabolites is still often based on trial and error approaches.

Most problems are usually encountered during the isolation of plant metabolites, although the major ingredients of most plant species are known. As plants do not have efficient excretion mechanisms, they tend to inactivate xenobiotics by storage in a polar but preferably insoluble form. Therefore, rather polar aqueous extractants are usually necessary to isolate the metabolites; these often also solubilize major plant ingredients, leading to a high ratio of more or less defined plant ingredients to metabolites in the extract. This situation often limits the efficiency of well-designed separation procedures.

Knowledge of the properties of the metabolites to be isolated is the most important input for the design of a successful strategy and methods to derive this information will be discussed in a separate section. In addition to yielding information on the stability of the metabolites with respect to hydrolysis, oxidation, etc., which might be important to exclude stressful procedures, it allows the selection of separation methods which utilize selectively the specific properties of the metabolites.

Having thus generally encompassed the thoughts which are applicable to the problem, a meaningful selection of methods and their proper sequence must be determined. Three method-specific features are a helpful guide in this respect: capacity, selectivity and resolution.

Capacity is the maximum amount of a given mixture which can be applied to a particular method while still allowing a defined separation efficiency. It is obvious that methods with high capacity are used in the earlier steps of the process and accordingly this feature is not important in the final purification processes.

Selectivity is the capability of a particular method to separate given single or groups of compounds on the basis of their distinct properties. Selective methods are also applied in the early steps of purification procedures for grouping the compounds on the basis of their solubility, charge, molecular weight, etc.

Resolution is the capability of a particular method to separate single compounds on the basis of the same property but with small differences in its numerical value. High resolution methods are required in the final purification steps because the processes selected for the earlier steps will usually result in groups of metabolites with a high degree of similarity regarding physical-chemical properties.

In addition to the proper use of the features mentioned above, it appears advisable to utilize different properties of the same metabolite(s) in different steps of the isolation process, as this will eventually lead to a highly effective overall separation effect. Methods, which are not compatible with certain properties of the metabolites to be isolated, might be utilized, however, by changing the properties of the metabolites in a desired way by derivatization. This technique is quite often used to enhance volatility for gas chromatography, reduce polarity for adsorption liquid chromatography, etc.

All strategic features mentioned here and in the forthcoming general sections are dedicated to the free or conjugated xenobiotic-related metabolites. Pesticide moieties covalently bound to biological macromolecules or those being—after degradation—reutilized by incorporation into endogenous compounds will not be addressed in this chapter.

Recently, various reviews on the isolation of drug metabolites, including methodological and strategic aspects have been published and are especially recommended for readers interested in animal metabolism studies (Coutts and Jones, 1980; Dieterle and Faigle, 1981; Martin and Reid, 1981).

SOLUBILIZATION AND ISOLATION

Solid samples

Various procedures have been used to isolate metabolites from solid biological samples, the most common being solvent extraction. In order to maximize extraction efficiency the sample must be thoroughly ground. It is common practice to remove the water from the sample, preferably by lyophilization, but other—more crude—procedures are also used, prior to grinding up the sample.

Care should be taken to avoid elevated temperatures during the grinding process to ensure the integrity of the metabolites present. Instead of removing the water, the moisture may be bound by mixing the sample thoroughly with a common drying agent, e.g. anhydrous sodium sulphate.

A common alternative to this procedure is to use water, water-miscible solvents or mixtures thereof and homogenize the sample during the extraction process with blenders. Using this procedure, Wheeler *et al.* (1979) compared the extraction efficiencies of three commonly used commercial blenders from carbofuran (**1.0**)† treated radishes and found very small differences.

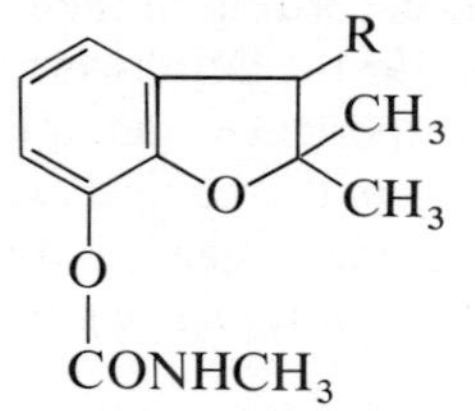

R: —H (**1.0**)
R: —OH (**1.1**)
R: —OOC—C(CH3)=CH—CH3 (**1.2**)

Besides the choice of the solvents, the procedure itself determines the efficiency of the extraction process. Thus, the extraction in a soxhlet extractor is generally very efficient and convenient, but stresses the sample due to the elevated temperature involved, so that preferably solvents with low boiling points should be selected. The solvent should also be frequently exchanged especially during the early phase of the extraction process to reduce the risk of artefact formation. Although less convenient, extraction at room temperature using shakers, mixers, blenders, ultrasonic agitation, etc. and separating the phases by filtration or preferably centrifugation appears to be generally the safer method. After thorough extraction by these methods the process might be continued in a soxhlet apparatus.

Another technique, which is preferred in our laboratory for faeces, is to place the lyophilized and thoroughly ground sample in a glass chromatography column and percolate the extracting solvent slowly through the column. The solvents for extraction have to be selected according to the nature of the metabolites of interest and the matrix to be extracted. They should be easily and readily removable (low boiling point), of chemical inertness and free of reactive contaminants (see section: Pitfalls, p. 345). The general purity of the solvents should be adequate for this early stage of the isolation and purification process.

† Parent structures are numbered consecutively with a sub-number 0 (e.g. **26.0**). Metabolites or derivatives of compound **26.0** are numbered **26.1**, etc.

Most metabolism studies are concerned with all metabolites present in a certain solid matrix, so that maximal extraction efficiency is desired. Consequently the solvents used are usually highly polar and water, methanol, ethanol, acetone, acetonitrile and their mixtures are usually employed. The choice of solvent is mostly arbitrary or based on experimental data obtained from preceding small-scale pilot experiments (Sonobe *et al.*, 1980a; b; Smith, 1981). Sometimes these polar solvents are used successively to ensure maximum extraction.

In some cases a sequence of solvents starting from very non-polar ones, like hexane, moving to very polar solvents or vice versa have been employed in order to group the metabolites according to their polarity or solubility (Larsen and Bakke, 1975; Bakke *et al.*, 1981a; Drinkwine and Fleeker, 1981). However, it appears to be the general experience that this approach usually failed, as overlap of the metabolite pattern in the various extracts was observed (Crayford and Hutson, 1980a). Hence a maximized extraction efficiency using polar solvents followed by grouping techniques, e.g. solvent partition, seems to be preferable.

After exhaustive extraction with solvents, hydrolytic enzymes have been applied to solubilize metabolites still attached to the solid matrix. Bradbury *et al.* (1981) incubated methanol extracted rat faeces with cellulase and thus improved the extractability from 60% to over 85%. Yu *et al.* (1980) could not significantly improve the extractability of buthidazole (**2**) metabolites from alfalfa plants upon incubation with sulphatase-contaminated β-glucuronidase.

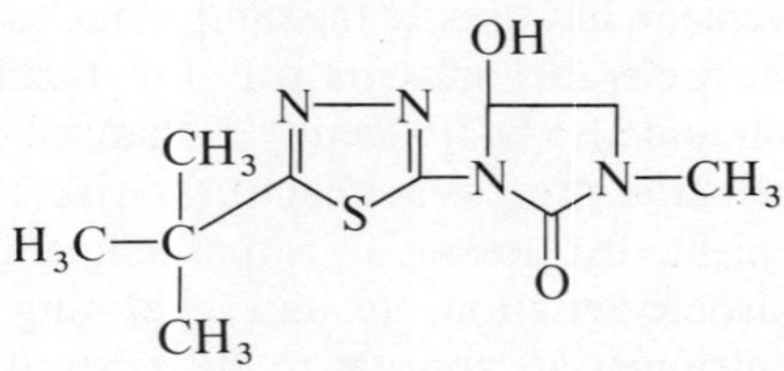

(**2**)

It is common practice to attempt finally the release of unextracted metabolites by acid or base treatment and to identify the released moieties. In some cases these unextractable compounds have been separated into fractions using biochemical methods and assigned as covalently bound or identified as endogenous compounds.

When the interest is dedicated to a defined group of metabolites a specific extraction solvent, taking into account the nature of the matrix, is preferred. Lipophilic conjugates, i.e. xenobiotics conjugated with fatty acids, diglycerides or similar endocons were extracted from spleen, liver, muscle, epididymal adipose tissue and carcass with chloroform or chloroform–methanol mixtures (Fears *et al.*, 1978; Quistad *et al.*, 1978a; Leighty *et al.*, 1980; Leighty, 1980; 1981).

However, a one-step procedure, originally developed by Bligh and Dyer (1959) for lipid extraction and purification from biological materials, separating metabolites into 'polar', 'non-polar' and 'non-extractable' fractions should be mentioned as it has been successfully used for the isolation of pesticide metabolites from animal tissues (Marco and Hermes, 1971), plant material (Marco and Hermes, 1971; Still and Mansager, 1972; 1973; 1975), faeces (Still and Mansager, 1975; Paulson *et al.*, 1975) and insects (Still and Leopold, 1978).

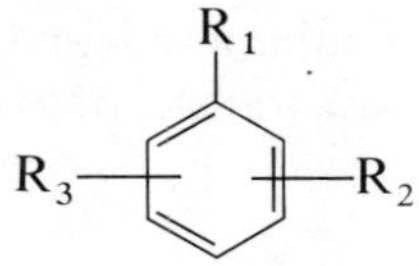

R_1	R_2	R_3	
$—NO_2$	$—Cl_5$	—	**(3.0)**
$—NH_2$	$—Cl_5$	—	**(3.1)**
—OH	$—Cl_5$	—	**(3.2)**
—H	$—Cl_5$	—	**(3.3)**
—SH	$—Cl_5$	—	**(3.4)**
$—SCH_3$	$—Cl_5$	—	**(3.5)**
$—SOCH_3$	$—Cl_5$	—	**(3.6)**
$—SCH_2COOH$	$—Cl_5$	—	**(3.7)**
—cyst.	$—Cl_5$	—	**(3.8)**
—N-acet. cyst.	$—Cl_5$	—	**(3.9)**
—N-mal. cyst.	$—Cl_5$	—	**(3.10)**
—glutath.	$—Cl_5$	—	**(3.11)**
$—NH_2$	$—Cl_4$	—H	**(3.12)**
—H	$—Cl_4$	—H	**(3.13)**
—cyst.	$—Cl_4$	—cyst.	**(3.14)**
$—NO_2$	$—Cl_4$	—N-mal. cyst.	**(3.15)**
$—NO_2$	$—Cl_4$	—glutath.	**(3.16)**
—glutath.	$—Cl_4$	—glutath.	**(3.17)**

Faecal and urinary metabolites from pentachloronitrobenzene (**3.0**) and hexachlorobenzene (**4**) treated animals were isolated by steam distillation (Renner *et al.*, 1978; Renner, 1980). The samples were adjusted to pH 1 and the steam-distilled metabolites continuously extracted from the condensate by organic solvents. Although this technique yields, besides separation from the non-volatile metabolites, a significant purification effect, its use in metabolism

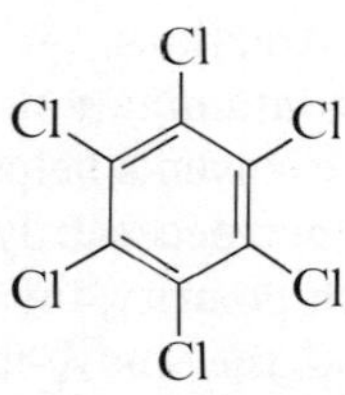

(**4**)

studies is rather limited and requires a careful evaluation of the metabolite stability, as the method is rather stressful. The technique is applicable for determining partial structures of bound or other unknown metabolites after hydrolysis (Hsu and Bartha, 1976; Hitchings and Roberts, 1980; O'Grodnick *et al.*, 1981) and has been used as common moiety approach in residue analysis of urea herbicides and other aniline containing pesticides (Baunok and Geissbühler, 1970).

A refined technique for the isolation of volatile compounds from biological material has been suggested by Zlatkis and Kim (1976). The procedure involves adsorption of the compound and transfer into glass wool followed by heat desorption.

Liquid samples

Metabolites present in aqueous samples like urine, bile, etc. and in water-reconstituted extracts of solid samples must be isolated, which implies concentration, clean-up and separation into suitable metabolite groups. Procedures specified by high capacity and selectivity, low resolution and convenient handling are required. Among the techniques employed, solvent partition and adsorption on different solid materials are the most common procedures used. As in most cases not all the metabolites will readily partition into the solvent, serious consideration has to be focused on the selectivity. It is therefore usually advantageous to select conditions which allow a clear separation into metabolite groups, rather than striving for maximum transfer into the organic phase. The most dominant variables in this context are the pH value of the aqueous phase and the polarity of the organic solvent. For extraction of acidic or basic compounds the pH is adjusted, so that they are present in a non-ionized form. Often the procedure is used sequentially, i.e. extraction at different pH and also with different solvents to effect fractionation (Muecke *et al.*, 1976; Lamoureux and Rusness, 1980) or back-extraction from the organic phase for purification purposes. Care must be taken for labile compounds which may form artefacts on pH changes.

A large variety of organic solvents has been used, e.g. hexane, dichloromethane, chloroform, diethyl ether, ethyl acetate, *n*-butanol, and mixtures like acetonitrile/diethyl ether, ethanol/diethyl ether, ethanol/hexane and ethyl acetate/ethanol. The solvents should have basically the same properties as those used for solid sample extractions. Additionally the water-solubility should be low to avoid cross contamination of the metabolites. The usual practice of water saturation of the solvents helps to maintain constant volumes during the procedure, but does not necessarily prevent carry-over effects. As the water solubility increases with polarity, this effect is especially pronounced with solvents like *n*-butanol and the above-mentioned mixtures containing high percentages of the polar partner. Although *n*-butanol is often employed as the 'ultima ratio' for the extraction of polar metabolites especially for glucuronides, the high water solubility and the high boiling point (b.p. 118 °C) precludes its general use. A high extraction strength can be obtained with the mixtures and a careful selection of the ratio may limit the water-solubility to an acceptable degree, thus providing an effective extractant for polar compounds (Irving *et al.*, 1967; Shafik and Enos, 1969). For an enhanced extractability the aqueous phase may be saturated with, for example, ammonium sulphate or sodium chloride, but this usually means the sacrifice of the metabolites remaining in the aqueous salt solution (Schwartz *et al.*, 1979; Frincke and Henderson, 1980; Belasco and Baude, 1981; Glickman *et al.*, 1981; Mihara *et al.*, 1981a).

Also in partitioning of metabolites it is usual—besides pH changes—to increase stepwise the polarity of the organic solvents. This is usually more successful than the sequential extraction from solid samples, as the procedure can be more easily controlled.

To separate large amounts of lipophilic contaminants from small amounts of lipophilic metabolites distribution between hexane and acetonitrile or heptane and methanol has been commonly used for, e.g. faeces (Yanagi *et al.*, 1976; Kaneko *et al.*, 1981a), eggs, milk, liver, kidney, muscle, fat and blood (Yanagi *et al.*, 1976; Ivie, 1978; 1980; Atallah *et al.*, 1980; Ivie and Hunt, 1980; Oehler and Ivie, 1980), rat liver microsomal incubate (Leighty *et al.*, 1976) and whole fish (Bruce, 1972).

There are some useful techniques for reducing large amounts of contaminants prior to solvent partition. Genuine liquid samples like urine and bile are lyophilized and the residue extracted with a polar solvent, e.g. methanol (Crawford *et al.*, 1980; Kaneko *et al.*, 1981b; Mihara *et al.*, 1981b). Arndts and Rominger (1978) reported removal of up to 90% of the physiological urinary ingredients without any metabolite loss applying this simple procedure. Similar effects are obtained by reducing the volume of extracts beyond the solubility of major coextracted contaminants. Using this procedure large amounts of natural ingredients precipitate, whereas the metabolites usually remain completely in solution. A similar technique is used to add water-miscible solvents, like acetone or methanol to the aqueous solution and precipitate unlabelled ingredients (Yanagi *et al.*, 1976; Lores and Bradway, 1977; Franklin *et al.*, 1981; Son *et al.*, 1980). As by far the major methanol-soluble solute in mammalian urine is urea Muecke *et al.* (1981) destroyed this compound by urease prior to evaporation and reextraction. However, the possibility of enzyme inhibition in studies with urea herbicides has to be considered (Cervelli *et al.*, 1975).

Gross *et al.* (1979) applied ultrafiltration using membranes with defined cut-offs to purify aqueous extracts from chlorotoluron (**5.0**) treated wheat.

Ion-pair extraction is a method recommended for the isolation of ionizable compounds by forming ion pairs with suitable counter ions of opposite charge, which can be readily extracted from aqueous solutions. Several excellent reviews on this subject have been published by Schill *et al.* (1977a; b; c). A

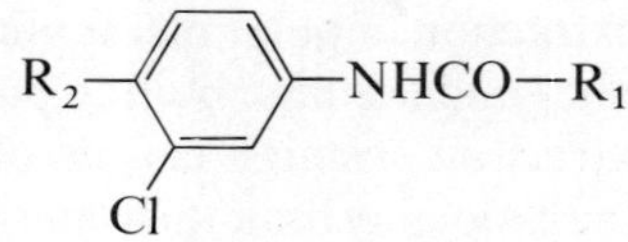

R_1	R_2	
$-N(CH_3)_2$	$-CH_3$	**(5.0)**
$-NH_2$	$-CH_3$	**(5.1)**
$-N(CH_3)_2$	$-CH_2OH$	**(5.2)**
$-NHCH_3$	$-CH_2OH$	**(5.3)**
$-NH_2$	$-CH_2OH$	**(5.4)**
$-N(CH_3)_2$	$-CH_2SCH_3$	**(5.5)**
$-NHCH_3$	$-CH_2SCH_3$	**(5.6)**
$-NH_2$	$-CH_2SCH_3$	**(5.7)**
$-N(CH_3)_2$	$-COOH$	**(5.8)**
$-N(CH_3)CHO$	$-COOH$	**(5.9)**
$-NHCH_3$	$-COOH$	**(5.10)**
$-NH_2$	$-COOH$	**(5.11)**
$-N(CH_3)_2$	$-CH_2O$-gluc.	**(5.12)**
$-NHCH_3$	$-CH_2O$-gluc.	**(5.13)**

suitable counter ion will form a strong ion bond with the metabolite ion and confers solubility in organic, water-immiscible solvents on the resulting ion pairs. Although apparently this method would be a suitable tool for the isolation of highly hydrophilic conjugates (Bakke, 1976; Paulson, 1976) and other ionized metabolites, it is rarely used by pesticide metabolism chemists. A certain counter ion might be ideal for one metabolite or one group of metabolites, but not for others. Furthermore, the amounts of counter ions necessary to maintain a desired equilibrium are usually high and remain after extraction in the aqueous phase, thus impeding the isolation of the remaining hydrophilic metabolites. After extraction and maybe further purification it is advantageous to cleave the ion pair to regain the original structure for appropriate derivatization and eventual spectroscopic elucidation. As the formation and the dissociation of the ion pair usually involves drastic pH changes, this method might be too stressful to some rather labile metabolites. In conclusion, ion pair extraction is not a tool for general metabolism studies dealing with a large number of unknown compounds, but a very useful method in residue analysis (Mundy and Machin, 1981) and probably for the isolation of a specific, known, partially known or speculated metabolite. Following this line, Baldwin and Hutson (1980) isolated 2-amino-4-chloro-5-fluorophenyl sulphate as the cetylpyridinium salt **(6.1)** from the urine of rats to develop a monitoring method for exposure to 3-chloro-4-fluoroaniline **(6.0)**.

A comprehensive review on the design of solvent extraction methods has been recently published by Bush (1981).

The introduction of non-ionic hydrophobic resins in metabolism studies provided a very powerful tool for the removal of metabolites from aqueous

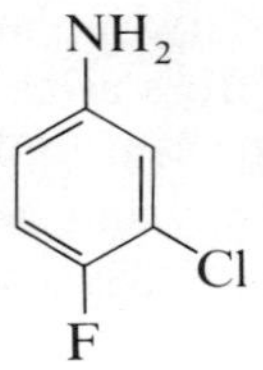

(6.0)

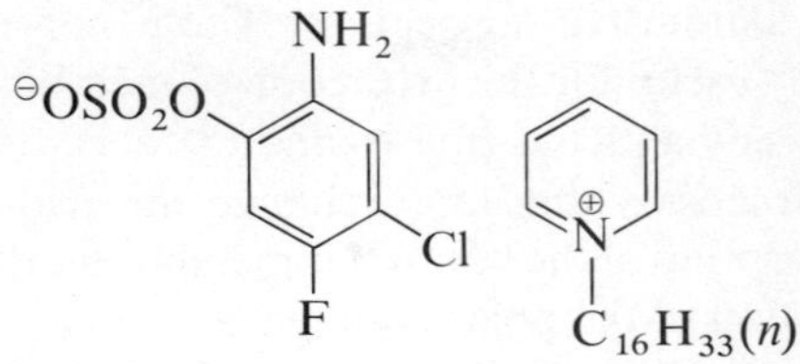

(6.1)

solutions and has been consequently used increasingly. The commonly used resins of this type are Amberlite XAD-2 and Amberlite XAD-4 and the corresponding analogues, e.g. Bio-Beads SM-2 and SM-4, Servachrom XAD-2 and XAD-4. They are styrene–divinylbenzene copolymers, XAD-2 having the smaller specific surface, but larger pores, compared to XAD-4. This implies, that with respect to capacity, XAD-4 is to be preferred for small and XAD-2 for larger molecules. An evaluation of Amberlite XAD-2 and XAD-4 for the recovery of glucuronides and sulphates from aqueous solution has been reported (White and Schwartz, 1980). The authors concluded that Amberlite XAD-4 is the better adsorbent and aqueous acetone a better desorbing solvent compared with methanol for the conjugates tested. The Fargo metabolism group introduced an ethylvinylbenzene-divinylbenzene copolymer (Porapak Q, originally designed as stationary phase in gas chromatography) as an extracting hydrophobic resin. The basic mechanisms of adsorption and desorption are assumed to be the same as those for XAD-2 and XAD-4. A review on the extraction of organic compounds from water with porous organic polymers including Amberlite and Porapak has been published by Dressler (1979).

These resins extract from aqueous phases compounds covering a wide range of polarity including polar conjugates, i.e. glucuronides, sulphates, mercapturic acids, etc. The compound to be adsorbed has to possess a hydrophobic moiety. Aromatic rings and alkyl groups increase adsorption, as does adjusting the pH of the aqueous solution to maintain the compounds in their non-ionized forms. Increasing the ionic strength of the aqueous phase by adding salts also usually enhances adsorption (e.g. Yanagi *et al.*, 1976). Conveniently, it has often been experienced that metabolites which have been adsorbed only by enhancing ionic strength and/or adjusted pH may not be desorbed from the resin by plain water thus allowing the removal of salts and buffers, prior to eluting the metabolites. When the metabolites are adsorbed, the resin is usually washed with water to remove highly polar materials like inorganic salts, urea, glucose and others and the metabolites desorbed with methanol, acetonitrile, acetone or other polar eluents. For the desorption of very lipophilic metabolites less polar solvents might be used. On loading the resin column, a small portion of the metabolites often appears in the eluate with the void volume, having at least qualitatively the same composition as the major portion being adsorbed. Reloading of this non-adsorbed portion on to the same column is advisable to approach quantitative adsorption. These non-ionic hydrophobic resins are not only very useful for the extraction of metabolites from urine, bile and comparatively crude aqueous phases and extracts, but excellent tools for desalting metabolite fractions after ion-exchange chromatography and similar methods. Besides adsorption of the non-polar metabolites the method is especially suitable for the isolation of the polar conjugates.

O-, *S*- and *N*-Glucuronides of various compounds including pesticides were isolated from bile, urine, serosal and mucosal fluids using XAD-2 (Lynn *et al.*, 1978; Schwartz *et al.*, 1979; Stierlin *et al.*, 1979; Kojima *et al.*, 1980; Mihara *et al.*, 1981a), XAD-4 (Hignite *et al.*, 1981) and Porapak Q (Bakke and Price, 1979; Pekas, 1979; Bakke *et al.*, 1981b; Larsen and Bakke, 1981).

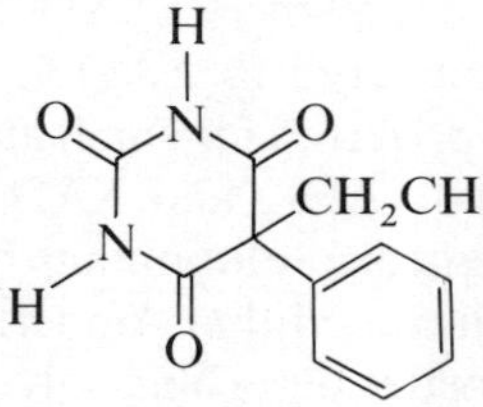

(7.0)

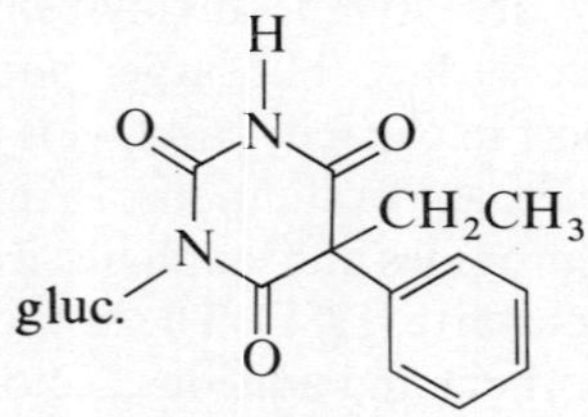

(7.1)

N-Glucosides (**7.1**, **8.1**) of phenobarbital (**7.0**) from human urine and of sulphamethazine (**8.0**) from swine tissue extracts were adsorbed by XAD-4 (Tang *et al.*, 1979) and XAD-2 (Paulson *et al.*, 1981). The glucoside (**9.1**) of maleic hydrazide (**9.0**) present in the methanol extract of treated tobacco plants

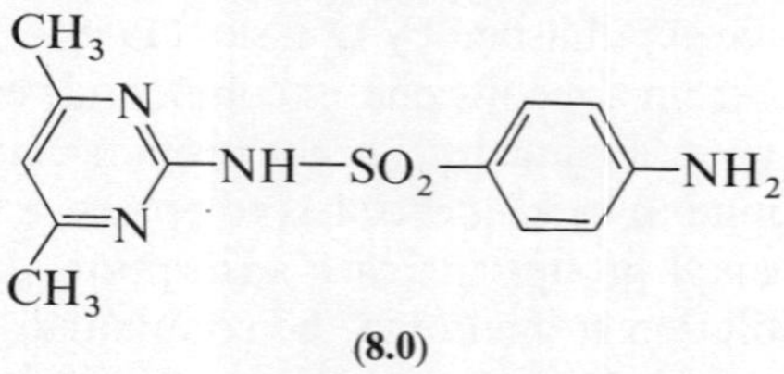

(8.0)

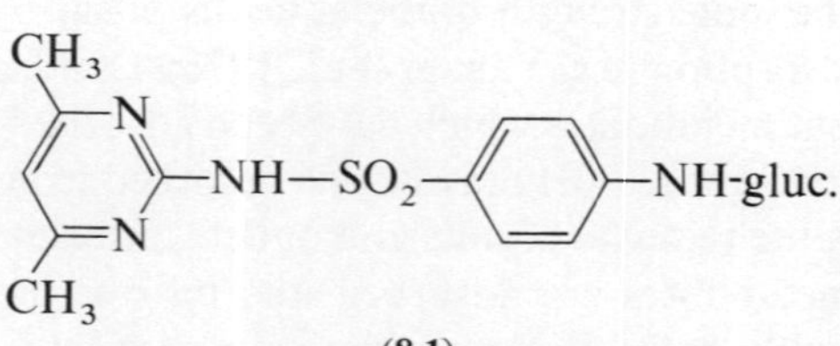

(8.1)

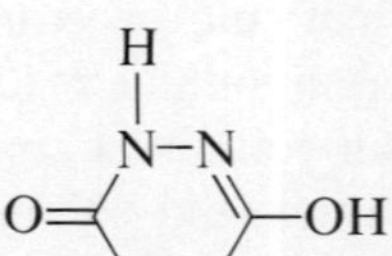

(9.0)

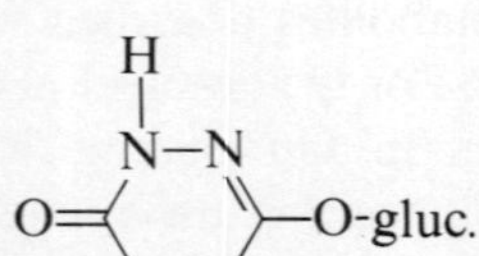

(9.1)

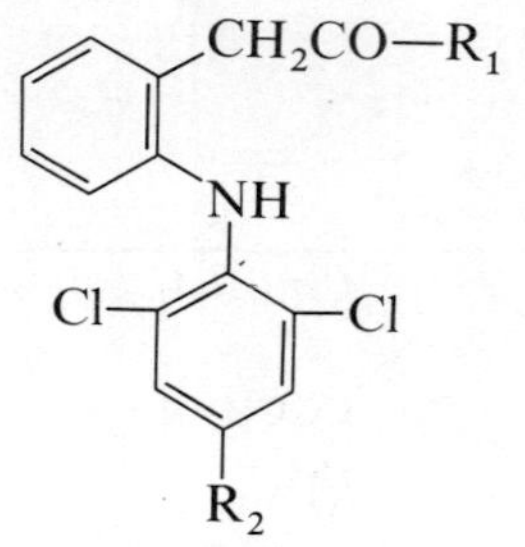

R_1	R_2	
—ONa	—H	**(10.0)**
—OH	—O-sulph. ac.	**(10.1)**
—taur.	—H	**(10.2)**

was recovered from aqueous solution using XAD-2 (Frear and Swanson, 1978). Sulphate conjugates (**10.1, 11.1, 11.2**) from the urine of diclofenac sodium (**10.0**) treated rats (Stierlin *et al.*, 1979) and from the urine of pindolol (**11.0**) treated rhesus monkeys (Maurer *et al.*, 1981) were efficiently adsorbed by XAD-2 and recovered by elution with methanol. Also the very polar glutathione (**12.15**) and cysteine (**12.13**) conjugates, as well as the corresponding mercapturic acid (**12.14**) of propachlor (**12.0**) could be extracted from urine, bile and

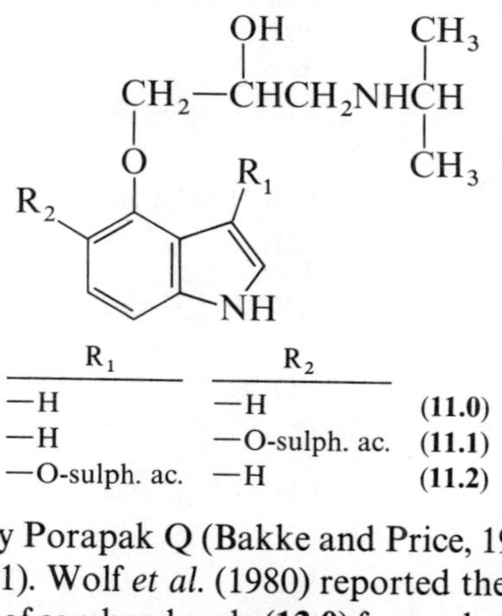

R_1	R_2	
—H	—H	**(11.0)**
—H	—O-sulph. ac.	**(11.1)**
—O-sulph. ac.	—H	**(11.2)**

faeces extracts of rats by Porapak Q (Bakke and Price, 1979; Bakke *et al.*, 1980; Larsen and Bakke, 1981). Wolf *et al.* (1980) reported the isolation of the glutathione conjugate (**13.1**) of cambendazole (**13.0**) from a hamster liver microsomal system with XAD-2. Even a taurine conjugate (**10.2**) could be adsorbed by XAD-2 from the urine of diclofenac sodium (**10.0**) treated beagle dogs (Stierlin *et al.*, 1979).

The information on the usefulness of these non-ionic resins for the isolation of amino acid conjugates is rather limited in the literature. But their general properties should allow efficient adsorption under adequate conditions. Franklin *et al.* (1981) demonstrated that hippuric acid (**65.1**) could be adsorbed on to XAD-2 from rat urine and recovered by elution with methanol.

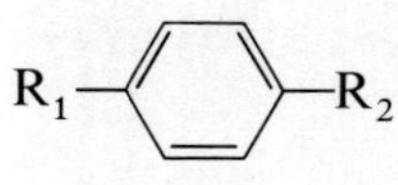

R_1	R_2	
—H	$-N(CH(CH_3)_2)COCH_2Cl$	(12.0)
—H	$-N(CH(CH_3)-CH_2OH)COCH_2SO_2CH_3$	(12.1)
—H	$-NHCOCH_2SO_2CH_3$	(12.2)
—OH	$-N(CH(CH_3)_2)COCH_2SO_2CH_3$	(12.3)
—OH	$-NHCOCH_2SO_2CH_3$	(12.4)
—OH	$-NHCOCH_3$	(12.5)
—H	$-N(CH(CH_3)_2)COCH_2S-X$	(12.6)
—H	$-N(CH(CH_3)-CH_2O\text{-gluc. ac.})COCH_2SO_2CH_3$	(12.7)
—H	$-N(CH(CH_3)-CH_2O\text{-gluc. ac.})COCH_3$	(12.8)
—H	$-N(CH(CH_3)-CH_2O\text{-gluc. ac.})H$	(12.9)
—H	$-N(CH(CH_3)_2)COCH_2O\text{-gluc. ac.}$	(12.10)
—O-gluc. ac.	$-N(CH(CH_3)_2)COCH_2SO_2CH_3$	(12.11)
—O-gluc. ac.	$-N(H)COOCH_3$	(12.12)

—H	—N($CH(CH_3)_2$)($COCH_2$-cyst.)	(**12.13**)
—H	—N($CH(CH_3)_2$)($COCH_2$-*N*-acet. cyst.)	(**12.14**)
—H	—N($CH(CH_3)_2$)($COCH_2$-glutath.)	(**12.15**)

These examples demonstrate that this method is applicable to the isolation not only of hydrophobic metabolites, but also of very hydrophilic metabolites. It is especially useful for polar compounds because alternative methods are rather limited as well as inefficient. The technique is easy and convenient to perform, has a high capacity, a high reproducibility and yields usually excellent recoveries. Furthermore, the resin columns are readily reconditioned and can thus be used repeatedly. making the approach very economical. For all these reasons, this approach has replaced to a certain extent conventional solvent extraction procedures.

The mechanism for adsorption and desorption of compounds on the non-ionized hydrophobic resins is essentially a reversed phase chromatographic process. Consequently all the stationary phases used in reversed phase h.p.l.c may be applied for this purpose. However, due to the comparatively high price and the low loading capacity these packing materials are not competitive for the general isolation of metabolites from biological liquids.

The application of ion-exchange resins for the extraction and isolation of metabolites from crude biological extracts is very limited and therefore not normally used in general metabolism studies. As these procedures are selectively dedicated to defined groups of metabolites, they are preferably used in exposure monitoring methods, for comparative purposes and in studies directed at one individual metabolite or a defined group of metabolites. Lores and Bradway (1977) used Amberlite CG-400, a strong anion-exchanger, to trap various alkylphosphates from urine samples of individuals exposed to organophosphorous pesticides.

Lamoureux and Davison (1975) compared the degree of mercapturic acid (**12.14**, **14.1**, **15.1**) formation in rats after oral ingestion of propachlor (**12.0**),

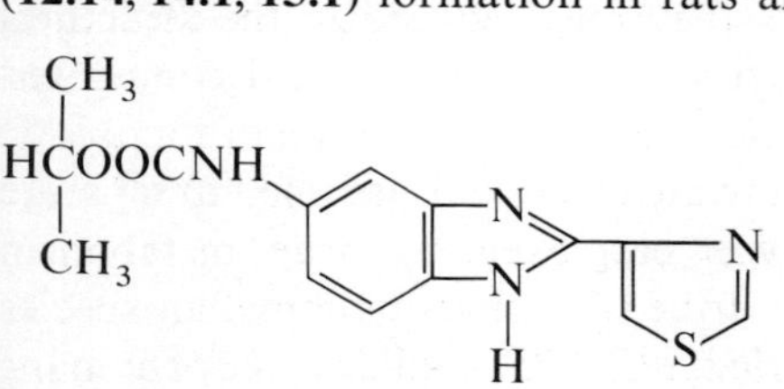

(**13.0**)

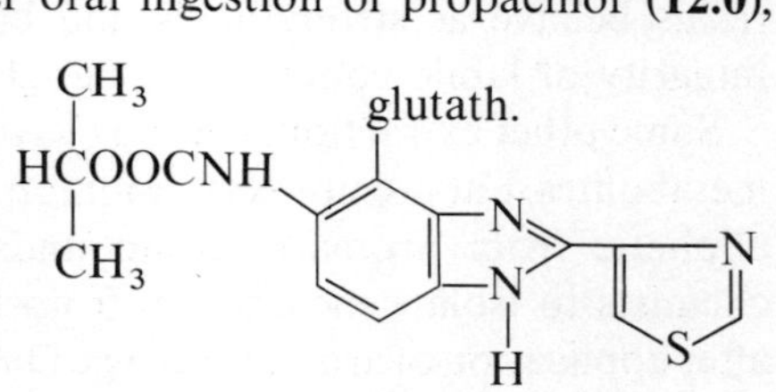

(**13.1**)

allidochlor (**14.0**) and fluorodifen (**15.0**). The urine was directly applied to a strong cation-exchange column Aminex AG 50W-X2 and the retarded mercapturic acid eluted with a pyridine-acetate buffer gradient. Axness and Fleeker (1979) investigated the metabolism of the butylcarbamoyl moiety of benomyl (**16.0**) in rats. The urine of treated animals was directly applied to a Dowex 50 × 8 column and the retained acidic metabolites including a presumed *N*-glucuronide (**16.1**) eluted by pyridine formate for further processing.

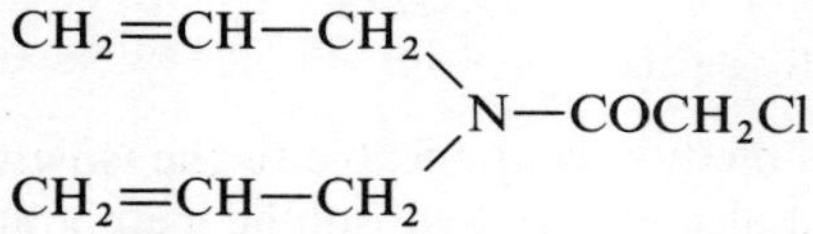

(14.0)

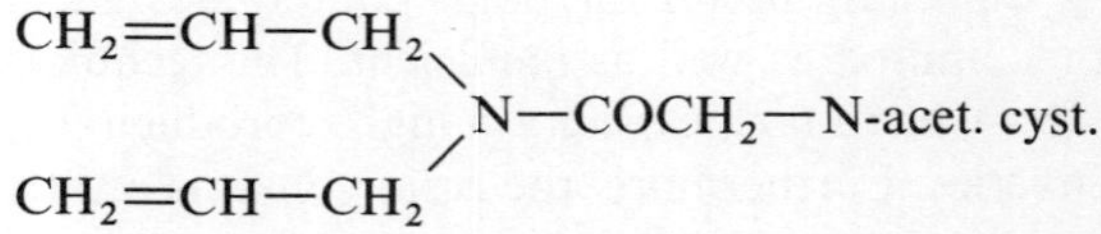

(14.1)

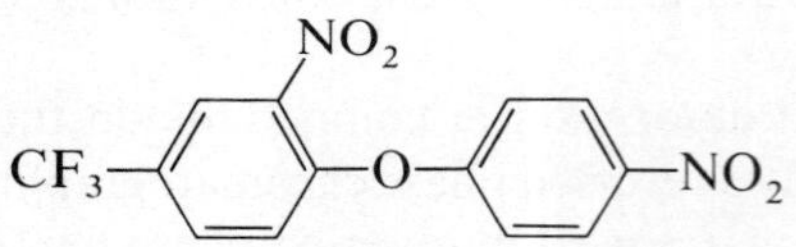

(15.0)

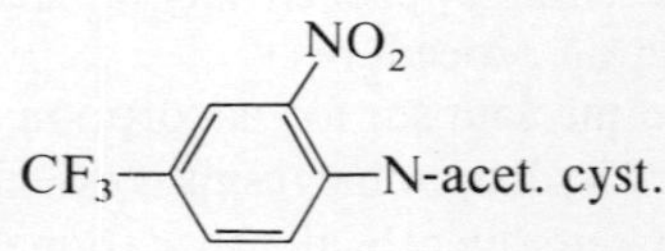

(15.1)

$CH_3CH_2CH_2CH_2NH{-}R$

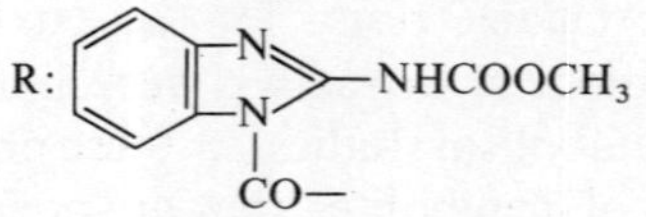

(16.0)

R: —gluc. ac. (**16.1**)
R: —CO-cyst. (**16.2**)
R: —CO-*N*-acet. cyst. (**16.3**)

For application in urine or other crude biological samples the use of strong ion-exchangers is preferred, as with weak ion exchangers the loading capacity might be impaired by accompanying natural electrolytes. As these exchange resins behave as strong acids and bases they may jeopardize the structural integrity of labile compounds, e.g. glucuronic and sulphuric acid conjugates.

Some other extraction methods have been used for the isolation of xenobiotic metabolites, but are not very common. Activated charcoal, assumed to separate aliphatic from aromatic compounds, was employed by drug metabolism chemists to isolate metabolites from the urine of various mammalian species after application of aromatic drugs. Dieterle *et al.* (1975) applied urine of humans (after the removal of the non-polar metabolites by 1,2-dichloroethane extraction) orally dosed with sulphinpyrazone (**17.0**) to charcoal. On desorption with methanol-ammonium hydroxide two highly polar metabolites, identified as *C*-glucuronides (**17.1**, **17.2**), were recovered. Ruelins *et al.* (1979) studied the metabolism of oxazepam (**18.0**) in swine. The urine of treated animals was pre-cleaned on Amberlite XAD-2 and then applied to activated charcoal. Removal of the metabolites with aqueous acetone yielded two diastereoisomeric *O*-glucuronides (**18.1**) with only 27 % recovery. The urine of monkeys treated with cartazolate (**19.0**) was concentrated to about 10 % of its original volume and then adsorbed on charcoal (Kripalani *et al.*, 1981). The metabolites including *O*-glucuronides (**19.1**, **19.2**, **19.3**) were recovered using sequentially pyridine/water and acetone/ethanol as eluent.

From these studies it appears that charcoal is a very strong adsorbent thus being prone to incomplete sample recovery. It has also been experienced that the results obtained are subject to high variations between different batches of the adsorbent and even within the same one. These features limit the use of activated charcoal in metabolism studies. However, it might be considered for selective purposes and its usefulness checked by appropriate pilot experiments.

The lipophilic properties of the hydroxypropylated dextran gel Sephadex LH-20 has been utilized for the isolation of compounds. The principal mechanisms of operation in using Sephadex LH-20 have been described as gel filtration,

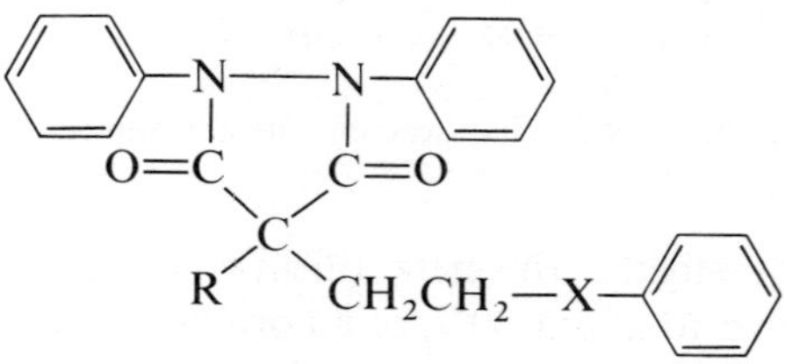

X	R	
—SO—	—H	(**17.0**)
—SO—	—gluc. ac.	(**17.1**)
$—SO_2—$	—gluc. ac.	(**17.2**)

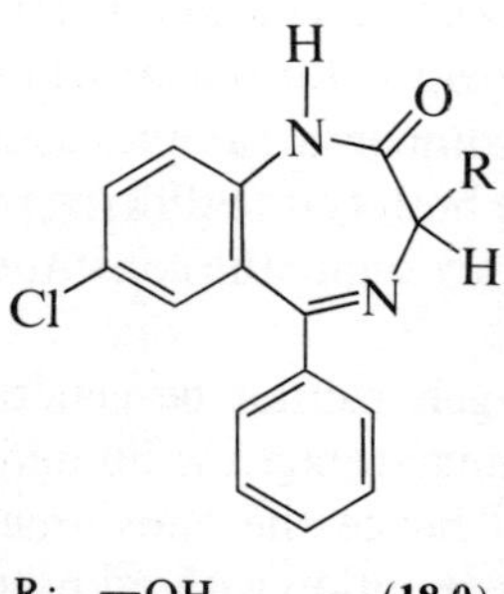

R: —OH (**18.0**)
R: —O-gluc. ac. (**18.1**)

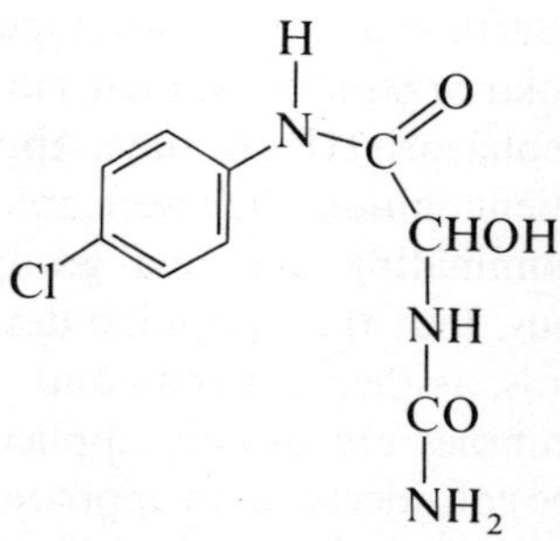

(**18.2**)

adsorption, partition, ion exclusion and retardation (Pharmacia, 1977). For its use as an extractant, partition and adsorption are the dominant features. Bakke *et al.* (1976) applied urine of propachlor (**12.0**) treated sheep and rats directly to a column of water-equilibrated Sephadex LH-20 and developed the chromatogram with water. The metabolites retained on the column included glucuronides (**12.7**, **12.11**), mercapturic acids (**12.14**) and cysteine conjugates (**12.13**), besides less polar propachlor metabolites.

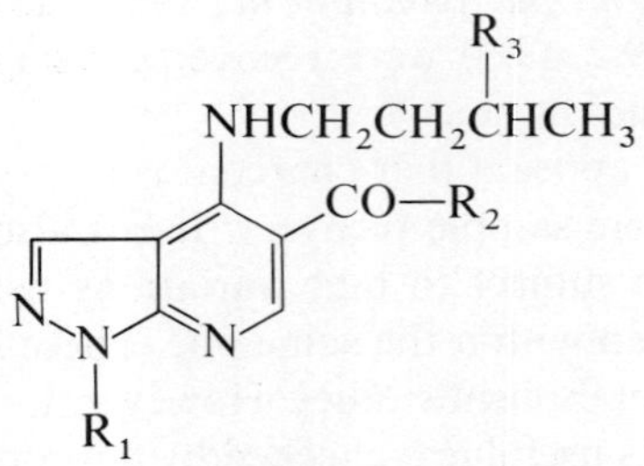

R_1	R_2	R_3	
$-C_2H_5$	$-OC_2H_5$	$-H$	(**19.0**)
$-H$	$-OH$	$-OH$	(**19.1**)
$-H$	$-OH$	$-H$	(**19.2**)
$-C_2H_5$	$-OH$	$-OH$	(**19.3**)

(**19.1**)–(**19.3**): gluc. ac. conj., site not defined

Son *et al.* (1980) investigated the metabolism of *o*-toluidine (**20.0**) in rats. The urine was concentrated to about 30% of its original volume, an equal volume of methanol was added, the mixture buffered to pH 9.2, the formed precipitate removed and the supernatant charged onto a Sephadex LH-20 column, equilibrated with methanol water (1/1 v/v) buffered to pH 9.2. The parent compound (**20.0**), various sulphuric acid (**20.12**, **20.13**, **20.14**) and glucuronic acid conjugates (**20.8**, **20.9**, **20.10**) were adsorbed on the column and afterwards eluted with the same solvent resulting in a remarkably good separation of the individual metabolites. Barth *et al.* (1981) developed a method for the enrichment of trace amounts of polar organic constituents in environmental samples using Sephadex LH-20 and Sephasorb HP ultrafine. Applying both gels in different modes using stepwise gradient elution, they were able to fractionate the constituents according to their dominating functional groups.

It is obvious, that the lipophilic dextran gels cannot be generally used as extracting tools, as their capacity and retaining strength is limited. Therefore, only small samples are usually applied and hence the above-mentioned examples may be considered as an approach to separation without prepurification, rather than as an isolation procedure.

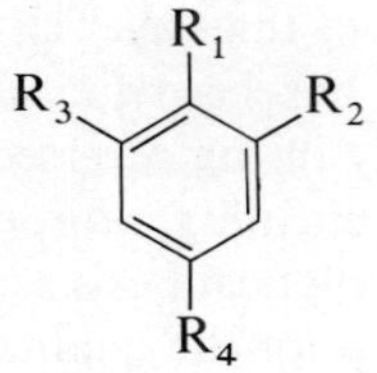

R_1	R_2	R_3	R_4	
$-NH_2$	$-CH_3$	$-H$	$-H$	**(20.0)**
$-NH_2$	$-CH_3$	$-H$	$-OH$	**(20.1)**
$-NH_2$	$-COOH$	$-H$	$-H$	**(20.2)**
$-NHCOCH_3$	$-CH_3$	$-H$	$-H$	**(20.3)**
$-NHCOCH_3$	$-CH_2OH$	$-H$	$-H$	**(20.4)**
$-NHCOCH_3$	$-COOH$	$-H$	$-H$	**(20.5)**
$-NHCOCH_3$	$-CH_3$	$-OH$	$-H$	**(20.6)**
$-NO$	$-COOH$	$-H$	$-H$	**(20.7)**
$-NH_2$	$-CH_3$	$-H$	—O-gluc. ac.	**(20.8)**
$-NHCOCH_3$	$-CH_2$—O-gluc. ac.	$-H$	$-H$	**(20.9)**
$-NHCOCH_3$	$-CH_3$	$-H$	—O-gluc. ac.	**(20.10)**
—NH-gluc. ac.	$-CH_3$	$-H$	—O-sulph. ac.	**(20.11)**
$-NH_2$	$-CH_3$	$-H$	—O-sulph. ac.	**(20.12)**
$-NH_2$	$-CH_3$	—O-sulf. ac.	$-H$	**(20.13)**
$-NHCOCH_3$	$-CH_3$	$-H$	—O-sulph. ac.	**(20.14)**

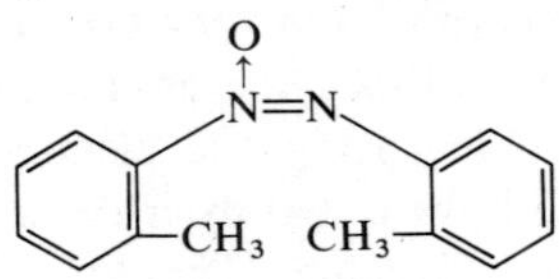

(20.15)

CHARACTERIZATION

The structure of the parent compound combined with a sound understanding of metabolic pathways will help to establish a hypothesis on the fate of a particular xenobiotic in the biological system under study. Consequently attempts will be made to substantiate this hypothesis and to design an adequate isolation strategy.

Knowledge of the number, distribution and the properties of the metabolites present in the sample provide the most important input for the selection of proper separation and purification methods and their sequence of application. In the literature the corresponding characterization methods used are often not stated, but may be deduced on the basis of the preparative procedure selected.

First, a simple and reliable analytical system needs to be established, which permits the efficient separation of the individual metabolites present and thus the determination of their number and distribution. This system should avoid stress on the metabolites and maintain the integrity of even labile metabolites during the analysis. In most cases thin-layer chromatography (t.l.c.), especially in the two-dimensional mode has been found useful for this purpose. It is expected that h.p.l.c., preferably in the reversed phase mode will increasingly replace t.l.c. due to its higher versatility, reproducibility, speed and its usually more gentle conditions. This analytical basis serves as reference for the characterization experiments as well as for the qualitative and quantitative measurement of the metabolites during the isolation process. The metabolite pattern obtained by t.l.c. or h.p.l.c. also yields information on the lipophilicity of the individual compounds.

It is advisable to start with weak hydrolysis experiments to gain an indication of the lability of individual metabolites at different pH and temperatures. Sensitivity of metabolites to air may be conveniently checked by two-dimensional t.l.c. The chromatogram is developed in the first dimension, kept for some time under aerated conditions and thereafter developed in the second dimension using the same mobile phase. Compounds not located on the diagonal, or tailing, are disintegrating during exposure to air or during the chromatographic process. Experiments may be carried out similarly to check sensitivity to other environmental factors, e.g. light, elevated temperature, etc.

One of the most important features to be utilized for the separation of metabolites is their electric charge. Although other techniques are feasible, the use of high voltage electrophoresis (h.v.e.) is very popular (Crawford *et al.*, 1980; Simoneaux *et al.*, 1980; Son *et al.*, 1980; Mayer *et al.*, 1981; Muecke *et al.*, 1981; Roberts and Wright, 1981). Measuring the electrophoretic mobility of the metabolites at different pH allows not only a determination of their charge, but also an estimation of their pK-values. This knowledge is a prerequisite for the evaluation of separation techniques such as ion exchange chromatography, pH-dependent extraction procedures, etc. The application of electrophoresis with special attention to metabolism of xenobiotics, including experimental procedures and general applications has been reviewed by Conway (1977). Paper (e.g. Whatman No. 1, Schleicher and Schuell 2043 B) is commonly used as carrier. Glass plates coated with silica gel, cellulose or other material are also used, but have certain disadvantages (More *et al.*, 1978; Schwartz *et al.*, 1979; Wright *et al.*, 1980; Roberts and Wright, 1981). The removal of the generated heat causes technical problems, so that usually only low potential gradients ($\leqslant 50$ V/cm) are applied, leading to longer running times. A short running time is an important feature as zone enlargement due to diffusion increases with the square root of time. Additionally most of the commercially available instruments permit a maximal migration distance of only about 20 cm.

The buffers are selected to allow high voltage at low current in order to minimize the running time for the reason mentioned above. The homogeneity of the field can be checked by scattering appropriate reference dyes, e.g. fuchsine or methylene blue over the starting line. Correction for electro-osmotic effects is made using an electro-neutral labelled reference compound, e.g. [^{14}C]glucose.

Two-dimensional electrophoresis at different pH values and chromatography in the first and h.v.e. in the second dimension (or vice versa) would in principle provide more detailed information, e.g. to attribute the electrophoretic and chromatographic properties to one individual fraction. However, due to technical difficulties (removal of buffer ions, limited resolution power, etc.) these methods lack versatility and are therefore rarely applied in metabolism studies.

It appears quite common to perform small scale partition experiments between aqueous phases at different pH using various organic solvents to define the properties of individual metabolites (e.g. Marsh *et al.*, 1981). This rather empirical approach yields an array of information from which the optimal conditions for a preparative procedure may be deduced. In the author's laboratory this experiment is often combined with the stability tests (pH, elevated temperature, etc.).

In addition to chromatographic, electrophoretic and solvent partition behaviour, which are commonly investigated, there are some other physical features of passing interest. In special cases, e.g. when incorporation into, or covalent binding to, large endogenous molecules is suspected, an estimation of the molecular size may be indicated. Dialysis, gel permeation chromatography and ultrafiltration have been utilized for characterization and eventually separation of pesticide related compounds (Gross *et al.*, 1979; Crayford and Hutson, 1980b; Hamboeck *et al.*, 1981).

Investigation of the volatility of metabolites might be useful in order to avoid losses during isolation procedures. Furthermore, this property can be utilized for isolation and purification purposes (see section: Solubilization and Isolation p. 289).

A wide array of enzymes has been used to investigate the presence of certain types of metabolites, especially conjugates, in crude biological samples. Although these methods are very helpful, one has to be aware that they are often non-specific and thus inconclusive. For example, a specific enzyme may completely cleave a metabolite in the pure state, but not, or only partially if present in a crude biological sample. Furthermore, selectivity, purity and the homogeneity of the commercial enzyme preparations are often crucial factors in the interpretation of these experiments. Quistad *et al.* (1978b) incubated undefined cycloprate (**21.0**) derived conjugates from apple leaves with various enzymes and determined the release of ether soluble products. Using β-glucuronidase, β-glucosidase, phosphatase, sulphatase, cellulase and mixtures thereof they report a cleavage—independent of the enzyme used—of 65 to 78 % compared to 12 % in the control experiment. Nevertheless, enzymes are often used to

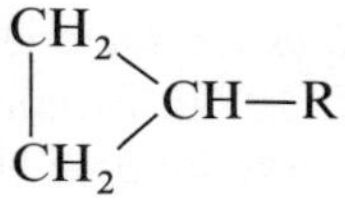

R: $—COO(CH_2)_{15}CH_3$ (**21.0**)
R: $—(CH_2)_{12}COOH$ (**21.1**)
R: $—(CH_2)_{14}COOH$ (**21.2**)
R: —CO-gly. (**21.3**)

cleave conjugates in crude biological samples on a preparative scale and to isolate the exocons (conjugate nomenclature according to Dorough, 1976). This approach has certain methodological drawbacks and the genuine structure of the conjugate may not be unequivocally deduced from that of the exocon. In many cases this approach cannot be avoided, as suitable non-destructive isolation and identification techniques for large molecules are not available. If necessary and appropriate, enzymatic cleavage is normally preferred to chemical hydrolysis as the conditions are milder and thus less prone to artefact formation (Roberts, 1977). In spite of these limitations, enzymes are used to gain information on the nature of conjugates present in crude samples in order to aid in the selection of adequate separation methods. They are also applied during the isolation procedure and are common and efficient tools—in combination with other techniques—for the characterization and identification of isolated pure metabolites as here the experimental conditions can be properly controlled.

β-Glucuronidase and arylsulphatase are the most common enzymes used for the characterization of metabolites in crude samples and also for their identification after purification (e.g. Baldwin and Hutson, 1980; Crawford *et al.*, 1980; 1981; Crayford and Hutson, 1980a; Son *et al.*, 1980; Belasco and Baude, 1981; Glickman *et al.*, 1981; Huckle *et al.*, 1981a; b; c; Hutson *et al.*, 1981; Kaneko *et al.*, 1981a; Mihara *et al.*, 1981a; Muecke *et al.*, 1981). Although both enzymes are commercially available from various biological sources in good purity, problems such as substrate selectivity, stability of compounds under incubation conditions and enzyme inhibition by ingredients present in the sample, prevent the unequivocal characterization of a metabolite based solely on its susceptibility towards an enzyme preparation. The significance of such characterization can be enhanced by a proper design of the experiment. This includes following the time course of cleavage, carrying out a negative control (using enzyme specific inhibitors), as well as a positive control (adding a suitable substrate to an identical blank sample). D-Glucaro-1,4-lactone and phosphates are usually applied as inhibitors of β-glucuronidase and arylsulphatase in the negative control sample, respectively. Suitable model substrates are commercially available and their cleavage is easily followed by photometric methods.

Other enzyme preparations have been used to release exocons and to characterize or even to identify conjugates, for example: *β-Glucosidases* (Still and Rusness, 1977; Frear *et al.*, 1978; Quistad *et al.*, 1978b; Gross *et al.*, 1979; Marquis *et al.*, 1979; Shimabukuro *et al.*, 1979; Huckle *et al.*, 1981a; b; Roberts and Wright, 1981; Scheel and Sandermann, 1981), *Hesperidinases* (Hodgson *et al.*, 1973; Frear and Swanson, 1978; Marquis *et al.*, 1979; Shimabukuro *et al.*, 1979), *Cellulases* (Roberts, 1977; Frear and Swanson, 1978, Quistad *et al.*, 1978b; Scheel and Sandermann, 1981), *Snail digestive juice* (Roberts, 1977), *Pectinases* (Still and Mansager, 1975; Otto *et al.*, 1979), *Proteinases* (Quistad *et al.*, 1978b; Scheel and Sandermann, 1981), *Phosphatases* (Quistad *et al.*,

1978b; Huckle *et al.*, 1981a; b), *Acylases* (Chen and Dorough, 1979; Jones and Gibson, 1980), L-*Cysteine C-S lyase* (Still and Rusness, 1977).

Combining various characterization methods a pool of information regarding stability, volatility, lipophilicity, molecular size, electric charge and solvent partition properties of the metabolites can be obtained as a guide in establishing an appropriate isolation procedure.

SEPARATION AND PURIFICATION TECHNIQUES

Pre-purification procedures

At this level the process usually consists of one rather coarse pre-purification step followed by several fine purification and separation procedures. For economic reasons it is advantageous to limit separation of metabolites during the pre-purification process and rather optimize removal of concomitantly isolated matrix ingredients, to avoid time-consuming purification of various subfractions obtained.

The procedures mentioned before (see section: Solubilization and Isolation, p. 286), e.g. extraction and solvent partitioning, adsorption on non-ionic hydrophobic resins, charcoal and other adsorbents and steam distillation may in some cases already be considered sufficient as pre-purification processes when the most adequate methods and the appropriate conditions have been selected.

Quite often, the same methods used for the final purification steps are applied for pre-purification purposes, only in a simpler mode to increase capacity and to reduce the time required to perform the procedure. For instance, adsorbents, etc. may be applied in a batch mode instead as column chromatography, or the chromatogram may be developed by simple step gradients rather than by sophisticated elution programmes. Also the dry-column chromatography technique (Loev and Goodman, 1967) has been successfully applied for pre-purification, as it combines high capacity with convenient handling and quite good predictability of separation parameters from simple thin-layer chromatography experiments. In the following subsections, methods generally used in separation and purification of pesticide metabolites will be discussed without special attention as to their use in pre- or final purification steps, as most of them may be modified to be used for either purpose.

Liquid column chromatography

Liquid column chromatography represents certainly the most versatile technique applied in metabolite isolation and purification. Although h.p.l.c. is only a variant thereof, this advanced technology will be the subject of a separate section. This section will deal with the application of what might be called conventional liquid column chromatography in pesticide metabolism studies. As the modes actually effecting a certain separation process are often multiple and difficult to assign, the subject will be dealt with according to the nature of the stationary phase as ion-exchange, partition, adsorption and size exclusion chromatography. For all general aspects of liquid chromatography, including basic concepts, separation mechanisms, modes of application, equipment, column packings, eluents, development of methods, the reader's attention is focused on the very recommendable monograph of Snyder and Kirkland (1979).

The development of h.p.l.c. has also been very stimulating for conventional liquid chromatography. The use of module units from h.p.l.c., such as pumps, gradient programmers, injection systems and continuous flow detectors, as well as the use of smaller and more uniform particles as stationary phase is now quite common. As some of the h.p.l.c. packing materials are also available as bulk material, the differences between both procedures are diminishing. Two recent reviews have been published dealing with this interconnecting area between h.p.l.c. and conventional liquid chromatography (Verzele and Geeraert, 1980; Nettleton, 1981).

All of the different modes of column chromatography have been successfully applied in the separation and purification of pesticide metabolites. In the following discussion special attention will be paid to the polar metabolites, as they present the major challenge in the isolation process.

Ion-exchange chromatography

Ion exchange chromatography has been intensively used not only for separation into neutrals, acids, bases and zwitterions, but also for the separation of individual metabolites within a certain group. During the last two decades, the formerly used polystyrene based resins have been increasingly replaced by those having a polyglycoside skeleton. Usually volatile buffers, such as carbonates, formates and acetates of pyridine and ammonium are used for their easy removal from the fractions eluted. Strong ion-exchange resins in their H^+ or OH^- forms, have also been applied to eliminate buffer ions from the eluate. The introduction of the non-ionic hydrophobic resins, i.e. Amberlite XAD-2 and XAD-4 and their analogues offered a convenient alternative method for removal of volatile and non-volatile buffer substances. As most of the polar pesticide metabolites have a negative rather than a positive electric charge, anion exchangers are most widely applied. Of the various cellulose and dextran based anion exchangers, the weakly basic diethylaminoethyl type (DEAE) is the most versatile and therefore the most commonly used.

In the mid 1960s a now classical paper was published by Knaak *et al.* (1965) demonstrating a fine separation of various *N*- and *O*-glucuronides (**22.4**, **22.5**, **22.6**, **22.7**) and sulphates (**22.2**, **22.3**) from the urine of guinea pigs and rats after intraperitoneal administration of carbaryl (**22.0**). The separation was accom-

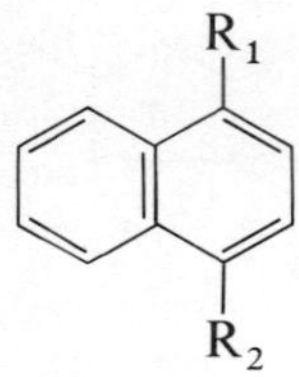

R_1	R_2	
—OCONHCH$_3$	—H	(**22.0**)
—OH	—H	(**22.1**)
—O-sulph. ac.	—H	(**22.2**)
—OCONHCH$_3$	—O-sulph. ac.	(**22.3**)
—OCON(CH$_3$)-gluc. ac.	—H	(**22.4**)
—O-gluc. ac.	—H	(**22.5**)
—OC(=NCH$_3$)O-gluc. ac.	—H	(**22.6**)
—OCONHCH$_3$	—O-gluc. ac.	(**22.7**)

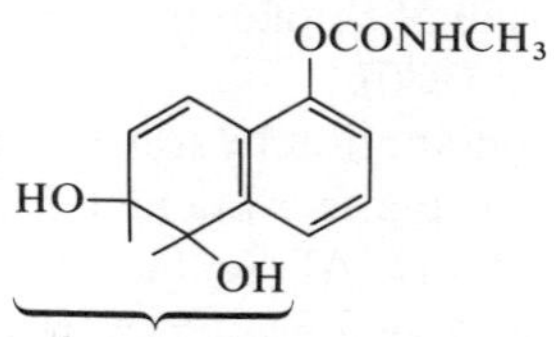

(**22.8**)

gluc. ac. conj. site not defined

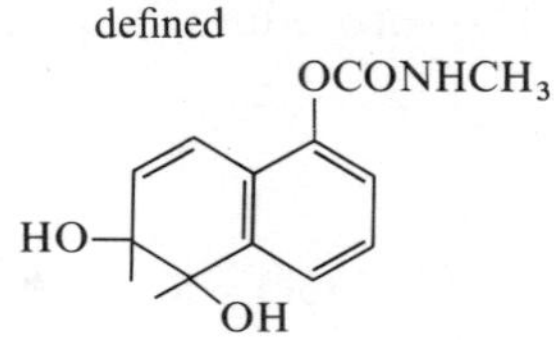

(**22.9**)

plished on a DEAE cellulose column eluting the metabolites with two subsequent linear tris buffers at pH 7.5 (0.01–0.05 M, 0.05–0.1 M). Later the tris buffer was replaced by ammonium formate and ammonium acetate and the method is still in use in carbaryl metabolism studies (Sullivan *et al.*, 1972; Chin *et al.*, 1979a; b).

Irving *et al.* (1967) investigated biliary metabolites of 2-acetylaminofluorene (AAF, **23.0**) excreted by rats and rabbits. On DEAE cellulose the *O*-glucuronide of *N*-hydroxy-2-acetyl-aminofluorene (**23.1**) was efficiently separated from the phenolic 3-, 5- and 7-hydroxy AAF glucuronides (**23.2**, **23.3**, **23.4**) and the 7-hydroxy AAF sulphate (**23.5**) using a stepwise elution with ammonium formate (pH 3.5, 0–1 M).

Frear *et al.* (1978) separated metabolites of chloramben (**24.0**) treated giant foxtail seedlings on DEAE cellulose using ammonium acetate as eluent in a

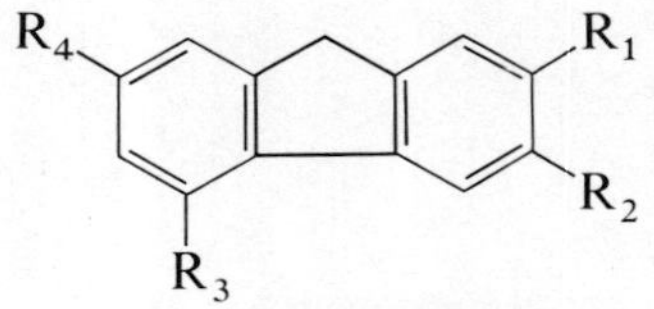

R_1	R_2	R_3	R_4	
$—NHCOCH_3$	—H	—H	—H	(**23.0**)
$—N(COCH_3)—$O-gluc. ac.	—H	—H	—H	(**23.1**)
$—NHCOCH_3$	—O-gluc. ac.	—H	—H	(**23.2**)
$—NHCOCH_3$	—H	—O-gluc. ac.	—H	(**23.3**)
$—NHCOCH_3$	—H	—H	—O-gluc. ac.	(**23.4**)
$—NHCOCH_3$	—H	—H	—O-sulph. ac.	(**23.5**)

neutral and an acidic fraction, thus separating the chloramben-*N*-glucoside (**24.1**) from the corresponding ester glucoside (**24.2**).

DEAE cellulose (ammonium formate elution) has also been used to divide the metabolites present in the urine of thiofanox (**25.0**) fed rats into a neutral and an acidic fraction (Chin *et al.*, 1980).

After removal of the lipophilic metabolites by solvent partitioning, Simoneaux *et al.* (1980) applied the aqueous phase obtained from tomatoes topically treated with methidathion (**26.0**) to a DEAE Sephadex A-25 column to separate the two major metabolites present. A zwitterionic cysteine (**26.5**) conjugate was eluted with water and the acidic monodesmethyl derivative (**26.1**) was recovered with a KBr gradient. This procedure had been originally used by

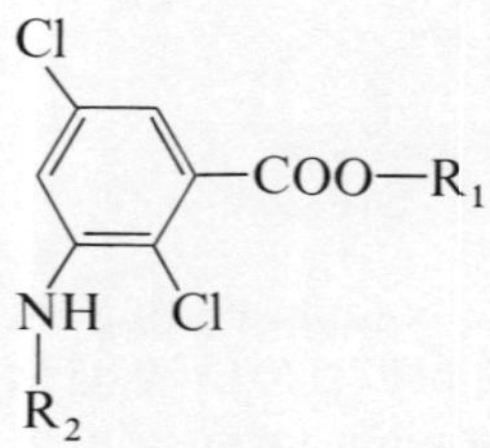

R_1	R_2	
—H	—H	(**24.0**)
—H	-gluc.	(**24.1**)
-gluc.	—H	(**24.2**)
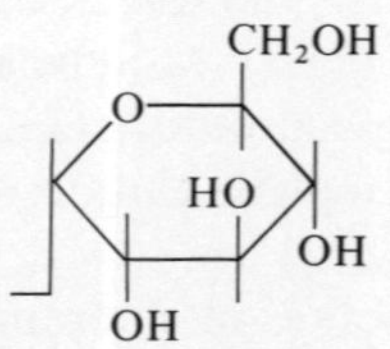	—H	(**24.3**)
$—CH_3$	—H	(**24.4**)

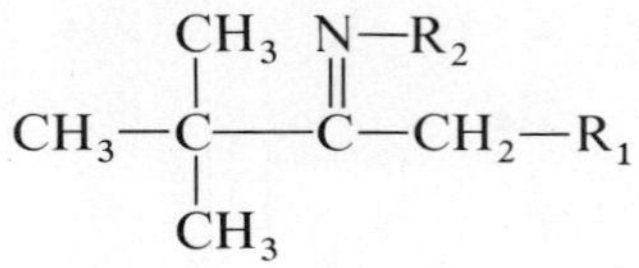

R_1	R_2	
$-SCH_3$	$-OCONHCH_3$	**(25.0)**
$-SOCH_3$	$-OCONHCH_3$	**(25.1)**
$-SO_2CH_3$	$-OCONHCH_3$	**(25.2)**
$-SCH_3$	$-OH$	**(25.3)**
$-SOCH_3$	$-OH$	**(25.4)**
$-SO_2CH_3$	$-OH$	**(25.5)**

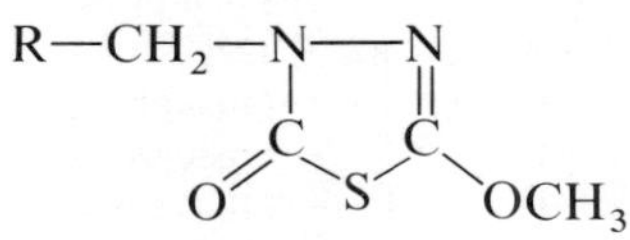

R: 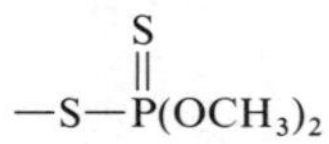 **(26.0)**

R: 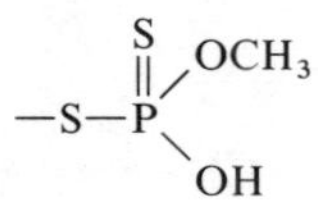 **(26.1)**

R:	$-SCH_3$	**(26.2)**
R:	$-SOCH_3$	**(26.3)**
R:	$-SO_2CH_3$	**(26.4)**
R:	-cyst.	**(26.5)**

Paulson *et al.* (1972) to separate metabolites from the excreta of propham **(27.0)** treated chicken. Applying a gradient of KBr (0–1 M) several sulphuric acid **(27.5, 27.6, 27.7, 27.8)** and two glucuronic acid conjugates **(27.9, 27.10)** were separated in a single run. The same method was also applied for the isolation of sulphates and glucuronides from the urine of sheep and rats fed on propham treated alfalfa (Paulson *et al.*, 1975). This procedure is still being used for the separation and purification of acidic compounds including glucuronides, cysteinates, glutathionates and cysteinyl glycinates (Pekas, 1979; Chopade *et al.*, 1981; Chopade and Dauterman, 1981).

DEAE Sephadex A-25 and isocratic elution with 0.2 M tris buffer was used for the purification of glutathionates **(12.15)**, cysteinates **(12.13)**, mercapturates **(12.14, 28.1)** and glucuronides **(12.7, 12.11)** from the excreta of rats treated with propachlor **(12.0)** and pentachloromethylthiobenzene **(28.0)** (Bakke *et al.*, 1981a; Larsen and Bakke, 1981). For the clean-up of various propachlor related glucuronides ECTEOLA cellulose, being a stronger base than the DEAE type, was used and the conjugates recovered with a 0.05 M acetate buffer (Bakke and Price, 1979).

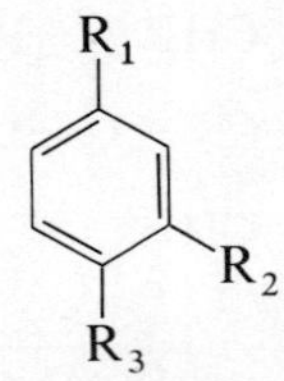

R_1	R_2	R_3	
$-NHCOOCH(CH_3)_2$	$-H$	$-H$	**(27.0)**
$-NHCOOCH(CH_3)_2$	$-OH$	$-H$	**(27.1)**
$-NHCOOCH(CH_3)_2$	$-H$	$-OH$	**(27.2)**
$-NHCOOCH(CH_3)_2$	$-OH$	$-OH$	**(27.3)**
$-NH_2$	$-H$	$-OH$	**(27.4)**
$-NHCOOCH(CH_3)_2$	$-H$	—O-sulph. ac.	**(27.5)**
$-NHCOOCH(CH_3)_2$	—O-sulph. ac.	$-OH$	**(27.6)**
$-NHCOOCH(CH_3)_2$	$-OCH_3$	—O-sulph. ac.	**(27.7)**
$-NH_2$	$-H$	—O-sulph. ac.	**(27.8)**
$-NHCOOCH(CH_3)_2$	$-H$	—O-gluc. ac.	**(27.9)**
$-NHCOOCH(CH_3)_2$	$-OH$	$-OH$	**(27.10)** gluc. ac. conj. site not defined

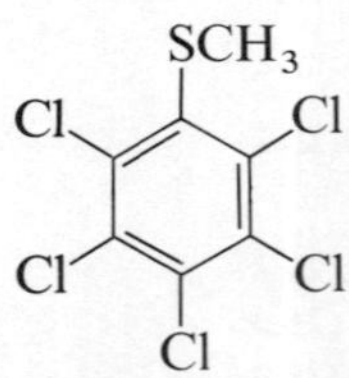

(28.0)

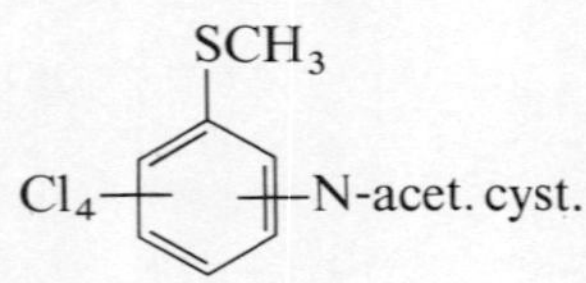

(28.1)

In contrast to the anion exchangers where the polyglycoside based types replaced to a high extent the resins based on polystyrene, the strongly acidic divinylbenzene cross-linked polystyrene based resins, carrying sulphonic acid groups as the active principle, are still the exchangers preferably used for cations. These resins, behaving as strong acids, have to be applied very carefully as they may easily produce artefacts by hydrolysis of labile metabolites. In the past these columns were often operated at elevated temperatures to enhance separation. As this increased dramatically the stress on the metabolites it is now usual to perform the separation at or even below room temperature. In spite of this rather severe drawback these resins are still occasionally used in pesticide metabolism studies.

For the most common resins, i.e. Dowex 50 and Dowex 1 and its equivalents, Schroeder (1972a; b) has recommended volatile developers for the separation of peptides which have been widely adopted for the separation of amino acid conjugates and other pesticide metabolites. Larsen *et al.* (1978) applied Chromabeads, Type A, to recover and to separate urinary metabolites of terbutryn (**29.0**) from rats and goats. The technique used was adapted from automatic

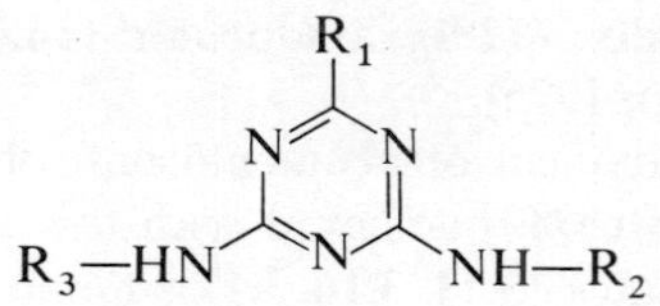

R_1	R_2	R_3	
$-SCH_3$	$-C(CH_3)_3$	$-CH_2CH_3$	**(29.0)**
$-SCH_3$	$-C(CH_3)_3$	$-CH_2COOH$	**(29.1)**
$-SCH_3$	$-C(CH_3)_2COOH$	$-CH_2CH_3$	**(29.2)**
$-SCH_3$	$-C(CH_3)_2COOH$	$-H$	**(29.3)**
$-SH$	$-C(CH_3)_3$	$-CH_2COOH$	**(29.4)**
$-SCH_3$	$-C(CH_3)_2CH_2O$-gluc. ac.	$-CH_2CH_3$	**(29.5)**
$-SCH_3$	$-C(CH_3)_2CH_2O$-gluc. ac.	$-H$	**(29.6)**
—S-gluc. ac.	$-C(CH_3)_3$	$-CH_2CH_3$	**(29.7)**
—S-gluc. ac.	$-C(CH_3)_2CH_2OH$	$-CH_2CH_3$	**(29.8)**
—S-gluc. ac.	$-C(CH_3)_3$	$-H$	**(29.9)**

amino acid analysis. The column was maintained at 60 °C and the metabolites eluted with different citrate buffers (pH 2.875 to pH 5.0) followed by 0.2 M NaOH containing 0.6 M NaCl. The metabolites isolated included various *S*- and *O*-glucuronides and other carboxy acids (**29.1**–**29.9**).

The water-soluble metabolites of atrazine (**30.0**) from smooth cordgrass were separated by Weete *et al.* (1980) on Aminex 50 W-X2 resin using the pyridine–acetate buffer gradient recommended by Schroeder (1972a) (pH 3.1, 0.2 M to pH 5, 2.0 M). The five metabolite fractions obtained were further separated on a strongly basic anion exchange resin (AG 1 × 2) eluting with an acetic acid gradient (0–0.5 M, 1–6 M). In addition to 2-hydroxyatrazine (**30.1**) and its deethylated derivative (**30.2**) evidence for the presence of various amino acid conjugates among the metabolites isolated was presented. This separation procedure is essentially the same as that used by Lamoureux *et al.* (1970; 1972) for the isolation of different amino acid conjugates of various 2-chloro-*s*-triazines from plants.

The same system, i.e. an Aminex 50 W-X2 column developed with a pyridine–acetate buffer (pH 3.1, 0.2 M to pH 5.0, 2.0 M), was also efficiently applied to isolate the mercapturic acids (**12.14**, **14.1**, **15.1**) from unprocessed urine of rats

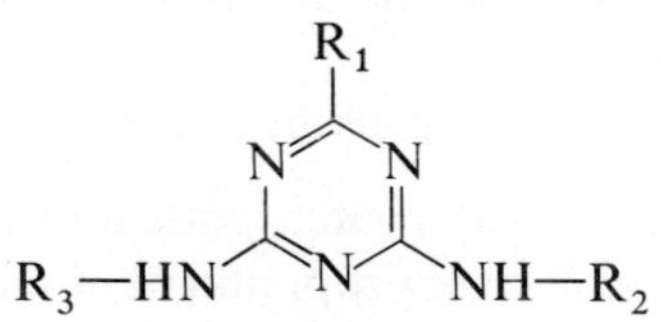

R_1	R_2	R_3	
$-Cl$	$-CH(CH_3)_2$	$-CH_2CH_3$	**(30.0)**
$-OH$	$-CH(CH_3)_2$	$-CH_2CH_3$	**(30.1)**
$-OH$	$-CH(CH_3)_2$	$-H$	**(30.2)**

fed with either propachlor (**12.0**), allidochlor (**14.0**) or fluorodifen (**15.0**) (Lamoureux and Davison, 1975).

Investigating the metabolism of pentachloronitrobenzene (**3.0**) in peanuts Lamoureux and Rusness (1980) achieved with this system the separation of mono and di-glutathionates (**3.11, 3.16, 3.17**), mono and di-cysteinates (**3.8, 3.14**) and *N*-malonyl cysteinates (**3.10, 3.15**). The weakly acidic nature of maleic hydrazide (**9.0**) has been utilized to separate this growth regulator from its major metabolite, i.e. the glucose conjugate (**9.1**), applying the plant extract to a strong anion exchange (AG 1 × 8 and Dowex 1 × 8) column (Frear and Swanson, 1978; Newsome, 1980).

Starratt and Bond (1981) used anion and cation exchange chromatography in their metabolism study of the fumigant methyl bromide (**31.0**) in granary weevils. The metabolites extracted with aqueous ethanol were purified on a strong acidic Amberlite CG-120 column and the metabolites recovered by elution with 1 M pyridine. Subsequently *S*-methyl cysteine (**31.1**), *S*-methyl glutathione (**31.2**) and *S*-methyl glutathione sulphoxide (**31.3**) could be separated on AG 1 × 2 columns using the gradient described by Schroeder (1972b).

CH_3—Br (**31.0**) CH_3-cyst. (**31.1**) CH_3-glutath. (**31.2**)

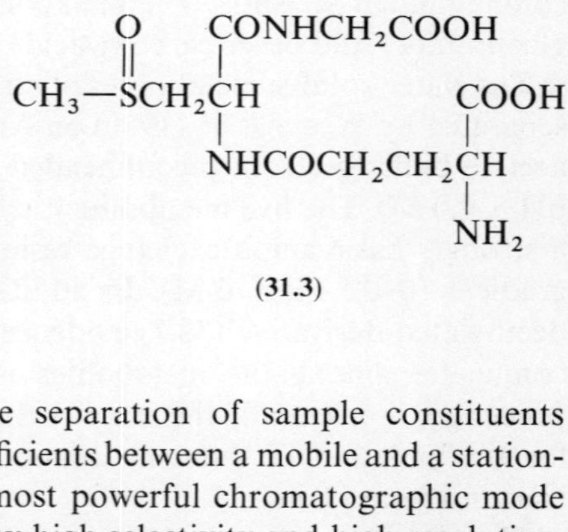

(**31.3**)

Partition chromatography

Partition chromatography involves the separation of sample constituents according to their different partition coefficients between a mobile and a stationary liquid phase. This is probably the most powerful chromatographic mode utilized at present, being characterized by high selectivity and high resolution. The fact that partition chromatography can be operated as normal and reversed phase systems, including the large number of partitioning phase pairs available, provides a high versatility covering metabolites to be separated from very polar to very non-polar in character.

Basically three different column packings have been used:

(a) Carrier (e.g. porous silica) coated with aqueous or water-immiscible liquids (e.g. buffers, β,β′-oxidipropionitrile, dimethylpolysiloxane).
(b) Organic phases covalently bound to porous silica (bonded phases, e.g. amino, diol, nitrile, alkyl).
(c) Non-ionic porous organic polymers (Amberlite XAD resins and analogues, Porapak).

The traditional technique of partition chromatography using stationary phases coated on usually deactivated carriers is very tedious. The major problem is the stripping of the stationary phase from the packing support. As a consequence of the changing amounts of stationary phases present, the surface of the carrier may become gradually exposed and eventually involved in the chromatographic process. Furthermore, the use of elution gradients is rather critical, as the solubility of the stationary phase will possibly change during the elution process. The capacity of this chromatography system is rather limited and scaling-up requires very careful and time-consuming pilot experiments, as enlarging the sample size may not only exhaust the chromatographic capacity, but can also destabilize the whole system. All these problems lead to a generally unstable system which is difficult to control. This is certainly the reason why, in pesticide metabolism studies, this technique had been rarely applied in the past, although high resolution can be achieved by well-elaborated systems. The development of alternative, more convenient to handle, and more stable partition systems, made this conventional technique almost obsolete.

In our study on the metabolism of chlorotoluron (**5.0**) in rats the technique was very successfully applied for neutral and acidic metabolites (Muecke *et al.*, 1976). Four benzoic acid derivatives (**5.8–5.11**) with different degrees of *N*-dealkylation in the urea moiety were efficiently separated on 0.01 M HCl coated Celite using two subsequent linear elution gradients (hexane/$CHCl_3$ 50/50 v/v to $CHCl_3$ to diethyl ether). The neutral chlorotoluron metabolites (**5.1–5.7**) were separated on H_2O coated Celite by elution with a linear gradient (hexane/CH_2Cl_2 80/20 v/v to CH_2Cl_2).

Apparently, quite often, when adsorption chromatography on porous adsorbents is performed, the dominating separation mechanism is partition rather than adsorption. The polar ingredients of an eluent (e.g. water, organic acids, etc.) are selectively absorbed on to the solid phase and may thus form a partition system *in situ*. This method has been occasionally intentionally utilized to establish a partition chromatography system.

In the course of the development of h.p.l.c., organic phases covalently bound to porous silica have become commercially available. Up to now more than a dozen different types of silica-bound materials for normal phase and for reversed phase chromatography are offered (Majors, 1980b). The vast majority of the materials available are designed for h.p.l.c.; i.e. the producers aim for small particle size ($\leq 10\ \mu m$) and narrow size distribution. The small particle size and the high price of these materials limit the use of bonded phases in conventional liquid chromatography. Recently, some of the bonded phases have become commercially available in larger particle sizes, so that acceptable inlet pressures for chromatography in conventional columns are obtained. As the particle size distribution is not as critical as in h.p.l.c., these preparations cover a wider range allowing more economical prices. Some of these materials are commercially available in prepacked columns.

However, looking through the recent pesticide metabolism literature it is evident that the use of bonded phases in conventional chromatography has not yet been exploited. The limited experience in our laboratory comparing Servachrom XAD-4 (50–100 μm) columns with prepacked LiChroprep RP-8 (40–63 μm) columns revealed that the bonded phase had the lower capacity, a lower retention, but better resolution. In some cases acidic fractions, including sulphates and glucuronides, could not be recovered from the resin using water-methanol gradients, but were well separated into individual metabolites on the bonded phase using the same eluent (Muecke and Bissig, 1981).

The lipophilic organic polymers, Amberlite XAD-2, XAD-4 and Porapak Q have been intensively utilized for extraction, pre-purification and desalting of pesticide metabolites. But even applying these resins in a rather simple mode, some separations of individual metabolites or metabolite groups were obtained. On Porapak Q, using water-methanol as eluent, Paulson *et al.* (1975) and Pekas (1979) fractionated to some degree metabolites of propham (**27.0**) from rat and sheep urine and of carbaryl (**22.0**) from serosal and mucosal fluids of rats, respectively. Earlier results on the fractionation of metabolites with special emphasis towards glucuronide conjugates on Porapak Q including stepwise elution with methanol–water gradients and sequential elution with different organic solvents, have been reviewed by Bakke (1976).

Glucuronides of drugs have also been grouped and separated on Amberlite XAD-2 using stepwise gradients of methanol in water (Hornke *et al.*, 1980; Kripalani *et al.*, 1981). White and Schwartz (1980) evaluated Amberlite XAD-2 and Amberlite XAD-4 for the recovery of model glucuronides and sulphates. From this study they conclude that the latter resin, using water–acetone mixtures as eluents, is the most suitable system for glucuronides and sulphates. However, it appears rather risky to generalize this conclusion, as the number of compounds tested and the parameters investigated were very limited.

Dieterle *et al.* (1979) recommended the use of micronized XAD-2 resins for the preparative separation of polar and non-polar metabolites. Applying a medium particle size of 12 μm and eluting with a ternary, isocratic system, typically containing methanol, water and small amounts of a non-polar solvent, e.g. toluene, leads to the *in situ* formation of a very powerful reversed phase partition system, as demonstrated with various urinary and biliary drug metabolites, including glucuronides and other acids.

Amberlite XAD-4 (50–100 μm) was used to separate the two major polar metabolites (**32.1**, **32.2**) of piperophos (**32.0**) from a rather crude rat urine fraction (Muecke *et al.*, 1981). Eluting with a linear water–methanol gradient, both metabolites were completely resolved and a vast purification was obtained.

At present it appears that partition chromatography on organic polymers or bonded phases is the mode with the highest potential for the pre-purification of crude metabolite extracts, prior to h.p.l.c.

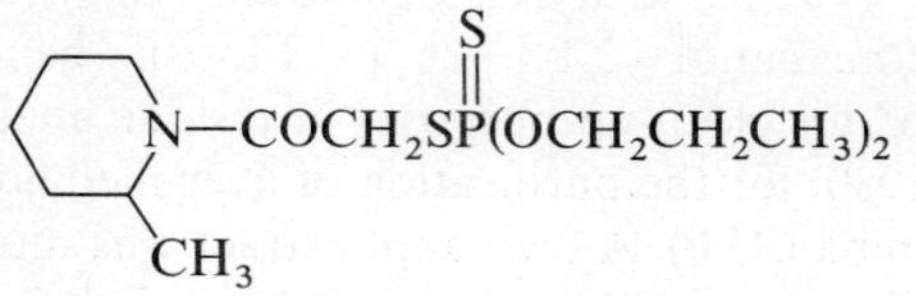

(32.0)

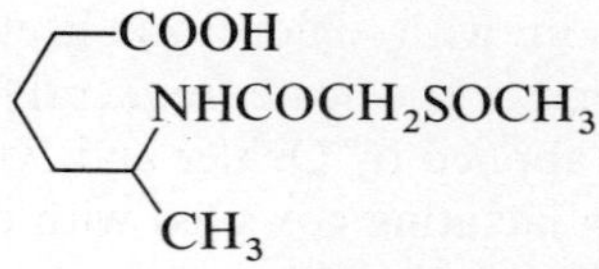

(32.1)

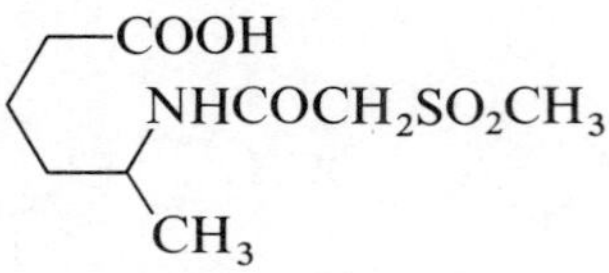

(32.2)

Adsorption chromatography

Adsorption chromatography is based on the interaction of solutes at high-surface porous particles such as silica or alumina. The degree to which different mechanisms are operative in liquid–solid chromatography is still obscure. The discussion on competition versus sorption in liquid–solid chromatography is not yet resolved and a generally accepted model has not yet been developed (Snyder and Poppe, 1980). Consequently, the separations cited in this section might be due to adsorption, partition or both, or other mechanisms may be involved.

The compounds in the sample are retained according to their different polarities depending on the presence and position of functional groups. The method combines high resolution, selectivity and simple operation. Optimal conditions may be chosen conveniently based on pilot t.l.c. experiments. The separation process is mainly controlled by the selection of the mobile phase, as only silica is commonly used as solid phase. With polar compounds adsorption to the solid phase may become irreversible, thus limiting application of the method to lipophilic metabolites. Derivatization of hydrophilic metabolites leading to stable and less polar products is frequently used to gain the advantages of this mode of chromatography for conjugates and other polar compounds (Blau and King, 1977).

However, in some cases column chromatography on silica has also been applied to very polar compounds. Consequently very hydrophilic eluents were required, which usually do not allow a defined chromatographic process. This approach, resulting usually in low recoveries and little or no separation of sample constituents, is not to be recommended; it fails to utilize the advantages of this method. However, if applied to polar compounds, it might be advisable to deactivate the stationary phase by the addition of water.

Axness and Fleeker (1979) applied the acidic fraction of benomyl (**16.0**) from rat urine to a silica dry column and recovered the metabolites, including a cysteinate (**16.2**), a mercapturate (**16.3**) and a *N*-glucuronide (**16.1**), upon elution with chloroform/acetone/methanol (2/1/1 v/v/v). Florisil, being a magnesia trisilicate frequently used for clean-up purposes in residue analysis, was applied by Oehler and Ivie (1980) for the purification of liver metabolites from lactating cows fed with dicamba (**33.0**). A methanol extract was directly applied to the column, the lipophilic ingredients eluted with hexane and ethyl acetate and the herbicide and its demethylated derivative (**33.1**) recovered with

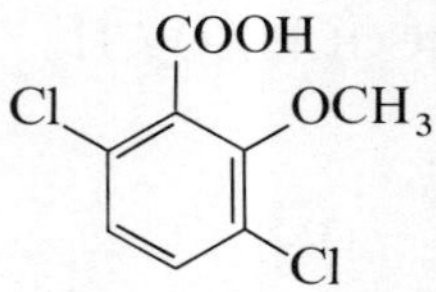

(33.0)

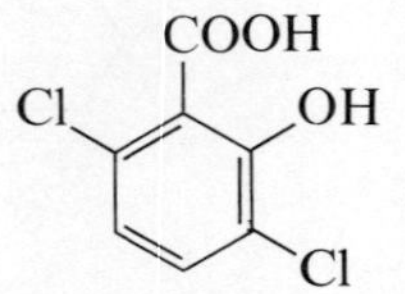

(33.1)

methanol containing 1% acetic acid. Sonobe *et al.* (1981) utilized Florisil columns for the purification of the recently discovered angelic acid ester conjugate (**1.2**) of 3-hydroxy carbofuran (**1.1**) isolated from carbofuran (**1.0**) treated carrots. Mixtures of ethyl ether and petroleum ether were used as eluent. An *O*-glucuronide (**22.8**) of carbaryl (**22.0**) isolated from the urine of treated rats by ion exchange chromatography on DEAE cellulose was purified on silica using an acetonitrile to methanol gradient (Sullivan *et al.*, 1972).

The most polar fraction from the excreta of rats treated orally or subcutaneously with terallethrin (**34.0**) was applied to a silica column and the metabolites eluted with methanol/chloroform (1/4 v/v) followed by pure methanol (Mihara *et al.*, 1981a). The metabolites were grouped into a fraction containing among several unknowns two defined carboxy acid derivatives (**34.1, 34.2**), a second

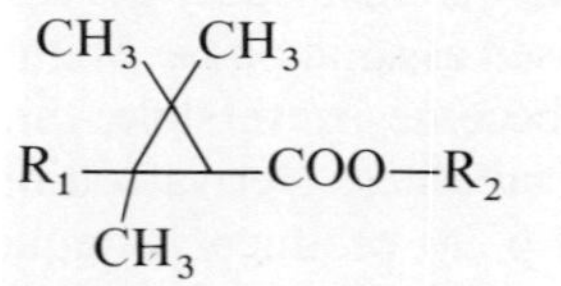

R_1	R_2	
$-CH_3$	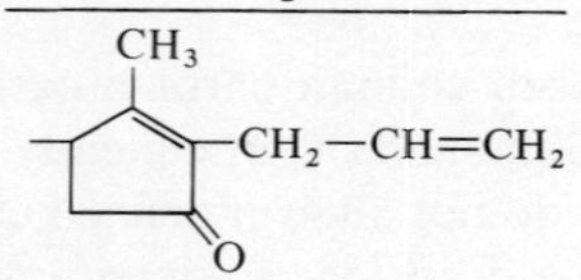	**(34.0)**
$-COOH$	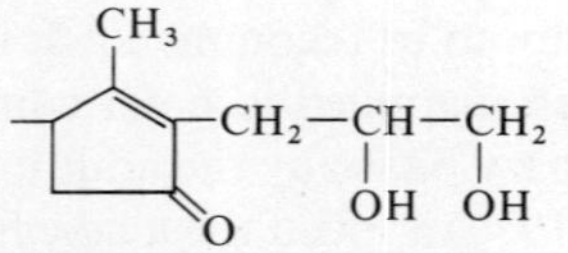	**(34.1)**
$-CH_2OH$	$-H$	**(34.2)**
$-CH_3$	—gluc. ac.	**(34.3)**

fraction containing only two unknowns, and a third one containing only an ester glucuronide (**34.3**).

Begum *et al.* (1979) used silica column chromatography as a pre-purification step in the study of the metabolism of pentachloronitrobenzene (**3.0**) in outdoor grown onions. The crude chloroform/methanol extracts of different plant parts were directly purified on silica using sequentially a series of different eluents, i.e. cyclohexane, chloroform, ethyl acetate, acetone and methanol. Separation and further purification was carried out with preparative t.l.c. Applying the crude extract from onion shoots to the silica column, a defined separation of metabolites was accomplished. Upon elution with cyclohexane three fractions were obtained, each containing two individual metabolites: pentachlorobenzene (**3.3**) and a non-identified metabolite, pentachloronitrobenzene (**3.0**) and pentachlorothioanisole (**3.5**), pentachloroaniline (**3.1**) and tetrachloroaniline (**3.12**). Continuing elution with more polar solvents recovered solely pentachlorophenylmethylsulphoxide (**3.6**).

The use of silica gel chromatography for the purification of rather lipophilic metabolites is numerous, e.g. pentachloronitrobenzene (**3.0**) metabolites from rhesus monkey urine and faeces (Koegel *et al.*, 1979), 3-chloro-4-fluoroaniline (**6.0**) metabolites from rat urine (Baldwin and Hutson, 1980); piperophos (**32.0**) metabolites from rat urine (Muecke *et al.*, 1981).

Lusby *et al.* (1981) investigated the metabolic fate of the *N*-nitrosoderivative (**35.1**) of pendimethalin (**35.0**) by an *in vitro* culture of soil-isolated streptomyces. The metabolite extract was pre-purified by filtration through a Celite pad and separated into individual metabolites on a silica column using benzene as eluent. Final separation and purification was achieved by preparative t.l.c. and h.p.l.c.

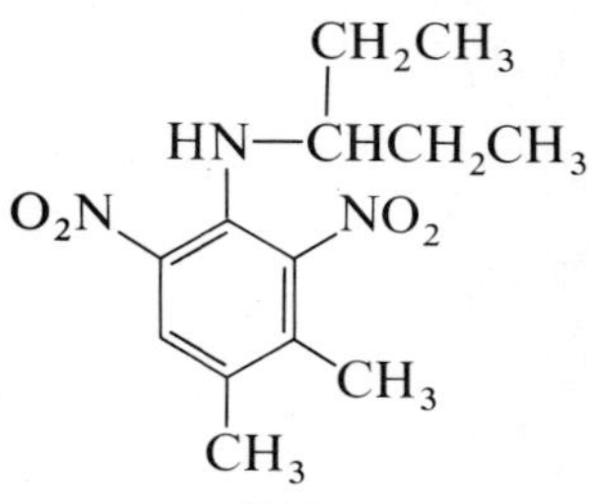

(**35.0**)

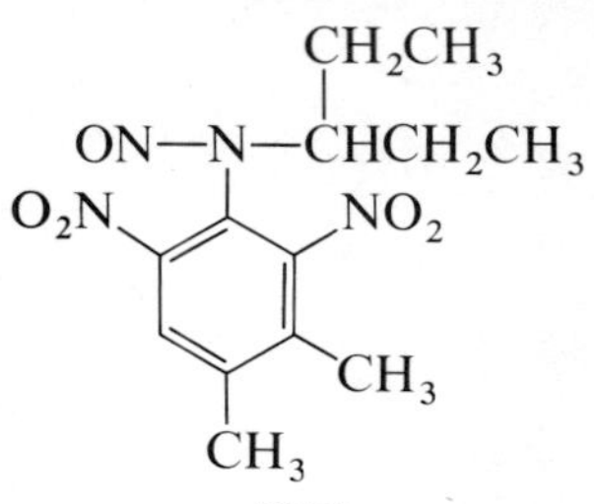

(**35.1**)

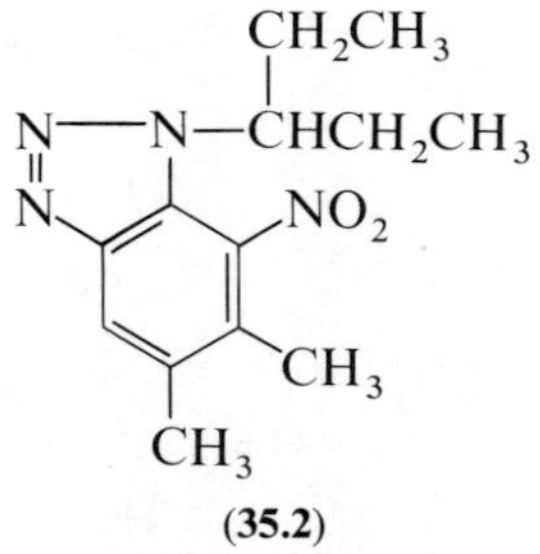

(**35.2**)

The *in vitro* metabolism of carbaryl (**22.0**) by human foetal kidney, lung and liver tissues was studied by Chin *et al.* (1979b). The non-ionic metabolite fraction obtained by DEAE cellulose chromatography was resolved into individual metabolites on silica, using a series of linear gradients. On isooctane to diisopropyl ether elution, 1-naphthol (**22.1**) separated from unchanged carbaryl, diisopropyl ether to acetonitrile isolated among several unknowns 5,6-dihydro-5,6-dihydroxy carbaryl (**22.9**), and finally acetonitrile to methanol recovered two non-identified metabolites. Based on a somewhat unsymmetric peak shape it was speculated that the dihydrodiol metabolite (**22.9**) may have been a mixture of the 3,4- and the 5,6-isomer. Including the eluted compounds not identified, ten metabolites were distinguished. Applying the same system, including pre-separation of the metabolites on a DEAE cellulose column, Chin *et al.* (1980) investigated the non-ionic fraction from the urine of rats treated with thiofanox (**25.0**). On elution with the first gradient the free oxime derivative (**25.3**) could be separated from the parent compound (**25.0**) and the free oxime sulphone (**25.5**); the second gradient yielded the sulphone derivative of the parent molecule (**25.2**) and the third one eluted separately the sulphoxide of the oxime (**25.4**) and that of the parent compound (**25.1**).

Paulson *et al.* (1977) examined the (pH 1) organic solvent extractable metabolites and the remaining water soluble metabolites after HCl hydrolysis from the excreta of rats and chickens orally dosed with perfluidone (**36.0**). For separation and purification of metabolites, silica column chromatography was

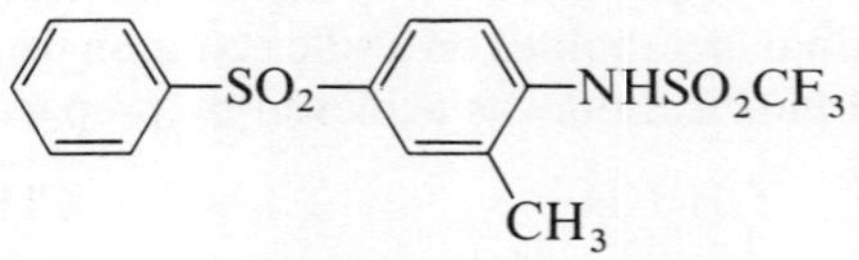

(36.0)

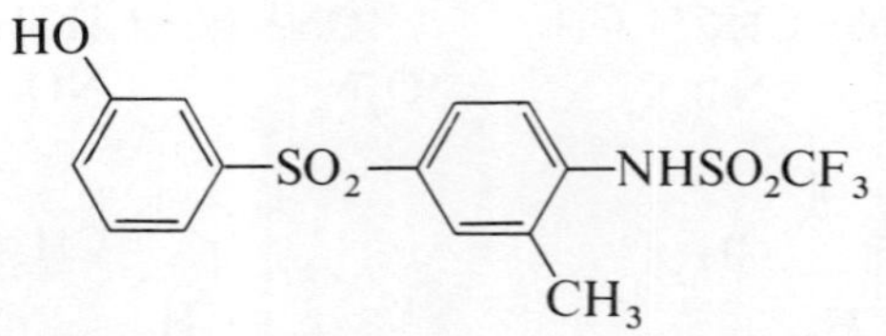

(36.1)

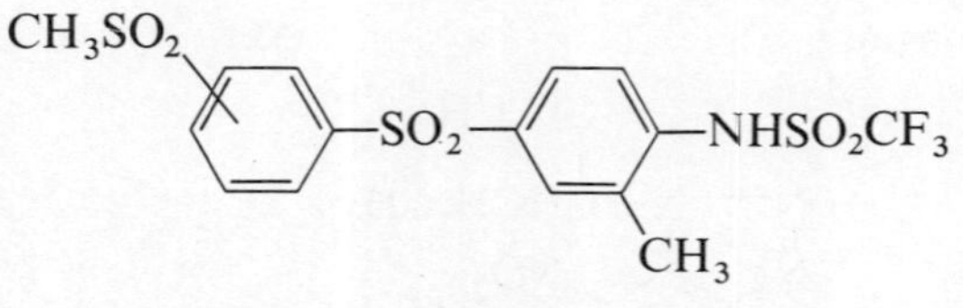

(36.2)

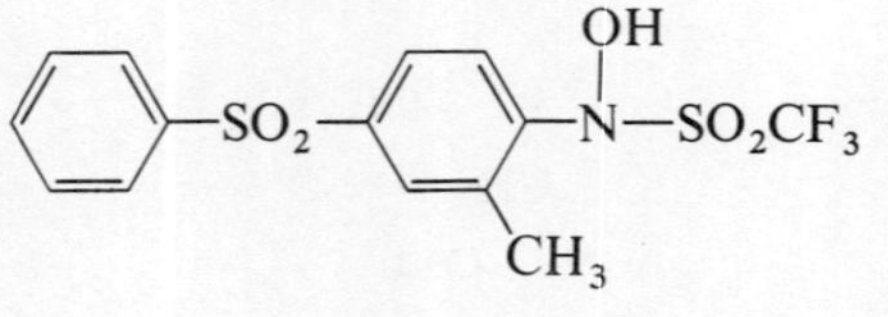

(36.3)

readily utilized. Columns eluted with petroleum ether/chloroform/acetic acid (35/60/5 v/v/v) effected the separation of the parent compound (**36.0**) from its 3′-hydroxy and its methylsulphonyl derivative (**36.1**, **36.2**) (chicken urine). From rat urine extracts, unchanged perfluidone (**36.0**) was separated from its *N*-hydroxy derivative (**36.3**). For further purification the polarity of the products was modified by methylation with diazomethane and silica column chromatography was again applied using chloroform as eluent.

As mentioned above, enhancing the lipophilicity of polar metabolites by adequate derivatization permits the extension of adsorption chromatography to rather polar compounds. Quite frequently alkylation of carboxy acids and phenols and acylation of amines and aliphatic hydroxy groups are used. Derivatization is preferably effected prior to the ultimate step of purification by t.l.c., g.c., h.p.l.c., etc. as changing the physico chemical properties of the metabolite may be an additional aid to the removal of final impurities. Furthermore, derivatization may be necessary anyhow to enhance vapour pressure of the metabolite for subsequent identification by m.s. and g.c.–m.s. (Paulson *et al.*, 1973; Bakke, 1976; Aschenbacher *et al.*, 1978; Tomasic, 1978; Fenselau and Johnson, 1980).

Kurihara *et al.* (1979) studying mercapturic acid formation from lindane (**37.0**) in rats, used derivatization for purification and eventual g.c.–m.s. identification. The mercapturic acids (**37.1**–**37.3**) were extracted from the urine with diethyl ether at pH 3 and hydrolytically deacetylated. The cysteine derivatives were *N*-trifluoroacetylated and esterified with *n*-butanol as recommended by Lamkin and Gehrke (1965). These derivatives were chromatographed on a silica column using hexane and hexane/dichloromethane (2/1 v/v) as eluent.

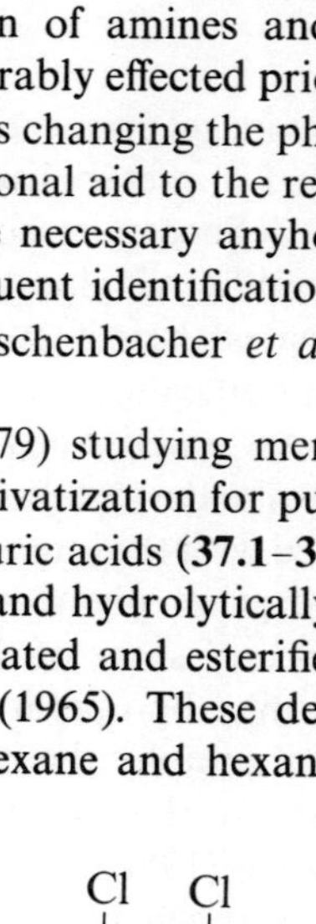

(**37.0**)

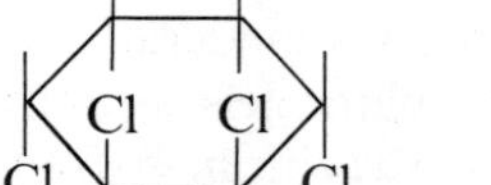

(**37.1**)

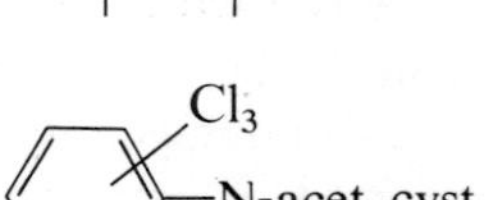

(**37.2**)

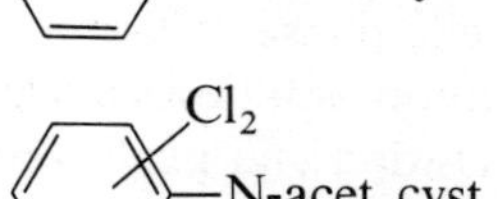

(**37.3**)

The compounds were separated and appeared in the order trichloro-, dichloro- and monochlorophenyl cysteine derivatives. The fractions consisted of mixtures of isomers.

For the determination of urinary dialkyl phosphate metabolites as a measure of human exposure to organophosphorus pesticides, Takade *et al.* (1979) introduced 1-(4-nitrobenzyl)-3-(4-tolyl)triazene as alkylating reagent for derivatization prior to g.c. analysis. The resulting 4-nitrobenzyl derivatives were purified on silica columns. The P=S derivatives were eluted with dichloromethane followed by acetone elution to recover the P=O derivatives.

Size exclusion chromatography

This technique, also named gel, gel filtration, gel permeation and steric exclusion chromatography has been widely applied in metabolism studies. The column packings used are bead form gels with defined pore sizes based on materials such as dextran, agarose, polyacrylamide, vinylpolymers, styrene divinyl benzene copolymers, silica and composites thereof. According to the nature of the material, or due to chemical modification of the surface, these gels behave hydrophilicly or lipophilicly. As stated above, in most separation and purification procedures used for pesticide metabolites the size exclusion principle is of lesser importance compared with the interactions of the molecules with the matrix by partition and adsorption phenomena. Additionally, the presence of charged groups at the surface of some of the gels would possibly suggest that ion exchange phenomena may be operative in some cases. For general aspects of size exclusion chromatography the reader is referred to the literature of the supplying companies and to the recently published textbooks of Kremmer and Boross (1979) and Fischer (1980).

This chromatographic mode is characterized by a high versatility (coping with polar and non-polar metabolites), a moderate resolution and very simple techniques. Separation on the hydrophilic gels is controlled by the pH and the ionic strength of the aqueous eluent; the chromatographic behaviour of metabolites on the lipophilic gels is mainly governed by the swelling characteristics and the polarity of the organic eluent. It has been a general experience with dextran gels, that especially aromatic ompounds interact specifically with the matrix and are therefore strongly retained. The presence of lower alcohols, i.e. methanol and ethanol, in the mobile phase enhances this interaction and the effect can be used for the optimization of a desired separation procedure. The most commonly used gels for separation and purification of pesticide metabolites have been the dextran based hydrophilic Sephadex G-10, G-15 and G-25 and the lipophilic Sephadex LH-20, as well as the hydrophilic polyacrylamide based Bio-Gel P.

Sephadex LH-20 has been for a long time a common tool for the purification of individual metabolites or groups of metabolites, e.g. propham (**27.0**) metabolites, including sulphates and glucuronides from the excreta of chicken, rat and sheep (Paulson *et al.*, 1972; 1975), atrazine (**30.0**) metabolites, including 2-hydroxy-*s*-triazines and amino acid conjugates, from smooth cordgrass (Weete *et al.*, 1980), chlorotoluron (**5.0**) metabolites, including carboxy acids, from rat urine (Muecke *et al.*, 1976), diclofop-methyl (**38.0**) metabolites, including carboxy acids, from wheat and wild oat (Shimabukuro *et al.*, 1979), terallethrin (**34.0**) metabolites, including glucuronides, 1,2-diol derivatives and carboxy acids, from rat excreta (Mihara *et al.*, 1981a), piperophos (**32.0**) metabolites from rat urine (Muecke *et al.*, 1981), terbutryn (**29.0**) metabolites from rat and goat faeces (Larsen *et al.*, 1978).

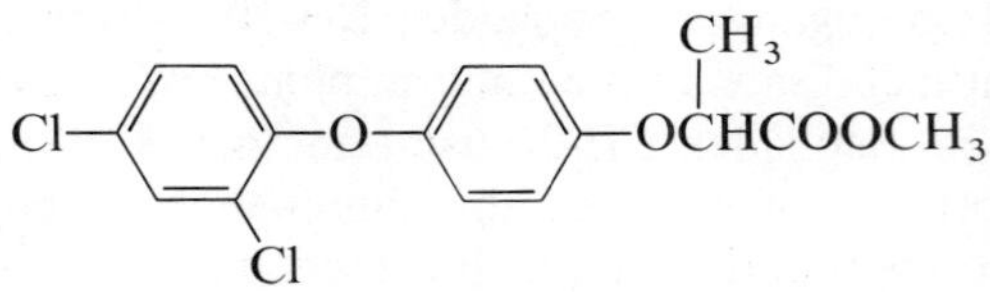

(**38.0**)

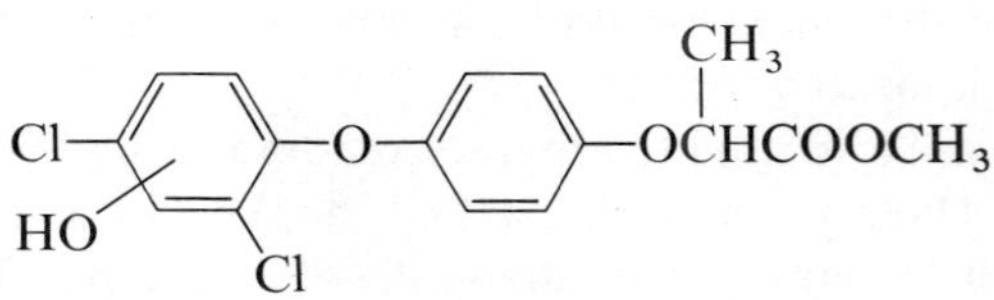

(**38.1**)

In the metabolism studies of propachlor (**12.0**) in animals Sephadex LH-20 was frequently used for purification as well as for separation of metabolite mixtures (Bakke *et al.*, 1976; 1981c; Bakke and Price, 1979; Larsen and Bakke, 1981). The metabolites from sheep urine could be separated, using water as eluent, into a fraction containing two methylsulphonyl derivatives (**12.1**, **12.3**) and two individual metabolites identified as a mercapturic acid (**12.14**) and a cysteine conjugate (**12.13**) of propachlor. Starting from rat urine, after pre-purification and fractionation on a Porapak Q column, three methylsulphonyl derivatives (**12.1**, **12.2**, **12.4**), a mercapturic acid conjugate (**12.14**) and a composite fraction consisting mainly of glucuronic acid conjugates were separately eluated with water. In continuation of their studies on the genesis of methylsulphonyl derivatives, the metabolism of pentachloromethylthiobenzene (**28.0**) and 2-acetamino-4-(chloromethyl)thiazole (**39**), an anti-inflammatory compound, was investigated in rats. Size exclusion chromatography on Sephadex LH-20 gave good separations of similar metabolites, including mercapturic acids, *S*-glucuronides and *S*-methyl derivatives (Bakke *et al.*, 1981a; b).

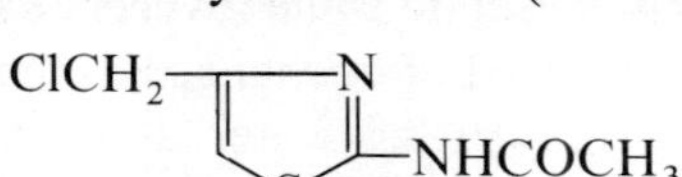

(**39**)

A fine separation of urinary metabolites from rats subcutaneously dosed with *o*-toluidine (**20.0**) was recently reported (Son *et al.*, 1980). A Sephadex LH-20 column was loaded with urine after removal of the ether extractable metabolites and developed with methanol/water/triethylamine bicarbonate buffer, 0.05 M, pH 9.2 (100/100/2 v/v/v). In the following order a glucuronide (**20.10**), a mixture of two glucuronides (**20.8, 20.9**), three sulphates (**20.12–20.14**) and the parent compound (**20.0**) were separately eluted. However, a double conjugate, i.e. the *N*-glucuronide of 4-amino-*m*-cresyl sulphate (**20.11**) was assumed to have decomposed during the procedure.

Shimada and Nambara (1979) investigated the chromatographic behaviour of sulphuric acid conjugates on Sephadex LH-20. It was demonstrated that different univalent metal salts of the same sulphate are recovered separately by chromatography on Sephadex LH-20 using methanol as eluent. This finding stresses the importance of ensuring that homogeneous populations of salts (or free acid) are present in the sample. Ion exchange prior to loading may help to solve such problems. When aqueous systems are used, addition of an excess of salts (buffer) to the eluent would be another way to create a homogeneous population of an ionizable species.

Lipophilic gels are also used to separate xenobiotic metabolites from natural occurring lipids. Harvey and Paton (1980) applied liver extracts from mice orally treated with Δ^6-tetrahydrocannabinol (**40**) to a Sephadex LH-20 column. On elution with choroform, triglycerides, cholesterol and free fatty acids were sequentially eluted and the metabolites recovered by a stepwise gradient of methanol in chloroform.

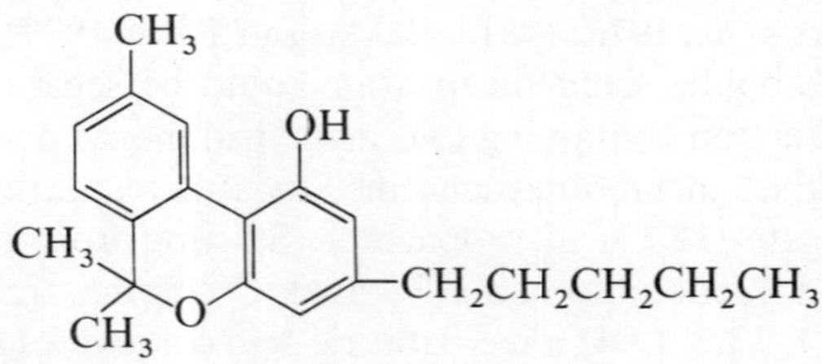

(**40**)

Crayford and Hutson (1980b) utilized μ-Styragel and tetrahydrofuran/acetic acid (99/1 v/v) for the separation of lipids in their metabolism study of orally applied 3-phenoxybenzoic acid (**41.2**) being an intermediate in the metabolism of cypermethrin (**41.0**). Skin extracts were applied to the gel column and separated in two fractions. The first fraction, eluted with the void volume, contained the triglycerides including 2- and 3-(3-phenoxybenzoyl) dipalmitin (**41.4, 41.5**) and the second fraction consisted of unchanged 3-phenoxybenzoic acid (**41.2**).

Biogel P-2 chromatography using water as eluent was used mainly for purification purposes rather than for separation of metabolites, e.g. a β-glucoside (**42.1**) and a β-gentiobioside (**42.2**) from diphenamide (**42.0**) treated

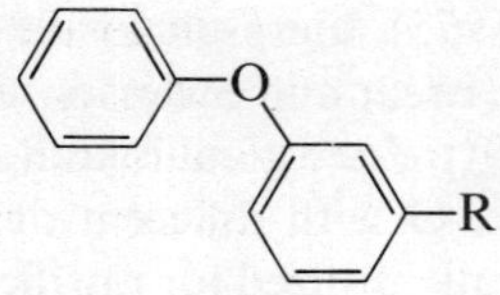

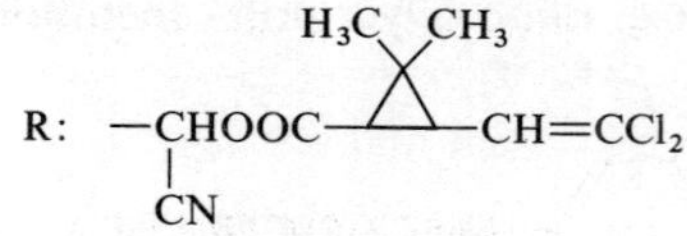

(41.0)

R: —CH_2OH (41.1)

R: —COOH (41.2)

R: —CH_2Cl (41.3)

$CH_2OOC(CH_2)_{14}CH_3$
R: —COOCH (41.4)
$CH_2OOC(CH_2)_{14}CH_3$

$CH_2OOC(CH_2)_{14}CH_3$
$CHCOOC(CH_2)_{14}CH_3$
R: —$COOCH_2$ (41.5)

R: —CO-taur. (41.6)

R: —CH_2—O-gluc. -arab. (41.7)

R: —CH_2—O-gluc. -xyl. (41.8)

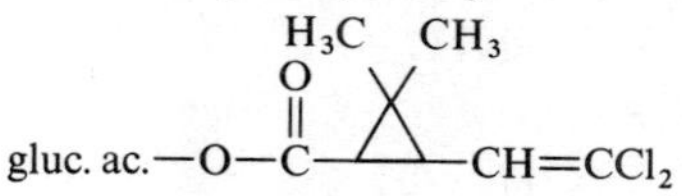

(41.9)

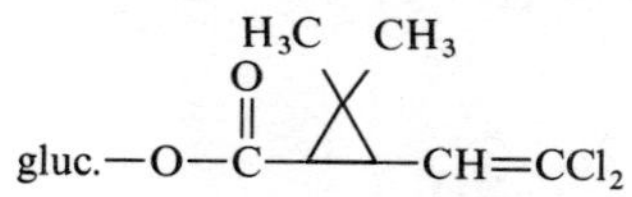

(41.10)

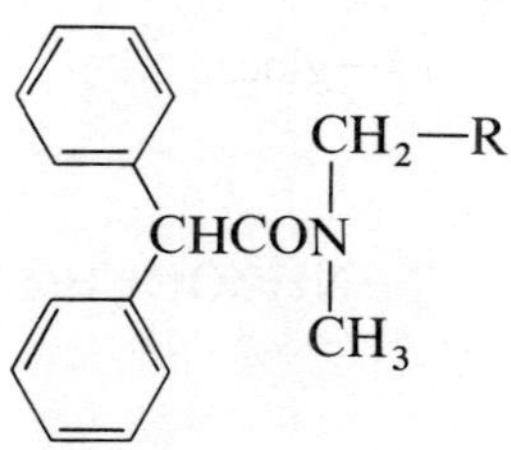

R: —H (42.0)

R: —O-gluc. (42.1)

R: —O-gluc. -gluc. (42.2)

tomatoes (Hodgson *et al.*, 1973), the β-glucoside (**9.1**) from maleic hydrazide (**9.0**) treated tobacco plants (Frear and Swanson, 1978), and the α-glucose ester (**24.3**) from chloramben (**24.0**) treated giant foxtail and barley (Frear *et al.*, 1978).

The hydrophilic Sephadex G with different degrees of cross-linking (G-10, G-15, G-25) have been typically utilized for purification of polar metabolites or polar metabolite fractions, e.g. Bligh–Dyer polar metabolites of chlorpropham

(43.0)

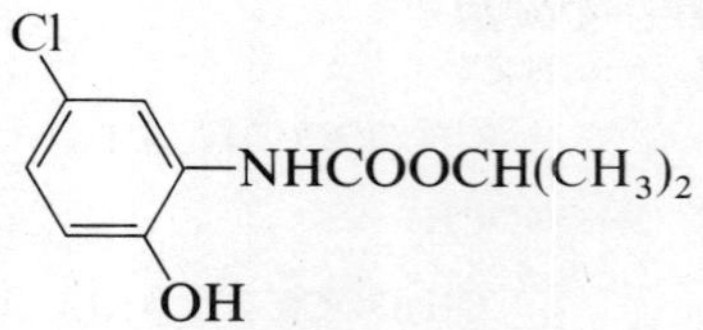

(43.1)

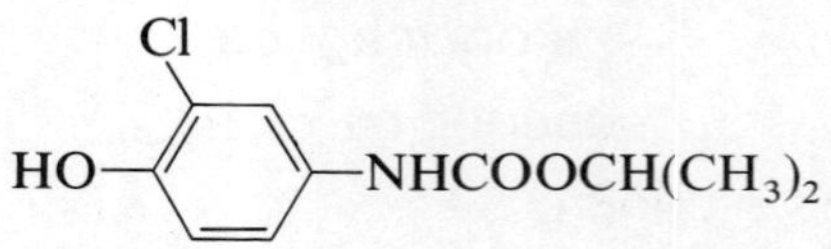

(43.2)

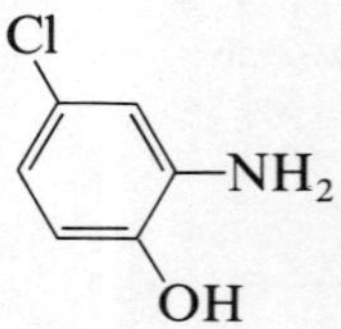

(43.3)

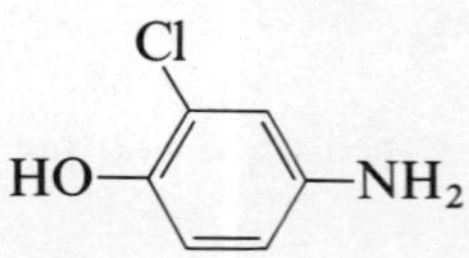

(43.4)

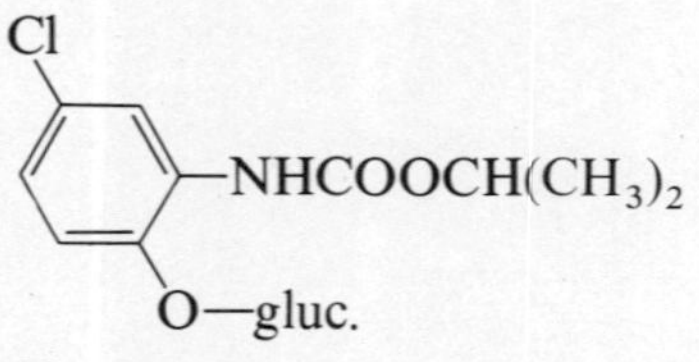

(43.5)

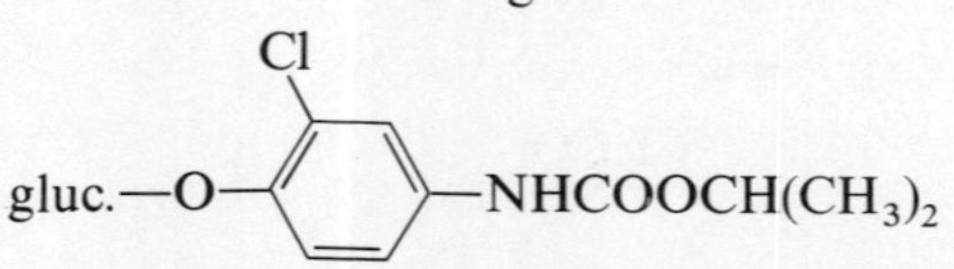

(43.6)

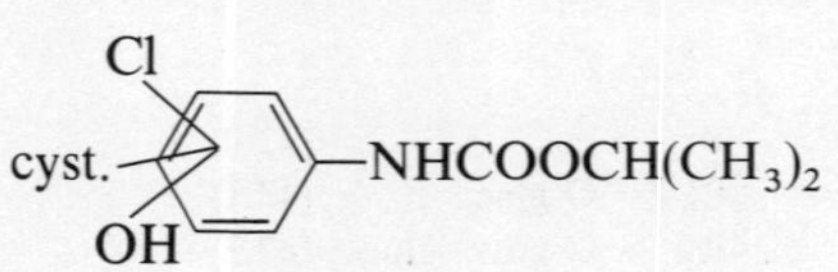

(43.7)

(**43.0**), including the β-glucosides of its 2- and 4-hydroxy derivatives (**43.5**, **43.6**) from soybeans and alfalfa (Still and Mansager, 1973; 1975), chlorotoluron (**5.0**) metabolites, including β-glucosides (**5.12**, **5.13**) from wheat (Gross *et al.*, 1979) and the glutathione (**31.2**) and the cysteine conjugate (**31.1**) of the fumigant methyl bromide (**31.0**) from granary weevils (Starrat and Bond, 1981). De Baun *et al.* (1979) applied Sephadex G-10 to separate the most polar metabolite fraction in the urine of rats orally dosed with molinate (**44.0**). Two fractions were obtained, one containing the glucuronides (**44.3**, **44.4**) and the other the corresponding hexahydroazepine derivatives (**44.1**, **44.2**).

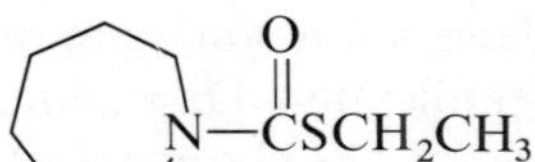

(**44.0**)

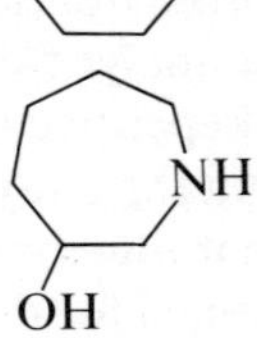

(**44.1**)

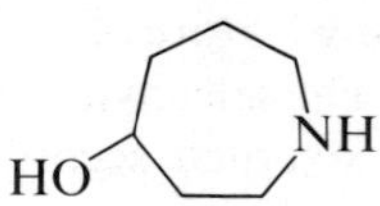

(**44.2**)

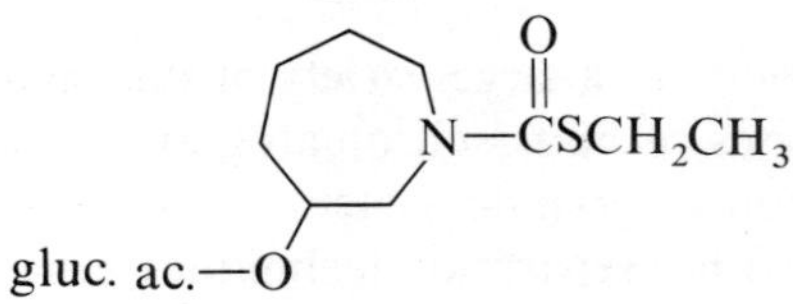

(**44.3**)

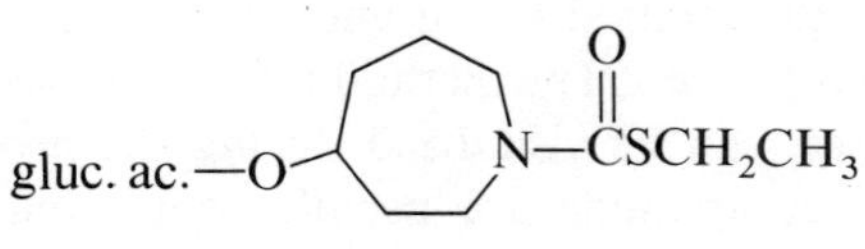

(**44.4**)

High performance liquid chromatography

The desire to improve the efficiency of traditional liquid chromatography had led to the use of stationary phases consisting of small, sperical particles ($\leq 10\ \mu$m) with narrow size distribution and high surface areas. The high back pressure of columns packed with these materials required the development of adequate equipment which ultimately led to the present stage of h.p.l.c. The great success and the still good commercial prospects of this technology have stimulated the development of a large variety of different column packings utilizing all modes of separation mechanisms known in solid–liquid and liquid–liquid chromatography. The technique and the application of h.p.l.c. has been described and discussed in numerous textbooks, e.g. Engelhardt (1977), Huber (1978), Hawk

(1979), Kaiser and Oelrich (1981), Pryde and Gilbert (1979). The reader's attention is again drawn to the monograph of Snyder and Kirkland (1979) and to a continuing publication edited by Horvath (starting 1980) dedicated to cover the advances and perspectives of high performance liquid chromatography. Two volumes of this series focusing on the fundamentals of h.p.l.c., written by the foremost experts in each individual field, are already published. Two recent publications deal with the general application of h.p.l.c. in pesticide biochemistry, analysis and metabolism (Schooley and Quistad, 1979; Harvey, 1980).

Although h.p.l.c. is basically an analytical tool, it has preparative potential. According to Verzele and Geeraert (1980), 'a chromatography is termed "preparative" when it isolates a fraction, pure or not so pure, from a mixture for further use', e.g. for identification. Depending on the nature of the sample, the separation goal desired and the chromatographic system used, samples of several mg may be applied to a typical analytical column (250 × 4.6 mm, i.d.) with an acceptable loss of resolution. Repetitive injections of the same sample and enhancing the column size are suitable ways to extend h.p.l.c. to larger sample sizes. However, it is obvious, that low capacity is an inherent feature of h.p.l.c. and hence, the method is predestined for the later steps in the separation and purification process. The dominating feature of h.p.l.c. is its high resolution required for final isolation procedures. The selectivity of the method depends on the operation mode and can thus be governed according to the sample composition.

The commercial availability of a large variety of stationary phases allowing the application of all modes of liquid chromatography and thus the almost unlimited number of chromatographic systems to choose from, provides a versatility unsurpassed by other separation techniques. Once the rather expensive equipment is at hand, the method is fast and economical regarding time, labour and cost of consumable items. The method is fast also, because the result of separation or purification can be visualized during the process, as radiolabelled and non-labelled compounds are usually monitored in the eluent simultaneously (Harvey, 1980). Furthermore, the parameters of the entire system can be readily controlled with the modern h.p.l.c. equipment, leading to a high degree of reproducibility. Being a closed system, usually operated under mild conditions and at ambient temperature, the method is not prone to artefact formation.

After a pre-purification process, h.p.l.c. is mostly applied as a sequence of several chromatographic steps to effect the ultimate purification. The purified metabolite is typically obtained in a small volume of eluent, thus limiting contamination with solvent impurities.

The same chromatographic modes outlined in the previous section on conventional liquid chromatography can be utilized, i.e. ion exchange, partition, adsorption and size exclusion. However, the mechanisms of solute retention, especially on the bonded phases, is obviously complex and still the subject of controversy. Therefore, the use of the terms adsorption h.p.l.c., partition h.p.l.c., etc. does not necessarily imply that this mode is the dominating one, operative in a given system.

Regarding isolation of pesticide metabolites, ion exchange and size exclusion h.p.l.c. is of limited use and therefore rarely used.

Ion exchange chromatography usually requires the use of aqueous buffers which—even if volatile—are potential sources for contamination of the purified metabolite. Furthermore, ion exchange packings are typically less efficient than those used for adsorption and partition, a fact which has also limited their use. In most cases partition chromatography in the reversed phase mode can be used instead for ionizable metabolites.

The limiting factor for size exclusion h.p.l.c. is its resolution. It must be assumed, that the differences in molecular weight or actual size of molecules in the sample at the penultimate stage of the purification process are rather small.

By far the most popular chromatographic mode applied is the reversed phase technique on bonded phases, usually on octadecyl silica (C_{18}, ODS). These and the other (e.g. dimethyl, octyl and phenyl silica) bonded reversed phase systems have some major advantages. With respect to polarity a wide spectrum of compounds can be separated including ionizable metabolites and polar conjugates. The mobile phase, commonly methanol/water and acetonitrile/water can be modified by addition of acids, buffer mixtures, salts, pairing ions, etc. to effect ion suppression, salting out effects and ion pair formation. The time necessary to establish equilibrium is short, thus allowing rapid analysis and a short reconditioning time between the runs after using gradient elution. Columns can be operated within a pH range of 2 and 8, separation performance is reproducible and the recoveries are usually high. Trace enrichment techniques can be conveniently applied, i.e. loading the sample in a large volume of a solvent, e.g. water, which does not elute the solute of interest. The solute will concentrate at the top of the column and may be recovered by gradient elution. The use of the so-called normal bonded phases, i.e. amino, diol, nitrile group carrying matrices has been rather limited. Practical aspects of bonded-phase h.p.l.c. have been comprehensively reviewed by Brown and Krstulovic (1979) and Majors (1980b).

In addition to reversed phase h.p.l.c., adsorption chromatography, preferably on silica, is the commonly used method in metabolite isolation. The advantages and limitations of this technique are essentially those discussed previously in the section on liquid column chromatography.

Ion pair h.p.l.c., conveniently performed on bonded phases, is recommended for compounds, e.g. sulphate conjugates, which are ionized within the pH range (pH 2–8) suitable for the operation of silica based columns (Schill *et al.*, 1977a; b; c; Tomlinson *et al.*, 1978; Hearn, 1980; Karger *et al.*, 1980). The technique has been successfully applied in analytical work, but rarely in preparative chromatography. The problems involved in removing the excess of counter ions and in cleavage of the ion pair to regain the genuine metabolites are considered prohibitory.

Final isolation and purification of individual metabolites is usually accomplished by two or more distinct chromatographic steps. It appears advisable—if appropriate—to utilize different separation modes for the individual steps or at least distinct chromatographic systems. This strategy also named multidimensional h.p.l.c., in preparative chromatography usually performed as an off-line technique, considerably enhances the efficiency of the separation process (Majors, 1980a). The sequence in which, for example, reversed phase and adsorption h.p.l.c. are applied depends on the pre-purification procedures. In our laboratory we generally prefer to use non-aqueous, low boiling eluents for the ultimate purification step as their fast and easy removal reduces the risk of recontamination of the usually small amounts (μg range) of metabolites obtained. This necessarily means applying reversed phase h.p.l.c. first and adsorption h.p.l.c. thereafter. Rather efficiently pre-purified metabolites of low polarity might be directly applied to a silica column and the eluent changed for the second run. Polar metabolites like carboxy acids and conjugates are primarily purified by reversed phase chromatography, applying ionization depression if necessary and thereafter derivatized to yield less polar products for purification on silica columns. Sulphate conjugates, known to be difficult to derivatize, can nevertheless be purified by adsorption chromatography on silica. In our laboratory it was empirically found that sulphuric acid conjugates are selectively desorbed from silica gel, when the mobile phase contains substantial amounts of acetone (Muecke and Ifflaender, 1976). This effect can be utilized for the purification (and isolation) of sulphates by chromatography on silica. In some cases polar metabolites are subjected for ultimate purification to reversed phase h.p.l.c., when proper chromatography of the genuine metabolite on silica is impossible and derivatization for one or the other reason is not appropriate.

Reversed phase h.p.l.c. on octadecyl silica has been widely utilized for the isolation of pesticide metabolites and its use is continually increasing. Many metabolites of different polarities, including exocons after hydrolytic or enzymatic cleavage have been purified on C_{18} columns, e.g. chlorpropham (**43.0**) metabolites from alfalfa, rat and sheep urine (Still and Mansager, 1975), cycloprate (**21.0**) metabolites from apples, oranges, rat tissues and faeces (Quistad *et al.*, 1978a; b), 3-phenoxybenzoic acid (**41.2**) metabolites from various plants (More *et al.*, 1978), propachlor (**12.0**) metabolites from urine, faeces and bile of rats (Larsen and Bakke, 1981), pentachloronitrobenzene (**3.0**) metabolites from peanuts (Rusness and Lamoureux, 1980), 2,4-D (**45**) and 2,4,5-T (**46**) metabolites from soil (McCall *et al.*, 1981), and piperophos (**32.0**) metabolites from rat urine (Muecke *et al.*, 1981).

Examples of the chromatography of some specific polar conjugates should be highlighted. Thus, Crawford *et al.* (1981) isolated a mixture of two stereo-

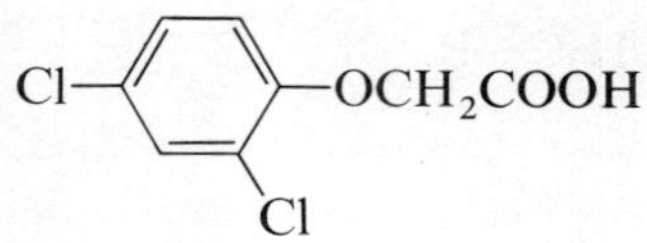

(45)

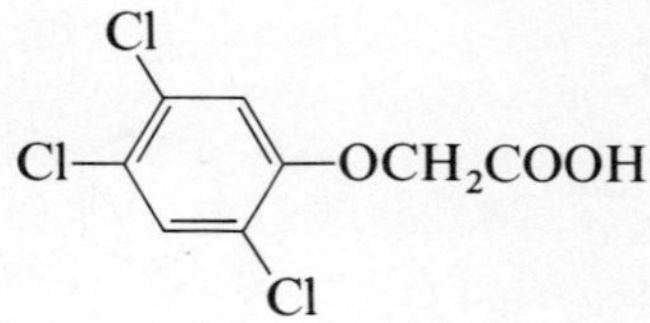

(46)

isomeric (*cis/trans*) β-glucuronides (**41.9**) from the urine of cypermethrin (**41.0**) dosed rats. Still and Rusness (1977) chromatographed a compound, presumably a β-glucoside (**43.6**), from chlorpropham (**43.0**) treated oat. A taurine conjugate (**41.6**) from the urine of cypermethrin (**41.0**) and 3-phenoxybenzoic acid (**41.2**) treated mice was purified on a C_{18} column (Hutson and Casida, 1978).

Cysteine conjugates (**43.7**, **3.8**, **3.14**) of chlorpropham (**43.0**) and pentachloronitrobenzene (**3.0**) were isolated from oat and peanuts (Still and Rusness, 1977; Lamoureux and Rusness, 1980; Lamoureux *et al.*, 1981). The same authors report also the isolation of a glutathione conjugate (**3.11**) and two malonyl-cysteine conjugates (**3.10**, **3.15**) from pentachloronitrobenzene (**3.0**) treated peanuts by reversed phase chromatography on C_{18} columns (Lamoureux and Rusness, 1980; Lamoureux *et al.*, 1981). Bakke *et al.* (1981a) purified a mercapturic acid conjugate (**28.1**) from the faeces of germ-free rats after oral application of pentachloromethylthiobenzene (**28.0**).

Verweij and Kientz (1981) investigated the metabolic fate of 1,2,3-thiadiazole-5-carboxaldoxime (**47.0**), an experimental antidote in organophosphate poisoning, in rabbits. Applying urine directly—without pre-purification—on to semi-preparative C_{18} columns (250 mm × 10 mm), mg amounts of the major metabolite, a glucuronide (**47.1**), were obtained for ^{1}H- and ^{13}C-n.m.r. spectroscopy. All of the authors cited above isolating acidic conjugates, used acidified aqueous methanol or acetonitrile as eluents for ion suppression, except Still and Rusness (1977) and Bakke *et al.* (1981a) who eluted with neutral solvents.

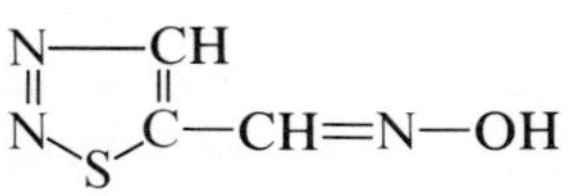

(47.0)

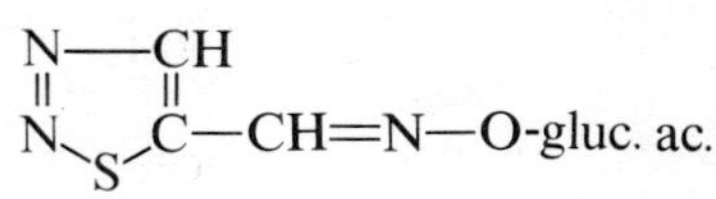

(47.1)

Wright *et al.* (1980) chromatographed the tetraacetyl derivative of an ester glucoside (**41.10**) derived from cypermethrin (**41.0**) treated lettuce on a C_{18} column using a acetonitrile/water mixture as eluent.

In a study on the metabolic conjugation of some carboxylic acids in the horse, Marsh *et al.* (1981) demonstrated the separation of hippuric acid (**65.1**), 2-naphthyl acetic acid (**48.0**) and its taurine (**48.1**), glutamine (**48.2**) and glycine conjugate (**48.3**) on C_{18} columns eluted with 20% methanol, adjusted to pH 3 with acetic acid. Illing and Wilson (1981a; b) and Sinclair and Caldwell (1982)

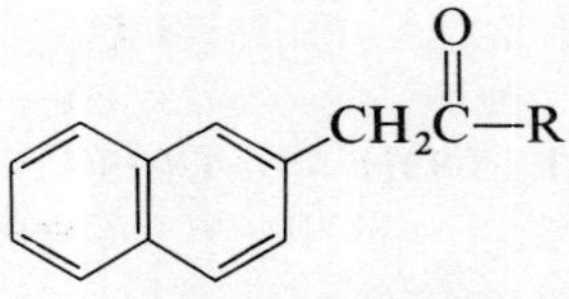

R:	—OH	(**48.0**)
R:	-taur.	(**48.1**)
R:	-glutam.	(**48.2**)
R:	-gly.	(**48.3**)

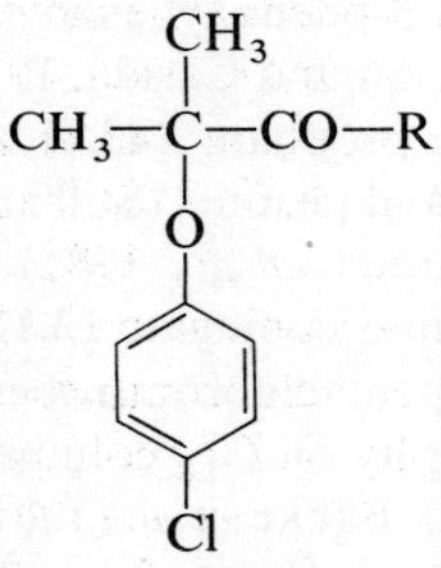

(**49**)

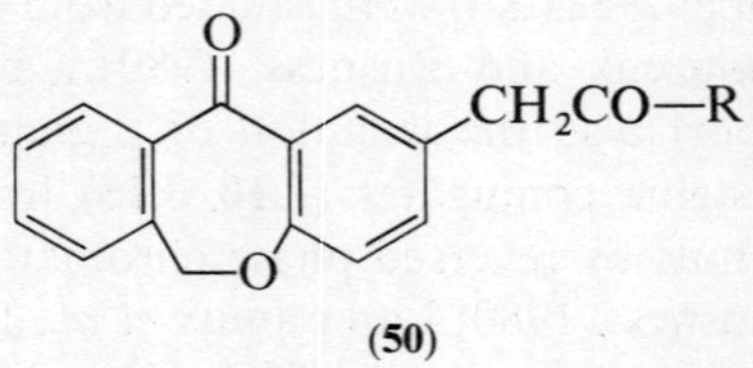

(**50**)

R: —OH (**49.0** or **50.0**)

R: 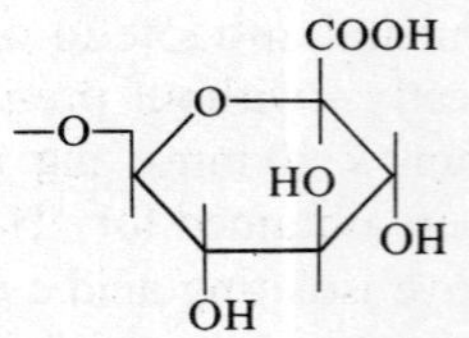

(**49.1** or **50.1**)

R: 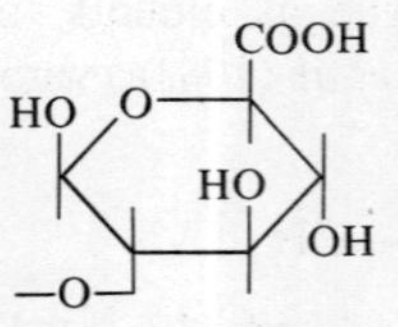

(**49.2** or **50.2**)

R: 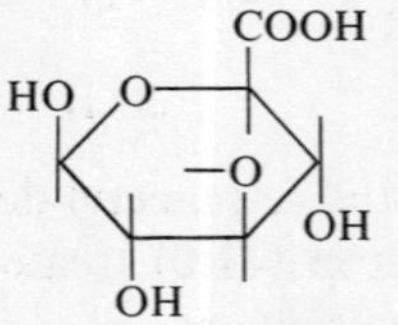

(**49.3** or **50.3**)

R: 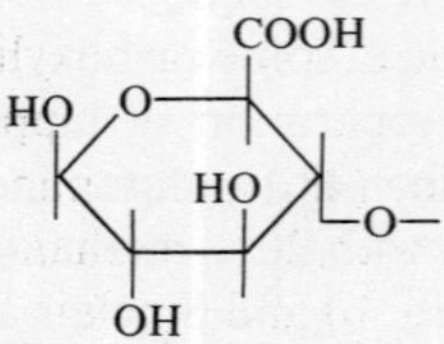

(**49.4** or **50.4**)

studied the acyl migration of ester glucuronides. Both groups were able to demonstrate separation of the four isomeric ester 'glucuronides' of clofibric acid (**49.0**, **49.1**–**49.4**) and isoxepac (**50.0**, **50.1**–**50.4**) respectively, on C_{18} columns using ion-suppressing aqueous eluents. The separation of two diastereoisomeric glucuronides [(*S*)-(+) and (*R*)-(−)] (**18.1**) of oxazepam (**18.0**) with 40% aqueous methanol, containing 0.025 M NaH_2PO_4 was reported by Ruelins *et al.* (1979). A general discussion of the separation of enantiomers by h.p.l.c. with the aid of chiral eluents has been published by Oelrich *et al.* (1980).

This rather random selection of recent publications demonstrates impressively the enormous versatility of octadecyl silica columns for separation and purification of pesticide metabolites.

The general experience with the bonded phases carrying functional groups, such as CN, NH_2, Diol is still rather limited and only a few examples of their utilization have been published. Hexane extracts from acidified urine and faeces of pentachloronitrobenzene (**3.0**) treated rats were chromatographed on a CN-column using methanol/water/acetic acid (49/49/2 v/v/v) as eluent (O'Grodnick *et al.*, 1981). A separation of the parent (**3.0**) and four potential metabolites including a mercapturic acid conjugate (**3.1**, **3.2**, **3.4**, **3.9**) was reported. Lusby *et al.* (1981) used a CN-column investigating the metabolism of *N*-nitrosopendimethalin (**35.1**) by soil isolated streptomyces cultures. Using in sequence a C-18 column and a CN-column (eluent: isooctane/propan-2-ol) a metabolite (**35.2**) was isolated and separated from the parent compound (**35.1**).

Crawford *et al.* (1981) applied a NH_2-column and hexane/dichloromethane/acetic acid (95/5/0.5 v/v/v) as eluent for the analysis of *cis*- and *trans*-cyclopropanecarboxylic acid derivatives in a study on the metabolites of cypermethrin (**41.0**) excreted by rats. Details were not given. Crayford and Hutson (1980b) isolated two major products from the skin of 3-phenoxybenzoic acid (**41.2**) treated rats. After separation by size exclusion chromatography, the starting material (**41.2**) and the mixture of 2- and 3-(3- phenoxybenzoyl)dipalmitin (**41.4**, **41.5**) were efficiently purified on a NH_2-column eluting with dichloromethane/methanol/acetic acid (95/4/1 v/v/v) and dichloromethane/hexane (10/90 v/v), respectively.

Adsorption h.p.l.c. on silica gel has been preferred for isolation and purification of metabolites with low or moderate polarity, e.g. oryzalin (**51**) metabolites from soil (Golab *et al.*, 1975). Son *et al.* (1980) separated seven ether

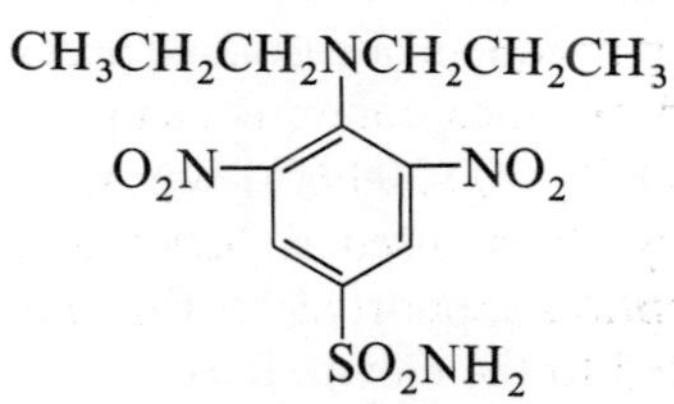

(**51**)

extractable urinary metabolites (**20.1, 20.3, 20.4, 20.6, 20.7, 20.15**) including the parent compound from *o*-toluidine (**20.0**) treated rats on silica eluted with hexane/ethanol (90/10 v/v). Upon acidifying the eluent by adding 0.2% of acetic acid two additional acidic metabolites (**20.2, 20.5**) were eluted separately. During the investigation on the nature of DDT (**52.0**) metabolites retained in liver and spleen of rats, h.p.l.c. on silica columns operated at 50 °C, using chloroform/heptane (80/20 v/v) as eluent, was applied to the purification of the newly discovered type of fatty acid conjugates (**52.1–52.4**) (Leighty, 1980; Leighty *et al.*, 1980).

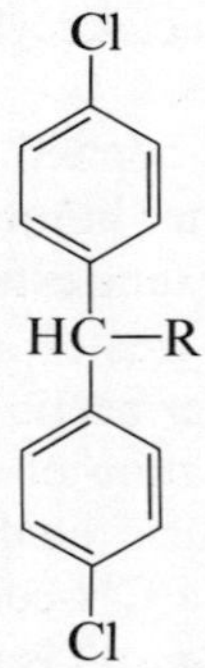

R: $-CCl_3$ (**52.0**)
R: $-CH_2OCO(CH_2)_{14}CH_3$ (**52.1**)
R: $-CH_2OCO(CH_2)_{16}CH_3$ (**52.2**)
R: $-CH_2OCO(CH_2)_7CH{=}CH(CH_2)_7CH_3$ (**52.3**)
R: $-CH_2OCO(CH_2)_7(CH{=}CHCH_2)_2(CH_2)_3CH_3$ (**52.4**)

Quistad *et al.* (1978a; b) investigated the metabolism of cycloprate (**21.0**) in rats, apples and oranges. Various ω-cyclopropyl fatty acids were released from plant and animal tissues by saponification; two of them, derived from rat liver (**21.1, 21.2**), were separated from the natural fatty acids and ultimately purified as their methyl esters on silica columns (eluent: ether/hexane 0.5/95.5 v/v) prior to identification by g.c.–m.s. The major urinary metabolite, a glycine conjugate (**21.3**) was finally purified as its *p*-phenylphenacyl ester by elution with ether/pentane (70/30 v/v) and identified by m.s.

Although dedicated to rather non-polar compounds, h.p.l.c. on silica may—under certain conditions—also be successfully applied for separation and purification of polar compounds. Various hydroxy-*s*-triazines (**53–58**), representing major degradation products of *s*-triazine herbicides in the environment, were efficiently separated on silica columns using an isocratic eluent (chloroform/methanol/water/H_3PO_4 700/300/60/1 v/v/v/v) (Ramsteiner and Hoermann, 1979). The ternary eluent used is believed to form a reversed-phase system *in situ*, an assumption supported by the elution order observed. The phosphoric acid was added to the eluent, in order to increase the solubility of the compounds.

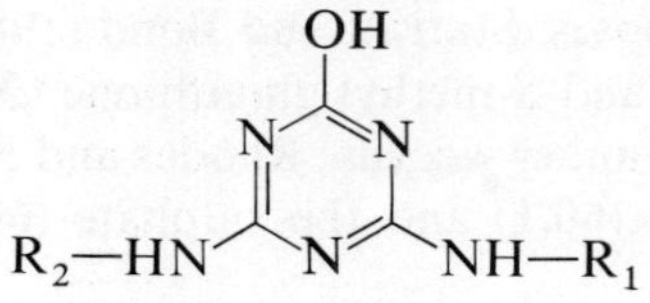

R_1	R_2	
$-CH(CH_3)_2$	$-CH(CH_3)_2$	**(53)**
$-CH(CH_3)_2$	$-CH_2CH_3$	**(54)**
$-CH_2CH_3$	$-CH_2CH_3$	**(55)**
$-H$	$-C(CH_3)_3$	**(56)**
$-H$	$-CH(CH_3)_2$	**(57)**
$-H$	$-CH_2CH_3$	**(58)**

Underivatized sulphuric acid conjugates can be successfully chromatographed on silica columns. Acetone has been found to show selective desorption properties for this type of conjugate (Muecke and Ifflaender, 1976). Studying the metabolism of the fungicide CGA 64 250 (**59.0**) in rats, this effect has been utilized to separate and purify urinary sulphuric acid conjugates, e.g. (**59.1**), using a gradient of acetone in chloroform (Muecke, 1981).

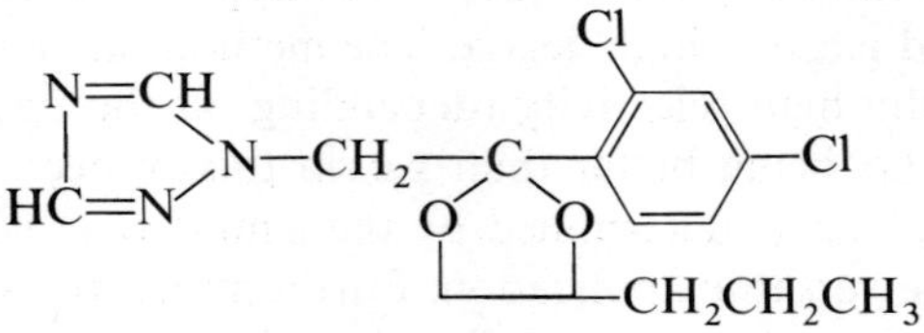

(**59.0**)

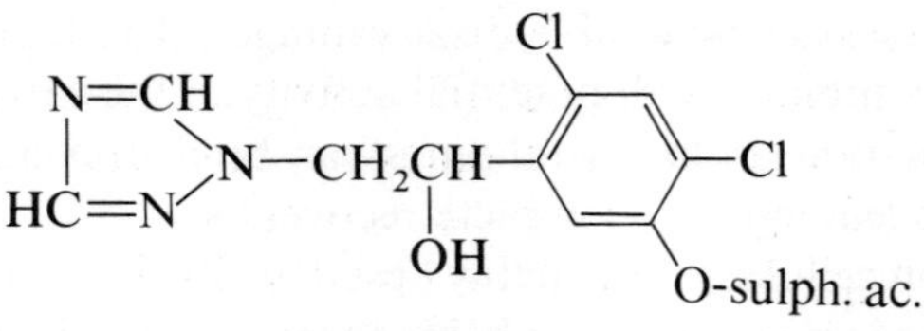

(**59.1**)

Ion-exchange h.p.l.c. is not commonly used for metabolite isolation, but occasionally for separation, detection and quantification in residue analysis, e.g. maleic hydrazide residues (Newsome, 1980) and arsenical residues (Woolson and Aharonson, 1980).

For reasons mentioned above, ion-pair h.p.l.c. has not been applied yet for metabolite isolation, although it represents an excellent separation tool. Fransson *et al.* (1976) demonstrated examples of separation for several classes of compounds, including phenols, carboxylic acids, glycinates, sulphates and glucuronides using straight phases as well as reversed phases. When utilized by metabolism chemists, ion-pair h.p.l.c. has been used as an analytical tool, but not for preparative purposes. Starratt and Bond (1981) analysed a mixture of *s*-methyl cysteine (**31.1**) and *S*-methyl glutathione (**31.2**) derived from methyl bromide (**31.0**) treated granary weevils; Rhodes and Houston (1981) quantified the urinary glucuronide (**60.1**) and the sulphate (**60.2**) of 1-naphthol (**60.0**) from treated rats.

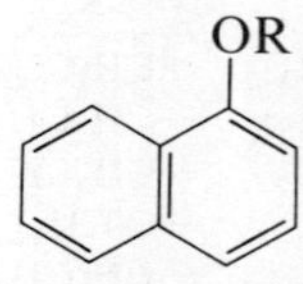

R: —H (**60.0**)
R: —gluc. ac. (**60.1**)
R: —sulph. ac. (**60.2**)

Thin-layer and paper chromatography

Both of these techniques have long been used for the analysis of metabolite mixtures and also for preparative purposes, i.e. separation, isolation and purification of metabolites. The methods still deserve their status as valuable tools in the repertoire of isolation techniques. This applies predominantly to t.l.c., which has replaced p.c. to a high degree. The methods are characterized by low capacity, potentially high selectivity (depending on the operation mode) and high resolution, t.l.c. being better than p.c. in this respect. The popularity of p.c. and especially t.l.c. is determined by the simple operation procedure, the high speed and the versatile application. Furthermore, the success of a certain separation objective can be easily predicted on the basis of analytical pilot runs in an identical system, taking into account the increased sample load. However, both methods possess considerable disadvantages. The high-surface exposure to aerial oxygen combined with potential activity of the carrier surface encourages artefact formation. Polar metabolites can bind irreversibly, especially on adsorptive phases leading to incomplete recoveries.

The usual p.c. on cellulose is probably best described as a partition chromatography system, the stationary phase being formed *in situ* by selective uptake of water from the solvent by the cellulose carrier. However, other effects, especially adsorption, will always be operative as well; consequently p.c. is especially used for isolating polar metabolites, including conjugates.

Preparative t.l.c. is commonly carried out on silica coated glass plates. The chromatographic mechanism is complex, but in the most traditional systems adsorption mechanisms play a significant role. Therefore, t.l.c. is, compared with p.c., more prone to irreversible binding of polar compounds and is hence preferred for metabolites with low and moderate polarity.

In addition to these standard types of p.c. and t.l.c., variants of the techniques have been developed, e.g. reversed phase systems by lipophilic impregnation or corresponding chemical modification of the carrier surface, and impregnation with buffer solutions or other chemicals to achieve selective effects. Furthermore, papers containing ion-exchange resins and adsorbents, like silica or alumina and plates coated with bonded phases, as used in h.p.l.c., or other phases like alumina or polyamide are commercially available. However, the various known special systems have seldom been utilized for preparative isolation of pesticide metabolites.

Whilst analytical t.l.c. is commonly performed on plates of 0.25 mm thickness, for preparative work, plates of up to 2 mm thickness are frequently used; this leads necessarily to reduced resolution. The decision on which plates to use must eventually be based on the need for capacity, resolution and consequently convenience regarding a certain separation. For use in a one-dimensional mode preparative samples are usually streaked rather than spotted onto the plate to increase capacity.

For the recovery of the separated metabolites from the carrier, the corresponding areas are usually cut out from the paper or scraped off the plate, desorbed and extracted with methanol, acetone or other polar solvents. Even when the paper or plate has been properly pre-washed, co-extraction of impurities from the carrier, can scarcely be avoided. I it is adequate with respect to the desorption of the metabolite, acetone has been found useful to minimize this effect; methanol or aqueous methanol dissolve considerable amounts of silicic acid from the carrier. It has often been found advantageous, to pass the t.l.c. or p.c. eluate through a commercial reversed-phase cartridge e.g. SEP-PAK to remove major contaminants. Nevertheless, in most cases an additional, ultimate purification step is necessary. Detailed information on theoretical and technical aspects and application of t.l.c. and p.c. in the field of xenobiotic metabolism have been published by Titus (1971), Touchstone and Dobbins (1978), Mikes (1979) and Nestler (1979).

Surveying the recent pesticide metabolism literature the importance of preparative t.l.c. becomes evident. Apart from some studies in which the isolation for identification often combined with g.c.–m.s., is solely based on t.l.c., there are few papers where no preparative t.l.c. was applied as an intermediate isolation step. Therefore, only those recent papers applying preparative t.l.c. to polar conjugates or utilizing a special technique will be cited here.

The following types of conjugates, isolated from various sources and related to the pesticides in parentheses were purified on silica coated plates in free form and/or as derivatives:

Glutathione conjugates (pentachloronitrobenzene (**3.0**)) in free form and after ethylation and acetylation (Lamoureux and Rusness, 1980; Lamoureux *et al.*, 1981),

Cysteine conjugates (pentachloronitrobenzene (**3.0**), chlorpropham (**43.0**), methidathion (**26.0**)) in free form (Still and Rusness, 1977; Rusness and Lamoureux, 1980), as *n*-butyl ester (Simoneaux *et al.*, 1980) and as *N*-trifluoroacetyl methyl ester (Lamoureux and Rusness, 1980; Rusness and Lamoureux, 1980; Lamoureux *et al.*, 1981),

N-Acetyl cysteine conjugates (pentachloronitrobenzene (**3.0**), 1,2-dichloropropane (**61**), propachlor (**12.0**), fluorodifen (**15.0**), allidochlor (**14.0**)) in free form (Lamoureux and Davison, 1975; Jones and Gibson, 1980; Renner, 1980) and as methyl ester (Lamoureux and Davison, 1975),

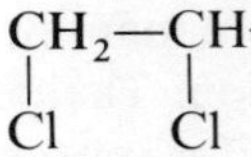

(61)

N-Malonyl cysteine conjugates (pentachloronitrobenzene (**3.0**)) as *n*-butyl or methyl ester (Lamoureux and Rusness, 1980; Lamoureux *et al.*, 1981),

3-*Thio lactic acid conjugates* (pentachloronitrobenzene (**3.0**)) as *n*-butyl ester (Rusness and Lamoureux, 1980),

β-D-*Glucose conjugates* (chlorpropham (**43.0**), 3-phenoxybenzoic acid (**41.2**), maleic hydrazide (**9.0**), diclofop-methyl (**38.0**), methazole (**62**)) in free form (Still and Rusness, 1977; Frear and Swanson, 1978; More *et al.*, 1978; Suzuki and Casida, 1981) and as acetyl derivative (Shimabukuro *et al.*, 1979; Suzuki and Casida, 1981),

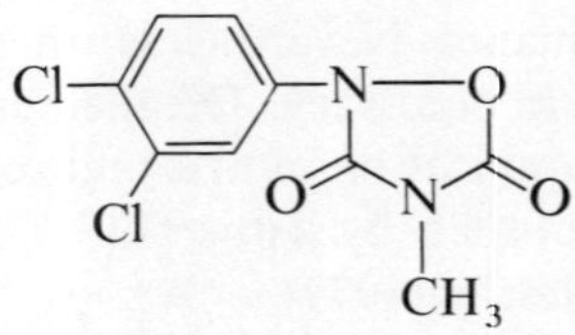

(62)

6-*O-Malonyl-β*-D-*glucose conjugates* (methazole (**62**)) in free form and after methylation and acetylation (Suzuki and Casida, 1981),

α-D-*Glucose conjugates* (chloramben (**24.0**)) in free form (Frear *et al.*, 1978),

Disaccharide conjugates, i.e. glucosylarabinose, glucosylxylose (3-phenoxybenzoic acid (**41.2**)) in free form and after acetylation (More *et al.*, 1978),

Glycine conjugates (toclofos-methyl (**63**)) in free form (Mihara *et al.*, 1981b),

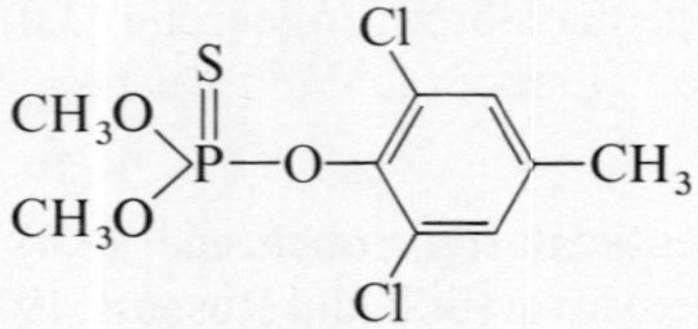

(63)

Glycylvaline conjugates (3-phenoxybenzoic acid (**41.2**)) in free form and after methylation (Huckle *et al.*, 1981c),

Taurine conjugates (cypermethrin (**41.0**), 3-phenoxy benzoic acid (**41.2**)) in free form and after methylation (Hutson and Casida, 1978).

Sulphuric acid and β-D-glucuronic acid conjugates are rarely isolated by preparative t.l.c. As more or less selective hydrolases are commercially available cleavage of the conjugates followed by identification of the exocons has been very common. Furthermore, sulphuric acid conjugates are usually rather labile under the acidic conditions commonly used for the chromatography of acidic conjugates, so that a high portion of the original conjugate may be lost during the process. Convenient derivatization techniques to maintain the integrity of sulphuric acid conjugates are not available. However, h.p.l.c. renders a powerful, fast and convenient alternative for the isolation of both types of conjugates.

Quistad *et al.* (1978a; b) utilized $AgNO_3$–impregnated silica plates for the separation of their ω-cyclopropyl fatty acids (**21.1**, **21.2**) as methyl ester derived from plant and animal tissue after treatment with cycloprate (**21.0**). This method, well established in lipid biochemistry, allows the separation of fatty acids according to their number of double bonds, as $AgNO_3$ forms π-complexes with unsaturated compounds and thus effecting selective retention on t.l.c. plates.

A few examples in which reversed phase t.l.c. systems have been used for preparative purposes have been published. Renner *et al.* (1978) separated pentachlorobenzene (**3.3**), 1,2,3,4- and 1,2,4,5-tetrachlorobenzene (**3.13**) on silica plates previously impregnated with paraffin oil. The compounds were recovered from the plates by vacuum sublimation from the corresponding areas. Gross *et al.* (1979) used aluminium sheets coated with polyamide for the isolation of an acidic metabolite (**5.8**) of chlorotoluron (**5.0**) from wheat plants. Kraft *et al.* (1981) recommended polyamide layers for a combined t.l.c.–m.s. system. The separated compounds were localized, the corresponding areas removed by means of a spatula and the material directly introduced into the ion source of a mass spectrometer, without prior elution of the compound.

Crayford and Hutson (1980b) ultimately purified the two isomeric dipalmitoyl-(3-phenoxybenzoyl)glycerols (**41.4**, **41.5**), isolated from the skin of rats dosed with 3-phenoxybenzoic acid (**41.2**) on silanized silica plates.

Roberts and Wright (1981) separated on C_{18} silica plates disaccharide conjugates isolated from cotton leaves treated with 3-phenoxybenzyl alcohol (**41.1**). Two fractions were obtained, the analytical results of which rendered evidence for the structure of a glucosylarabinose (**41.7**) and a glucosylxylose (**41.8**) conjugate, respectively.

Paper chromatography was and is still used for the isolation of metabolites, especially polar ones including conjugates, e.g. urinary rabbit metabolites of dinoseb acetate (**64.0**) including a glucuronide (**64.1**) (Ernst, 1969), urinary rat

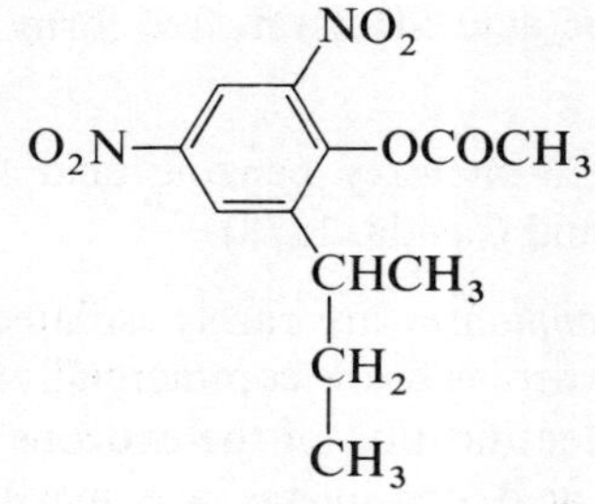

(64.0)

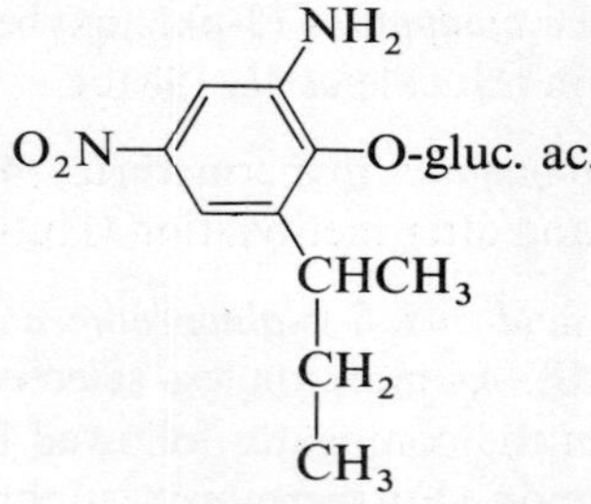

(64.1)

metabolites of dichlorvos (**65.0**) including a glycine (**65.1**) and a glucuronic acid (**65.2**) conjugate (Hutson *et al.*, 1971), urinary and biliary rat metabolites of cyanazine (**66.0**) including a mercapturic acid (**66.1**) and two glutathione conjugates (**66.2, 66.3**) (Crayford and Hutson, 1972), faecal rat metabolites of cyprazine (**67.0**), including polar 2-hydroxy-*s*-triazines (**67.7, 67.8**) (Larsen and Bakke, 1975), urinary and faecal rat and goat metabolites of terbutryn (**29.0**), including several *S*- and *O*-glucuronides (**29.5–29.9**) (Larsen and Bakke, 1978), two rat urinary mercapturic acid conjugates (**68.1, 68.2**) of cyanatryn (**68.0**) (Crawford *et al.*, 1980), urinary rat metabolites of propachlor (**12.0**) including several glucuronides (**12.7–12.12**) (Bakke and Price, 1979), metabolites from soybean and wheat cell suspension cultures of 2,4-D (**45**), including numerous amino acid conjugates (Scheel and Sandermann, 1981).

Roberts and Wright (1981) used chromatography on paper pretreated with sodium acetate solution for the separation of sugars obtained after chemical hydrolysis of the 3-phenoxybenzyl alcohol disaccharide conjugates (**41.7, 41.8**).

$CCl_2{=}CHOP(O)(OCH_3)_2$

(65.0)

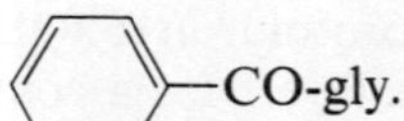

(65.1)

$CHCl_2CH_2$—O-gluc. ac.

(65.2)

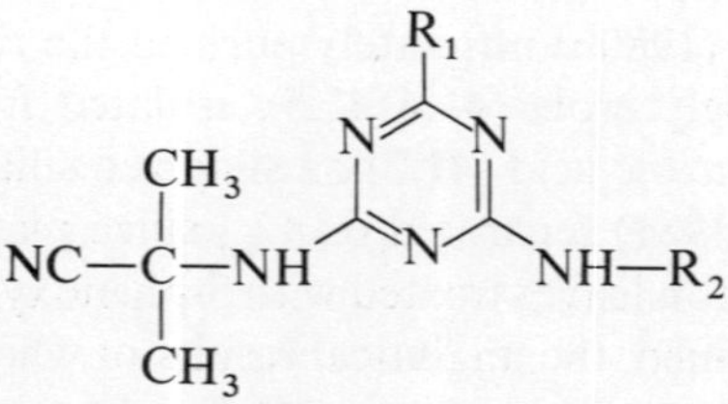

R_1	R_2	
—Cl	$—CH_2CH_3$	(66.0)
—N-acet. cyst.	$—CH_2CH_3$	(66.1)
-glutath.	$—CH_2CH_3$	(66.2)
-glutath.	—H	(66.3)

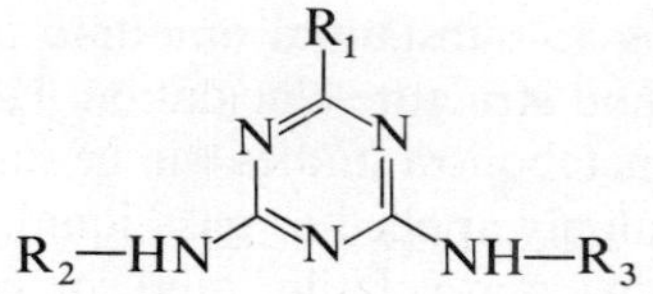

R_1	R_2	R_3	
—Cl	—(cyclopropyl)	$-CH(CH_3)_2$	**(67.0)**
—Cl	—(cyclopropyl)	$-CH(CH_3)CH_2OH$	**(67.1)**
—Cl	—(cyclopropyl)	$-CH(CH_3)COOH$	**(67.2)**
—Cl	—(cyclopropyl)	—H	**(67.3)**
—Cl	—H	$-CH(CH_3)_2$	**(67.4)**
—Cl	—H	$-CH(CH_3)CH_2OH$	**(67.5)**
—Cl	—H	—H	**(67.6)**
—OH	—H	$-CH(CH_3)_2$	**(67.7)**
—OH	—H	—H	**(67.8)**

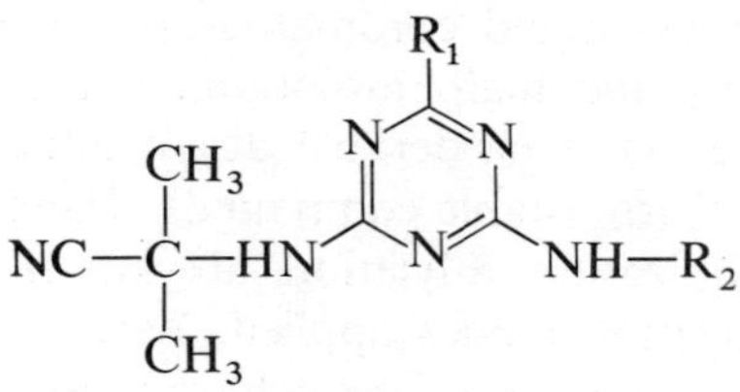

R_1	R_2	
$-SCH_3$	$-CH_2CH_3$	**(68.0)**
-*N*-acet. cyst.	$-CH_2CH_3$	**(68.1)**
-*N*-acet. cyst.	—H	**(68.2)**

Gas chromatography

Gas chromatography has been used for a long time as an almost indispensable tool in the qualitative and quantitative analysis of complex mixtures. Combining a gas chromatograph with a mass spectrometer has extended an analytical high-resolution method to a system which affords structural elucidation of unknown compounds. Although this technology has considerable limitations, its application in biochemistry and xenobiotic metabolism has greatly increased. A comprehensive series on the analysis of drugs and metabolites with g.c.–m.s. is being edited by Gudzinowicz and Gudzinowicz (commenced 1977); seven of the nine volumes planned have already been published. G.c.–m.s. has also received much attention in pesticide metabolism studies. Nevertheless, this technique will not be covered, because this chapter is dedicated to preparative isolation methods leading to substantial amounts of a pure metabolite for subsequent multiple method structure elucidation. Hence only the use of preparative g.c. in pesticide metabolism studies will be surveyed.

This technique is commonly applied as gas–liquid chromatography, i.e. the stationary phase consists of a non-volatile liquid coated on an inert carrier with large surface areas and uniform particle size, or on the inner wall of a narrow-bore capillary. The sample is injected into the gas chromatograph, enters the injection port, is volatilized at elevated temperature, passes into a stream of an inert gas and is carried over the stationary phase. The mobile phase (the gas stream) leaving the chromatographic column is fed completely (with non-destructive detectors) or partially, after splitting, (with destructive detectors) into a detecting device. The principal retention mechanism is partition between the liquid and the gaseous phase. The separation is controlled by the nature, surface area and the load of the stationary phase on the support and by the temperature of the system.

The application of g.c. in drug metabolism has been recently reviewed by VandenHeuvel and Zacchei (1978) and by Coutts and Jones (1980). Preparative aspects have been especially covered by Zlatkis and Pretorius (1971). Generally, preparative gas–liquid chromatography is characterized by high capacity, high selectivity and high resolution. In contrast to these positive aspects, preparative g.c. has considerable disadvantages. The application is limited to volatile and thermostable compounds. These features can be introduced by proper derivatization. Silylation, alkylation and acylation are the most common derivatization modes applied. Various recent comprehensive reviews of derivatization techniques for g.c. have been published, e.g. Blau and King (1977); Nicholson (1978); Pierce (1979; 1981) and Drozd (1981). The state of the art regarding gas chromatography and derivatization of glucuronides were summarized by Bakke (1976). Fenselau and Johnson (1980) reviewed recent derivatization techniques for intact glucuronides for mass spectroscopy and g.c.–m.s. Tomasic (1978) discussed g.c. of glucuronic acid conjugates and published a list of g.c. data of various derivatives. Even when adequate thermostability and volatility have been accomplished, the derivatizing groups may have to be removed after g.c. for identification and/or further isolation procedures. This complicates the whole process and makes it rather laborious. Furthermore, the trapping of μg amounts of a compound usually present as an aerosol from a rather fast gas stream causes a technical problem which can often not be solved satisfactorily and low collection efficiencies may result. Differently shaped collection tubes immersed in coldbaths, and occasionally adsorption on solids, e.g. resins, are commonly used trapping methods.

Preparative g.c. has been used for enzymatically derived exocons of chlorpropham (**43.0**) in plants and animals (Still and Mansager, 1972; 1973; 1975). Four metabolites, 2- and 4-hydroxy chlorpropham (**43.1**, **43.2**) and the two corresponding anilines (**43.3**, **43.4**) were purified on OV 1 columns for spectroscopic structure elucidation. All exocons were chromatographed as acetyl derivatives; the carbanilates (**43.1**, **43.2**) also separated as trimethylsilyl derivatives.

Paulson *et al.* (1972) investigated the fate of the analogue propham (**27.0**) in chicken. One free metabolite (**27.2**) as its methylated and acetylated derivative, and several others derived from conjugates after chemical or enzymatic cleavage (**27.1–27.4**) followed by acetylation, methylation or both were purified on OV 1 or on Dexsil columns. Larsen and Bakke (1975) investigated the metabolism of cyprazine (**67.0**) in rats and ultimately purified all metabolites identified (**67.1–67.8**) as the corresponding trimethyl silyl derivatives on SE-30 columns prior to mass spectroscopy. Larsen *et al.* (1978) preparatively gas chromatographed numerous metabolites, including five glucuronic acid conjugates (**29.5–29.9**) from the excreta of rats and goats orally dosed with terbutryn (**29.0**). The conjugates were purified on SE-30 columns in the form of their trimethylsilyl derivatives and analysed by m.s. With another portion, the trimethylsilyl groups were removed by hydrolysis with aqueous methanol, the conjugates cleaved with β-glucuronidase and the exocons again trimethylsilylated for structure confirmation by m.s.

The β-D-glucoside (**9.1**) of maleic hydrazide (**9.0**) isolated from tobacco plants was finally purified as its trimethylsilyl derivative on an OV 1 column (Frear and Swanson, 1978). As the derivatization appeared incomplete, the derivative was methanolysed to yield the free conjugate, which in turn was acetylated for m.s. analysis.

Shimabukuro *et al.* (1979) purified non-polar metabolites and some not defined derivatized samples, including a methylated hydroxy derivative (**38.1**) of diclofop-methyl (**38.0**) derived from wheat and wild oat on OV 1 columns. The position of the hydroxy group in the 2,4-dichlorophenyl ring was not determined.

Various chloroform-soluble metabolites from pentachloronitrobenzene (**3.0**) in peanuts, including a cysteine conjugate (**3.8**) were purified on a SP-2100 column (Rusness and Lamoureux, 1980). The non-polar metabolites (**3.1**, **3.5**, **3.6**) were chromatographed in the genuine form, (**3.4**) as acetyl, (**3.7**) as methyl or butyl and (**3.8**) as *N*-trifluoroacetyl methyl derivative.

Preparative g.c. was used for the separation of pentachloromethylthiobenzene (**28.0**) metabolites from rat excreta, but details were not given (Bakke *et al.*, 1981a).

The method was readily utilized in the study of propachlor (**12.0**) metabolism in rats (Bakke and Price, 1979; Larsen and Bakke, 1981). SE-30 columns were used to separate the metabolites, including several glucuronides (**12.7**, **12.9**, **12.11**, **12.8** or **12.10**) and a mercapturic acid conjugate (**12.14**). The glucuronides and the mercapturates were chromatographed as trimethylsilyl derivatives, as were all the non-conjugated metabolites (**12.1**, **12.2**, **12.4**, **12.5**). One metabolite (**12.6**) was acetylated. It was assumed, that the *S*-acetyl derivative obtained was an artefact of the derivatization procedure and that the metabolite was an unknown *S*-conjugate.

From the various metabolites of diflubenzuron (**69.0**) identified in cow and sheep excreta, including milk and bile, a urinary glycine conjugate (**69.1**) was purified as methyl ester in large amounts (about 10 mg) on a SE-30 column (Ivie, 1978).

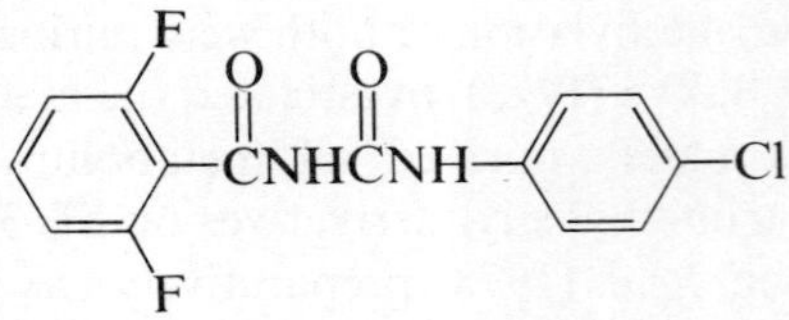

(69.0)

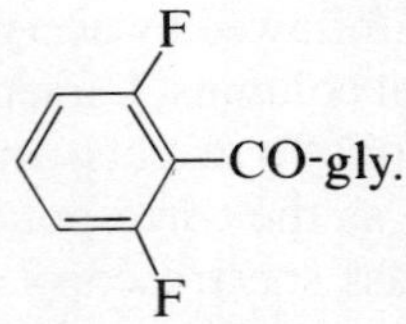

(69.1)

Other techniques

In addition to the major isolation techniques discussed in the previous sections, there are some which have been only occasionally applied to pesticide metabolites.

Counter-current distribution (c.c.d.) is a support-free, multiple liquid–liquid partition method, the commercial instruments allowing up to several hundred distinct partitioning steps. The separation of the sample components depends on the differences between their partition coefficients in the solvent system selected and the number of partition steps applied. The method is applicable to polar, as well as to non-polar metabolites by the proper selection of an adequate pair of immiscible solvents. C.c.d. is characterized by a high capacity and a high selectivity, but low resolution. Being a closed and support-free system, the method is free of stress to the sample, which is generally recovered quantitatively. However, c.c.d. is very cumbersome, the apparatus is expensive and complicated, the separation process is sometimes disturbed by unexpected formation of long-lasting emulsions, large solvents volumes have to be handled and, as a consequence, the method is rather slow. The fact, that the result of the process cannot be continuously monitored with a detector is an additional time-consuming disadvantage. Theoretical and practical aspects of c.c.d. have been described by Titus (1971) and Prochazka (1979). Considering these above-mentioned characteristics of the methods, it is not surprising that the few examples of its application in pesticide metabolism are in the earlier literature.

Esser *et al.* (1968) purified and separated two major urinary rat metabolites (**26.3**, **26.4**) of methidathion (**26.0**) using a 60-step apparatus and applying an ethyl acetate-water system. With a 30-step apparatus Shuntner and Thompson (1966) separated diazoxon (**70.1**) from diazinon (**70.0**) from cattle tick larvae applying methanol/water (80/20 v/v) as lower and toluene/*n*-hexane (2/1 v/v) as upper phase.

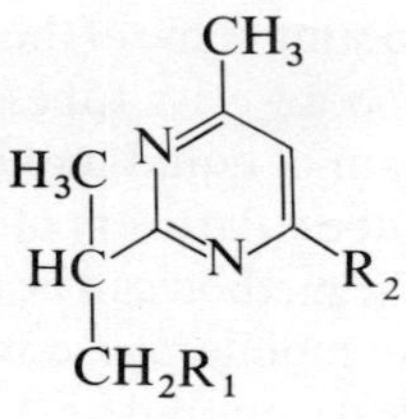

R_1	R_2	
—H	—OP(=S)(OCH$_2$CH$_3$)$_2$	**(70.0)**
—H	—OP(=O)(OCH$_2$CH$_3$)$_2$	**(70.1)**
—OH	—OH	**(70.2)**

Muecke *et al.* (1970) purified one of the urinary rat metabolites of diazinon (**70.2**) on a 80-step system with ethyl acetate-methanol/water (30/70 v/v). Tomasic (1978) listed examples of various glucuronides mainly of steroids and drugs, which had been purified or isolated by c.c.d. But here too, the references cited cease in the early 1970s.

Lin *et al.* (1977) reported the separation of albuterol (**71.0**), a *β*-adrenergic bronchodilator, and its sulphuric acid conjugate (**71.1**) from human urine in 250 steps using *n*-butanol-potassium phosphate buffer (0.1 M, pH 10).

Assandri and Perazzi (1974) studied with model compounds the separation of phenolic *O*-glucuronides and phenolic sulphates by c.c.d. recommending as phases *n*-propanol/*n*-butanol/ethyl acetate (1/1/3 v/v/v)-sodium phosphate buffer (0.01 M, pH 6.8). In their studies, instead of a conventional, discontinuous apparatus according to Craig, a continuous flow counter-current system was used. This technique, named droplet counter-current chromatography

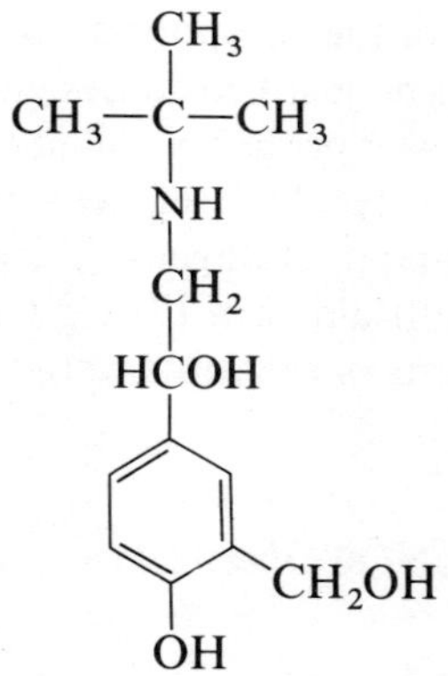

(71.0)

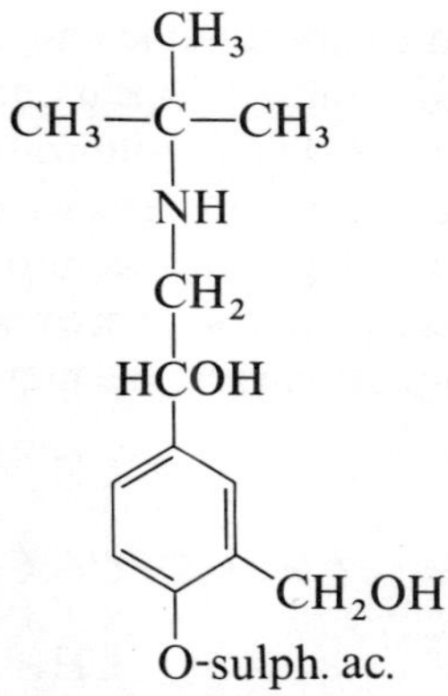

(71.1)

(d.c.c.c.), is a modification eliminating some of the disadvantages of c.c.d. Several vertical interconnected narrow bore glass tubes are filled with the stationary liquid phase. A continuous stream of liquid mobile phase is, in form of discrete droplets, pumped through the tubes. Partition of the solutes occurs between the two immiscible liquid phases. The method can be employed in the ascending and in the descending mode, i.e. the mobile phase may be lighter or heavier than the stationary phase. Comparing d.c.c.c. and c.c.d. the resolution is improved, the separation time is reduced, the instrument is easier and more convenient to operate, the solvent volumes and the bench space required are smaller, and the effluent can be continuously monitored. Meanwhile d.c.c.c. apparatus has become commercially available from different companies, but no application of this method for pesticide metabolite isolation has yet been published. However, it is assumed, that this method has potential and its application is expected in the future, especially for crude samples of polar character, e.g. conjugates. A comprehensive paper on this method including its application for the isolation of natural products, e.g. glycosides, has been published by Hostettmann (1980).

In some rare cases high voltage electrophoresis (h.v.e.) has been utilized for the isolation of charged pesticide metabolites, e.g. Paulson *et al.* (1972) purified a glucuronide (**27.9**) of propham (**27.0**) on paper using an acidic pyridine acetate buffer. The metabolite was eluted from the support with methanol. Although it would generally appear, that for acidic and amphoteric conjugates, e.g. glucuronic acid, sulphuric acid, mercapturic acid and amino acid derivatives, preparative h.v.e. represents a useful tool for penultimate purification (Hutson, 1976), it obviously lacks in resolution power. As alternative and superior methods for this purpose, e.g. h.p.l.c. are readily available, preparative h.v.e. remains restricted to a few very selected problems.

In some studies, where several mg of a particular metabolite had been isolated, recrystallization was carried out as the ultimate purification step (e.g. Esser *et al.*, 1968; Ernst, 1969; Muecke *et al.*, 1970; Ivie, 1978; Kojima *et al.*, 1980). Although being a very effective method, it will not be further considered here, as it is restricted to appreciable amounts of the compounds, which are usually not available. Generally it appears not to be justified to design an experiment so as to be able to apply recrystallization, as it probably will not be economical. There are some exceptions, e.g. when the labelled segment of a pesticide is incorporated to a high degree into a major endogenous compound. Field grown sugar beet plants were treated with aldicarb (**72**), ^{14}C-labelled in the methylthio group. At harvest the ingredients of the roots were fractionated and

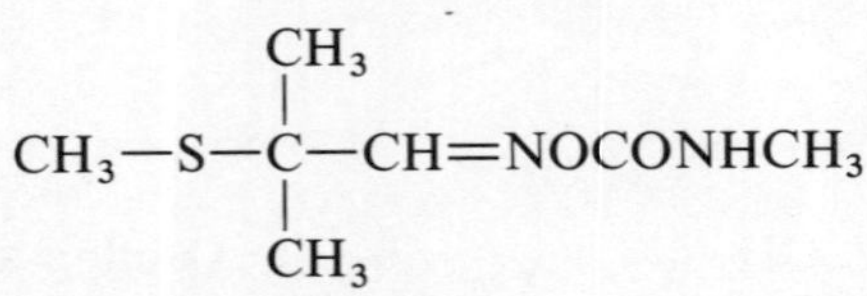

(72)

the isolated labelled sugar was recrystallized several times (Rouchaud *et al.*, 1981).

Kenndler and Kaniansky (1981) separated paraquat (**73**) and diquat (**74**) by isotachophoresis from the typical cations present in plant extracts and thereafter identified the compounds by m.s. as their chlorides.

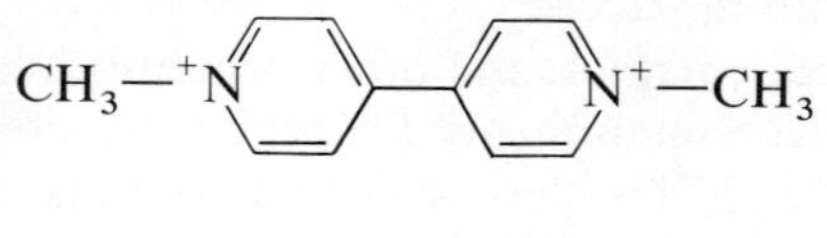

(**73**)

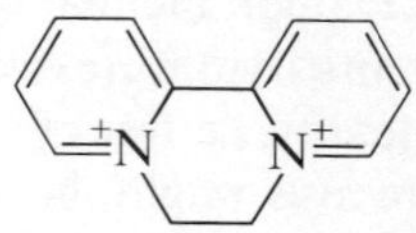

(**74**)

PITFALLS IN ISOLATION AND PURIFICATION PROCEDURES

The matrices from which metabolites have to be isolated, e.g. animal excreta, plant and animal tissue and soil, are complex structures containing numerous ingredients with a wide variety of physico-chemical properties. In all biotic or non-biotic systems being investigated for their potential to degrade pesticides, a wide array of physical, chemical and biochemical transformation principles are operative leading usually to an appreciable number of different transformation products. Occasionally these products are still chemically reactive being thus prone to artefact formation upon access to physical or chemical reactants. The complexity of the whole system and the disproportionate ratio between the amount of metabolites to that of the matrix, makes the control of artefact formation, especially in the early stages of the isolation process, rather difficult. Sometimes adequate analytical techniques (e.g. t.l.c., h.p.l.c.) can only be meaningfully applied after the metabolites have been released from the matrix and pre-purified to some degree. After analytical methods are applied to determine the (hopefully still genuine) metabolite pattern, qualitative and quantitative control after each isolation, purification or derivatization procedure is indispensable to ensure the entirety of the metabolites finally identified. During the isolation and purification of metabolites various common and uncommon pitfalls have been experienced, some of which will be discussed below. A very systematic paper on pitfalls in drug metabolism methodology, including pitfalls in the design of experiments and in structure elucidation has been published by Beckett and Cowan (1978).

Sampling is by definition an act by which the process under study is stopped. This necessarily implies that the sample has to be worked up immediately or stored under conditions suppressing any further reaction mediated by physical (e.g. light), chemical (e.g. oxygen) or biological (e.g. microbes) means. Storage of samples at low temperatures, in the dark, preventing the access of oxygen might be adequate. Antimicrobial agents are sometimes added to prevent bacterial growth. However, employing these precautions may help to maintain, but do not ensure integrity of the sample.

The sampling process itself may be important, e.g. a deficient separation device in a rat metabolism study has caused formation of an artefact. A urinary sulphoxide metabolite (**26.3**) of methidathion (**26.0**) was in contact with faeces reduced to the corresponding sulphide (**26.2**) (Esser *et al.*, 1968). Renner (1980) reported the reduction of pentachloronitrobenzene (**3.0**) to pentachloroaminobenzene (**3.1**) upon incubation with rat faeces.

The solvents used for extraction, partition and other procedures need to be of defined quality, i.e. free of reactive contaminants. Proper purification immediately before use might be advisable for very sensitive solvents. Adequate purification procedures are available in the literature on the subject. The problems caused by ether containing peroxides are well-known and documented, i.e. oxidation of sulphenyl and sulphinyl derivatives, primary and secondary amines and others (e.g. Beckett and Cowan, 1978; Gemborys and Mudge, 1981). Using dichloromethane, containing cyanogen chloride, as extraction solvent gave rise to artefacts by reaction with primary and secondary amines, e.g. 4-benzamidopiperidine (**75**) was transformed to 4-benzamido-1-cyanopiperidine (**76**). Analysis of a number of different batches of dichloromethane from various suppliers revealed that cyanogen chloride was present in most of the samples (Franklin *et al.*, 1978). Apparently this reaction had led to the discovery of a 'new salivary metabolite' (**78**) of morpholine (**77**) (Wishnok and Tannenbaum, 1977).

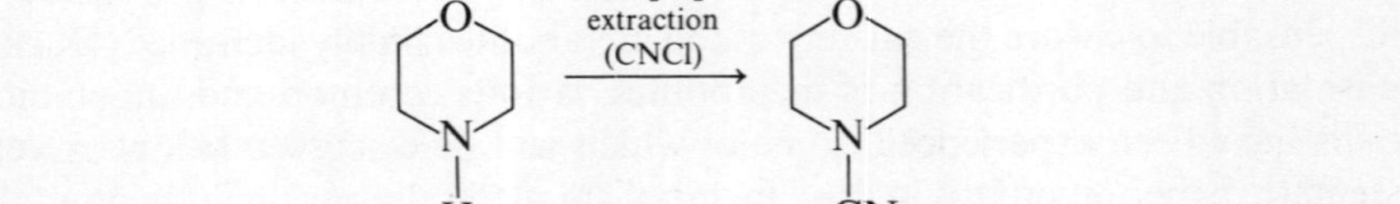

(**75**) (**76**)

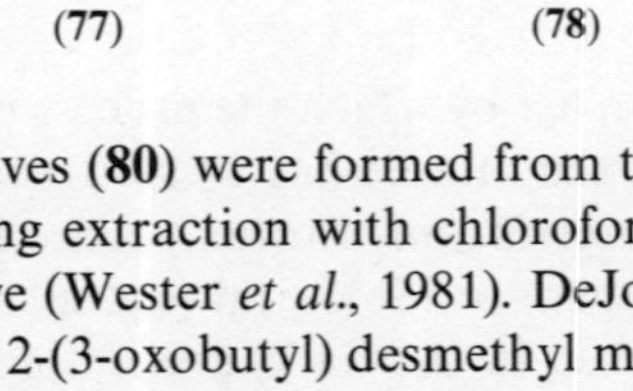

(**77**) (**78**)

Carbamate derivatives (**80**) were formed from tricyclic anti-depressants, e.g. octriptyline (**79**) during extraction with chloroform containing phosgene and ethanol as preservative (Wester *et al.*, 1981). DeJongh *et al.* (1981) claimed an artificial formation of 2-(3-oxobutyl) desmethyl mianserin (**82**) from desmethyl mianserin (**81**) extracting urine at pH 8 or above with ethyl acetate. No artefact was formed when ethyl acetate was replaced by *n*-hexane, dichloromethane, dichloroethane, isopropanol, chloroform or methyl acetate.

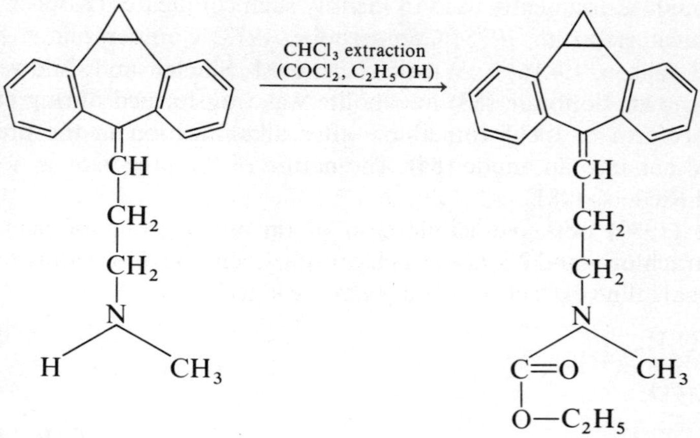

(79) (80)

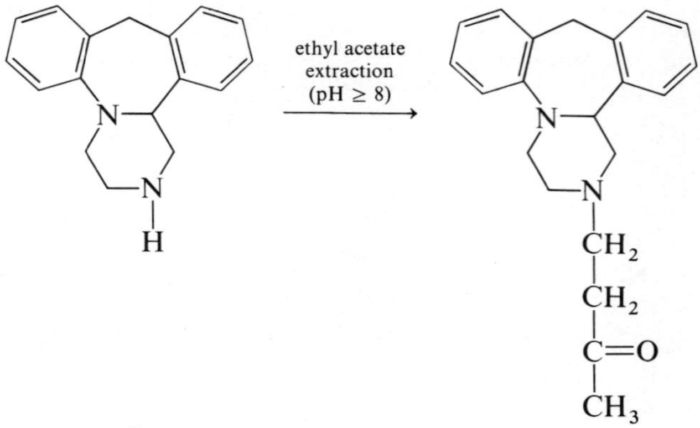

(81) (82)

The loss of volatile metabolites during evaporation of extracts and other procedures are occasionally observed (e.g. Son *et al.*, 1980), but may be reduced by applying adequate equipment.

Special precautions have to be taken, when the pH of the sample is changed in order to effect ion suppression for efficient extraction or other reasons as transformation by hydrolysis and other mechanisms may readily occur (e.g. Shono *et al.*, 1978; Glickman *et al.*, 1981). Especially labile conjugates, e.g. sulphuric acid, glucuronic acid and glucose conjugates may be cleaved under acidic and alkaline conditions (Frear, 1976; Paulson, 1976; Frear *et al.*, 1978; Kadlubar *et al.*, 1978; Tomasic, 1978; Pekas, 1979; Hornke *et al.*, 1980).

Additionally, at pH above seven, ester glucuronides may undergo rearrangement, i.e. migration of the exocon to the adjacent hydroxy group of the glucuronic acid, leading to the 2-, 3- and 4-*O*-isomers which are not substrates for β-glucuronidase, frequently used to identify such conjugates (Kopcewicz *et al.*, 1974; Blanckaert *et al.*, 1978; Compernolle, 1978; Compernolle *et al.*, 1978; Illing and Wilson, 1981a; b; Wilson *et al.*, 1981; Sinclair and Caldwell, 1982).

An unknown clofibrate (**83**) metabolite was transformed during extraction with chloroform or dichloromethane after alkalinization in the presence of ammonia ions into an amide (**84**). The nature of the precursor is not known (Siek and Rieders, 1981).

Renner (1980) observed elimination of the nitro group of pentachloro-, 2,3,5,6-tetrachloro- and 2,3,4,6-tetrachloro-nitrobenzene (**85**) leading to phenols (**86**) upon alkaline extraction of their hexane solution.

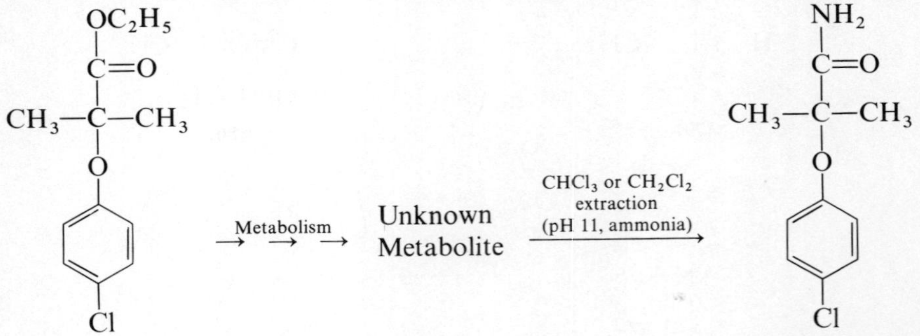

(**83**) (**84**)

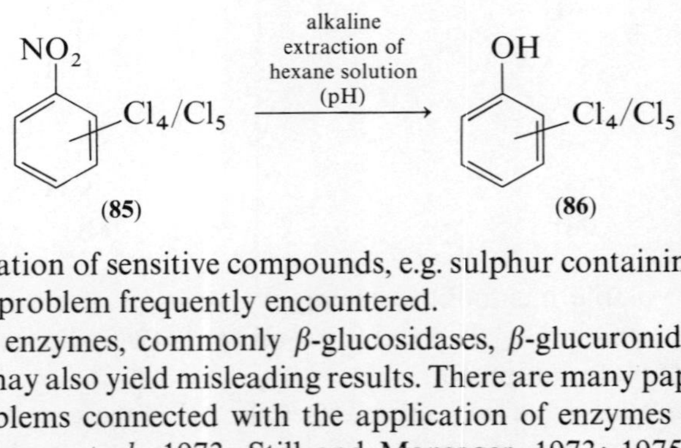

(**85**) (**86**)

Aerial oxidation of sensitive compounds, e.g. sulphur containing metabolites, is a common problem frequently encountered.

The use of enzymes, commonly β-glucosidases, β-glucuronidases and aryl-sulphatases, may also yield misleading results. There are many papers published, in which problems connected with the application of enzymes have been described (Hodgson *et al.*, 1973; Still and Mansager, 1973; 1975; Frear, 1976; Paulson, 1976; Lin *et al.*, 1977; Dorough, 1979; Hutson *et al.*, 1981; Huckle *et al.*, 1981a; b). Most frequent problems are incomplete cleavage and 'false positive' cleavage. The latter, mainly due to contaminated enzyme preparations, can be overcome to some extent by proper control experiments as outlined above (section: characterization). The former cases may be mainly explained by different substrate specificities of enzymes from different biological sources and by enzyme inhibitors present in the sample. *C*-, *S*- and *N*-Glucuronides may not all be suitable substrates for β-glucuronidases (Dieterle *et al.*, 1975;

Tomasic, 1978; Heirwegh and Compernolle, 1979). Also *S*-glucosides and *O*-(6-*O*-malonyl) glucosides may resist enzymatic hydrolysis by β-glucosidase (Heirwegh and Compernolle, 1979; Frear, 1976, cited by Suzuki and Casida, 1981). Susceptibility of substrates may depend on stereoselectivity of the enzyme (Ruelins *et al.*, 1979). Regarding ester glucuronides, alkaline conditions may lead to sequential acyl migration, the products formed being resistant to enzymatic hydrolysis. Enzyme cleavage of stable conjugates leads occasionally to labile exocons which in turn give rise to artefact formation. Without special protection during the cleavage procedure with β-glucuronidase, *S*-glucuronide (**88**) exocons may be obtained as their disulphide derivatives (**89**) by spontaneous oxidation, e.g. with dimethametryn (**87**) (Donzel and Muecke, 1978).

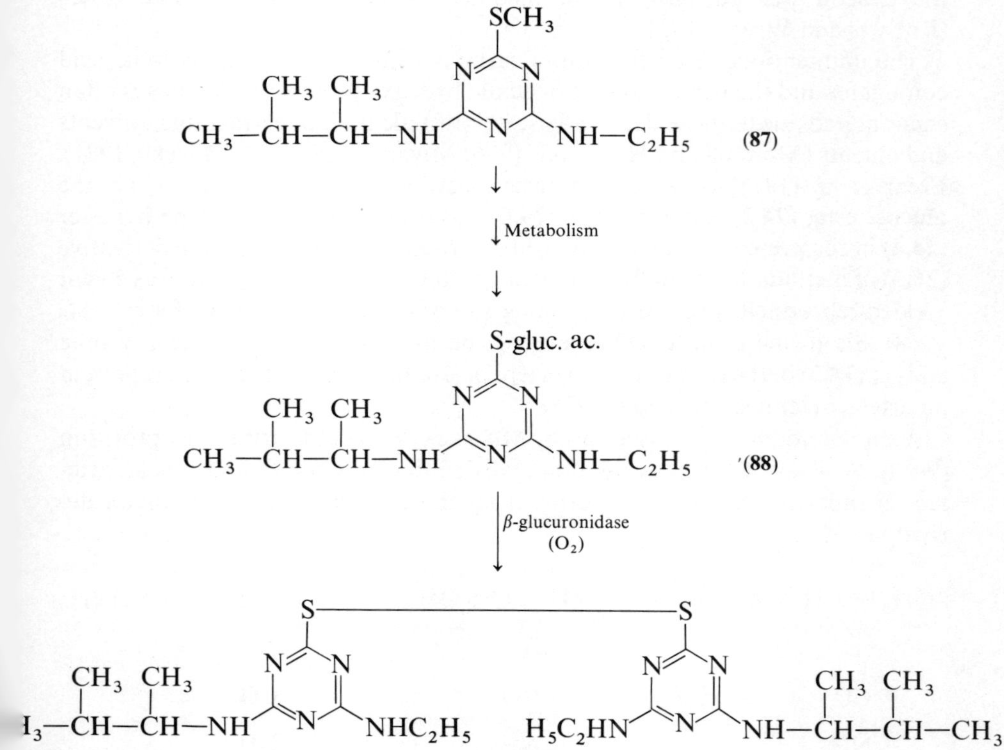

(**87**) (**88**) (**89**)

Even when conjugates can be completely and selectively cleaved by enzymes an additional problem may arise. The exocons released may be multifunctional, so that a decision on the structure of the original conjugate will be difficult or even impossible (Fenselau and Johnson, 1980; Kirkpatrick *et al.*, 1981). This fact adds another credit to the rationale for the isolation and identification of whole conjugates as far as is feasible.

The frequently applied chemical hydrolysis of conjugates is even more prone to artefact formation than is enzymatic cleavage; the usually drastic conditions may cleave the conjugate and degrade the exocon and endocon as well (Heirwegh and Compernolle, 1979; Paulson and Portnoy, 1970). The *N*-methyl moiety of diphenamid (**42.0**) is metabolically oxidized to the *N*-methylol derivative followed by conjugation (**42.1**, **42.2**). Upon hydrolysis the exocon lost formaldehyde being transformed to a primary amine (Hodgson *et al.*, 1973; Frear, 1976). *N*-Methylol derivatives (of various urea herbicides) decomposed readily upon t.l.c. at room temperature, but could be chromatographed at 0–5 °C (Suzuki and Casida, 1981). Upon prolonged hydrolysis of the disaccharide conjugates (**41.7**, **41.8**) of 3-phenoxybenzyl alcohol (**41.1**) with 4 M HCl the exocon was partially transformed to 3-phenoxybenzyl chloride (**41.3**) (Roberts and Wright, 1981).

The unintentional esterification of carboxy acids, including glucuronic acid conjugates and the derivatization of acidic hydroxy groups by alcohols is often encountered due to the widespread use of lower alcohols as extractants, solvents and eluents (Mumma and Hamilton, 1976; Miyano *et al.*, 1981, Muecke, 1981). Frear *et al.* (1978) observed transesterification on Florisil columns, i.e. the glucose ester (**24.2**) of chloramben (**24.0**) was transformed into the methyl ester (**24.4**) in the presence of methanol. Chromatography of a sulphoxide derivative (**26.3**) of methidathion (**26.0**) on a silica gel column using ethyl acetate as eluent yielded reproducibly the corresponding sulphone derivative (**26.4**) (Esser *et al.*, 1968). Sisenwine *et al.* (1972) reported the discovery of a new urinary urea adduct (**18.2**) of the tranquillizer oxazepam (**18.0**), which is believed to represent an artefact (Jenner and Testa, 1978).

A similar adduct (**92**) was found by Muecke (1975) in the urine of isoproturon (**90**) treated rats. Careful evaluation revealed that this compound was an artefact chemically formed in the urine from the corresponding *N*-methylol derivative (**91**).

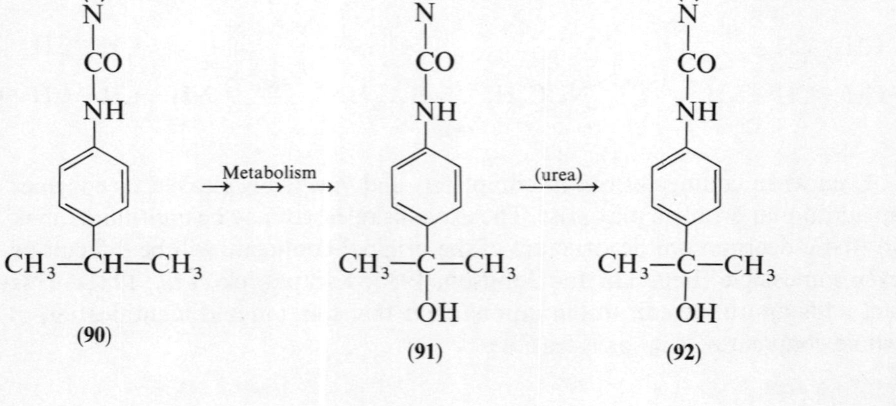

(**90**) (**91**) (**92**)

The rather drastic conditions previously used in ion-exchange chromatography often gave rise to hydrolytic artefacts, e.g. Larsen and Bakke (1975) assumed that some of the 2-hydroxy-*s*-triazine derivatives (**94**) from cyprazine (**67.0**) were formed by hydrolysis of the corresponding 2-chloro-*s*-triazines (**93**).

Also during derivatization procedures unexpected reactions may occur which may lead to an equivocal or incorrect interpretation of the metabolite structure. Treatment of glucuronides with diazomethane may lead not only to methylation of the carboxy group, but also to a partial methylation of the glucuronic acid hydroxy groups (Knaak *et al.*, 1967; Muecke, 1981). Even destruction of glucuronic acid conjugates by diazomethane has been reported (Huckle *et al.*, 1981c). Furthermore, diazomethane may, upon prolonged treatment at room temperature, react with an aliphatic double bond (**97**) as

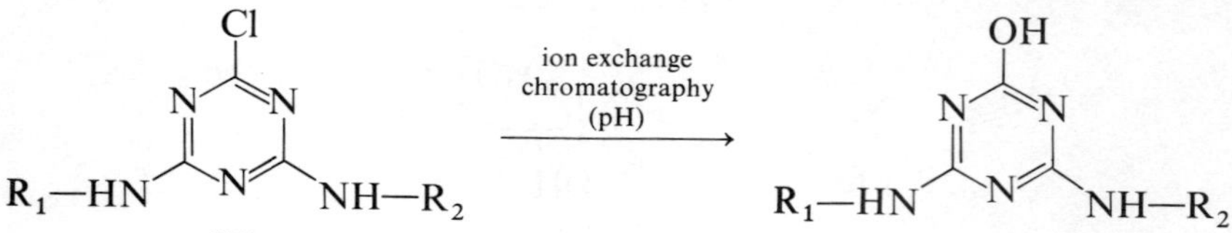

(**93**) (**94**)

R_1 and R_2: various groups.

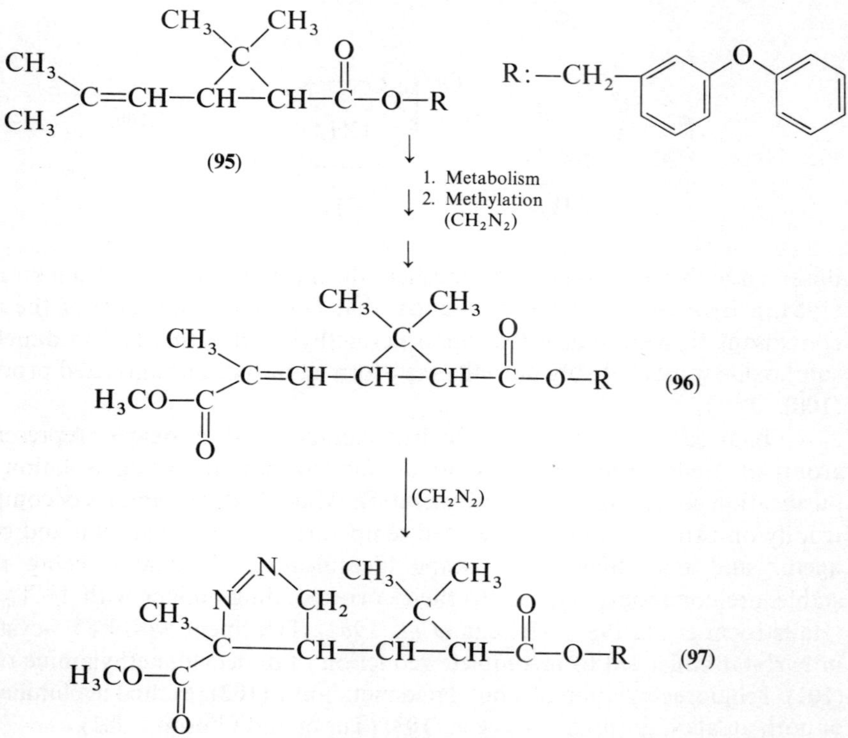

(**95**) (**96**) (**97**)

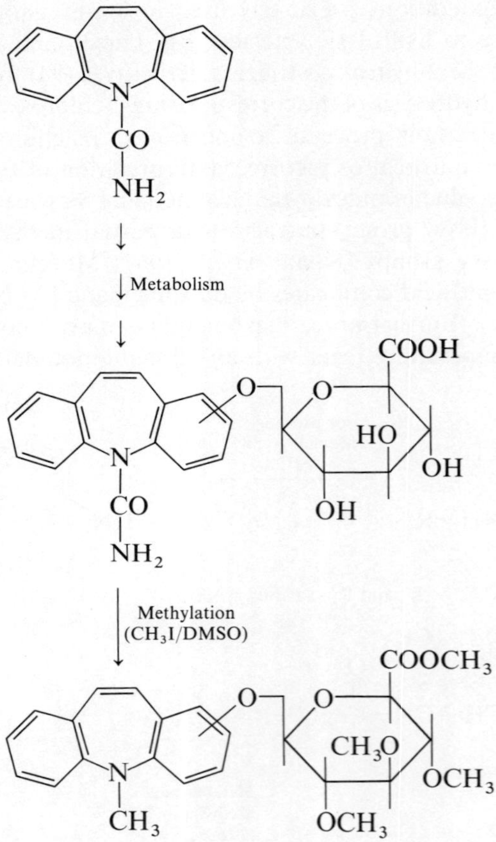

(98)

(99)

(100)

observed with a metabolite (**96**) of phenothrin (**95**) reported by Kaneko *et al.* (1981a). Lynn *et al.* (1978) mentioned, that *O*-glucuronides (**99**) of the anticonvulsant carbamazepine (**98**) upon permethylation with CH_3I in dimethylsulphoxide were probably partially transformed into decarbamylated products (**100**).

N-Oxidized metabolites, e.g. hydroxylamines and *N*-oxides, represent a group of labile products prone to artefact formation during isolation and purification as well as during derivatization. Many hydroxylamines decompose readily on exposure to light, elevated temperature, aerial oxidation and other agents and may thus often escape identification. *N*-Oxides, being more stable, are commonly reduced to their corresponding amines with $TiCl_3$/HCl (Hallstroem *et al.*, 1981; McLean *et al.*, 1981). Treatment with Fe^{++}-cysteine in barbital buffer led to incomplete reduction of dodecyldimethylamine oxide (**101**). Trifluoracetylation of a butylated metabolite (**102**) resulted in elimination of both an alkyl group and oxygen (**103**) (Turan and Gibson, 1981).

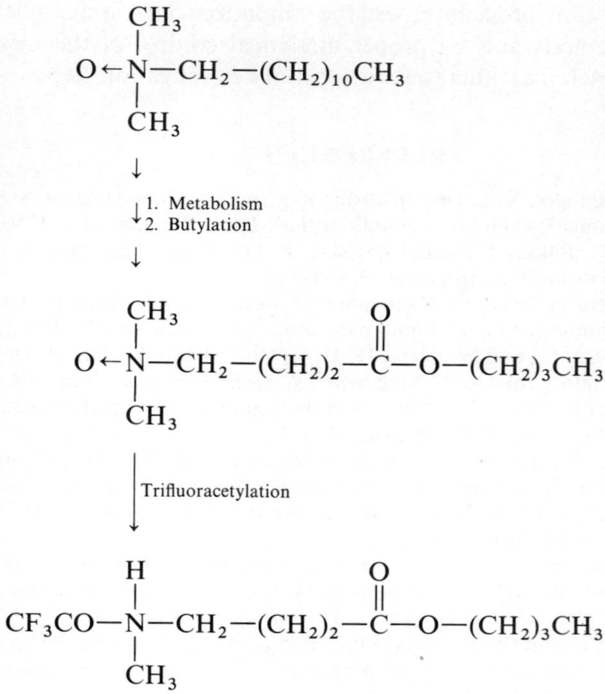

(101)

(102)

(103)

CONCLUSIONS

The separation and purification of polar pesticide metabolites, such as conjugates, still represent a challenge to the metabolism chemist. The ACS symposium held in Vail, 1975 (Kaufman *et al.*, 1976) presented the state of the art on bound and conjugated pesticide metabolites, including isolation methods, of the mid 1970s.

Since then, considerable progress has been made regarding the improvement of the methods for the isolation of polar metabolites. Noteworthy is the increasing utilization of h.p.l.c. with its various application modes for these purposes. However, early methods, especially conventional column chromatography and preparative thin-layer chromatography maintain their value as efficient and convenient tools.

Of the new separation techniques appearing on the horizon, only d.c.c.c. is considered to have a potential for the isolation of polar pesticide metabolites.

As a considerable array of suitable and efficient methods is already available, emphasis should be placed on the development of an adequate strategy based on metabolite characterization so as to accomplish the isolation goals in the most economical way.

Pitfalls in isolation procedures can be minimized by careful selection of materials and methods and by proper analytical control of the metabolites before and after each individual separation and derivatization step.

REFERENCES

Arndts, D., and Rominger, K. L. (1978). 'Isolierung von Metaboliten neuer Arzneimittel zur massenspektrometrischen Strukturaufklärung', *Drug. Res.*, **28**, 1951–1960.

Aschenbacher, P. W., Bakke, J. E., and Zaylskie, R. G. (1978). 'Mass spectra of aniline glucuronides', *Biomed. Mass Spectrom.*, **5**, 409–412.

Assandri, A., and Perazzi, A. (1974). 'Separation of phenolic *O*-glucuronides and phenolic sulphate esters by multiple liquid–liquid partition', *J. Chromatogr.*, **95**, 213–221.

Atallah, Y. H., Yu, Ch. C., and Whitacre, D. M. (1980). 'Metabolite fate of the herbicide buthidazole in lactating cows and laying hens', *J. Agric. Food Chem.*, **28**, 278–286.

Axness, M. E., and Fleeker, J. R. (1979). 'Metabolism of the butylcarbamoyl moiety of benomyl in rat', *Pestic. Biochem. Physiol.*, **11**, 1–12.

Bakke, J. E. (1976). 'Recent advances in the isolation and identification of glucuronide conjugates', in *Bound and Conjugated Pesticide Residues* (eds D. D. Kaufman, G. G. Still, G. D. Paulson, and S. K. Bandal), ACS Symposium Series 29, pp. 55–67, American Chemical Society, Washington, DC.

Bakke, J. E., Aschenbacher, P. W., Feil, V. J., and Gustafsson, B. E. (1981a). 'The metabolism of pentachloromethylthiobenzene in germ-free and conventional rats', *Xenobiotica*, **11**, 173–178.

Bakke, J. E., Feil, V. J., and Price, C. E. (1976). 'Replacement of a chlorine with a methylsulfonyl group in the metabolism of propachlor (2-chloro-*N*-isopropylacetanilide', *Biomed. Mass Spectrom.*, **3**, 226–229.

Bakke, J. E., Gustafsson, J. A., and Gustafsson, B. E. (1980). 'Metabolism of propachlor by the germfree rat', *Science*, **210**, 433–435.

Bakke, J. E., and Price, C. E. (1979). 'Metabolism of 2-chloro-*N*-isopropylacetanilide (propachlor) in the rat', *J. Environ. Sci. Health*, **14**, 427–441.

Bakke, J. E., Rafter, J., Larsen, G. L., Gustafsson, J. A., and Gustafsson, B. E. (1981c). 'Enterohepatic circulation of mercapturic acid and cysteine conjugates of propachlor', *Drug Metab. Dispos.*, **9**, 525–528.

Bakke, J. E., Rafter, J. J., Lindeskog, P., Feil, V. J., Gustafsson, J. A., and Gustafsson, B. E. (1981b). 'Metabolism of 2-acetamido-4-chloromethyl)-thiazole in germfree and conventional rats', *Biochem. Pharmacol.*, **30**, 1839–1844.

Baldwin, M. K., and Hutson, D. H. (1980). 'The metabolism of 3-chloro-4-fluoro-aniline in dog and rat', *Xenobiotica*, **10**, 135–144.

Barth, T., Tjessem, K., and Aaberg, A. (1981). 'Fractionation of polar organic constituents in environmental samples using the lipophilic dextran gels Sephadex LH 20 and Sephasorb HP ultrafine', *J. Chromatogr.*, **214**, 83–93.

Baunok, I., and Geissbühler, H. (1970). 'Specific determination of urea herbicide residues by EC gas chromatography after hydrolysis and iodine derivative formation', *Bull. Environ. Contam. Toxicol.*, **3**, 7–17.

Beckett, A. H., and Cowan, D. A. (1978). 'Pitfalls in drug-metabolism methodology', in *Drug Metabolism in Man* (eds J. W. Gorrod and A. H. Beckett), Taylor and Francis, London.

Begum, S., Scheunert, I., Haque, A., Klein, W., and Korte, F. (1979). 'Conversion of [^{14}C]pentachloronitrobenzene in onions', *Pestic. Biochem. Physiol.*, **11**, 189–200.

Belasco, I. J., and Baude, F. J. (1981). 'The metabolism of [^{14}C]cymoxanil in the rat', *Pestic. Sci.*, **12**, 27–36.

Blanckaert, N. T., Compernolle, F., Leroy, P., Van Houtte, R., Fevery, J., and Heirwegh, K. P. M. (1978). 'The fate of bilirubin-IXα glucuronide in cholestasis and during storage in-vitro', *Biochem. J.*, **171**, 203–214.

Blau, K., and King, G. S. (eds) (1977). *Handbook of Derivatives for Chromatography*, Heyden, London–Bellmawr–Rheine.

Bligh, E. G., and Dyer, W. J. (1959). 'A rapid method of total lipid extraction and purification', *Can. J. Biochem. Physiol.*, **37**, 911–917.

Bradbury, A., Powell, G. M., Curtis, C. G., and Rhodes, C. (1981). 'The enhanced biliary secretion of a taurine conjugate in the rat after intraduodenal administration of high doses of fenclozic acid', *Xenobiotica*, **11**, 665–674.

Brown, P. R., and Krstulovic, A. M. (1979). 'Review: Practical aspects of reversed-phase liquid chromatography applied to biochemical and biomedical research', *Anal. Biochem.*, **99**, 1–21.

Bruce, W. N. (1972). 'Separation and identification of carbofuran, its metabolites and conjugates found in fish exposed to ring ^{14}C-labelled carbofuran using ITLC silica gel strips', *J. Econ. Entomol.*, **65**, 1738–1740.

Bush, M. T. (1981). 'Design of solvent extraction methods', in *Methods in Enzymology* (ed. W. B. Jacoby), Vol. 77, pp. 353–372, Academic Press, New York.

Cervelli, S., Nannipieri, P., Giovannini, G., and Perna, A. (1975). 'Jack bean urease inhibition by substituted ureas', *Pestic. Biochem. Physiol.*, **5**, 221–225.

Chen, K.-Ch., and Dorough, H. W. (1979). 'Glutathione and mercapturic acid conjugations in the metabolism of naphthalene and 1-naphthyl-*N*-methylcarbamate (Carbaryl)', *Drug. Chem. Toxicol.*, **2**, 331–354.

Chin, B. H., Eldridge, J. M., Anderson, J. H., and Sullivan, L. J. (1979a). 'Carbaryl metabolism in the rat. A comparison of in-vivo, in-vitro (tissue explant), and liver perfusion technique', *J. Agric. Food Chem.*, **27**, 716–720.

Chin, B. H., Sullivan, L. J., Eldridge, J. M., and Tallant, M. J. (1979b). 'Metabolism of carbaryl by kidney, liver, and lung from human postembryonic fetal autopsy tissue', *Clin. Toxicol.*, **14**, 489–498.

Chin, B. H., Tallant, M. J., Duane, W. C., and Sullivan, L. J. (1980). 'Metabolism of carbamate insecticide thiofanox in rats', *J. Agric. Food Chem.*, **28**, 1085–1090.

Chopade, H. M., and Dauterman, W. C. (1981). 'Studies on the in-vitro metabolism of methidathion by rat and mouse liver', *Pestic. Biochem. Physiol.*, **15**, 105–119.

Chopade, H. M., Dauterman, W. C., and Simoneaux, B. J. (1981). 'Studies on residues and the metabolic pathway of methidathion in tomato fruit', *Pestic. Sci.*, **12**, 17–26.

Compernolle, F. (1978). 'Structure revision of disaccharidic conjugates of bilirubin-IXα in human bile and identification of phenylazo derivatives B_4, B_5 and B_6 as 2-, 3- and 4-O-acyl-glucuronides', *Biochem. J.*, **175**, 1095–1101.

Compernolle, F., Van Hees, G. P., Blanckaert, N., and Heirwegh, K. P. M. (1978). 'Glucuronic acid conjugates of bilirubin-IXα in normal bile compared with post-obstructive bile', *Biochem. J.*, **171**, 185–201.

Conway, W. D. (1977). 'Electrophoresis', in *Drug Fate and Metabolism* (eds E. R. Garret and J. L. Hirtz), Methods and Techniques, Vol. 1, pp. 65–134, Marcel Dekker, New York and Basle.

Coutts, R. T., and Jones, G. R. (1980). 'Significance of analytical techniques in drug metabolism studies', in *Concepts in Drug Metabolism* (eds P. Jenner and B. Testa), Part A, pp. 1–51, Marcel Dekker, New York and Basle.

Crawford, M. J., Croucher, A., and Hutson, D. H. (1981). 'The metabolism of the pyrethroid insecticide cypermethrin in rats; excreted metabolites', *Pestic. Sci.*, **12**, 399–411.

Crawford, M. J., Hutson, D. H., and Stoydin, G. (1980). 'The metabolic fate of a herbicidal methylmercapto-*s*-triazine (cyanatryn) in the rat', *Xenobiotica*, **10**, 169–185.

Crayford, J. V., and Hutson, D. H. (1972). 'The metabolism of the herbicide, 2-chloro-4-(ethylamino)-6-(1-cyano-1-methylamino)-*s*-triazine in the rat', *Pestic. Biochem. Physiol.*, **2**, 295–307.

Crayford, J. V., and Hutson, D. H. (1980a). 'The metabolism of the herbicide flamprop-methyl in rats and some comparative data for dogs, mice and rabbits', *Pestic. Sci.*, **11**, 573–590.

Crayford, J. V., and Hutson, D. H. (1980b). 'Xenobiotic triglyceride formation', *Xenobiotica*, **10**, 349–354.

DeBaun, J. R., Bova, D. L., Tseng, C. K., and Menn, J. J. (1979). 'Metabolism of the rice herbicide, molinate (Ordram®) in the rat', in *Advances in Pesticide Science* (ed. H. Geissbühler), Vol. 3, pp. 547–550, Pergamon Press, Oxford.

DeJongh, G. D., Van den Wildenberg, H. M., Nienwenhuyse, H., and Van der Veen, F. (1981). 'The metabolism of mianserin in women, rabbits, and rats', *Drug. Metab. Dispos.*, **9**, 48–53.

Dieterle, W., and Faigle, J. W. (1981). 'Preparative liquid chromatographic methods in drug metabolism studies', presented at the 4th International Bioanalytical Forum, University of Surrey, Sept. 7–10.

Dieterle, W., Faigle, J. W., and Mory, H. (1979). 'Preparative reversed phase chromatography of polar and non-polar metabolites on column packed with micronized XAD-2 resin', *J. Chromatogr.*, **168**, 27–34.

Dieterle, W., Faigle, J. W., Mory, H., Richter, W. J., and Theobald, W. (1975). 'Biotransformation and pharmacokinetics of sulfinpyrazone (Anturan®) in man', *Europ. J. Clin. Pharmacol.*, **9**, 135–145.

Donzel, B., and Muecke, W. (1978). 'Metabolism of C 18 898 in the rat', unpublished results.

Dorough, H. W. (1976). 'Biological activity of pesticide conjugates', in *Bound and Conjugated Pesticide Residues* (eds D. D. Kaufman, G. G. Still, G. D. Paulson, and S. K. Bandal), ACS Symposium Series 29, pp. 11–34, American Chemical Society, Washington, DC.

Dorough, H. W. (1979). 'Metabolism of insecticides by conjugation mechanisms', *Pharmac. Ther.*, **4**, 433–471.

Dressler, M. (1979). 'Extraction of trace amounts of organic compounds from water with porous organic polymers', *J. Chromatogr.*, **165**, 167–206.

Drinkwine, A. D., and Fleeker, J. R. (1981). 'Metabolism of 2,5-dichloro-4-hydroxyphenoxyacetic acid in plants', *J. Agric. Food Chem.*, **29**, 763–766.

Drozd, J. (1981). *Chemical Derivatization in Gas Chromatography* (Journal of Chromatography Library: Volume 19), Elsevier Science Publishers, Amsterdam.

Engelhardt, H. (1977). *Hochdruck-Flüssigkeits-Chromatographie*, Springer, Berlin.

Ernst, W. (1969). '10. Metabolism of substituted dinitrophenols and ureas in mammals and methods for the isolation and identification of metabolites', *J.S. AFR. Chem. Inst.*, **22**, (Special Issue), 79S–88S.

Esser, H. O., Muecke, W., and Alt, K. O. (1968). 'Der Abbau des Insectizids GS 13 005 in der Ratte: Strukturaufklärung der wichtigsten Metabolite [1]', *Helv. Chim. Acta.*, **51**, 513–517.

Fears, R., Baggaley, K. H., Alexander, R., Morgan, B., and Hindley, R. M. (1978). 'The participation of ethyl 4-benzyloxybenzoate (BRL 10894) and other aryl-substituted acids in glycerolipid metabolism', *J. Lipid Res.*, **19**, 3–11.

Fenselau, C., and Johnson, L. P. (1980). 'Analysis of intact glucuronides by mass spectrometry and gas chromatography—mass spectrometry—A. Review', *Drug. Metab. Dispos.*, **8**, 274–283.

Fischer, L. (1980). *Gel Filtration Chromatography*, Elsevier/North-Holland Biomedical Press, Amsterdam.

Franklin, R. A., Heatherington, K., Morrison, B. J., Sherren, P., and Ward, T. J. (1978). 'Detection of cyanogen chloride as an impurity in dichloromethane and its significance in drug and metabolite analysis', *Analyst*, **103**, 660–662.

Franklin, R. A., Heatherington, K., Morrison, B., and Ward, T. (1981). 'Metabolism of 4-benzamido-1-[4(indol-3-yl)-4-oxobutyl] piperidine in rats and monkeys', *Xenobiotica*, **11**, 159–165.

Fransson, B., Wahlund, K.-G., Johansson, I. M., and Schill, G. (1976). 'Ion-pair chromatography of acid drug metabolites and endogenic compounds', *J. Chromatogr.*, **125**, 327–344.

Frear, D. S. (1976). 'Pesticide conjugates-glycosides', in *Bound and Conjugated Pesticide Residues* (eds D. D. Kaufman, G. G. Still, G. D. Paulson, and S. K. Bandal), ACS Symposium Series 29, pp. 35–54, American Chemical Society, Washington, DC.

Frear, D. S., and Swanson, H. R. (1978). 'Behaviour and fate of [^{14}C] maleic hydrazide in tobacco plants', *J. Agric. Food Chem.*, **26**, 660–666.

Frear, D. S., Swanson, H. R., Mansager, E. R., and Wien, R. G. (1978). 'Chloramben metabolism in plants: isolation and identification of glucose ester', *J. Agric. Food Chem.*, **26**, 1347–1351.

Frincke, J. M., and Henderson, G. L. (1980). 'The major metabolite of fentanyl in the horse', *Drug. Metab. Dispos.*, **8**, 425–427.

Gemborys, M. W., and Mudge, G. H. (1981). 'Formation and disposition of the minor metabolites of acetaminophen in the hamster', *Drug. Metab. Dispos.*, **9**, 340–351.

Glickman, A. H., Hamid, A. A. R., Rickert, D. E., and Lech, J. J. (1981). 'Elimination and metabolism of permethrin isomers in rainbow trout', *Toxicol. Appl. Pharmacol.*, **57**, 88–98.

Golab, T., Bishop, C. E., Donoho, A. L., Manthey, J. A., and Zornes, L. L. (1975). 'Behaviour of ^{14}C-oryzalin in soil and plants', *Pestic. Biochem. Physiol.*, **5**, 196–204.

Gross, D., Laanio, T., Dupuis, G., and Esser, H. O. (1979). 'The metabolic behaviour of chlorotoluron in wheat and soil', *Pestic. Biochem. Physiol.*, **10**, 49–59.

Gudzinowicz, B. J., and Gudzinowicz, M. J. (1977–1980). *Analysis of Drugs and Metabolites by Gas Chromatography-Mass Spectrometry*, Vols 1–7, Marcel Dekker, Basle.

Hallstroem, G., Lindecke, B., and Anderson, E. (1981). 'Metabolism of *N*-(5-pyrrolidinopent-3-ynyl)-succinimide (BL 14) in rat liver preparations. Characterization of four oxidative reactions', *Xenobiotica*, **11**, 459–471.

Hamboeck, H., Fischer, R. W., Di Iorio, E. E., and Winterhalter, K. H. (1981). 'The binding of *s*-triazine metabolites to rodent hemoglobins appears irrelevant to other species', *Mol. Pharmacol.*, **20**, 579–584.

Harvey, J., jun. (1980). 'Modern high performance liquid chromatography in pesticide metabolism studies', *ACS Symposium Series, 136* (*CHO1*), 1–14.

Harvey, D. J., and Paton, W. D. M. (1980). 'Identification of in-vivo liver metabolites of Δ^6-tetrahydrocannabinol produced by the mouse', *Drug. Metab. Dispos.*, **8**, 178–186.

Hawk, G. L. (ed.) (1979). *Biological/Biomedical Applications of Liquid Chromatography*, Marcel Dekker, New York, Basle.

Hearn, M. T. W. (1980). 'Ion-pair chromatography on normal- and reversed phase systems', in *Advances in Chromatography* (eds J. C. Giddings, E. Grushka, J. Cazes, Ph. R. Brown), Vol. 18, pp. 59–100, Marcel Dekker, New York, Basle.

Heirwegh, K. P. M., and Compernolle, F. (1979). 'Microanalytic detection and structure elucidation of ester-glycosides', *Biochem. Pharmacol.*, **28**, 2109–2114.

Hignite, Ch. E., Tschanz, Ch., Lemons, S., Wiese, H., Azarnoff, D. L., and Huffman, D. H. (1981). 'Glucuronic acid conjugates of clofibrate: four isomeric structures', *Life Sci.*, **28**, 2077–2081.

Hitchings, E. J., and Roberts, T. R. (1980). 'Degradation of the herbicide flamprop-methyl in soil', *Pestic. Sci.*, **11**, 591–599.

Hodgson, R. H., Frear, D. S., Swanson, H. R., and Regan, L. A. (1973). 'Alteration of diphenamid metabolism in tomato by ozone', *Weed Sci.*, **21**, 542–549.

Hornke, I., Fehlhaber, H.-W., Girg, M., and Jantz, H. (1980). 'Metabolism of nomifensine (Alival®, Merital®): isolation and identification of the conjugates of nomifensine-^{14}C from human urine', *Br. J. Clin. Pharmac.*, **9**, 255–264.

Horvath, C. (ed.) (1980). *High Performance Liquid Chromatography Advances and Perspectives*, Vols 1, 2, Academic Press, New York and London.

Hostettmann, K. (1980). 'Droplet counter-current chromatography and its application to the preparative scale separation of natural products', *Planta Medica*, **39**, 1–18.

Hsu, T.-S., and Bartha, R. (1976). 'Hydrolyzable and Nonhydrolyzable 3,4-dichloroaniline-humus complexes and their respective rates of biodegradation', *J. Agric. Food Chem.*, **24**, 118–122.

Huber, J. F. K. (ed.) (1978). *Instrumentation for High-Performance Liquid Chromatography*, Journal of Chromatography Library: Volume 13, Elsevier Scientific Publishers, Amsterdam.

Huckle, K. R., Chipman, J. K., Hutson, D. H., and Millburn, P. (1981b). 'Metabolism of 3-phenoxybenzoic acid and the enterohepatorenal disposition of its metabolites in the rat', *Drug Metab. Dispos.*, **9**, 360–368.

Huckle, K. R., Climie, I. J. G., Hutson, D. H., and Millburn, P. (1981c). 'Dipeptide conjugation of 3-phenoxybenzoic acid in the mallard duck', *Drug. Metab. Dispos.*, **9**, 147–149.

Huckle, K. R., Hutson, D. H., and Millburn, P. (1981a). 'Species differences in the metabolism of 3-phenoxybenzoic acid', *Drug. Metab. Dispos.*, **9**, 352–359.

Hutson, D. H. (1976). 'Glutathione conjugates', in *Bound and Conjugated Pesticide Residues* (eds D. D. Kaufman, G. G. Still, G. D. Paulson, and S. K. Bandal), ACS Symposium Series 29, pp. 103–131, American Chemical Society, Washington, DC.

Hutson, D. H., and Casida, J. E. (1978). 'Taurine conjugation in metabolism of 3-phenoxybenzoic acid and the pyrethroide insecticide cypermethrin in mouse', *Xenobiotica*, **8**, 565–571.

Hutson, D. H., Gaughan, L. C., and Casida, J. E. (1981). 'Metabolism of the cis- and trans-isomers of cypermethrin in mice', *Pestic. Sci.*, **12**, 385–398.

Hutson, D. H., Hoadley, E. C., and Pickering, B. A. (1971). 'The metabolic fate of [vinyl-1-^{14}C] dichlorvos in the rat ofter oral and inhalation exposure', *Xenobiotica*, **1**, 593–611.

Illing, H. P. A., and Wilson, I. D. (1981a). 'pH Dependent formation of β-glucuronidase resistant conjugates from the biosynthetic ester glucuronide of isoxepac', presented at the 4th International Bioanalytical Forum, Guildford, Sept. 7–10.

Illing, H. P. A., and Wilson, I. D. (1981b). 'pH Dependent formation of β-glucuronidase resistant conjugates from the biosynthetic ester of isoxepac', *Biochem. Pharmacol.*, **30**, 3381–3384.

Irving, Ch. C., Wiseman, R., jur., and Hill, J. T. (1967). 'Biliary excretion of the *O*-glucuronide of *N*-hydroxy-2-acetylaminofluorene by the rat and rabbit', *Cancer Res.*, **21**, 2309–2317.

Ivie, G. W. (1978). 'Fate of diflubenzuron in cattle and sheep', *J. Agric. Food Chem.*, **26**, 81–89.

Ivie, G. W. (1980). 'Fate of the plant growth regulator mefluidide [*N*-[2,4-dimethyl-5[[(trifluoromethyl)sulfonyl]amino]phenyl]acetamide] in a cow and sheep', *J. Agric. Food Chem.*, **28**, 1286–1288.

Ivie, G. W., and Hunt, L. M. (1980). 'Metabolites of cis- and trans-permethrin in lactating goats', *J. Agric. Food Chem.*, **28**, 1131–1138.

Jenner, P., and Testa, B. (1978). 'Novel pathways in drug metabolism', *Xenobiotica*, **8**, 1–25.

Jones, A. R., and Gibson, J. (1980). '1,2-Dichloropropane: metabolism and fate in the rat', *Xenobiotica*, **10**, 835–846.

Kadlubar, F., Flammang, T., and Unruh, L. (1978). 'The role of *N*-hydroxy arylamine *N*-glucuronides in arylamine-induced urinary bladder carcinogenesis: metabolite profiles in acidic, neutral and alkaline urines of 2-naphthylamine and 2-nitronaphthalene treated rats', in *Conjugation Reactions in Drug Biotransformation* (ed. A. Aitio), pp. 443–454, Elsevier/North Holland Biomedical Press.

Kaiser, R. E., and Oelrich, E. (1981). *Optimation in HPLC*, Dr. Alfred Hüthing Verlag GmbH., Heidelberg.

Kaneko, H., Ohkawa, H., and Miyamoto, J. (1981a). 'Absorption and metabolism of dermally applied phenothrin in rats', *J. Pestic. Sci.*, **6**, 169–182.

Kaneko, H., Ohkawa, H., and Miyomoto, J. (1981b). 'Metabolism of tetramethrin isomers in rats', *J. Pestic. Sci.*, **6**, 425–435.

Karger, B. L., LePage, J. N., and Tanaka, N. (1980). 'Secondary chemical equilibria in high-performance liquid chromatography', in *High Performance Liquid Chromatography Advances and Perspectives* (ed. C. Horvath), Vol. 1, pp. 113–206, Academic Press, New York.

Karger, B. L., Snyder, L. R., and Horvath, C. (1973). *An Introduction to Separation Science*, John Wiley, New York.

Kaufman, D. D., Still, G. G., Paulson, G. D., and Bandal, S. K. (eds), (1976). *Bound and Conjugated Pesticide Residues*, ACS Symposium Series 29, American Chemical Society, Washington, DC.

Kenndler, E., and Kanianski, D. (1981). 'Off-line combination of isotachophoresis and mass spectrometry', *J. Chromatogr.*, **209**, 306–309.

Kirkpatrick, D., Finn, Ch. M., Conway, B., Hawkins, D. R., Honda, T., Ishida, M., and Powell, G. P. (1981). 'Biokinetics and metabolism of *N*-(2,3-dichlorophenyl)-3,4,5,6-tetrachlorophthalamic acid in rats', *J. Agric. Food Chem.*, **29**, 608–614.

Knaak, J. B., Eldridge, J. M., and Sullivan, L. J. (1967). 'Systematic approach to preparation and identification of glucuronic acid conjugates', *J. Agric. Food Chem.*, **15**, 605–609.

Knaak, J. B., Tallant, M. J., Bartley, W. J., and Sullivan, L. J. (1965). 'The metabolic fate of cabaryl in the rat, guinea pig, and man', *J. Agric. Food Chem.*, **13**, 537–543.

Koegel, W., Mueller, W. F., Coulston, F., and Korte, F. (1979). 'Fate and effects of pentachloronitrobenzene in rhesus monkeys', *J. Agric. Food Chem.*, **27**, 1181–1185.

Kojima, S., Muruyama, K., and Babasaki, T. (1980). 'Identification of urinary metabolites of 2-isopropyl-naphthalene in rats', *Drug. Metab. Dispos.*, **8**, 463–466.

Kopcewicz, J., Ehmann, A., and Bandurski, R. S. (1974). 'Enzymatic esterification of indole-3-acetic acid to myoinositol and glucose', *Plant Physiol.*, **54**, 846–851.

Kraft, R., Otto, A., Makower, A., and Etzold, G. (1981). 'Combined thin-layer chromatography/mass spectroscopy without substance elution', *Anal. Biochem.*, **113**, 193–196.

Kremmer, T., and Boross, L. (1979). *Gel Chromatography: Theory, Methodology, Applications*, John Wiley, Chichester, New York, Brisbane, Toronto.

Kripalani, K. J., Dreyfuss, J., Nemec, J., Cohen, A. I., Meeker, F., and Egli, P. (1981). 'Biotransformation in the monkey of cartazolate (SQ 65396), a substituted pyrazolopyridine having anxiolytic activity', *Xenobiotica*, **11**, 481–488.

Kurihara, N., Tanaka, K., and Nakajima, M. (1979). 'Mercapturic acid formation from lindane in rats', *Pestic. Biochem. Physiol.*, **10**, 137–150.

Lamkin, W. M., and Gehrke, Ch. W. (1965). 'Quantitative gas chromatography of amino acids', *Anal. Chem.*, **37**, 383–389.

Lamoureux, G. L., and Davison, K. L. (1975). 'Mercapturic acid formation in the metabolism of propachlor, CDAA, and fluorodifen in the rat', *Pestic. Biochem. Physiol.*, **5**, 497–506.

Lamoureux, G. L., Gouot, J.-M., Davies, D. G., and Rusness, D. G. (1981). 'Pentachloronitrobenzene metabolism in peanut. 3. Metabolism in peanut cell suspension cultures', *J. Agric. Food Chem.*, **29**, 996–1002.

Lamoureux, G. L., and Rusness, D. G. (1980). 'Pentachloronitrobenzene metabolism in peanut. 1. Mass spectral characterization of seven glutathione-related conjugates produced in-vivo or in-vitro', *J. Agric. Food Chem.*, **28**, 1057–1070.

Lamoureux, G. L., Shimabukuro, R. H., Swanson, H. R., and Frear, D. S. (1970). 'Metabolism of 2-chloro-4-ethylamino-6-isopropylamino-*s*-triazine (atrazine) in excised sorghum leaf sections', *J. Agric. Food Chem.*, **18**, 81–86.

Lamoureux, G. L., Stafford, L. E., and Shimabukuro, R. H. (1972). 'Conjugation of 2-chloro-4,6-bis(alkylamino)-*s*-triazines in higher plants', *J. Agric. Food Chem.*, **20**, 1004–1010.

Larsen, G. L., and Bakke, J. E. (1975). 'Metabolism of 2-chloro-4-cyclopropylamino-*s*-triazine (cyprazine) in the rat', *J. Agric. Food Chem.*, **23**, 388–392.

Larsen, G. L., and Bakke, J. E. (1978). 'Mass spectral characterization of the glucuronide conjugates of terbutryn (2-(*t*-butylamino)-4-(ethylamino)-6-(methylthio)-*s*-triazine) metabolites from rats and goat', *Biomed. Mass Spectrom.*, **5**, 391–394.

Larsen, G. L., and Bakke, J. E. (1981). 'Enterohepatic circulation in formation of propachlor (2-chloro-*N*-isopropyl-acetanilide) metabolites in the rat', *Xenobiotica*, **11**, 473–480.

Larsen, G. L., Bakke, J. E., and Feil, V. J. (1978). 'Metabolism of [^{14}C]-terbutryn (2-(*t*-butylamino)-4-(ethylamino)-6-(methylthio)-*s*-triazine) by rats and goats', *Biomed. Mass Spectrom.*, **5**, 382–390.

Leighty, E. G. (1980). 'In-vivo and in-vitro conjugation of a fatty acid to DDT', *Fed. Proc. Fed. Am. Soc. Exp. Biol.*, **39**, 625.

Leighty, E. G. (1981). 'Decreased retention of fatty acid conjugated DDT metabolites in rats given injections of heparin, bile salts or lecithin', *Res. Commun. Chem. Pathol. Pharmacol.*, **31**, 69–76.

Leighty, E. G., Fentiman, A. F., jr., and Foltz, R. L. (1976). 'Long-retained metabolites of Δ^9 and Δ^8-tetrahydrocannabinols identified as novel fatty acid conjugates', *Res. Commun. Chem. Pathol. Pharmacol.*, **14**, 13–28.

Leighty, E. G., Fentiman, A. F., jr., and Thompson, R. M. (1980). 'Conjugation of fatty acids to DDT in the rat: possible mechanism for retention', *Toxicology*, **15**, 77–82.

Lin, C., Li, Y., McGlotten, J., Morton, J. B., and Symchowicz, S. (1977). 'Isolation and identification of the major metabolite of albuterol in human urine', *Drug. Metab. Dispos.*, **5**, 234–238.

Loev, B., and Goodman, M. M. (1967). 'Dry-column chromatography: a preparative chromatography technique with the resolution of thin-layer chromatography', *Chem. & Ind.*, 2026–2032.

Lores, E. M., and Bradway, D. E. (1977). 'Extraction and recovery of organophosphorous metabolites from urine using an ion exchange resin', *J. Agric. Food Chem.*, **25**, 75–79.

Lusby, W. R., Oliver, J. E., Smith, R. H., jr., Kearney, P. C., and Finegold, H. (1981). 'Metabolism of *N*-nitrosopendimethalin and *N*-nitropendimethalin by a streptomyces isolated from soil', *J. Agric. Food Chem.*, **29**, 245–250.

Lynn, R. K., Smith, R. G., Thompson, R. M., Deinzer, M. L., Griffin, D., and Gerber, N. (1978). 'Characterization of glucuronide metabolites of carbamazepine in human urine by gas chromatography and mass spectroscopy', *Drug. Metab. Dispos.*, **6**, 494–501.

Majors, R. E. (1980a). 'Multidimensional high-performance-liquid chromatography', *J. Chromatogr. Sci.*, **18**, 571–579.

Majors, R. E. (1980b). 'Practical operation of bonded-phase columns in high-performance liquid chromatography', in *High-Performance Liquid Chromatography* (ed. C. Horvath), Vol. 1, Academic Press, New York, London, Toronto, Sidney, San Francisco.

Marco, G., and Hermes, P. (1971). 'Biphasic extraction of radioactive metabolites from treated biological material', unpublished results.

Marquis, L. Y., Shimabukuro, R. H., Stolzenberg, G. E., Feil, V. J., and Zaylskie, R. G. (1979). 'Metabolism and selectivity of fluchloralin in soybean roots', *J. Agric. Food Chem.*, **27**, 1148–1156.

Marsh, M. V., Caldwell, J., Smith, R. L., Horner, M. W., Houghton, E., and Moss, M. S. (1981). 'Metabolic conjugation of some carboxylic acids in the horse', *Xenobiotica*, **11**, 655–663.

Martin, L. E., and Reid, E. (1981). 'Isolation of drug metabolites', in *Progress in Drug Metabolism* (eds J. W. Bridges and L. F. Chasseaud), Vol. 6, pp. 197–248, John Wiley, Chichester, New York, Brisbane, Toronto.

Maurer, G., Donatsch, P., Galliker, H., Kiechel, J. R., and Meier, J. (1981). 'Pindolol: disposition and metabolism in rhesus monkeys after chronic treatment', *Xenobiotica*, **11**, 33–41.

Mayer, P., Kriemler, H. P., Hamboeck, H., and Laanio, T. L. (1981). 'Metabolism of O,O-dipropyl-S-[2(2′-methyl-1′-piperidinyl)-2-oxo-ethyl]phosphorodithioate (C 19 490) in paddy rice', *Agric. Biol. Chem.*, **45**, 335–360.

McCall, P. J., Vrona, S. A., and Kelley, S. S. (1981). 'Fate of uniformly carbon-^{14}ring labeled 2,4,5-trichlorophenoxyacetic acid and 2,4-dichlorophenoxyacetic acid', *J. Agric. Food Chem.*, **29**, 100–107.

McLean, S., Davies, N. W., Watson, H., Fauretto, W. A., and Bignall, J. C. (1981). '*N*-Hydroxyphenacetin, a new urinary metabolite of phenacetin in the rat', *Drug. Metab. Dispos.*, **9**, 255–260.

Mihara, K., Ohkawa, H., and Miyamoto, J. (1981a). 'Metabolism of terallethrin in rats', *J. Pestic. Sci.*, **6**, 211–222.

Mihara, K., Ohkawa, H., and Miyamoto, J. (1981b). 'Metabolism of toclofos-methyl in rats and mice', *J. Pestic. Sci.*, **6**, 65–74.

Mikes, O. (ed.) (1979). *Chromatographic and Allied Methods*, Ellis Horwood, Chichester.

Miyano, K., Ota, T., and Toki, S. (1981). 'Stereo-selective formation of glucuronides in metabolism of hexobarbital enantiomers in-vivo. Isolation and quantitation of glucuronides in rabbit urine', *Drug. Metab. Dispos.*, **9**, 60–64.

More, J. E., Roberts, T. R., and Wright, A. N. (1978). 'Studies of the metabolism of 3-phenoxybenzoic acid in plants', *Pestic. Biochem. Physiol.*, **9**, 268–280.

Muecke, W. (1975). 'Metabolism of CGA 18 731 in the rat', unpublished results.

Muecke, W. (1981). 'Metabolism of CGA 64 250 in the rat', unpublished results.

Muecke, W., Alt, K. O., and Esser, H. O. (1970). 'Degradation of ^{14}C-labelled diazinon in the rat', *J. Agric. Food Chem.*, **18**, 208–212.

Muecke, W., and Bissig, R. (1981). 'Metabolism of metolachlor in the rat', unpublished results.

Muecke, W., and Ifflaender, U. (1976). 'Movement of sulfuric acid conjugates on stationary phases of silica using acetone containing eluents', unpublished results.

Muecke, W., Kriemler, H. P., Hug, P., and Alt, K.O. (1981). 'Metabolism of *O,O*-dipropyl-S-[2(2′-methyl-1′-piperidinyl)(2-oxo-ethyl]phosphorodithioate (C 19 490) in the rat', *Agric. Biol. Chem.*, **45**, 43–51.

Muecke, W., Menzer, R. E., Alt, K. O., Richter, W., and Esser, H. O. (1976). 'The metabolism of chlorotoluron in the rat', *Pestic. Biochem. Physiol.*, **6**, 430–441.

Mumma, R. O., and Hamilton, R. H. (1976). 'Amino acid conjugates', in *Bound and Conjugated Pesticide Residues* (eds D. D. Kaufman, G. G. Still, G. D. Paulson, and S. K. Bandal), ACS Symposium Series 29, pp. 68–85, American Chemical Society, Washington, DC.

Mundy, D. E., and Machin, A. F. (1981). 'Determination of pentachlorophenol and related compounds in animal materials by high-performance liquid chromatography and gas chromatography', *J. Chromatogr.*, **216**, 229–238.

Nestler, H. J. (1979). 'Paper chromatography', in *Drug Fate and Metabolism Methods and Techniques* (eds E. R. Garrett and J. L. Hirtz), Vol. 3, pp. 69–131, Marcel Dekker, New York and Basle.

Nettleton, D. E., jr. (1981). 'Preparative liquid chromatography, I. Approaches utilizing highly compressed beds', *J. Liq. Chromatogr.*, **4**, 141–173.

Newsome, W. H. (1980). 'A method for the determination of maleic hydrazide and its β-D-glucoside in foods by high-pressure anion-exchange liquid chromatography', *J. Agric. Food Chem.*, **28**, 270–272.

Nicholson, J. D. (1978). 'Derivative formation in the quantitative gas-chromatographic analysis of pharmaceuticals', A. Review, *Analyst*, **103**, part. 1, pp. 1–28, part. 2, pp. 193–222.

Oehler, D. D., and Ivie, G. W. (1980). 'Metabolic fate of the herbicide dicamba in a lactating cow', *J. Agric. Food Chem.*, **28**, 685–689.

Oelrich, E., Preusch, H., and Wilhelm, E. (1980). 'Separation of enantiomers by high performance liquid chromatography using chiral eluents', *J. of High Resol. Chrom.*, **3**, 269–272.

O'Grodnick, J. S., Adamovics, J. A., Blake, S. H., and Wedig, J. (1981). 'The metabolic fate of ^{14}C-labeled pentachloronitrobenzene in osborne-mendell rats', *Chemosphere*, **10**, 67–72.

Otto, S., Beutel, P., Drescher, N., and Huber, R. (1979). 'Investigations into the degradation of bentazon in plant and soil', in *Advances in Pesticide Sciences* (ed. H. Geissbuehler), Vol. 3, pp. 551–556, Pergamon Press, Oxford.

Paulson, G. D. (1976). 'Sulfate ester conjugates—their synthesis, purification, hydrolysis and chemical and spectral properties', in *Bound and Conjugated Pesticide Residues* (eds D. D. Kaufman, G. G. Still, G. D. Paulson, and S. K. Bandal), ACS Symposium Series 29, pp. 86–102 American Chemical Society, Washington, DC.

Paulson, G. D., Docktor, M. M., Jacobsen, A. M., and Zaylskie, R. G. (1972). 'Isopropyl carbanilate (propham) metabolism in the chicken: balance studies and isolation and identification of excreted metabolites', *J. Agric. Food Chem.*, **20**, 867–876.

Paulson, G. D., Giddings, J. M., Lamoureux, C. H., Mansager, E. R., and Struble, C. B. (1981). 'The isolation and identification of ^{14}C-sulfamethazine (4-amino-*N*-(4,6-dimethyl-2-pyrimidinyl)[^{14}C]benzenesulfonamide) metabolites in the tissues and excreta of swine', *Drug. Metab. Dispos.*, **9**, 142–146.

Paulson, G. D., Jacobsen, A. M., and Still, G. G. (1975). 'Animal metabolism of propham (isopropylcarbanilate): the fate of residues in alfalfa when consumed by the rat and sheep', *Pestic. Biochem. Physiol.*, **5**, 523–535.

Paulson, G. D., Jacobsen, A. M., and Zaylskie, R. G. (1977). 'Herbicide perfluidone (1,1,1-trifluoro-*N*-[2-methyl-4-(phenylsulfonyl)phenyl]methanesulfonamide): its metabolism in rats and chickens', *Pestic. Biochem. Physiol.*, **7**, 62–72.

Paulson, G. D., and Portnoy, C. E. (1970). 'Sulfate ester conjugates: a one step method for replacing the sulfate with an acetyl group', *J. Agric. Food Chem.*, **18**, 180–181.

Paulson, G. D., Zaylskie, R. G., and Docktor, M. M. (1973). 'Characterization of aryl glucuronic acid conjugates by derivatization and mass spectral analysis', *Anal. Chem.*, **45**, 21–27.

Pekas, J. C. (1979). 'Further metabolism of naphthyl *N*-methylcarbamate (carbaryl) by the intestine', *Pestic. Biochem. Physiol.*, **11**, 166–175.

Pharmacia (1977). *Sephadex LH-20. Chromatography in Organic Solvents*, Pharmacia, Uppsala, Sweden.

Pierce, A. E. (1979). *Sylilation of Organic Compounds*, Pierce Chemical Co. Publisher.

Pierce, A. E. (1981). *Pierce Handbook and General Catalog 1981–82*, Pierce Eurochemie BV, Rotterdam, Holland.

Prochazka, Z. (1979). 'Countercurrent distribution', in *Chromatographic and Allied Methods* (ed. O. Mikes), pp. 621–648, Ellis Horwood, Chichester.

Pryde, A., and Gilbert, M. T. (1979). *Applications of High Performance Liquid Chromatography*, Chapman and Hall, London.

Quistad, G. B., Staiger, L. E., and Schooley, D. A. (1978a). 'Environmental degradation of the miticide cycloprate (hexadecyl cyclopropanecarboxylate). 1. Rat metabolism', *J. Agric. Food Chem.*, **26**, 60–66.

Quistad, G. B., Staiger, L. E., and Schooley, D. A. (1978b). 'Environmental degradation of the miticide cycloprate. 2. Metabolism by apples and oranges', *J. Agric. Food Chem.*, **26**, 66–70.

Ramsteiner, K. A., and Hoermann, W. D. (1979). 'High pressure liquid chromatographic determination of hydroxy-*s*-triazine residues of plant material', *J. Agric. Food Chem.*, **27**, 934–938.

Renner, G. (1980). 'Metabolic studies on pentachloronitrobenzene (PCNB) in rats', *Xenobiotica*, **10**, 537–550.

Renner, G., Schuster, K. P., Kuhnhen, H. B., and Stachel, H.-D. (1978). 'Isolierung von Stoffwechselprodukten der Pestizide Hexachlorbenzol and Pentachlornitrobenzol aus Harn und Faeces von Ratten und Kaninchen', *Chemosphere*, **12**, 943–947.

Rhodes, J. C., and Houston, J. B. (1981). 'Quantification of naphthyl conjugates. Comparison of high performance liquid chromatography and selective enzyme methods', *Xenobiotica*, **11**, 63–70.

Roberts, T. R. (1977). 'The metabolism of the herbicide flamprop-isopropyl in barley', *Pestic. Biochem. Physiol.*, **7**, 378–390.

Roberts, T. R., and Wright, A. N. (1981). 'The metabolism of 3-phenoxybenzyl alcohol a pyrethroid metabolite in plants', *Pestic. Sci.*, **12**, 161–169.

Rouchaud, J., Moons, C., and Meyer, J. A. (1981). 'The metabolism of [^{14}C]aldicarb in the root of sugar beet plants', *Pestic. Sci.*, **12**, 548–556.

Ruelius, H. W., Tio, C. O., Knowles, J. A., McHugh, S. L., Schillings, R. T., and Sisenwine, S. F. (1979). 'Diastereo-isomeric glucuronides of oxazepam isolation and stereoselective enzymic hydrolysis', *Drug. Metab. Dispos.*, **7**, 40–43.

Rusness, D. G., and Lamoureux, G. L. (1980). 'Pentachloro-nitrobenzene metabolism in peanut. 2. Characterisation of $CHCl_3$-soluble metabolites produced in vivo', *J. Agric. Food Chem.*, **28**, 1070–1077.

Scheel, D., and Sandermann, H., jr. (1981). 'Metabolism of 2,4-dichlorophenoxyacetic acid in cell suspension cultures of soybean (glycine max L.) and wheat (triticum aestivum L.)', *Planta*, **152**, 248–252.

Schill, G., Borg, K. O., Modin, R., and Persson, B.-A. (1977c). 'Ion-pair extraction methods', in *Progress in Drug Metabolism* (eds J. W. Bridges and L. F. Chasseaud), Vol. 2, pp. 219–258, John Wiley, Chichester, New York, Brisbane, Toronto.

Schill, G., Modin, R., Borg, K. O., and Persson, B.-A. (1977a). 'Ion-pair extraction and chromatography', in *Drug Fate and Metabolism, Methods and Techniques* (eds E. R. Garrett and J. L. Hirtz), Vol. 1, pp. 135–185, Marcel Dekker, New York and Basle.

Schill, G., Modin, R., Borg, K. O., and Persson, B.-A. (1977b). 'Ion-pair extraction and ion-pair chromatography', in *Handbook of Derivatives for Chromatography* (eds K. Blau and G. S. King), pp. 500–529, Heyden, London–Bellmawr–Rheine.

Schooley, D. A., and Quistad, G. B. (1979). 'High-pressure, high-resolution liquid chromatography and its application to pesticide analysis and biochemistry', in *Progress in Drug Metabolism* (eds J. W. Bridges and L. F. Chasseaud), Vol. 3, pp. 1–113, John Wiley, Chichester, New York, Brisbane, Toronto.

Schroeder, W. A. (1972a). 'Separation of peptides by chromatography on columns of Dowex 50 with volatile developers', in *Methods in Enzymology* (eds C. H. W. Hirs and S. N. Timasheff), Vol. 25, pp. 203–213, Academic Press, New York.

Schroeder, W. A. (1972b). 'Separation of peptides by chromatography on columns of Dowex 1 with volatile developers', in *Methods in Enzymology* (eds C. H. W. Hirs and J. N. Timasheff), Vol. 25, pp. 214–221, Academic Press, New York.

Schwartz, D. E., Jordan, J. C., Vetter, W., and Oesterhelt, G. (1979). 'Metabolic studies of ornidazole in the rat, in the dog and in man', *Xenobiotica*, **9**, 571–581.

Shafik, M. T., and Enos, H. F. (1969). 'Determination of metabolic and hydrolytic products of organophosphorus pesticide chemicals in human blood and urine', *J. Agric. Food Chem.*, **17**, 1186–1189.

Shimabukuro, R. H., Walsh, W. C., and Hoerauf, R. A. (1979). 'Metabolism and selectivity of diclofop-methyl in wild oat and wheat', *J. Agric. Food Chem.*, **27**, 615–623.

Shimada, K., and Nambara, T. (1979). 'Studies on steroids, CXLVI. Chromatographic behaviour of organic sulphates on Sephadex LH 20', *J. Chromatogr.*, **174**, 440–442.

Shono, T., Unai, T., and Casida, J. C. (1978). 'Metabolism of permethrin isomers in American cockroach adults, house fly adults and cabbage looper larvae', *Pestic. Biochem. Physiol.*, **9**, 96–106.

Shuntner, C. A., and Thompson, P. G. (1966). 'Detection and estimation of [^{14}C]diazoxon in-vivo by countercurrent distribution and inhibition assay', *Biochem. Biophys. Acta*, **115**, 225–227.

Siek, T. S., and Rieders, E. F. (1981). 'Formation of 2-(4-chlorophenoxy)-2-methylpropanamide from a clofibrate metabolite during extraction of urine', *J. Anal. Toxicol.*, **5**, 194–197.

Simoneaux, B. J., Martin, G., Cassidy, J. E., and Ryskiewich, D. P. (1980). 'Isolation and identification of the major polar metabolites of methidathion in tomatoes', *J. Agric. Food Chem.*, **28**, 1221–1224.

Sinclair, K. A., and Caldwell, J. (1982). 'The formation of β-glucuronidase resistant glucuronides by the intramolecular rearrangement of glucuronic acid conjugates at mild alkaline pH', *Biochem. Pharmacol.*, **31**, 953–957.

Sisenwine, S. F., Tio, C. O., Shrader, St. R., and Ruelius, H. W. (1972). 'The biotransformation of oxazepam (7-chloro-1,3-dihydro-3-hydroxy-5-phenyl-2,H-1,4-benzo-diazepin-2-one) in man, miniature swine and rat', *Arzneim.-Forsch.* (*Drug Res.*), **22**, 682–687.

Smith, A. E. (1981). 'Comparison of solvent systems for the extraction of atrazine, benzoylprop, flamprop, and trifluralin from weathered field soils', *J. Agric. Food Chem.*, **29**, 111–115.

Snyder, L. R., and Kirkland, J. J. (1979). *Introduction to Modern Liquid Chromatography*, John Wiley, New York.

Snyder, L. R., and Poppe, H. (1980). 'Mechanism of solute retention in liquid–solid chromatography and the role of the mobile phase in affecting separation, Competition versus sorption', *J. Chromatogr.*, **184**, 363–413.

Son, O. S., Everett, D. W., and Fiala, E. S. (1980). 'Metabolism of *O*-[methyl-^{14}C]toluidine in the F 344 rat', *Xenobiotica*, **10**, 457–468.

Sonobe, H., Carver, R. A., Krause, R. T., and Kamps, L. R. (1980a). 'Extraction of biologically incorporated ^{14}C-phorate residues from root crops', *Abstr. Papers, Am. Chem. Soc. 2 Congr. N. Amer. Pest, 112.*

Sonobe, H., Carver, R. A., Krause, R. T., and Kamps, L. R. (1980b). 'Extraction of biologically incorporated ^{14}C-carbofuran residues from root crops', *Abstr. Papers Am. Chem. Soc. 2 Congr. N. Amer. Pest, 113.*

Sonobe, H., Kamps, L. R., Mazzola, E. P., and Roach, J. A. G. (1981). 'Isolation and identification of a new conjugated carbofuran metabolite in carrots: angelic acid ester of 3-hydroxycarbofuran', *J. Agric. Food Chem.*, **29**, 1125–1129.

Starratt, A. N., and Bond, E. J. (1981). 'Metabolism of methylbromide by susceptible and resistant strains of the granery weevil, sitophilus granarius (L.)', *Pestic. Biochem. Physiol.*, **15**, 275–281.

Stierlin, H., Faigle, J. W., Sallmann, A., Kueng, W., Richter, W. J., Kriemler, H. P., Alt, K. O., and Winkler, T. (1979). 'Biotransformation of diclofenac sodium (Voltaren®) in animals and in man, 1. Isolation and identification of principal metabolites', *Xenobiotica*, **9**, 601–610.

Still, G. G., and Leopold, R. A. (1978). 'The elimination of (*N*-[[4-chlorophenyl)amino]-carbonyl]-2,6-difluorobenzamide by the boll weevil', *Pestic. Biochem. Physiol.*, **9**, 304–312.

Still, G. G., and Mansager, E. R. (1972). 'Aryl hydroxylation of isopropyl-3-chlorocarbanilate by soybean plants', *Phytochem.*, **11**, 515–520.

Still, G. G., and Mansager, E. R. (1973). 'Soybean shoot metabolism of isopropyl-3-chlorocarbanilate: ortho and para aryl hydroxylation', *Pestic. Biochem. Physiol.*, **3**, 87–95.

Still, G. G., and Mansager, E. R. (1975). 'High performance liquid and gas liquid chromatography applied to the purification of pesticide plant and animal metabolites', *Chromatographia*, **8**, 129–135.

Still, G. G., and Rusness, D. G. (1977). '*S*-Cysteinylhydroxychlorpropham: formation of the *S*-cysteinyl conjugate of isopropyl-3′-chloro-4′-hydroxycarbanilate in oat (Avena sativa L.)', *Pestic. Biochem. Physiol.*, **7**, 210–219.

Sullivan, L. L., Eldridge, J. M., Knaak, J. B., and Tallant, M. J. (1972). '5,6-Dihydro-5,6-dihydroxycarbaryl glucuronide as a significant metabolite of carbaryl in the rat', *J. Agric. Food Chem.*, **20**, 980–985.

Suzuki, T., and Casida, J. E. (1981). 'Metabolites of diuron, linuron, and methazole formed by liver microsomal enzymes and spinach plants', *J. Agric. Food Chem.*, **29**, 1027–1033.

Takade, D. Y., Reynolds, J. M., and Nelson, J. H. (1979). '1-(4-Nitrobenzyl)-3-(4-tolyl)triazene as derivatizing reagent for the analysis of urinary dialkyl phosphate metabolites of organophosphorus pesticides by gas chromatography', *J. Agric. Food Chem.*, **27**, 746–752.

Tang, B. K., Kalow, W., and Grey, A. A. (1979). 'Metabolic fate of phenobarbital in man, *N*-glucoside formation', *Drug. Metab. Dispos.*, **7**, 315–318.

Titus, E. O. (1971). 'Paper, column and thin layer chromatography, counter current distribution and electrophoresis', in *Concepts in Biochemical Pharmacology* (eds B. B. Brodie, and R. J. Gilette), Part 2, *Handbook of Pharmacology*, Vol. 28/2, pp. 123–159, Springer-Verlag, Berlin–Heidelberg–New York.

Tomasic, J. (1978). 'Analysis of glucuronic acid conjugates', in *Drug Fate and Metabolism, Method and Techniques* (eds E. R. Garret and J. L. Hirtz), Vol. 2, pp. 281–335, Marcel Dekker, New York and Basle.

Tomlinson, E., Jefferies, T. M., and Riley, C. M. (1978). 'Ion-pair high-performance liquid chromatography', *J. Chromatogr.*, **159**, 315–358.

Touchstone, J. C., and Dobbins, M. F. (1978). *Practice of Thin Layer Chromatography*, John Wiley, New York.

Turan, T. S., and Gibson, W. B. (1981). 'A comparison of the elimination and biotransformation of dodecyldimethylamine oxide (DDAO) by rats, rabbits, and man', *Xenobiotica*, **11**, 447–458.

VandenHeuvel, W. J. A., and Zacchei, A. G. (1978). 'Gas liquid chromatography', in *Drug Fate and Metabolism, Methods and Techniques* (eds E. R. Garrett and J. L. Hirtz), Vol. 2, pp. 49–142, Marcel Dekker, New York and Basle.

Verweij, A., and Kientz, Ch. E. (1981). 'High-performance liquid chromatographic separation, isolation and identification of 1,2,3-thiadiazole-5-carboxy aldoxime glucuronide in rabbit urine', *J. Chromatogr.*, **226**, 165–173.

Verzele, M., and Geeraert, E. (1980). 'Preparative liquid chromatography', *J. Chromatogr. Sci.*, **18**, 559–570.

Weete, J. D., Pillai, P., and Davies, D. D. (1980). 'Metabolism of atrazine by spartina alterniflora, 2. Water-soluble metabolites', *J. Agric. Food Chem.*, **28**, 636–640.

Wester, R., Noonan, P., Markos, Ch., Bible, R., jun., Aksamit, W., and Hribar, J. (1981). 'Identification of carbamate derivatives formed during chloroform extraction of tricyclic antidepressants from urine', *J. Chromatogr.*, **209**, 463–466.

Wheeler, W. B., Thompson, N. P., Edelstein, R. L., and Krause, R. T. (1979). 'Comparison of blenders for the extraction of carbofuran from radishes', *Bull. Environm. Contam. Toxicol.*, **23**, 387–390.

White, J. D., and Schwartz, D. P. (1980). 'Evaluation of Amberlite XAD-4 for recovery of glucuronides and sulfates from aqueous solution', *J. Chromatogr.*, **196**, 303–308.

Wilson, I. D., Illing, H. P. A., Bryce, T., and Chamberlain, J. (1981). 'Artefactual conjugates of isoxepac', 4th International Bioanalytical Forum, Guildford, Sept. 7–10.

Wishnok, J. S., and Tannenbaum, S. R. (1977). 'An unknown salivary mopholine metabolite', *Anal. Chem.*, **49**, 715A–718A.

Wolf, D. E., Vandenheuvel, W. J. A., Tyler, T. R., Walker, R. E., Koninszy, F. R., Gruber, V., Arison, B. H., Rosegay, A., Jacob, T. A., and Wolf, F. J. (1980). 'Identification of a glutathione conjugate of cambendazole formed in the presence of liver microsomes', *Drug. Metab. Dispos.*, **8**, 131–136.

Woolson, E. A., and Aharonson, N. (1980). 'Separation and detection of arsenical pesticide residues and some of their metabolites by high pressure liquid chromatography–graphite furnace atomic absorption spectrometry', *J. Assoc. off. Anal. Chem.*, **63**, 523–528.

Wright, A. N., Roberts, T. R., Dutton, A. J., and Doig, M. V. (1980). 'The metabolism of cypermethrin in plants: the conjugation of the cyclopropyl moiety', *Pestic. Biochem. Physiol.*, **13**, 71–80.

Yanagi, Y., Haga, F., Endo, M., and Kitagawa, S. (1976). 'Comparative metabolic study in nimetazepam and its desmethyl derivative (nitrazepam) in dogs', *Xenobiotica*, **6**, 101–112.

Yu, C. C., Atallah, Y. H., and Whitacre, D. M. (1980). 'Metabolism of the herbicide buthidazole in corn seedlings and alfalfa plants', *J. Agric. Food Chem.*, **28**, 1090–1095.

Zlatkis, A., and Kim, K. (1976). 'Column elution and concentration of volatile compounds in biological fluids', *J. Chromatogr.*, **126**, 475–485.

Zlatkis, A., and Pretorius, V. (eds), (1971). *Preparative Gas Chromatography*, Wiley, Interscience, New York.

Progress in Pesticide Biochemistry and Toxicology, Volume 3
Edited by D. H. Hutson and T. R. Roberts

CHAPTER 7

The role of soft ionization mass spectral techniques in pesticide metabolite identification

D. E. Games

INTRODUCTION

Mass spectrometric methods have had extensive utility in the identification of pesticides and their metabolites (Biros, 1971; Alford, 1975; 1977; 1978; Safe, 1979; Sphon and Brumley, 1980). This is because of the techniques' ability to provide valuable structural information from very small amounts of sample which have been isolated by thin-layer chromatography (TLC) or high-performance liquid chromatography (LC). A further factor is the techniques' ability to examine crude biological extracts by gas chromatography–mass spectrometry (GC–MS) and to provide valuable information as to the nature of metabolic products, without the necessity of a time-consuming and tedious separation. Putative metabolites can subsequently be synthesized and the structures of the metabolic products confirmed by chromatographic and mass spectral comparison. Because of the complexity of the traces often produced, particularly if capillary GC has been used, peaks of interest have to be identified by GC using a radioactivity detector or if the pesticide is not chlorinated or brominated a cold labelling technique whereby the pesticide of interest and its cold labelled analogue are fed to the system of interest and metabolites can be readily identified from their characteristic isotope patterns (Pohl *et al.*, 1975).

The vast majority of studies of pesticide metabolism using mass spectrometry reported to date have relied on electron impact (EI) ionization and direct probe or gas chromatographic introduction of the sample into the mass spectrometer. Although this approach will suffice for many studies, as will be explained later, it can lead to anomalous results, non-availability of valuable structural information and the overlooking of metabolic products.

It is the aim of this chapter to illustrate the advantages of using a combination of mass spectral ionization methods and to show their unique ability to provide valuable structural information. In addition, it is hoped that the reader will become aware of the advantages of using combined high-performance liquid chromatography–mass spectrometry (LC–MS) in studies of pesticide metabolites. Emphasis will be placed on commercially available ionization methods but some other less accessible techniques will be discussed.

PROBLEMS ASSOCIATED WITH ELECTRON IMPACT MASS SPECTROMETRY AND GC–MS

Before discussing the various soft ionization mass spectral techniques and LC–MS it is necessary to enquire into the limitations of EI mass spectrometry and GC–MS. In order to obtain an EI spectrum of a compound, a sample has to be vaporized prior to ionization. This can result in thermal degradation of the sample if it is volatile and thermally labile and results in the obtaining of the spectra of the thermal degradation products; this can lead to incorrect identification. Other compounds of low volatility either cannot be vaporized or are thermally degraded. Metabolic products and particularly conjugated metabolites often fall into one or other of these classes of compound. Some of these problems can be overcome by chemical derivatization and extensive use of this approach has been made in metabolic studies to obtain mass spectral information. However, one often does not know what functionalities are available for derivatization; the chemical reactions involved can produce unexpected products leading to difficulty in characterization; a mixture of products may be produced; and usually only very small amounts of sample are available, making chemical manipulation difficult. Alternatively, a variety of 'tricks' can be employed by the mass spectrometrist, e.g. the use of an inert surface in the direct introduction probe, with or without rapid heating, or the introduction of the sample close to the electron ion beam. These approaches have recently been reviewed (Doyle Daves, Jr., 1979). However, even these approaches often fail to provide satisfactory data. A further problem in EI mass spectrometry is that conventionally 70 eV is the electron beam energy used; this can result in the presence of low abundance or no molecular ions for compounds with labile structures. The former problem can cause considerable difficulties in the relative molecular mass assignment if a sample is not pure, since there may be other peaks due to impurities in the molecular ion region. Often assignment of molecular weight can be assisted in this situation by lowering the electron beam energy. However, this results in loss of sensitivity and is of no assistance if a molecular ion is not present.

Gas chromatographic examination of compounds requires that the sample be vaporized before chromatography can be effected; this again can result in thermal degradation and many compounds do not have sufficient vapour pressure for gas chromatographic study. There are solutions to this problem, such as derivatization and the advent of fused silica capillary columns has improved our ability to handle more difficult compounds. However, the rapid expansion in the use of high-performance liquid chromatography (LC) in the last decade testifies to the limitations of GC for studying many classes of compound and in particular metabolites because of their polar nature.

The purpose of this section is not to decry the use of EI mass spectrometry or GC–MS but to indicate their limitations. Let us now proceed to see how we can overcome the problems outlined above. It should be emphasized at this point that an approach to structural determination which relies on any one technique is not advocated and the more information that one can obtain about a particular sample from whatever source, the better. Since mass spectrometry is in general a destructive technique one often leaves it to the last resort for structural studies if only minute amounts of sample are available.

SOFT IONIZATION MASS SPECTRAL TECHNIQUES

Soft ionization mass spectral methods can be broadly divided into two types: those which are used with conventional mass spectral introduction methods, i.e. direct insertion probe, hot box or GC, and those which utilize an alternative type of sample supply, usually where ionization is effected on a surface and vaporization of the sample prior to ionization is not necessary. This means that these latter techniques can handle compounds not amenable to the former methods. Chemical ionization (CI) and field ionization (FI) fall into the former class; desorption CI occupies an intermediate position and the remaining ionization methods discussed fall into the latter class. This section is not intended as a fully comprehensive survey of all ionization methods; they have been selected because of their commercial availability and ability or potential ability to assist in the identification of metabolic products. For more comprehensive coverage the reader is referred to general reviews in the area (Doyle Daves, Jr., 1979; Milne and Lacey (1974).

Chemical ionization mass spectrometry

Chemical ionization (CI) occurs in a mass spectrometer when a sample is ionized by an ion–molecule reaction. Ionization is effected by reaction of reagent ions (which are generated from a reactant gas in a high pressure source), and are usually present in large excess over the vaporized sample molecules. Depending on the reagent gas used, different types of reaction are involved in the production of CI spectra. If, for example, methane is used as the reagent gas then in the positive mode the reactant species are mainly $[CH_5]^+$ and $[C_2H_5]^+$ with smaller amounts of $[C_3H_5]^+$. These ions undergo mainly acid-base reactions with the sample molecule resulting in the production of a protonated molecular ion. In addition the $[C_2H_5]^+$ and $[C_3H_5]^+$ ions also function as Lewis acids and form collision-stabilized complexes $[M(C_2H_5)]^+$ and $[M(C_3H_5)]^+$ resulting in the presence of $[M + 29]^+$ and $[M + 41]^+$ ions in the spectrum. The $[M + 29]^+$ ion is usually of the order of 10% of the abundance of the $[M + 1]^+$ ion and the $[M + 41]^+$ ion is of lower abundance. The internal energy content of the protonated molecular ion is usually much less than that in EI and, in addition, they are even electron ions which are usually more stable to fragmentation than are the odd electron ions produced by EI. Thus relative molecular mass assignment is much easier than in EI spectra and the fragment ions produced can often provide complementary structural information to that obtained from EI mass spectra. Figure 1 shows the EI and methane CI spectra of dimethoate (**1**) and illustrates these features.

$$(CH_3O)_2\overset{\overset{\displaystyle S}{\|}}{P}SCH_2\overset{\overset{\displaystyle O}{\|}}{C}NHCH_3$$

(**1**)

The energy transfer in CI is dependent on the differences in proton affinity between the molecule and the conjugate base of the relevant reagent ion. In addition the proton affinity of the molecule under study must exceed that of the conjugate base if proton transfer is to occur. This is generally observed when methane is used as reagent gas; however, with ammonia, collision-stabilized complexes $[M(NH_4)]^+$ can be exclusively formed or protonation resulting in the formation of $[M + H]^+$ ions can occur because the conjugate base of ammonia $[NH_4]^+$ has one of the highest proton affinities of the reagent gases most commonly used. In the former case the proton affinity of the molecule

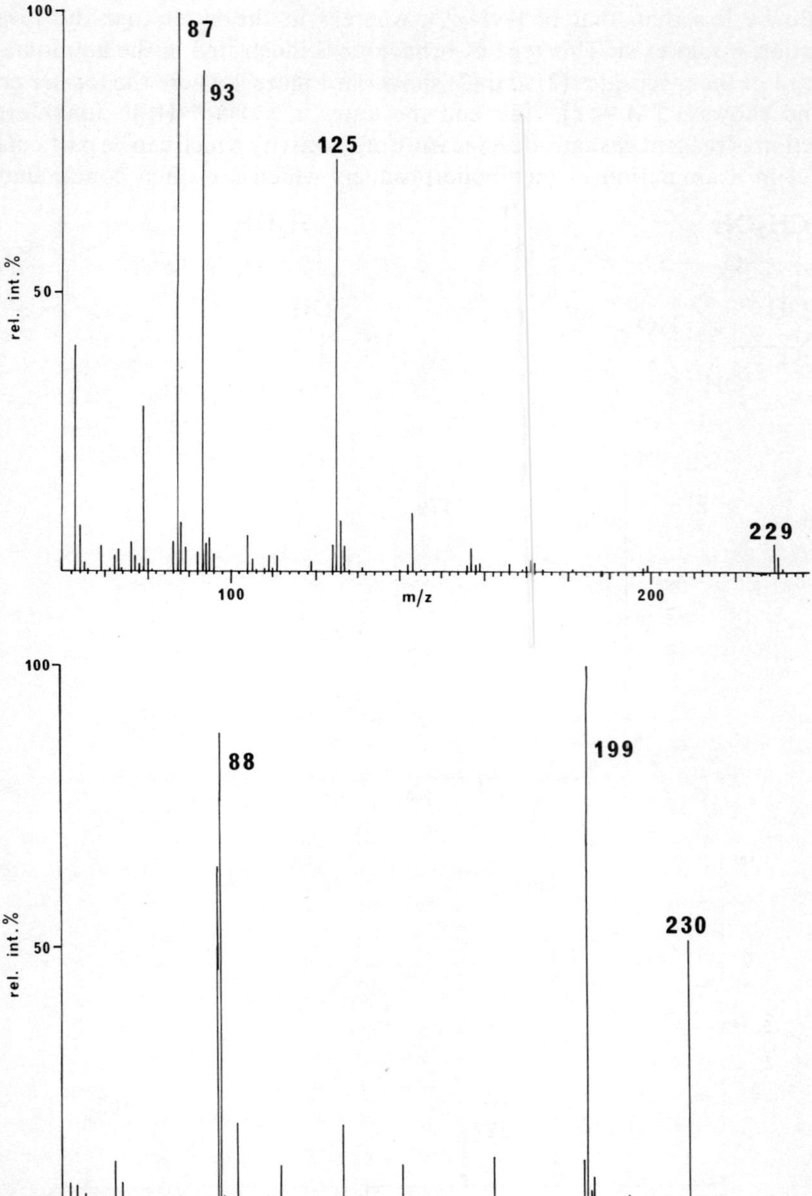

Figure 1 EI and methane CI mass spectra of dimethoate (**1**)

would be less than that of $[NH_4]^+$, whereas in the latter case the reverse situation would exist. This type of behaviour is illustrated in the ammonia CI spectra of the glycosides (**2**) and (**3**) shown in Figure 2 where the former compound shows a $[M + 1]^+$ ion and the latter, a $[M + NH_4]^+$ ion. Careful selection of reagent gas can add a measure of selectivity which can be particularly useful in examination of metabolic products which are often contaminated.

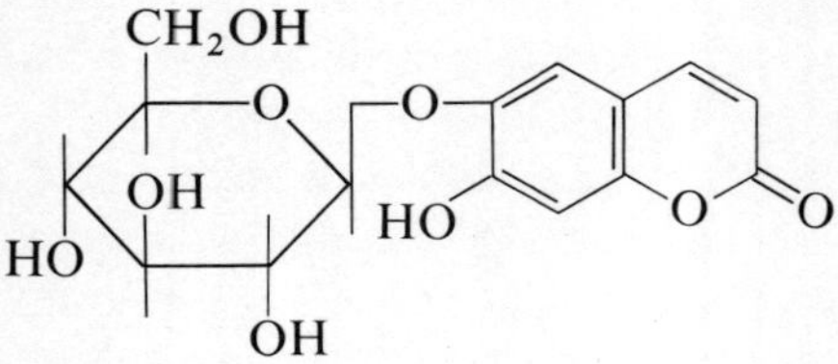

(**2**)

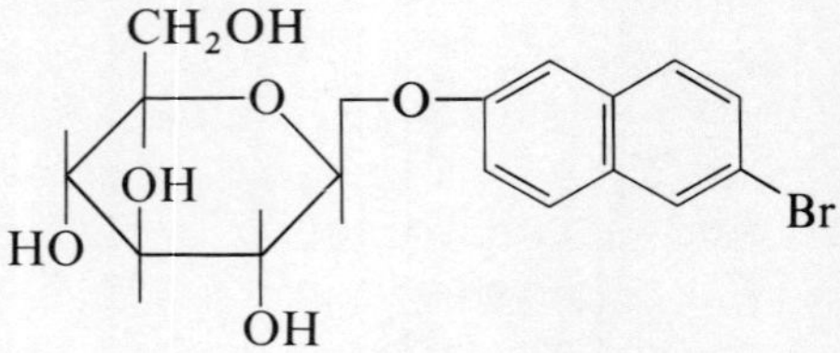

(**3**)

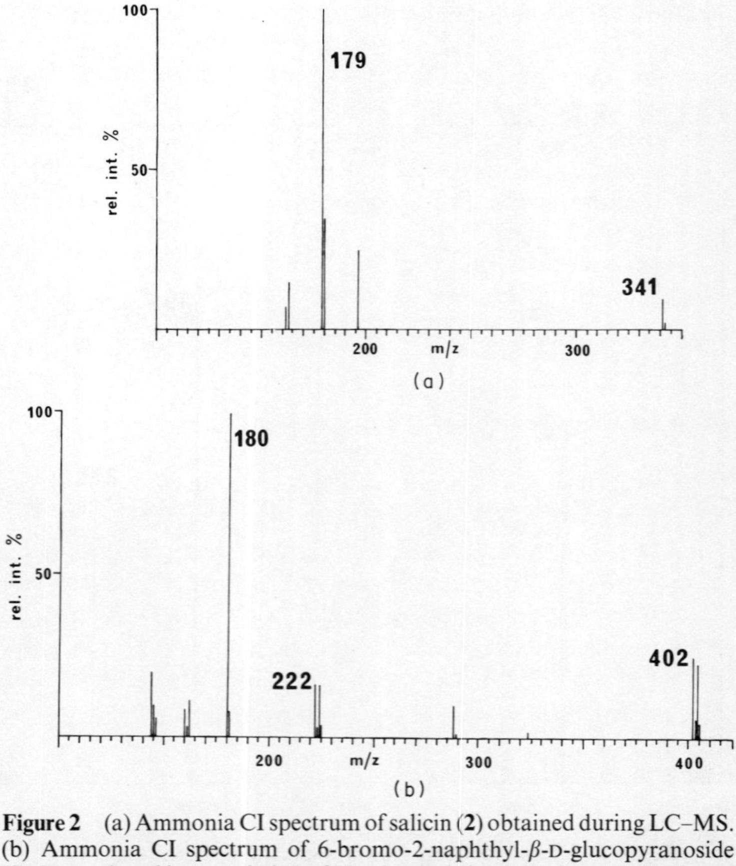

Figure 2 (a) Ammonia CI spectrum of salicin (**2**) obtained during LC–MS. (b) Ammonia CI spectrum of 6-bromo-2-naphthyl-β-D-glucopyranoside (**3**). Reproduced by permission of Heyden and Son Ltd., from D. E. Games and E. Lewis, *Biomedical Mass Spectrometry*, 1980, **7**, 433

Some common reagent gases which undergo acid-base type reactions in CI are listed in order of their increasing proton affinities: methane, water, methanol, isobutane and ammonia.

For more detailed discussion of the selection of reagent gases for CI, the reader is referred to relevant reviews (Munson, 1971; 1977; Field, 1972; Arsenault, 1972; Milne and Lacey, 1974; Richter and Schwarz, 1978).

Ionization occurs with other reagent gases by charge exchange resulting in the formation of $M^{+\cdot}$ ions. In this case the energy of the molecular ion is controlled by the difference between the recombination energy of the reagent ions and the ionization potential of the molecule. Reagent gases which undergo this type of reaction are listed in order of decreasing recombination energy: helium, neon, argon, nitrogen, krypton and xenon.

The spectra produced are similar to EI spectra. Some reagent gases exhibit a combination of mechanisms, e.g. hydrogen, and combinations of reagent gases, e.g. argon and water have been used to produce spectra exhibiting EI and CI characteristics.

Of particular importance in the pesticide area is negative ion chemical ionization mass spectrometry. This is because the technique enables higher sensitivities to be obtained for compounds with high electron affinities. Additional selectivity can be introduced; fragment ions formed can provide complementary structural information to that obtained in EI and positive CI and finally chloride attachment negative CI can be a very powerful technique for providing relative molecular mass data. There are three important ion-forming reactions which occur in negative ion CI:

(a) Resonance capture of a thermalized electron

$$M + e^- \longrightarrow M^{-\cdot}$$

(b) Dissociative resonance capture of a low energy electron

$$AB + e^- \longrightarrow A^{\cdot} + B^-$$

(c) Ion–molecule reactions between ions in the ion source and neutrals.

A large number of reactions of this type can occur including charge transfer, proton transfer, hydride transfer, oxygen exchange reactions with halogen or hydrogen and anion-molecule adduct formation. For a more detailed exposition of the technique the reader is referred to reviews on the subject (Jennings, 1977; Brandenberger, 1978; 1979; Hunt and Crow, 1979; Hass, 1980; Dougherty, 1981a). Its possible utility in the identification of metabolic products will be discussed later.

CI assists in providing relative molecular mass information for compounds which do not provide this data in conventional EI mass spectrometry, because the molecules readily fragment. The technique does not, however, necessarily help for thermally labile or low volatility compounds if conventional methods of sample supply, i.e. direct insertion probe or gas chromatography, are used.

As a result a number of different methods of sample supply have been developed which reduce the likelihood of thermal decomposition of the sample and hence enable more difficult compounds to be handled. These include rapid heating from an inert surface (Beuhler *et al.*, 1974), extension of the probe so that the sample is in the ion plasma and use of a variety of inert surfaces on the probe (Baldwin and McLafferty, 1973a; Cotter, 1979; Cotter and Fenselau, 1979; Thenot *et al.*, 1979; Carroll *et al.*, 1979a; Hansen and Munson, 1978; 1980), introduction of the sample in solution and using the solvent as a CI reagent gas (Baldwin and McLafferty, 1973b) and desorption chemical ionization (DCI) (Hunt *et al.*, 1977). The latter technique involved loading a solution of the sample on to an activated (Hunt *et al.*, 1977) or a non-activated emitter (Arpino and Devant, 1979; Rapp *et al.*, 1980; Sphon *et al.*, 1980; Bruins, 1980a; b; Beaugrand and Devant, 1980), similar to the type used in field desorption (FD) mass spectrometry and rapid heating of the emitter. Impressive data from difficult molecules including glycosides and glucuronides have been obtained in both the positive and negative CI modes (Arpino and Devant, 1979; Rapp *et al.*, 1980; Sphon *et al.*, 1980; Bruins, 1980a; b; 1981; Beaugrand and Devant, 1980; Hostettmann *et al.*, 1981; Gower *et al.*, 1981; Esmans *et al.*, 1980). The technique considerably extends the range of compounds which provide relative molecular mass data from CI and the technique should have considerable utility in metabolic studies particularly in providing spectra from conjugated metabolites.

Field ionization and field desorption mass spectrometry

Ionization of a vaporized sample may be effected by use of an intense electric field (of the order of 10^9–10^{10} Vm^{-1}). This technique is called field ionization (FI) (Beckey, 1969; 1971; Milne and Lacey, 1974; Beckey, 1977); in this form of ionization the quantity of excess energy imparted to the molecular ion is usually small resulting in the presence of abundant molecular ions in the spectra produced. The sample is supplied by conventional methods including gas chromatography and hence, although the technique is an excellent method for confirmation of relative molecular mass, it provides relatively few fragment ions and it suffers from the same problems as do EI and conventional CI with thermally labile and/or low volatility compounds. A further problem is that the technique is relatively insensitive due to poor sample ionization efficiency. With the advent of other forms of ionization the technique is of little utility in areas relevant to this review. However, the use of FI GC–MS may have utility in some metabolic studies where positive or negative CI has failed to provide satisfactory relative molecular mass data (Damico and Barron, 1971; Games *et al.*, 1973; 1974; 1976; 1978; Milberg and Cook, 1979; Zune *et al.*, 1976; Knight *et al.*, 1980). Figure 3 shows the EI and FI spectra obtained from lindane (γ-HCH, **4**) during GC–MS and illustrates the type of spectra obtained using the technique.

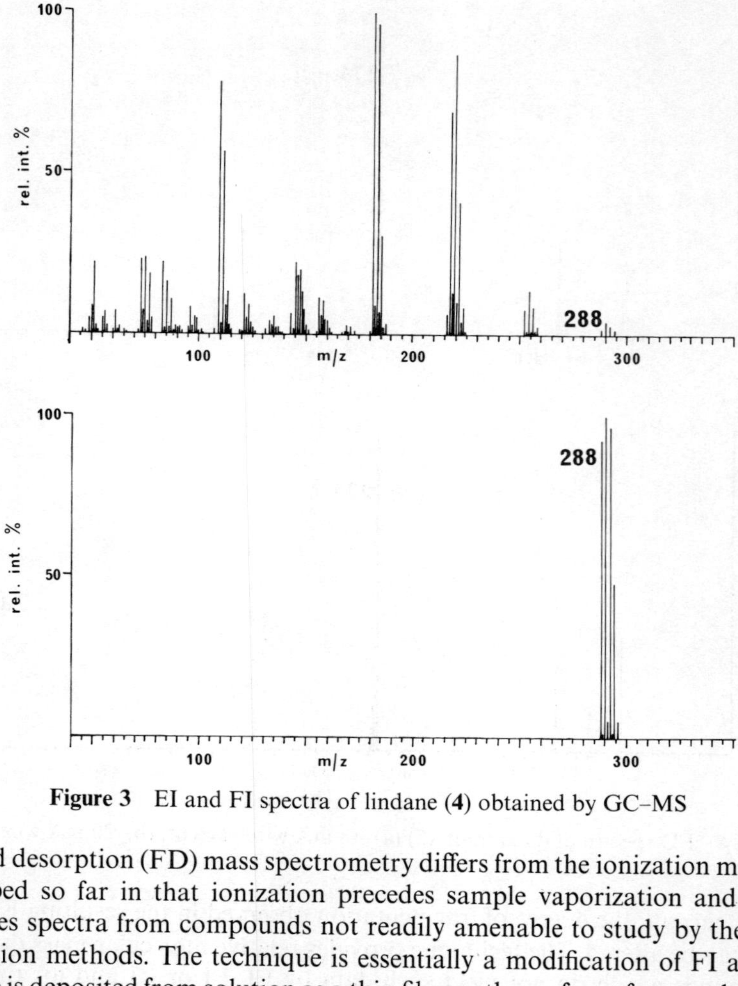

Figure 3 EI and FI spectra of lindane (**4**) obtained by GC–MS

Field desorption (FD) mass spectrometry differs from the ionization methods described so far in that ionization precedes sample vaporization and hence provides spectra from compounds not readily amenable to study by the other ionization methods. The technique is essentially a modification of FI and the sample is deposited from solution as a thin film on the surface of an anode which is normally a thin tungsten wire activated so that it has a coating of carbon needles. The anode is then inserted into a FI source through a vacuum lock and the field is applied. Ionization occurs and the cations formed are strongly repelled by the anode resulting in desorption of the ion. A current is normally passed through the wire to assist the desorption process and, by careful control

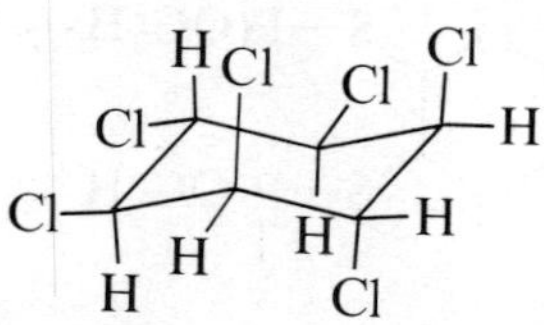

(**4**)

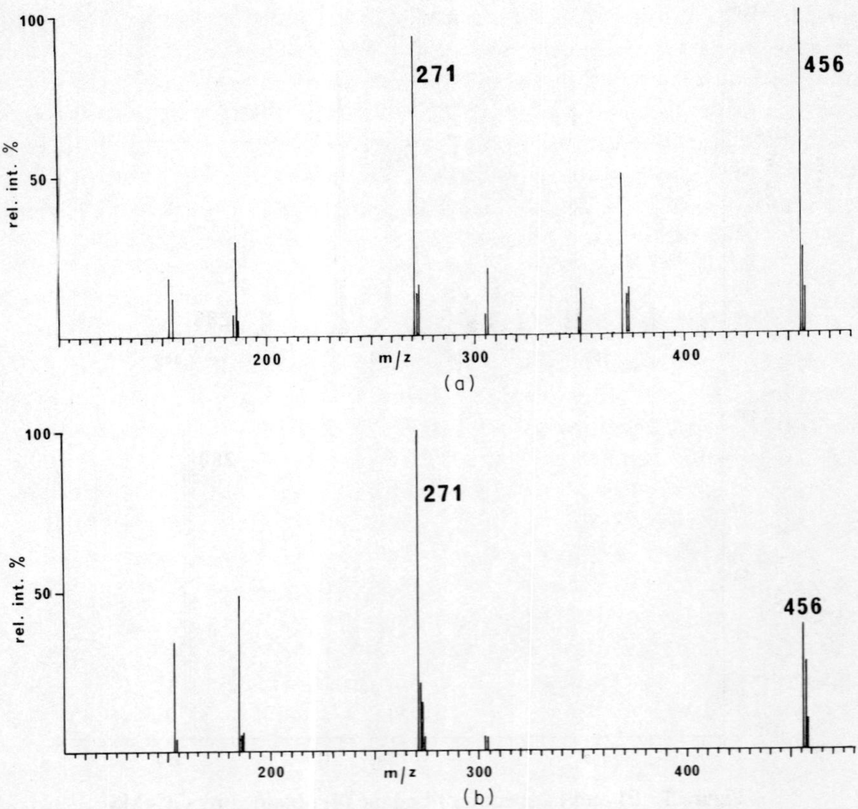

Figure 4 FD spectra of dioxathion (**5**) (a) 18 mA wire current, (b) 20 mA wire current

of this current, the degree of fragmentation observed in the resulting FD spectrum can be altered. The technique provides relative molecular mass data from compounds which do not yield such data by EI, FI or CI and for thermally labile compounds, structurally useful fragment ions are present. Figure 4 shows FD spectra obtained from dioxathion (**5**) and illustrates the extra fragmentation which is induced at higher wire currents in FD spectra. This compound fails to

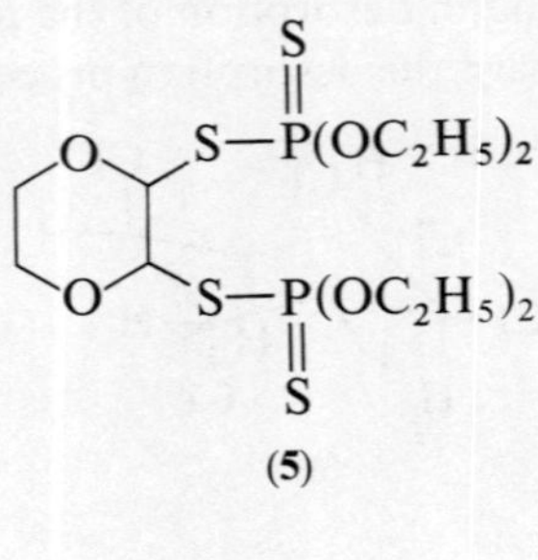

(**5**)

provide relative molecular mass data in its EI and CI spectra (Fales *et al.*, 1975). Although the technique works very well for standard samples problems are often encountered with polar molecules isolated from biological samples unless considerable care has been taken during work-up to remove excess alkali metal ions. Interpretation of spectra can sometimes be complicated since $M^{+\cdot}$, $[M + 1]^+$ and cationized molecular ions can occur. The reader is referred to reviews on FD for a detailed discussion of the finer points of the technique (Beckey, 1977; Beckey and Schulten, 1975; 1979; Schulten, 1979).

^{252}Cf-Plasma desorption mass spectrometry

Extremely rapid heating can avoid thermal decomposition of heat-sensitive compounds, since there is insufficient time for the thermal decomposition of the sample to occur before it is vaporized. This principle is exploited in ^{252}Cf-plasma desorption mass spectrometry (Macfarlane, 1980). The sample to be analysed is deposited on a thin metal foil which is then placed in a ^{252}Cf fission fragment source. When a ^{252}Cf nucleus undergoes spontaneous fission it splits into two ions with $\sim$ 100 MeV energy. The ions have sufficient energy to pass through the thin foil. When the fission fragment passes through the metal foil extremely rapid local heating occurs. This results in the desorption of the molecules with production of positive and negative ions which are then accelerated out of the ion source into a time-of-flight analyser system. A number of systems of this type have been built but the technique is not commercially available. Until recently it had advantages in terms of the ability to provide relative molecular mass data from compounds not amenable to studies by other ionization methods. However, fast atom bombardment mass spectrometry appears to be equally able to handle many of these problem molecules. The current utility of ^{252}Cf-plasma desorption mass spectrometry lies in its ability to handle molecules of relatively high mass (10 000 daltons) because of the use of time-of-flight mass analysis. It has recently been shown that low-energy ion bombardment using kiloelectronvolt K^+ and Cs^+ ions provides similar spectra to those obtained with ^{252}Cf-plasma desorption mass spectrometry (Ens *et al.*, 1981).

Secondary ion and fast atom bombardment mass spectrometry

Secondary ion mass spectrometry (SIMS) has long been utilized for the detection of metals on surface layers of metallic and non-metallic materials. In recent years it has been shown also to be useful for the provision of mass spectra from low volatility, thermally labile polar molecules (Benninghoven, 1979; Day *et al.*, 1980). The sample is deposited on a metal surface and is then bombarded with ions, usually argon ions. Data reported to date are of molecules of comparatively low relative molecular mass and FD mass spectrometry would appear to be a superior technique, since it has been demonstrated to be capable of providing relative molecular mass data from a wider range of compounds and from much larger molecules (Morris *et al.*, 1981). This may, however, be a somewhat unfair criticism since the instrumentation used to date for SIMS is of limited performance. The main utility of SIMS appears to lie in its ability to selectively ionize ionic compounds (Unger *et al.*, 1981) and its possible utility with combined high-performance liquid chromatography–mass spectrometry (LC–MS) (Benninghoven *et al.*, 1980; Smith *et al.*, 1981). Preliminary data have appeared showing that SIMS can be used with LC–MS. However, the data are of a rather crude nature and it is difficult to ascertain, if this approach to LC–MS is in fact viable.

The most exciting recent development in mass spectrometry has been the advent of fast atom bombardment (FAB). The technique is closely related to SIMS, but uses a beam of neutral argon atoms to bombard the sample. There is also a difference in sample loading techniques, in that the sample is in a matrix system, usually of glycerol. Spectacular results have been obtained with the technique from low volatility, labile polar molecules (Barber *et al.*, 1981). In addition to providing relative molecular mass data from compounds not amenable to FD study, the technique provides an abundance of structurally useful fragment ions. It is easy to use and the technique has also been shown to be capable of providing mass spectral data from large biological molecules in excess of 3000 daltons. Although in its infancy, the rapidity with which it has become commercially available testifies as to its importance. In the pesticide metabolism area it should have considerable utility for the characterization of conjugated metabolites and preliminary data indicates that LC–MS with FAB is a real possibility (Hunt *et al.*, 1981).

Other ionization methods

The ionization methods already discussed should be capable of providing mass spectral data from most types of pesticide metabolic products. Three other ionization methods are worthy of some mention particularly in view of their potential utility with LC–MS.

Laser ionization

The use of a pulsed laser to ionize compounds on a metal surface has been shown to be a very effective method for providing spectra from low volatility thermally labile polar molecules (Posthumus *et al.*, 1978; Heinen, 1981). With this approach, either electro-optical ion detection or photoplate detection have to be used if a magnetic instrument is used or a time-of-flight instrument has to be used. A repetitive pulsed laser enables a magnetic instrument with conventional electrical recording to be used (Heresch *et al.*, 1980) or a continuous wave laser can be used with a quadrupole mass spectrometer (Stoll and Röllgen, 1979). Recently laser ionization has been used with a moving stainless steel belt (Hardin and Vestal, 1981). Although good spectra have been obtained, they are not yet sufficiently reproducible for the system to be used for LC–MS at the present time.

An alternative approach has been to use a pulsed laser to produce neutrals which are subsequently ionized using a CI reagent gas (Cotter, 1980). This approach has also been used with a moving belt LC–MS interface (Hunt *et al.*, 1980). However, as yet no on-line LC–MS data have been reported. Improved FD mass spectra can also be obtained by use of a laser (Schulten *et al.*, 1978).

Rapid heating of a liquid solution

As a spin-off of studies directed towards development of a LC–MS interface a new form of ionization method has been discovered (Blakely *et al.*, 1980). In order to effect vaporization of the LC–MS eluent, it is rapidly heated in a metal tube. This results in the formation of a particle beam which is impacted on a mildly heated surface and results in the formation of ions from the solute. Addition of ionic components to the solution is required and excellent mass spectral data from difficult compounds have been reported. This approach to LC–MS appears to be a very promising one for the handling of more difficult compounds.

Atmospheric pressure ionization

Atmospheric pressure ionization (API) can be an extremely sensitive technique for the detection of organic compounds with high electron affinities when used in the negative ionization mode (Carroll, 1979b). Until recently the technique has been limited to studies of compounds of relatively high volatility. Two approaches have been described which enable more difficult samples to be handled. One involves ionization of the samples in the liquid phase or in solution using argon, helium or nitrogen ions formed by a corona discharge (Tsuchiya and Taira, 1980). Although no particularly difficult molecules have yet been studied the technique shows promise as an LC–MS method, particularly since a method for inducing more fragmentation into the API spectra is also described. The other approach (Thomson and Iribarne, 1979) which involves field induced evaporation of ions from liquid surfaces in an API source shows promise as a technique for handling of more difficult compounds and would also appear to be amenable to use for LC–MS.

COMBINED HIGH-PERFORMANCE LIQUID CHROMATOGRAPHY–MASS SPECTROMETRY (LC–MS)

The reasons for the requirement of LC–MS for studies in this area were discussed earlier. A variety of different types of liquid chromatographic–mass spectrometric interface have been described and the reader is referred to recent reviews on the subject (McFadden, 1979; 1980; Arpino and Guiochon, 1979; Games, 1980a; b; 1981a; b; 1982) for full discussion of the various methods used. We will confine ourselves here to the two types of approach which are currently commercially available.

Direct liquid introduction systems

In this approach a portion of the LC eluent is fed directly into the ion source of a mass spectrometer configured for CI (Arpino *et al.*, 1974). The solvent is ionized and acts as a CI reagent gas for the solute, producing CI spectra from the solute. Polar solvents have to be used if spectra resembling conventional CI spectra of the type formed from acid-base reactions are required. Recently it has been shown that ammonia can be used as a CI reagent gas with systems of this type (Arpino *et al.*, 1981). The amount of eluent from the LC which is passed to the mass spectrometer depends on the flow rates used for LC and the pumping system of the mass spectrometer. If conventional columns are used, of the order of 5% of the eluent can be fed into the mass spectrometer ion source. This of course places a restriction on the sensitivity that can be obtained. However, use of microbore LC enables all the eluent to be handled resulting in considerably improved sensitivity (Henion and Maylin, 1980; Henion, 1981; Schafer and Levsen, 1981). Although very simple interfaces have been used in this approach they encounter problems in handling even moderately volatile materials, resulting in blocking of the interface. In order to overcome this problem and to take full advantage of the 'direct chemical ionization' effect observed when samples are introduced in solution (Baldwin and McLafferty, 1973b) a small diameter diaphragm has to be introduced at the end of the probe to enable nebulization of the eluent to occur (Arpino *et al.*, 1981). Use of this type of interface enables quite difficult molecules, e.g. glycosides to be readily handled. Systems of this type are commercially available from Hewlett-Packard (Melera, 1980) and Nermag (Arpino *et al.*, 1981). They have been shown to work well and should be of considerable assistance in metabolic studies.

Moving belt systems

With systems of this type the LC eluent is fed on to a continuous moving belt made of Kapton®. Solvent is removed by an infra-red heater between two vacuum locks. The residual solute is then flash vaporized into the source of a mass spectrometer, where EI or CI spectra are obtained. A clean-up heater is included to remove residual solute. Systems of this type are commercially available from Finnigan (McFadden *et al.*, 1976; 1977) Finnigan MAT (Brunnee *et al.*, 1981) and V. G. Analytical (Millington *et al.*, 1980). The former system was the first to be developed and the interface was sited outside the ion source. The latter systems enter the ion source resulting in improved sensitivity and in the case of the Finnigan MAT system improved ability in the handling of more difficult compounds. Data reported using this system shows its ability in this context, to be equivalent to desorption CI, whereas from data available to date, the other systems do not appear to be as good at handling thermally labile low volatility compounds as desorption CI. As with the direct liquid introduction system, the use of microbore LC leads to improved sensitivity (Westwood *et al.*, 1982). One of the major problems with this type of interface is the difficulty experienced in handling mobile phases with high percentage water contents. A number of solutions to the problem have been developed and the reader is referred to a recent review (Alcock *et al.*, 1982) for further details. Systems of this type have been effectively used in a wide range of areas and should have considerable utility in studies of pesticide metabolism (Alcock *et al.*, 1982).

APPLICATIONS AND POTENTIAL APPLICATIONS OF SOFT IONIZATION METHODS

Compared with the utilization of EI mass spectrometry, soft ionization mass spectrometric techniques have had comparatively little application. In addition to describing areas where the techniques have been applied in metabolite identification, studies which indicate areas of potential utility will also be included. The techniques have often proved to be most effective when used in combination with EI mass spectrometry.

Chemical ionization

This has been the most widely utilized technique and has proved to be particularly effective when used with GC–MS.

Organophosphorous compounds

The methane CI mass spectra of 29 organophosphorous insecticides and their metabolites have been reported (Holmstead and Casida, 1974). In many cases the $[M + H]^+$ was of significant abundance enabling relative molecular masses to be more readily assigned than with the EI spectra, which often fail to show molecular ions. Structurally useful fragment ions were also produced in the CI spectra complementing those observed in the EI spectra. Variation in structural types produced marked changes in this fragmentation behaviour. Comparison of the EI, methane, isobutane, and methanol CI spectra obtained during capillary GC–MS of 23 pesticides of the phosphate, phosphorothiolate, phosphorothionate and phosphorodithioate types (Stan, 1977a) indicated that isobutane was the CI reagent gas of choice for routine identification, providing abundant protonated molecular ions and significant fragmentations. This approach has been applied in the detection of organophosphorous pesticide residues in food at the ppb level (Stan, 1977b). In another comparative study (Busch *et al.*, 1978), the EI, methane positive CI and methane, methane–oxygen mixture and oxygen negative ion CI spectra of 16 organophosphorous pesticides were studied. Methane positive CI proved to be the method of choice for provision of relative molecular mass data in most cases except for azinphos-methyl, azinphos-ethyl and carbophenothion. The negative ion CI spectra with all reagent gases showed intense peaks for phosphate and phenate anions which complement the fragment ions observed in EI and positive CI. Thus a combination of positive and negative CI should be of considerable assistance in assignments of structure to metabolic products. The sensitivity of the negative ion methods was greater than either the EI or methane positive CI and in selected cases can be up to 800 times more sensitive. Chloride attachment negative CI mass spectra of organophosphorous pesticides also show abundant phosphate anions and phenate anions, but have the advantage of also having $[M + Cl]^-$ ions present in most cases, which provide relative molecular mass data (Dougherty and Wander, 1980).

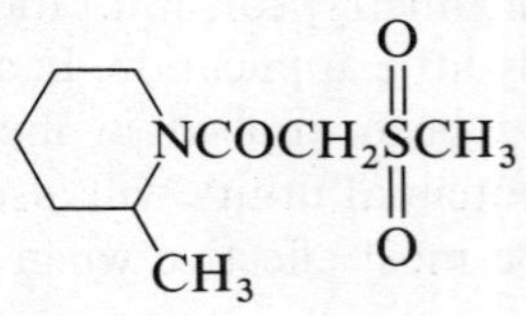

(6)

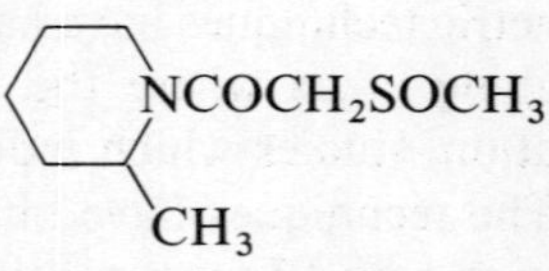

(7)

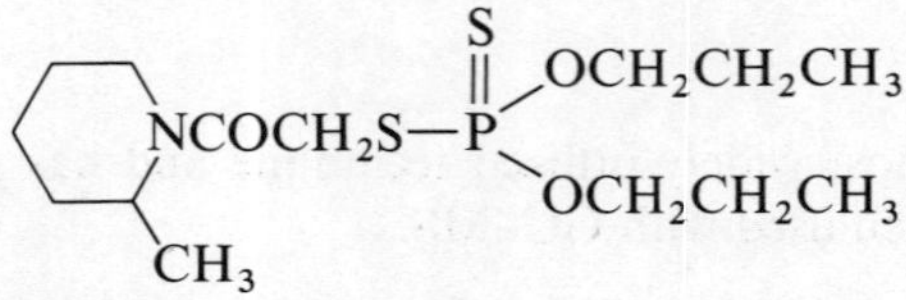

(8)

Methane CI has been used in the assignment of the relative molecular masses of the metabolites (**6**) and (**7**) of the organophosphorous pesticide (**8**) (Mayer *et al.*, 1981a). Compound (**6**) showed a highest mass ion at *m/z* 204 $[M - CH_3]^+$ in EI whereas in CI it exhibited an $[M + 1]^+$ ion at *m/z* 220. In the case of (**7**) the highest mass ion in its EI spectrum was *m/z* 140 $[M - SOCH_3]^+$, but the CI spectrum showed as $[M + 1]^+$ ion at *m/z* 204. Other examples of the use of methane CI include the identification of metabolites, photoproducts and oxidative degradation products of dioxathion (Harned and Casida, 1976) and the identification of the cysteine conjugate and demonomethyl methidathion as major polar metabolites of methidathion in tomatoes (Simoneaux *et al.*, 1980).

In most metabolic studies, radioactively labelled pesticides are fed to the biological system and the radioactivity is followed to enable recovery of metabolic products. Studies of the photodegradation of the systemic insecticide cytrolane (mephosfolan) (**9**) where approximately equimolar mixtures of either $^{12}C/^{13}C$-imido-labelled (**9a**) or $^{12}C/^{13}C$-ethyl-labelled material (**9b**) were used

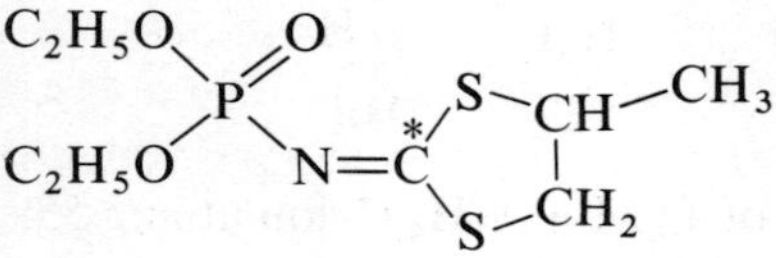

(**9a**)

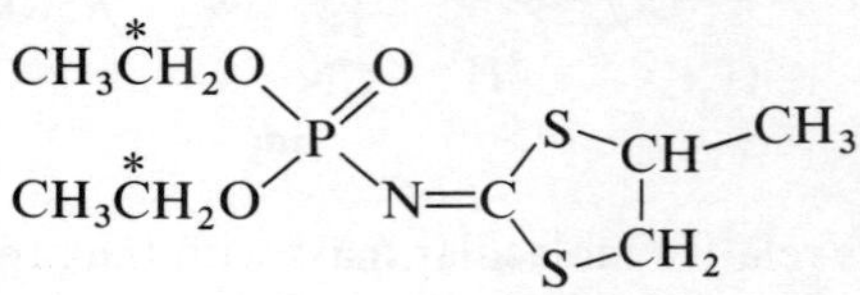

(**9b**)

illustrate an alternative approach (Ku *et al.*, 1979). Use of these mixtures facilitated identification of the photo-products by GC-methane CI mass spectrometry because the molecular and fragment ions in the mass spectra appeared as characteristic doublets. Endogenous plant materials gave only singlet ions in their mass spectra and thus could be ignored. Using this approach, the major photodegradation products were identified as cyclic *S*,*S*-propylene dithiocarbonate, 2-imino-5-methyl-1,3-dithiolane and diethylphosphate, with ethyl phosphate, phosphoric acid and diethyl methyl phosphate being minor products.

Carbamates

The CI mass spectra of a series of *N*-methylcarbamate insecticides and their metabolites have been reported (Holmstead and Casida, 1975), and the EI and isobutane CI spectra of some aromatic carbamates and their corresponding phenols have been studied (Jankowski and Paré, 1980). One of the major problems with compounds of this class is that they undergo thermal decomposition on GC. Hence they either have to be derivatized prior to GC–MS study or examined by LC–MS. An example of the former approach is the determination of carbofuran and its metabolites as heptafluorobutyryl derivatives by methane CI GC–MS in crop residues (Chapman and Robinson, 1977). The other approach will be discussed later.

Pyrethrins

The pyrethrins, natural and synthetic often exhibit weak molecular ions in their EI spectra. Isobutane CI has been shown to give protonated molecular ions for six naturally occurring pyrethrin insecticides (Holmstead and

Soderlund, 1977). Methane CI has assisted in the studies of the photodecomposition products of resmethrin (Ueda *et al.*, 1974) and kadethrin (Ohsawa and Casida, 1979). In studies of the metabolism of cypermethrin (**10**) in plants, ammonia CI of a tetraacetyl-β-D-glucopyranosyl ester of the acid (**11**) confirmed

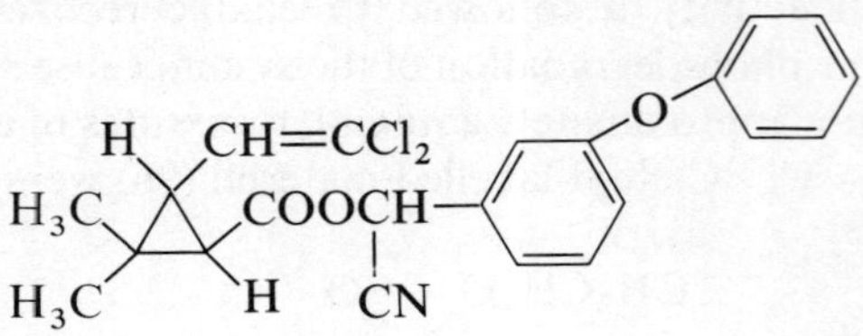

(**10**)

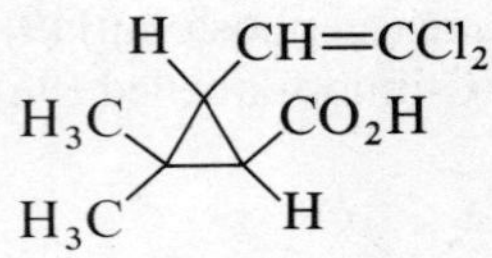

(**11**)

its relative molecular mass with the presence of a $[M + NH_4]^+$ ion at m/z 556 and assisted in indicating the presence of disaccharide conjugates when the metabolism of the acid (**11**) a metabolite of cypermethrin was studied in separate experiments (Wright *et al.*, 1980). Ammonia CI has also assisted in the identification of β-D-glucopyranose 3-phenoxybenzyl ether as a metabolite of 3-phenoxybenzyl alcohol, a metabolite of a number of synthetic pyrethroids including cypermethrin, in abscised cotton leaves (Roberts and Wright, 1981).

Chlorinated hydrocarbons

Since molecular ions in the EI spectra of many of these classes of compound and their metabolites are of low abundance or are absent, CI has been extensively used for their identification.

The positive methane CI spectra of polychlorinated hydrocarbons and their metabolites of the cyclodiene types are less complex than their EI spectra (Biros *et al.*, 1972; Dougherty *et al.*, 1974). Protonated molecular ions were present for all oxygenated molecular species and those with unsubstituted double bonds. Other structural types gave molecular ions. In general the most abundant ions in the spectra were due to elimination of Cl or OH from the molecular ion and ions due to retro-Diels–Alder fragmentation, which can be particularly useful in metabolite identification (McKinney *et al.*, 1974; Oswald *et al.*, 1974a). Isobutane CI yields similar spectra to that reported with methane, but with less fragmentation (McKinney *et al.*, 1974; Oswald *et al.*, 1974a). In their negative ion CI spectra using methane, isobutane or methylene chloride as enhancement gas, all the spectra gave an $[M + Cl]^-$ ion as the base peak and usually molecular anions of substantial relative abundance (Dougherty *et al.*, 1972; 1974).

Methane and isobutane CI with GC–MS have been very useful in elucidating the metabolic products of endrin in pine mice (Petrella *et al.*, 1977; Oswald, *et al.*, 1974a) and the metabolic products of aldrin and dieldrin (Oswald *et al.*, 1974a; McKinney *et al.*, 1974). A similar approach was used in the identification of the metabolites of *cis*- and *trans*-chlordane by a soil microorganism (Beeman and Matsumura, 1981). Examination of the retro-Diels–Alder fragment ions was particularly useful in elucidation of the nature of the dehydrochlorinated and oxygenated metabolites. Methane CI with GC–MS has been used to identify the degradation products of mirex (**12**) in the environment (Carlson *et al.*, 1976). Compounds derived from mirex by successive losses of chlorine were found together with kepone (**13**). Using similar techniques, mirex was found to

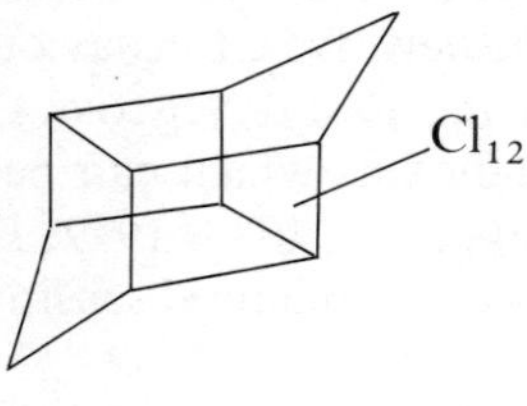

(**12**)

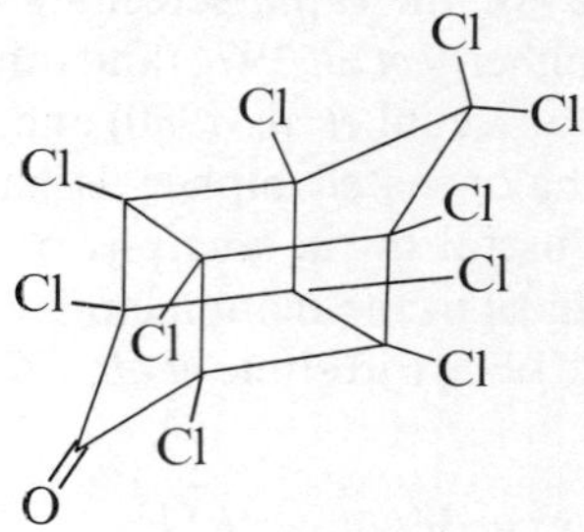

(**13**)

undergo reductive dechlorination with an iron(II) porphyrin model system (Holmstead, 1976). Kepone (**13**) exhibits a weak $M^{+\cdot}$ ion at *m/z* 486 in its EI spectrum; its methane CI spectrum shows an abundant $[M + 1]^+$ ion at *m/z* 487. GC–MS with methane CI was used to identify kepone in environmental samples together with monohydrokepone, one of its degradation products, and assisted in the identification of kepol (the reduction product of kepone in which the ketone has been reduced to a secondary alcohol) in a human faeces extract (Harless *et al.*, 1978).

Chlorinated hydrocarbon pesticides and their metabolites of the DDT type have been the subject of a number of detailed CI mass spectrometric studies (McKinney *et al.*, 1974; Dougherty *et al.*, 1974; 1975; Oswald *et al.*, 1974a). Studies in the positive CI mode with methane as the reagent gas showed no appreciable advantages for the saturated and unsaturated DDT types over EI and relative molecular mass assignment was complicated since $M^{+\cdot}$ ions were usually observed. However, for the oxidative metabolites the use of CI was advantageous enabling relative molecular mass to be assigned which would not have been possible from the EI spectra (McKinney *et al.*, 1974; Oswald *et al.*, 1974a). Isobutane CI in the positive mode provides more abundant ions in the molecular ion region for these compounds than EI or CI and negative CI with isobutane was particularly useful for relative molecular mass assignment and for screening of samples since $[M + Cl]^-$ ions were usually the base peak in the spectra (Dougherty *et al.*, 1974; 1975).

Although polychlorinated biphenyls are not within the scope of this chapter, they have been extensively studied and are worthy of some discussion. They yield protonated molecular ions in their positive CI spectra (Dougherty *et al.*,

1974; Oswald *et al.*, 1974a; b). The EI spectra are structurally much more useful than the CI spectra and preferable for structural studies of the oxygenated metabolites of the chlorinated biphenyls with lower numbers of chlorine atoms. However, for studies of the metabolites of the higher polychlorinated biphenyls, CI is useful for relative molecular mass assignment and the technique has been used in studies of the metabolism of 2,4,4′-trichlorobiphenyl by *Acinetobacter* species P6 (Furukawa *et al.*, 1979). Negative CI has been advocated for the rapid screening of tissue extracts for polychlorinated biphenyls (Dougherty *et al.*, 1974) and other classes of chlorinated compounds (Dougherty, 1981b; Keuhl *et al.*, 1980) and a detailed study of the negative CI behaviour of polyhalogenated biphenyls has appeared (Crow *et al.*, 1981). Negative CI is also useful in the analysis of tetrachlorodibenzo-*p*-dioxins which can be by-products in the manufacture of some pesticides (Hass *et al.*, 1978; 1979; Hunt *et al.*, 1975, Firestone *et al.*, 1979) particularly in view of its enhanced sensitivity.

Other classes of pesticides

Using a combination of CI mass spectrometry with feeding of equal amounts of unlabelled and ^{13}C-labelled warfarin to a rat liver microsomal preparation, a new metabolite of this rodenticide was located (Pohl *et al.*, 1975). Subsequent feeding of ^{2}H-labelled warfarin in the benzylic position together with unlabelled material, which resulted in the loss of this label, together with analysis of the mass spectrum of the ^{13}C-labelled metabolite, strongly indicates (**14**) as the

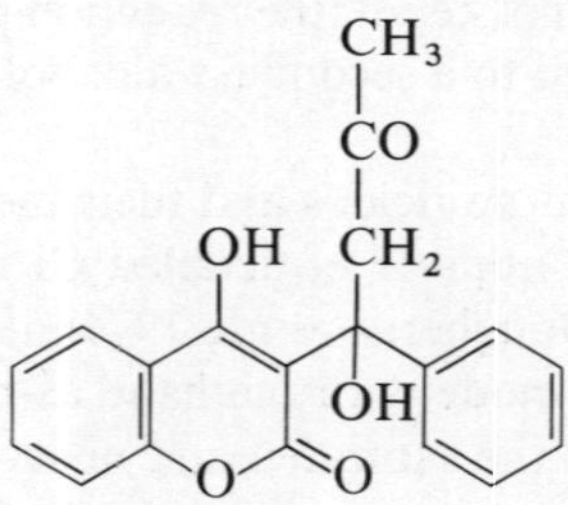

(**14**)

structure of the metabolite. Further studies using ^{2}H-labelled material confirmed the presence of this metabolite together with other known metabolites.

The EI and CI isobutane mass spectra of the TMS derivatives of six suspected simazine metabolites have been studied by GC–MS (Lusby and Kearney, 1978). EI mass spectra proved to be structurally more useful, with the CI spectra providing confirmation of relative molecular masses. For the identification of the sulphoxide and sulphone metabolites of 5,6-dihydro-2-methyl-1,4-oxathiin-3-carboxanilide (**15**), EI mass spectra were preferable to methane or hydrogen CI spectra (Onuska *et al.*, 1976). However, the CI mass spectrum of the metabolite (**16**) is structurally more informative than its EI spectrum.

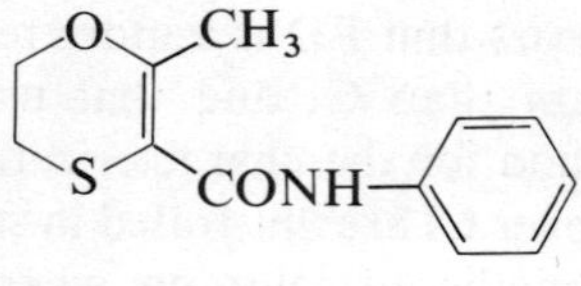

(15)

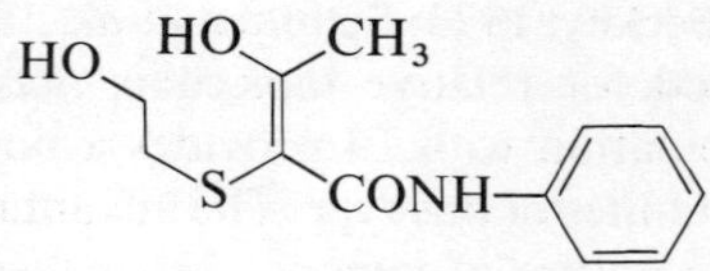

(16)

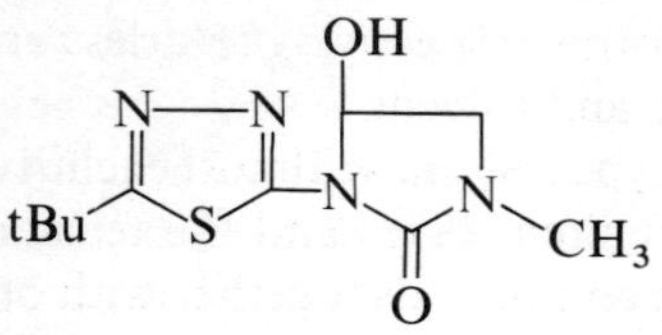

(17)

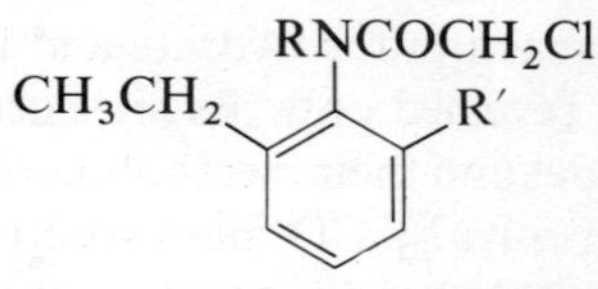

(18a) R = $CH_2COOCH_2CH_3$, R′ = CH_3CH_2
(18b) R = $CH(CH_3)CH_2COCH_3$, R′ = CH_3

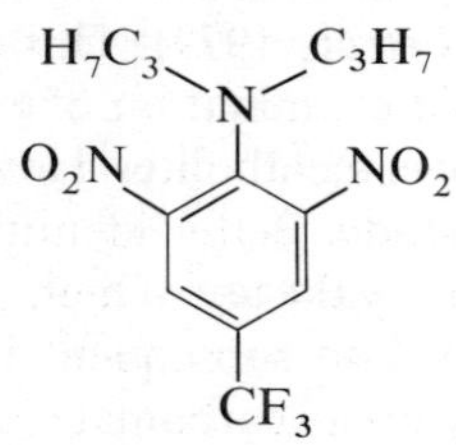

(19)

In most studies, CI mass spectrometry has been used to confirm the relative molecular masses of metabolic products. Examples of the techniques used in this context include metabolic studies of the herbicide buthidazole (**17**) in corn seedlings and alfalfa (Yu *et al.*, 1980), the herbicides diethatyl-ethyl (**18a**) and metolachlor (**18b**) by a soil fungus (McGahen and Tiedje, 1978), the insect repellent, *N*,*N*-diethyl-*m*-toluamide (DEET) in man (Wu *et al.*, 1979), the selective pre-emergence herbicide trifluralin (**19**) in soil (Wheeler *et al.*, 1979), the herbicide 2,6-dinitro-4(trifluoromethyl)benzenamine by *Streptomyces* isolated from soil (Lusby *et al.*, 1980) and the identification of the sensitized photo-oxidation products of bromacil in water (Acher and Dunkelblum, 1979).

Field ionization and field desorption mass spectrometry

The EI and FI mass spectra of some pesticides have been compared (Damico *et al.*, 1969). FI gave almost exclusively molecular ions. FI has been used for the quantification of the herbicide trifluralin in rat tissues and excreta (Heck *et al.*, 1977) and FI with GC–MS has been shown to be useful in providing relative molecular mass data from mixtures of polychlorinated biphenyls and in a study of the metabolism of a phosphorodithioate pesticide (Milberg and Cook, 1979).

Polychlorinated hydrocarbons and their metabolites of the cyclodiene types give $M^{+\cdot}$ ions by FD mass spectrometry with little fragmentation (Schulten and Beckey, 1973; Schulten *et al.*, 1973). It appears that FD is a more reliable method for relative molecular mass assignment than CI and that used in combination with EI provides a powerful method for the characterization of metabolites of this type. The advantages of FD over CI are illustrated in studies of the nature of kepone (**13**) in benzene and methanol solutions where FD studies confirmed the infra-red and NMR analyses, which indicated that kepone existed in the hydrate and hemiketal forms respectively (Harless *et al.*, 1978). CI revealed only the presence of kepone and solvent.

Pesticides and their metabolites of the DDT type also show almost exclusively molecular ions by FD mass spectrometry (Schulten, 1976) and the technique has been used to characterize mixtures of these compounds together with other chlorinated pesticides in extracts from environmental samples (Suzuki and Yamoto, 1980) and to identify the herbicides benthiocarb, CNP and oxadiazin in aqueous extracts (Yamoto *et al.*, 1979). This approach may be useful as a screening method for the rapid examination of extracts of biological materials for the possible presence of novel metabolites. However, structural identification will require a more detailed study. Better identification is provided for all the presence of known compounds by the use of high-performance liquid chromatography, collection of fractions and subsequent FD study. This approach has been used to identify phenylurea, carbamate and thiocarbamate pesticides (Schulten and Stöber, 1978; Stöber and Schulten, 1980). However, this type of study would be better if EI spectra were also taken since they are structurally more informative and the combination of all three sets of data would provide better confirmatory evidence.

An example of the utility of FD mass spectrometry in combination with other techniques for the characterization of conjugated metabolites is the identification of (**20**) as a metabolite of the wild oat herbicide flamprop-methyl (**21**) (Dutton *et al.*, 1976). Figure 5 shows the FD mass spectrum of the methylated metabolite and provides confirmation of its relative molecular mass. A similar

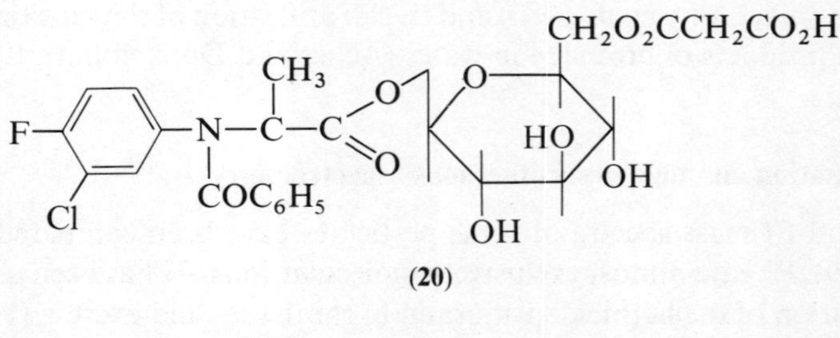

(20)

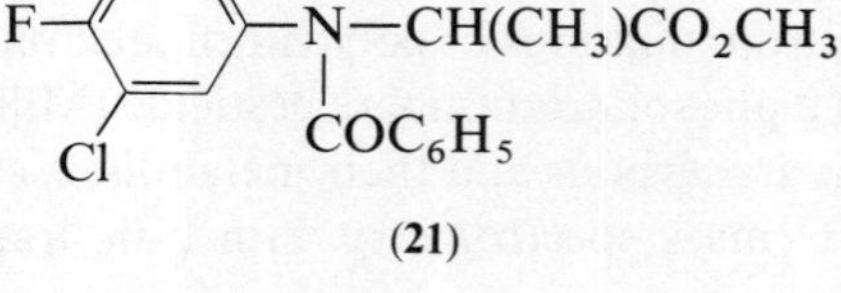

(21)

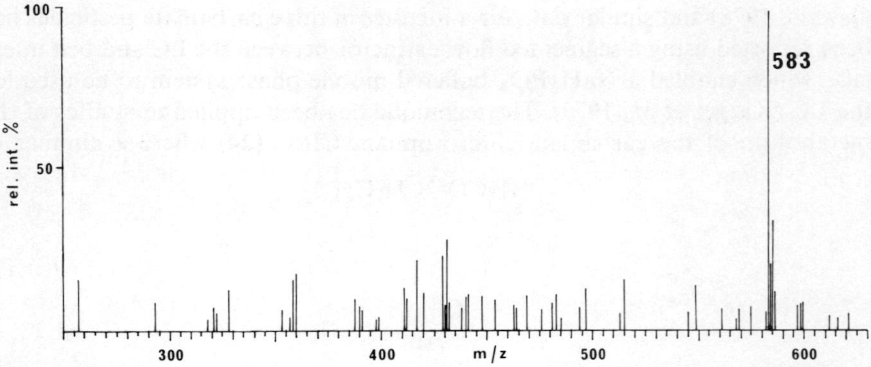

Figure 5 FD spectrum of methylated metabolite (**20**) of flamprop-methyl (**21**). Reproduced by permission of Pergamon Press, from A. J. Dutton, T. R. Roberts and A. N. Wright, *Chemosphere*, 1976, **3**, 195

approach has assisted in the identification of sugar conjugates in plants of 3-phenoxybenzoic acid (More *et al.*, 1978), and 3-phenoxybenzyl alcohol (Roberts and Wright, 1981) where CI failed to provide relative molecular mass information. Other examples of FD mass spectrometry assisting in the identification of metabolites are studies of piperophos (**8**) in rat (Muecke *et al.*, 1981), 3-phenoxybenzoic acid in the mouse (Hutson and Casida, 1978), dimethametryn (**22**) in paddy rice (Mayer, 1981b) and demeton-*S*-methyl sulphoxide (**23**) by soil microorganisms (Ziegler *et al.*, 1980).

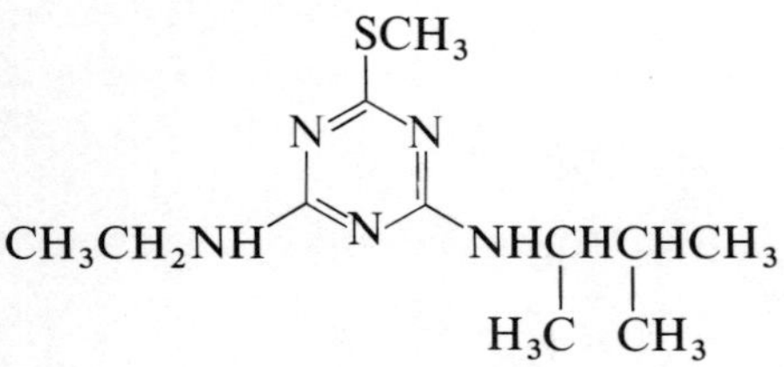

(22)

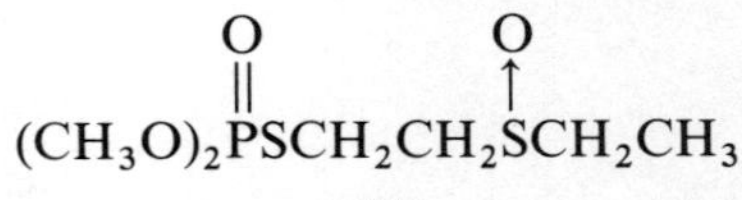

(23)

APPLICATIONS OF COMBINED HIGH-PERFORMANCE LIQUID CHROMATOGRAPHY–MASS SPECTROMETRY

The potential of LC–MS for the studies of pesticides was indicated in the first paper describing a moving belt interface (McFadden *et al.*, 1976), where LC–MS data obtained from a mixture of propoxur, carbaryl, malathion and aldrin were shown and the detection limit for carbaryl in the selected ion monitoring mode was shown to be in the low ng region. A detailed study of a mixture of twenty carbamate pesticides with this system showed that full spectra could be obtained if between 200 and 400 ng of sample were injected on column (Wright and

Oswald, 1978) and similar data for a mixture of three carbamate pesticides has been reported using a segmented flow extractor between the LC and belt interface, which enabled a NaH_2PO_4-buffered mobile phase system to be used for the LC (Karger *et al.*, 1979). The technique has been applied in studies of the metabolism of the carbamate chlorpropham CIPC (**24**) where a number of

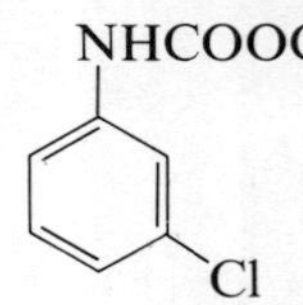

(24)

metabolic products were identified in rat urine extracts and paddy rice extracts (Games and Weerasinghe, 1980; Eckers *et al.*, 1980; Games, 1980a; b). Figure 6 shows the reconstructed total ion current trace obtained from one of these extracts by LC–MS and Figure 7 shows the isobutane CI spectrum

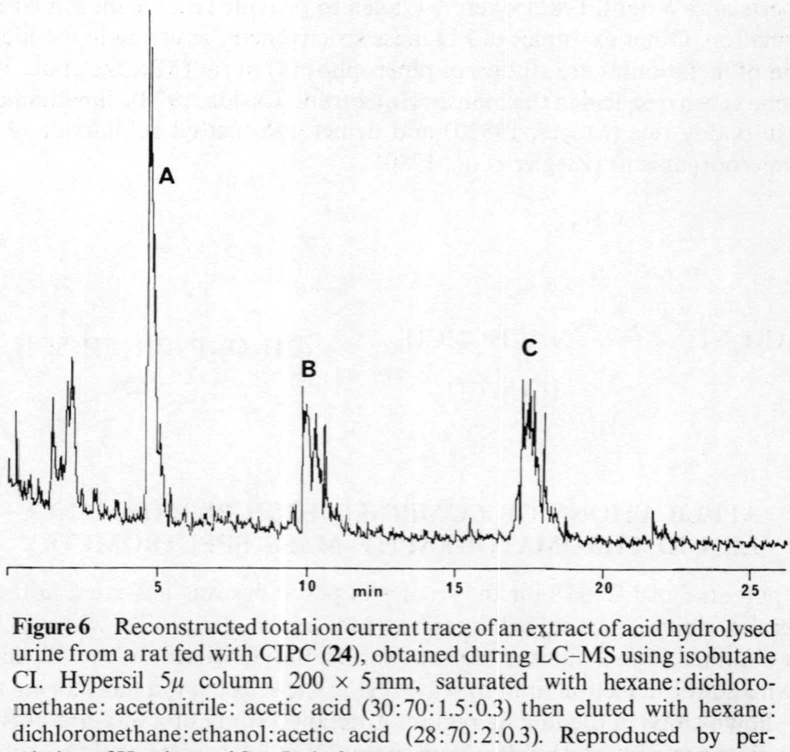

Figure 6 Reconstructed total ion current trace of an extract of acid hydrolysed urine from a rat fed with CIPC (**24**), obtained during LC–MS using isobutane CI. Hypersil 5μ column 200 × 5 mm, saturated with hexane:dichloromethane: acetonitrile: acetic acid (30:70:1.5:0.3) then eluted with hexane: dichloromethane:ethanol:acetic acid (28:70:2:0.3). Reproduced by permission of Heyden and Son Ltd., from D. E. Games, *Biomedical Mass Spectrometry*, 1981, **8**, 454

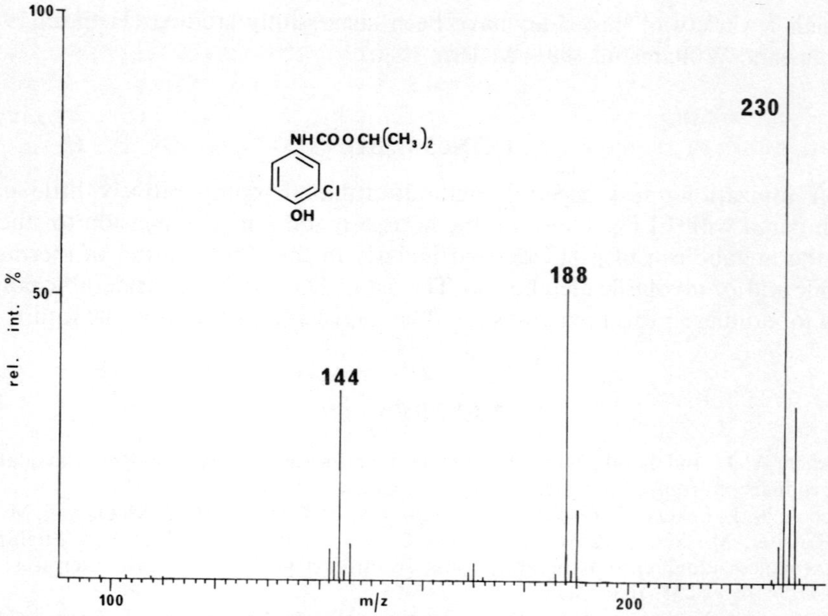

Figure 7 Isobutane CI spectrum obtained from component A (Figure 6). Reproduced by permission of Heyden and Son Ltd., from D. E. Games, *Biomedical Mass Spectrometry*, 1981, **8**, 454

obtained from one of the peaks. EI spectra were also obtained from the components and structural assignments were made on the basis of the LC–MS data and comparison with synthetic samples. Further examples of the utility of this approach are provided by the identification of chlorophenols in human urine (Wright *et al.*, 1981) and the identification of the metabolites of polychlorinated biphenyls (Dymerski *et al.*, 1979). Negative CI with a moving belt LC–MS interface has also been used to identify the photolysis products of the pre-emergence herbicide oryzalin (**25**) (Skinner *et al.*, 1980).

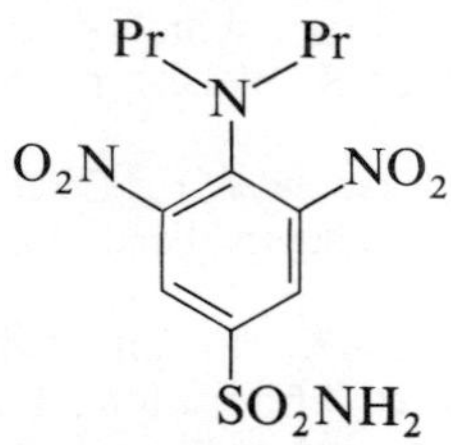

(**25**)

Although there are as yet no examples of the application of other types of LC–MS interfaces in studies of pesticide metabolites. The potential of the direct liquid introduction system in this context is shown by a number of reports in which a variety of pesticides have been successfully studied (Henion, 1978; Evans and Williamson, 1981; Melera, 1980).

CONCLUSIONS

Soft ionization mass spectral methods although comparatively little used compared with EI mass spectral methods can add a new dimension to studies of the metabolism of pesticides particularly in the identification of thermally labile and/or involatile metabolites. The use of LC–MS has considerable potential for studies in this area and should be increasingly utilized in the future.

REFERENCES

Archer, A. J., and Dunkelblum, E. (1979). 'Identification of sensitized photooxidation products of bromacil in water', *J. Agr. Food Chem.*, **27**, 1164–1167.

Alcock, N. J., Eckers, C., Games, D. E., Games, M. P. L., Lant, M. S., McDowall, M. A., Rossiter, M., Smith, R. W., Westwood, S. A., and Wong, H.-Y. (1982). 'High-performance liquid chromatography–mass spectrometry with transport interfaces', *J. Chromatogr.*, **251**, 165–174.

Alford, A. (1975). 'Environmental applications of mass spectrometry', *Biomed. Mass Spectrom.*, **2**, 229–253.

Alford, A. (1977). 'Environmental applications of mass spectrometry', *Biomed. Mass Spectrom.*, **4**, 1–22.

Alford, A. (1978). 'Environmental applications of mass spectrometry', *Biomed. Mass Spectrom.*, **5**, 259–286.

Arpino, P., Baldwin, M. A., and McLafferty, F. W. (1974). 'Liquid chromatography–mass spectrometry. II—continuous monitoring', *Biomed. Mass Spectrom.*, **1**, 80–82.

Arpino, P. J., and Devant, B. (1979). 'Analysis of non-volatile organic substances by desorption/chemical ionization mass spectrometry', *Analysis*, **7**, 348–354.

Arpino, P. J., and Guiochon, G. (1979). 'LC/MS coupling', *Anal. Chem.*, **51**, 682A–701A.

Arpino, P. J., Krien, P., Vajta, S., and Devant, G. (1981). 'Optimization of the instrumental parameters of a combined liquid chromatograph–mass spectrometer, coupled by an interface for direct liquid introduction. II. Nebulization of liquids by diaphragms,' *J. Chromatogr.*, **203**, 117–130.

Arsenault, G. P. (1972). 'Chemical ionization mass spectrometry', in *Biochemical Applications of Mass Spectrometry* (ed. G. R. Waller), pp. 817–832, Wiley–Interscience, New York.

Baldwin, M. A., and McLafferty, F. W. (1973a). 'Direct chemical ionization of relatively involatile samples. Application to underivatized oligopeptides', *Org. Mass Spectrom.*, **7**, 1353–1356.

Baldwin, M. A., and McLafferty, F. W. (1973b). 'Liquid chromatography–mass spectrometry interface—1: The direct introduction of liquid solutions into a chemical ionization mass spectrometer', *Org. Mass Spectrom.*, **7**, 1111–1112.

Barber, M., Bordoli, R. S., Sedgwick, R. D., and Tyler, A. N. (1981). 'Fast atom bombardment of solids as an ion source in mass spectrometry', *Nature*, **293**, 270–275.

Beaugrand, C., and Devant, G. (1980). 'Chemical ionization desorption probe for non-volatile organic compound analysis', in *Advances in Mass Spectrometry* (ed. A. Quayle), Vol. 8, pp. 1806–1811, Heyden, London.

Beckey, H. D. (1969). 'Determination of the structures of organic molecules and quantitative analysis with the field ionization mass spectrometer', *Angew. Chem.*, **8**, 623–639.

Beckey, H. D. (1971). *Field Ionization Mass Spectrometry*, Pergamon, Oxford.

Beckey, H. D. (1977). *Principles of Field Ionization and Field Desorption Mass Spectrometry*, Pergamon, London.

Beckey, H. D., and Schulten, H.-R. (1975). "Field desorption mass spectrometry', *Angew. Chem.*, **14**, 403–415.

Beckey, H. D., and Schulten, H.-R. (1979). 'Field ionization and field desorption mass spectrometry in analytical chemistry', in *Mass Spectrometry, Part A, Vol. 3 in Practical Spectroscopy*' (eds C. Merritt, Jr., and C. N. McEwen), pp. 145–264, Marcel Dekker, New York.

Beeman, R. W., and Matsumura, F. (1981). 'Metabolism of *cis*- and *trans*-chlordane by a soil microorganism', *J. Agr. Food Chem.*, **29**, 84–89.

Benninghoven, A. (1979). 'Organic secondary ion mass spectrometry', in *Trace Organic Analysis: A New Frontier in Analytical Chemistry* (eds H. S. Hertz and S. N. Chesler), pp. 627–635, N.B.S. Washington DC.

Benninghoven, A., Eicke, A., Junack, M., Sichtermann, W., Krizek, J., and Peters, H. (1980). 'Application of a secondary ion mass spectrometer as a detector in liquid chromatography', *Org. Mass Spectrom.*, **15**, 459–462.

Beuhler, R. J., Flanigan, E., Greene, L. J., and Friedman, L. (1974). 'Proton transfer mass spectrometry of peptides. A rapid heating technique for underivatized peptides containing arginine'. *J. Amer. Chem. Soc.*, **96**, 3990–3999.

Biros, F. J. (1971). 'Recent applications of mass spectrometry and combined gas chromatography–mass spectrometry to pesticide residue analyses', *Residue Review*, **40**, 1–63.

Biros, F. J., Dougherty, R. C., and Dalton, R. (1972). 'Positive chemical ionization mass spectra of polycyclic chlorinated pesticides', *Org. Mass Spectrom.*, **6**, 1161–1169.

Blakely, C. R., Carmody, J. J., and Vestal, M. L. (1980). 'A new soft ionization technique for mass spectrometry of complex molecules', *J. Amer. Chem. Soc.*, **102**, 5931–5933.

Brandenberger, H. (1978). 'Negative ion mass spectrometry—a new tool for the forensic toxicologist' in *Instrumental Applications in Forensic Drug Chemistry* (eds M. Klein, A. V. Kruegel and S. P. Sobol), pp. 48–59. US Government Printing Office, Washington, DC.

Brandenberger, H. (1979). 'Present state and future trends of negative ion mass spectrometry in toxicology, forensic and environmental chemistry' in *Recent Developments in Mass Spectrometry in Biochemistry and Medicine* (ed. A. Frigerio), Vol. 2, pp. 227–255, Plenum, New York.

Bruins, A. P. (1980a). 'Negative ion chemical ionization mass spectrometry of steroid glycosides by 'In Beam' sample introduction on glass rods and thermal desorption from a platinum wire' in *Advances in Mass Spectrometry* (ed. A. Quayle), Vol. 8, pp. 246–254, Heyden, London.

Bruins, A. P. (1980b). 'Probe for the introduction of non-volatile solids into a chemical ionization source by thermal desorption from a platinum wire', *Anal. Chem.*, **52**, 605–607.

Bruins, A. P. (1981). 'Negative ion desorption chemical ionization mass spectrometry of some underivatized glucuronides', *Biomed. Mass Spectrum.*, **8**, 31–35.

Brunnee, C., Delgmann, L., Dielmann, G., Meyer, W., and Thorenz, P. (1981). LC/MS Interface for EI and DCI technique', *Twenty-ninth Conference on Mass Spectrometry and Allied Topics*, Minneapolis, 280.

Busch, K. L., Bursey, M. M., Hass, J. R., and Sovocool, G. W. (1978). 'Comparison of five ionization methods for producing mass spectra of typical organophosphorous pesticides', *Applied Spectroscopy*, **32**, 388–399.

Carlson, D. A., Konyha, K. D., Wheeler, W. B., Marshall, G. P., and Zaylskie, R. G. (1976). Mirex in the Environment. Its degradation to kepone and related compounds', *Science*, **194**, 939–941.

Carroll, D. I., Dzidic, I., Horning, M. G., Montgomery, F. E., Nowlin, J. G., Stillwell, R. N., Thenot, J.-P., and Horning, E. C. (1979a). 'Chemical ionization mass spectrometry of nonvolatile organic compounds', *Anal. Chem.*, **51**, 1858–1860.

Carroll, D. I., Dzidic, I., Stillwell, R. N., and Horning, E. C. (1979b). 'Atmospheric pressure ionization mass spectrometry' in *Trace Organic Analysis: A New Frontier in Analytical Chemistry*, eds. H. S. Hertz and S. N. Chesler, pp. 655–671, N.B.S. Washington, DC.

Chapman, R. A., and Robinson, J. R. (1977). 'Simplified method for the determination of residues of carbofuran and its metabolites in crops using gas–liquid chromatography–mass fragmentography', *J. Chromatogr.*, **140**, 209–218.

Cotter, R. J. (1979). 'Probe for direct exposure of solid samples to the reagent gas in a chemical ionization mass spectrometer', *Anal. Chem.*, **51**, 317–318.

Cotter, R. J. (1980). 'Laser desorption chemical ionization mass spectrometry', *Anal. Chem.*, **52**, 1767–1770.

Cotter, R. J., and Fenselau, C. (1979). 'The effects of heating rate and sample size on the direct exposure/chemical ionization mass spectra of some biological conjugates', *Biomed. Mass Spectrom.*, **6**, 287–293.

Crow, F. W., Bjorseth, A., Knapp, K. T., and Bennett, R. (1981). 'Determination of polyhalogenated hydrocarbons by glass capillary gas chromatography–negative ion chemical ionization mass spectrometry', *Anal. Chem.*, **53**, 619–625.

Damico, J. N., and Barron, R. P. (1971). 'Application of field ionization to gas liquid chromatography–mass spectrometry (GLC–MS) studies', *Anal. Chem.*, **43**, 17–21.

Damico, J. N., Barron, R. P., and Sphon, J. A. (1969). 'Field ionization spectra of some pesticidal and other biologically significant compounds', *Int. J. Mass Spectrom. Ion Phys.*, **2**, 161–182.

Day, R. J., Unger, S. E., and Cooks, R. G. (1980). 'Molecular secondary ion mass spectrometry', *Anal. Chem.*, **52**, 557A–572A.

Dougherty, R. C. (1981a). 'Negative chemical ionization mass spectrometry', *Anal. Chem.*, **53**, 625A–636A.

Dougherty, R. C. (1981b). 'Negative chemical ionization mass spectrometry: applications in environmental analytical chemistry', *Biomed. Mass Spectrom.*, **8**, 283–292.

Dougherty, R. C., Dalton, R., and Biros, F. J. (1972). 'Negative chemical ionization mass spectra of polycyclic chlorinated pesticides', *Org. Mass Spectrom.*, **6**, 1171–1181.

Dougherty, R. C., Roberts, J. D., and Biros, F. J. (1975). 'Positive and negative chemical ionization mass spectra of some aromatic chlorinated pesticides', *Anal. Chem.*, **47**, 54–59.

Dougherty, R. C., Roberts, J. D., Tannenbaum, H. P., and Biros, F. J. (1974). 'Positive and negative chemical ionization mass spectra of polychlorinated pesticides' in *Mass Spectrometry and NMR Spectroscopy in Pesticide Chemistry* (eds R. Haque and F. J. Biros), pp. 33–47, Plenum, New York.

Dougherty, R. C., and Wander, J. D. (1980). 'Chloride attachment negative chemical ionization mass spectra of organophosphate pesticides', *Biomed. Mass Spectrom.*, **7**, 401–404.

Doyle Daves, Jr. G. (1979). 'Mass spectrometry of involatile and thermally unstable molecules', *Acc. Chem. Research.*, **12**, 359–365.

Dutton, A. J., Roberts, T. R., and Wright, A. N. (1976). 'Characterization of acidic conjugates of flamprop in wheat', *Chemosphere*, **3**, 195–200.

Dymerski, P., Kennedy, M., and Kaminsky, L. (1979). 'Analysis of individual polychlorinated biphenyls (PCBs) and their hepatic microsomal metabolites by HPLC-MS' in *Trace Organic Analysis: A New Frontier in Analytical Chemistry* (eds H. S. Hertz and S. N. Chesler), pp. 685–690. N.B.S., Washington, DC.

Eckers, C., Games, D. E., Lewis, E., Rao, K. R. N., Rossiter, M., and Weerasinghe, N. C. A. (1980). 'Studies of natural products and pesticides and their metabolites by LC/MS and other mass spectral methods', in *Advances in Mass Spectrometry* (ed. A. Quayle), pp. 1396–1404, Heyden, London.

Ens, W., Standing, K. G., Chait, B. T., and Field, F. H. (1981). 'Comparison of mass spectra obtained with low-energy ion and high-energy californium-252 fission fragment bombardment', *Anal. Chem.*, **53**, 1241–1244.

Esmans, E. L., Freyne, E. J., Vanbroeckhoven, J. H., and Alderweireldt, F. C. (1980). 'Chemical ionization desorption mass spectrometry as an additional tool for the structure elucidation of nucleosides', *Biomed. Mass Spectrom.*, **7**, 377–380.

Evans, N., and Williamson, J. E. (1981). 'The construction and use of simple interfaces for combined liquid chromatography mass spectrometry', *Biomed. Mass Spectrom.*, **8**, 316–321.

Fales, H. M., Milne, G. W. A., Winkler, H. U., Beckey, H. D., Damico, J. N., and Barron, R. (1975). 'Comparison of mass spectra of some biologically important compounds as obtained by various ionization techniques', *Anal. Chem.*, **47**, 207–218.

Field, F. H. (1972). 'Chemical ionization mass spectrometry', in *MTP International Review of Sciences* (ed. A. Maccoll), Vol. 5, pp. 133–181, Butterworth, London.

Firestone, D., Clower, Jr., M., Borsetti, A. P., Teske, R. H., and Long, P. E. (1979). 'Polychlorodibenzo-*p*-dioxin and pentachlorophenol residues in milk and blood of cows fed technical pentachlorophenol', *J. Agric. Food Chem.*, **6**, 1171–1177.

Furukawa, K., Tonomura, K., and Komibayashi, A. (1979). 'Metabolism of 2,4,4′-trichlorobiphenyl by Acinetobacter sp. P6', *Agric. Biol. Chem.*, **43**, 1577–1583.

Games, D. E. (1980a). 'Combined high-performance liquid chromatography–mass spectrometry', *Analytical Proceedings*, **17**, 110–116.

Games, D. E. (1980b). 'Applications of combined high-performance liquid chromatography—mass spectrometry', *Analytical Proceedings*, **17**, 322–326.

Games, D. E. (1981a). 'Combined high-performance liquid chromatography–mass spectrometry', *Biomed. Mass Spectrom.*, **8**, 454–462.

Games, D. E. (1981b). 'Combined high-performance liquid chromatography–mass spectrometry', in *Biological Mass Spectrometry: Soft Ionization* (ed. H. R. Morris), pp. 54–68, Heyden, London.

Games, D. E. (1982). 'High-performance liquid chromatography/mass spectrometry', in *Advances in Chromatography* (ed. J. Calvin Giddings), Vol. 20, in press, Marcel Dekker, New York.

Games, D. E., Hirter, P., Kuhnz, W., Lewis, E., Weerasinghe, N. C. A., and Westwood, S. A. (1981). 'Studies of combined liquid chromatography–mass spectrometry with a moving-belt interface', *J. Chromatogr.*, **203**, 131–138.

Games, D. E., Jackson, A. H., and Millington, D. S. (1973). 'Applications of field ionization mass spectrometry in the analysis of organic mixtures', *Tetrahedron Letters*, 3063–3066.

Games, D. E., Jackson, A. H., Millington, D. S., and Rossiter, M. (1974). 'Field ionization with gas chromatography/mass spectrometry in structural studies of natural products', in *Advances in Mass Spectrometry* (ed. A. R. West), Vol. 6, pp. 137–142, Academic Press, London.

Games, D. E., Jackson, A. H., Taylor, K. T., and Haskins, N. J. (1976). 'Field desorption, field ionization and electron impact mass spectrometry of barbiturates' in *Advances in Mass Spectrometry in Biochemistry and Medicine* (eds A. Frigerio and N. Castagnoli), Vol. 1, pp. 383–390, Spectrum Publications, New York.

Games, D. E., Lee, M. G., Lewis, I. A. S., and Rossiter, M. (1978). 'Studies of drugs and their metabolites by FI GC/MS', in *Advances in Mass Spectrometry* (ed. N. Daly), Vol. 7, pp. 1616–1621, Heyden, London.

Games, D. E., and Weerasinghe, N. C. A. (1980). 'Analysis of the metabolites of CIPC (chloropropham)', *J. Chromatogr. Sci.*, **18**, 106–107.

Gower, J. L., Beaugrand, C., and Sallot, C. (1981). 'Positive and negative ion desorption chemical ionization of some penicillins', *Biomed. Mass Spectrum.*, **8**, 36–40.

Hansen, G., and Munson, B. (1978). 'Surface chemical ionization mass spectrometry', *Anal. Chem.*, **50**, 1130–1134.

Hansen, G., and Munson, B. (1980). 'Chemical ionization mass spectrometry of thermally labile compounds', *Anal. Chem.*, **52**, 245–248.

Hardin, E. D., and Vestal, M. L. (1981). 'Laser ionization mass spectrometry of non-volatile samples', *Anal. Chem.*, **53**, 1492–1497.

Harless, R. L., Harris, D. E., Sovocool, G. W., Zehr, R. D., Wilson, N. K., and Oswald, E. O. (1978). 'Mass spectrometric analyses and characterization of kepone in environmental and human samples', *Biomed. Mass Spectrom.*, **5**, 232–237.

Harned, W. H., and Casida, J. E. (1976). 'Dioxathion Metabolites, photoproducts and oxidative degradation products', *J. Agr. Food Chem.*, **24**, 689–699.

Hass, J. R. (1980). 'Negative ion chemical ionization mass spectrometry', in *Progress in Spectroscopy Series* (eds C. Merritt, Jr. and C. N. McEwen), Vol. 13, Part B, pp. 287–375, Marcel Dekker, New York.

Hass, J. R., Friesen, M. D., Harvan, D. J., and Parker, C. E. (1978). 'Determination of polychlorinated dibenzo-*p*-dioxins in biological samples by negative chemical ionization mass spectrometry', *Anal. Chem.*, **50**, 1474–1479.

Hass, J. R., Friesen, M. D., and Hoffmann, M. K. (1979). 'The mass spectrometry of polychlorinated dibenzo-*p*-dioxins', *Org. Mass Spectrom.*, **14**, 9–16.

Heck, H. d'A., Dyer, R. L., Scott, A. C., and Anbar, M. (1977). 'Determination and disposition of trifluralin in the rat. Separation by Sequential high-pressure liquid chromatography and quantitation by field ionization mass spectrometry', *J. Agr. Food Chem.*, **25**, 901–908.

Heinen, H. J. (1981). 'On ion formation in laser desorption mass spectrometry with LAMMA', *Int. J. Mass Spectrom. Ion. Phys.*, **38**, 309–322.

Henion, J. D. (1978). 'Drug analysis by continuously monitored liquid chromatography/mass spectrometry with a quadrupole mass spectrometer', *Anal. Chem.*, **50**, 1687–1693.

Henion, J. D. (1981). 'Continuous monitoring of total micro LC eluant by direct liquid introduction LC/MS', *J. Chromatogr. Sci.*, **19**, 57–64.

Henion, J. D., and Maylin, G. A. (1980). 'Drug analysis by direct liquid introduction micro liquid chromatography mass spectrometry', *Biomed. Mass Spectrom.*, **7**, 115–121.

Heresch, F., Schmid, E. R., and Huber, J. F. K. (1980). 'Repetitive laser desorption mass spectrometry for nonvolatile organic compounds', *Anal. Chem.*, **52**, 1803–1807.

Holmstead, R. L. (1976), 'Studies of the degradation of mirex with an Iron(II) porphyrin model system', *J. Agr. Food Chem.*, **24**, 620–624.

Holmstead, R. L., and Casida, J. E. (1974). 'Methane chemical ionization spectra of 29 organophosphorous pesticides', *J. Assoc. Off. Anal. Chem.*, **57**, 1050–1055.

Holmstead, R. L., and Casida, J. E. (1975), 'Chemical ionization mass spectrometry of *N*-methylcarbamate insecticides, some of their metabolites and related compounds', *J. Assoc. Off. Anal. Chem.*, **58**, 541–547.

Holmstead, R. L., and Soderlund, D. M. (1977). 'Separation and analysis of pyrethrins by combined gas–liquid chromatography–mass spectrometry. Chemical ionization mass spectrometry', *J. Assoc. Off. Anal. Chem.*, **60**, 685–689.

Hostettmann, K., Doumas, J., and Hardy, M. (1981). 'Desorption/chemical ionization mass spectrometry of naturally occurring glycosides', *Helv. Chim. Acta.*, **64**, 297–303.

Hunt, D. F., Bone, W. M., and Shabanowitz, J. (1980). 'Laser desorption CI mass spectrometry', *28th Annual Conference on Mass Spectrometry and Allied Topics, New York*, 620.

Hunt, D. F., Bone, W. M., Shabanowitz, J., Rhodes, J., and Ballard, J. M. (1981). 'Sequence analysis of oligopeptides by secondary ion/collision activated dissociation mass spectrometry', *Anal. Chem.*, **53**, 1704–1706.

Hunt, D. F., and Crow, F. W. (1979). 'Electron capture negative ion chemical ionization mass spectrometry', in *Trace Organic Analysis: A New Frontier in Analytical Chemistry* (eds H. S. Hertz and S. N. Chesler), pp. 601–607. NBS, Washington, DC.

Hunt, D. F., Harvey, T. M., and Russell, J. W. (1975). 'Oxygen as a reagent gas for the analysis of 2,3,7,8-tetrachlorodibenzo-*p*-dioxin by negative ion chemical ionization mass spectrometry, *J.C.S. Chem. Comm.*, 151–152.

Hunt, D. F., Shabanowitz, J., Botz, F. K., and Brent, D. A. (1977). 'Chemical ionization mass spectrometry of salts and thermally labile organics with field desorption emitters as solids probes', *Anal. Chem.*, **49**, 1160–1163.

Hutson, D. H., and Casida, J. E. (1978). 'Taurine conjugation in metabolism of 3-phenoxybenzoic acid and the pyrethroid insecticide cypermethrin in mouse', *Xenobiotica*, **8**, 565–571.

Jankowski, K., and Paré, J. R. J. (1980). 'Mass spectrometric study of carbamates', *Eur. J. Mass Spectrom. Biochem. Med. Environ. Res.*, **1**, 141–147.

Jennings, K. R. (1977). 'Negative chemical ionization mass spectrometry', in *Mass Spectrometry* (ed. R. A. W. Johnstone), Vol. 4, pp. 203–216, Chemical Society, London.

Karger, B. L., Kirby, D. P., Vouros, P., Foltz, R. L., and Hidy, B. (1979). 'On-line reversed phase liquid chromatography–mass spectrometry', *Anal. Chem.*, **51**, 2324–2328.

Knight, D. W., Knight, M. E., Rossiter, M., Jones, A., and Games, D. E. (1980). 'Mass spectral methods in structural studies of naturally-occurring volatiles', in *Advances in Mass Spectrometry* (ed. A. Quayle), Vol. 8, pp. 1116–1126, Heyden, London.

Ku, C. C., Kapoor, I. P., Stout, S. J., and Rosen, J. D. (1979). 'Photodegradation of cytrolane (mephosfolan) systemic insecticide in the aquatic environment using C-13 as a mass tracer', *J. Agr. Food Chem.*, **27**, 1046–1050.

Kuehl, D. W., Whitaker, M. J., and Dougherty, R. C. (1980). 'Micromethods for toxic residue screening by negative chemical ionization mass spectrometry', *Anal. Chem.*, **52**, 935–940.

Lusby, W. R., and Kearney, P. C. (1978). 'Gas chromatographic–mass spectral analyses of *s*-triazine metabolites', *J. Agric. Food Chem.*, **26**, 635–638.

Lusby, W. R., Oliver, J. E., and Kearney, P. C. (1980). 'Metabolism of 2,6-dinitro-4-(trifluoromethyl)benzamine by a *Streptomyces* isolated from soil', *J. Agric. Food Chem.*, **28**, 641–644.

Mayer, P., Kriemler, H.-P., Hambock, H., and Laanio, T. L. (1981a). 'Metabolism of *O,O*-dipropyl-*s*-[2-(2′-Methyl-1′-piperidinyl)-2-oxo-ethyl]phosphorodithioate (C19490) in paddy rice', *Agric. Biol. Chem.*, **45**, 335–360.

Mayer, P., Kriemler, H.-P., and Laanio, T. L. (1981b). 'Metabolism of *N*-(1,′2′-dimethylpropyl)-*N′*-ethyl-6-methylthio-1,3,5-triazine-2,4-diamine (C18898) in paddy rice', *Agric. Biol. Chem.*, **45**, 361–368.

Macfarlane, R. D. (1980). '^{252}Cf-plasma desorption mass spectrometry (PDMS)', in *Biochemical Applications of Mass Spectrometry* (eds G. R. Waller and O. C. Dermer), First Supplementary Volume, pp. 1209–1218, Wiley, New York.

McFadden, W. H. (1979). 'Interfacing chromatography and mass spectrometry', *J. Chromatogr. Sci.*, **17**, 2–16.

McFadden, W. H. (1980). 'Liquid chromatography/mass spectrometry systems and applications', *J. Chromatogr. Sci.*, **18**, 97–115.

McFadden, W. H., Bradford, D. C., Games, D. E., and Gower, J. L. (1977). 'Applications of combined liquid chromatography/mass spectrometry', *American Laboratory*, 55–64.

McFadden, W. H., Schwartz, H. L., and Evans, S. (1976). 'Direct analysis of liquid chromatographic effluents', *J. Chromatogr.*, **122**, 389–396.

McGahen, L. L., and Tiedje, J. M. (1978). 'Metabolism of two new acylanilide herbicides, Antor herbicide (H-22234) and Dual (Metolachlor) by the soil fungus *Chaetomium globosum*', *J. Agr. Food Chem.*, **26**, 414–419.

McKinney, J. D., Oswald, E. O., dePaul Palarszek, S. M., and Corbett, B. J. (1974). 'Characterization of chlorinated hydrocarbon pesticide metabolites of the DDT and polycyclodiene type by electron impact chemical ionization mass spectrometry' in *Mass Spectrometry and NMR Spectroscopy in Pesticide Chemistry* (eds R. Haque and F. J. Biros), pp. 5–32, Plenum, New York.

Melera, A. (1980). 'Design, operation and applications of a novel LCMS CI Interface' in *Advances in Mass Spectrometry* (ed. A. Quayle), Vol. 8, pp. 1517–1615, Heyden, London.

Milberg, R. M., and Cook, Jr., J. C. (1979). 'Some applications of high sensitivity combined field ionization gas chromatography–mass spectrometry', *J. Chromatogr. Sci.*, **17**, 43–47.

Millington, D. S., Yorke, D. A., and Burns, P. (1980). 'A new liquid chromatography-mass spectrometry interface' in *Advances in Mass Spectrometry* (ed. A. Quayle), Vol. 8, pp. 1819–1825, Heyden, London.

Milne, G. W. A., and Lacey, M. J. (1974). 'Modern ionization techniques in mass spectrometry', *Critical Reviews in Analytical Chemistry*, **4**, 45–104.

More, J. E., Roberts, T. R., and Wright, A. N. (1978). 'Studies of the metabolism of 3-phenoxybenzoic acid in plants', *Pesticide Biochemistry and Physiology*, **9**, 268–280.

Morris, H. R., Dell, A., and McDowell, R. A. (1981). 'Extended performance using a high field magnet mass spectrometer', *Biomed. Mass Spectrom.*, **8**, 463–473.

Muecke, W., Kriemler, H.-P., Hug, P., and Alt, K. O. (1981). 'Metabolism of *O,O*-dipropyl-S-[2-(2′-methyl-1′-piperidinyl)-2-oxo-ethyl]phosphorodithionate (C19490 Piperophos) in rat', *Agric. Biol. Chem.*, **45**, 43–51.

Munson, B. (1971). 'Chemical ionization mass spectrometry', *Anal. Chem.*, **43**, 28A–42A.

Munson, B. (1977). 'Chemical ionization mass spectrometry', *Anal. Chem.*, **49**, 772A–778A.

Ohsawa, K., and Casida, J. E. (1979). 'Photochemistry of the potent knockdown pyrethroid kadethrin', *J. Agric. Food Chem.*, **27**, 1112–1120.

Onuska, F. I., Comba, M. E., and Harrison, A. G. (1976). 'Chemical ionization and high resolution electron impact mass spectra of 5,6-dihydro-2-methyl-1,4-oxathiin-3-carboxanilide and three of its metabolites', *Biomed. Mass Spectrom.*, **3**, 248–254.

Oswald, E. O., Albro, P. W., and McKinney, J. D. (1974a). 'Utilization of gas–liquid chromatography coupled with chemical ionization and electron impact mass spectrometry for the investigation of potentially hazardous environmental agents and their metabolites', *J. Chromatogr.*, **98**, 363–448.

Oswald, E. O., Levy, L., Corbett, B. J., and Walker, M. P. (1974b). Differentiation and characterization of isomeric polychlorinated diphenyls by gas–liquid chromatography coupled with electron impact and chemical ionization mass spectrometry', *J. Chromatogr.*, **93**, 63–90.

Petrella, V. J., McKinney, J. D., Fox, J. P., and Webb, R. E. (1977). 'Identification of metabolites of endrin. Metabolism in endrin susceptible and resistant strains of pine mice', *J. Agric. Food Chem.*, **25**, 393–398.

Pohl, L. R., Nelson, S. D., Garland, W. A., and Trager, W. F. (1975). 'The rapid identification of a new metabolite of warfarin via a chemical ionization mass spectrometry ion doublet technique', *Biomed. Mass Spectrom.*, **2**, 23–30.

Posthumus, M. A., Kistemaker, P. G., Meuzelaar, H. L. C., and Ten Noever de Brauw, M. C. (1978). 'Laser desorption–mass spectrometry of polar nonvolatile bioorganic molecules', *Anal. Chem.*, **50**, 985–991.

Rapp, U., Dielmann, G., Games, D. E., Gower, J. L., and Lewis, E. (1980). 'The interface of desorption chemical ionization in mass spectral ionization methods', in *Advances in Mass Spectrometry* (ed. A. Quayle), Vol. 8, pp. 1660–1668, Heyden, London.

Richter, W. J., and Schwarz, H. (1978). 'Chemical ionization—a mass spectrometric analytical procedure of rapidly increasing importance', *Angew. Chem. Int. Ed.*, **17**, 424–439.

Roberts, T. R., and Wright, A. N. (1981). 'The metabolism of 3-phenoxybenzyl alcohol, a pyrethroid metabolite, in plants', *Pestic. Sci.*, **12**, 161–169.

Safe, S. (1979). 'Environmental applications of mass spectrometry' in *Specialist Periodical Reports, Mass Spectrometry* (ed. R. A. W. Johnstone), Vol. 5, pp. 234–249, Chemical Society, London.

Schafer, K. H., and Levsen, K. (1981). 'Direct coupling of a micro high-performance liquid chromatograph and a mass spectrometer', *J. Chromatogr.*, **206**, 245–252.

Schulten, H.-R. (1976). 'Field desorption mass spectrometry of commercial pesticides and mixtures of pesticides', *J. Agric. Food Chem.*, **24**, 743–750.

Schulten, H.-R. (1979). 'Biochemical, medical and environmental applications of field-ionization and field-desorption mass spectrometry', *Int. J. Mass Spectrom. Ion Phys.*, **32**, 97–283.

Schulten, H.-R., and Beckey, H. D. (1973). 'Field desorption mass spectrometry of pesticides and their metabolites', *J. Agric. Food Chem.*, **21**, 372–375.

Schulten, H.-R., Lehmann, W. D., and Haaks, D. (1978). 'Laser assisted field desorption mass spectrometry', *Org. Mass Spectrom.*, **13**, 361–365.

Schulten, H.-R., and Stöber, I. (1978). 'Determination and identification of biocides of the phenylurea, carbamate and thiocarbamate type in surface waters by HPLC and FDMS', *Fresenius Z. Anal. Chem.*, **293**, 370–376.

Schulten, H.-R., Prinz, H., Beckey, H. D., Tomberg, W., Klein, W., and Korte, F. (1973). 'Felddesorptions-massenspektrometrie, eine neue methode zur identifizierung von pestiziden und deren metaboliten', *Chemosphere*, **1**, 23–30.

Simoneaux, B. J., Martin, G., Cassidy, J. E., and Ryskiewch, D. P. (1980). 'Isolation and identification of the major metabolites of methidathion in tomatoes', *J. Agr. Food Chem.*, **28**, 1221–1224.

Skinner, R. F., Thomas, Q., Giles, J., and Crosby, D. G. (1980). 'The determination of herbicide photolysis products by LC/MS', *J. Chromatogr. Sci.*, **18**, 108–109.

Smith, R. D., Burger, J. E., and Johnson, A. L., (1981). 'Liquid chromatography–mass spectrometry with electron impact and fast ion bombardment with a ribbon storage interface', *Anal. Chem.*, **53**, 1603–1611.

Sphon, J. A., and Brumley, W. C. (1980). 'Pesticides', in *Biochemical Applications of Mass Spectrometry* (eds G. R. Waller and O. C. Dermer), 1st Supplementary Volume, pp. 713–749, Wiley, New York.

Sphon, J., Andrzejewski, D., Brumley, W., Dreifuss, P., and Roach, J. (1980). 'The application of mass spectrometry to the analysis of food toxicants', in *Advances in Mass Spectrometry* (ed. A. Quayle), Vol. 8, pp. 1490–1498, Heyden, London.

Stan, H.-J. (1977a). 'Chemical ionization mass spectra of organophosphorous pesticides with various reagent gases', *Fresenius Z. Anal. Chem.*, **287**, 104–111.

Stan, H.-J. (1977b). 'Detection of organophosphorous pesticide residues in food at the ppb level with open tubular column gas chromatography—mass spectrometry', *Z. Lebensm.-Unters.-Forsch.*, **164**, 153–159.

Stöber, I., and Schulten, H.-R. (1980). 'Combined application of high-pressure liquid chromatography and field desorption mass spectrometry for the determination of biocides of the phenylurea and carbamate type in surface water', *Sci. Total Environ.*, **16**, 249–262.

Stoll, R., and Röllgen, F. W. (1979). 'Laser desorption mass spectrometry of thermally labile compounds using a continuous wave CO_2 laser', *Org. Mass Spectrom.*, **14**, 642–645.

Suzuki, M., and Yamato, Y. (1980). 'Application of field desorption mass spectrometry to the positive identification of pesticidal contaminants in the environment' in *Recent Developments in Mass Spectrometry in Biochemistry and Medicine* (eds A. Frigerio and M. McCamish), Vol. 6, pp. 543–550. Elsevier, Amsterdam.

Thenot, J. P., Nowlin, J., Carroll, D. I., Montgomery, F. E., and Horning, E. C. (1979). 'Coated glass probe tips for the determination of thermally labile compounds by mass spectrometry', *Anal. Chem.*, **51**, 1101–1104.

Thomson, B. A., and Iribarne, J. V. (1979). 'Field induced ion evaporation from liquid surfaces at atmospheric pressure', *J. Chem. Phys.*, **71**, 4451–4463.

Tsuchiya, M., and Taira, T. (1980). A new method of ionization for organic compounds in the liquid phase under atmospheric pressure', *Int. J. Mass Spectrom. Ion Phys.*, **34**, 351–359.

Ueda, K., Gaughan, L. C., and Casida, J. E. (1974). 'Photodecomposition of resmethrin and related pyrethroids', *J. Agr. Food Chem.*, **22**, 212–216.

Unger, S. E., Vincze, A., Cooks, R. G., Chrisman, R., and Rothman, L. D. (1981). 'Identification of quaternary alkaloids in mushroom by chromatography/secondary ion mass spectrometry', *Anal. Chem.*, **53**, 976–981.

Westwood, S. A., Games, D. E., Lant, M. S., and Woodhall, B. J. (1982). 'Micro-column high-performance liquid chromatography', *Analytical Proceedings*, **19**, 121–123.

Wheeler, W. B., Stratton, G. D., Twilley, R. R., Ou, L.-T., Carlson, D. A., and Davidson, J. M. (1979). 'Trifluralin degradation and binding in soil', *J. Agr. Food Chem.*, **27**, 702–706.

Wright, A. N., Roberts, T. R., Dutton, A. J., and Doig, M. V. (1980). 'The metabolism of cypermethrin in plants: the conjugation of the cyclopropyl moiety', *Pesticide Biochemistry and Physiology*, **13**, 71–80.

Wright, L. H., Edgerton, T. R., Arbes, Jr., S. J., and Lores, L. M. (1981). 'The determination of underivatized chlorophenols in human urine by combined high performance liquid chromatography mass spectrometry and selected ion monitoring', *Biomed. Mass Spectrom.*, **8**, 475–479.

Wright, L. H., and Oswald, E. O. (1978). 'Liquid chromatography/mass spectrometric methods of analyses for carbamate pesticides', *26th Annual Conference on Mass Spectrometry and Allied Topics, St. Louis*, 47–48.

Wu, A., Pearson, M. L., Shekoski, D. L., Soto, R. J., and Stewart, R. D. (1979). 'High resolution gas chromatographic/mass spectrometric characterization of urinary metabolites of *N,N*-diethyl-*m*-toluamide (DEET) in man', *J. High Resolut. Chromatogr. Chromatogr. Commun.*, **2**, 558–562.

Yamato, Y., Suzuki, M., and Watanabe, T. (1979). 'Field desorption mass spectrometric analyses of organic compound residues in the environment. II-Herbicides'. *Biomed. Mass Spectrom.*, **6**, 205–207.

Yu, C. C., Atallah, Y. H., and Whitacre, D. M. (1980). 'Metabolism of the herbicide buthidazole in corn seedlings and alfalfa plants', *J. Agric. Food Chem.*, **28**, 1090–1095.

Ziegler, W., Engelhardt, G., Wallnöfer, P. R., Oehlmann, L., and Wagner, K. (1980). 'Degradation of demeton S-methyl sulphoxide (Metasystox R) by soil micro-organisms, *J. Agric. Food Chem.*, **28**, 1102–1106.

Zune, A., Dobberstein, P., Maurer, K. H., and Rapp, U. (1976). 'Identification of drugs using a gas chromatography–mass spectrometry system equipped with electron impact–chemical ionization and electron impact–field ionization–field desorption combination sources, *J. Chromatogr.*, **122**, 365–371.

Progress in Pesticide Biochemistry and Toxicology, Volume 3
Edited by D. H. Hutson and T. R. Roberts

CHAPTER 8

Metabolism of pyrethrins and pyrethroids in insects

D. M. Soderlund, J. R. Sanborn and P. W. Lee

INTRODUCTION

During the past decade the synthetic pyrethroids have emerged as a major class of agriculturally useful organic insecticides. The impact of pyrethroids was enhanced by the discovery (Elliott *et al.*, 1973) that photosensitive moieties of the naturally-occurring pyrethrins and the early synthetic analogues could be isosterically replaced with groups resistant to photooxidative attack. The resulting photostabilized esters retained both the desirable insecticidal potency and the insect–mammal selectivity of the earlier compounds. Since that time, intensive efforts have been directed at the discovery and development of pyrethroid insecticides for use in agriculture. Numerous recent reviews have dealt with the development of pyrethroids and their application in insect control (Elliott, 1976; 1977; Elliott *et al.*, 1978; Elliott, 1980), pyrethroid chemistry

(Elliott and Janes, 1978), structure–activity relationships (Elliott *et al.*, 1974a; Elliott and Janes, 1979) and the metabolism of pyrethroids (Casida *et al.*, 1975/76; Miyamoto, 1976; Ruzo and Casida, 1977; Hutson, 1979; Casida *et al.*, 1979; Chambers, 1980; Casida and Ruzo, 1980). Reviews of metabolism have concentrated on the extensive literature concerning the metabolic fate of pyrethroids in mammals, but have dealt with studies of the metabolic fate of pyrethroids in insects briefly if at all. In this chapter, we review studies of the metabolic fate of both the natural pyrethrins and synthetic pyrethroids in insects *in vivo* and in relevant insect-derived enzyme preparations *in vitro*. We also discuss the contribution of metabolism to structure–toxicity relationships and to insect resistance to pyrethroids.

Problems and limitations in insect metabolism studies

Since knowledge of the metabolic fate of an insecticide in insects is not required for registration, fewer metabolism studies are performed with insects than with mammalian species. In many cases, insect metabolism studies are initiated only to answer specific questions concerning structure–activity relationships or resistance mechanisms or to provide data for comparison with concurrent metabolism studies in mammals. This situation has led to many obvious gaps in the literature, so that there is little or no information at all on the metabolism in insects of some pyrethroids that represent critical stages in the development of this insecticide class.

The exceptional insecticidal potency of pyrethroids has also exerted a practical limitation on metabolism studies in insects. Radiolabelled compounds are essential in order to identify metabolites, and significant progress in understanding the metabolic fate of pyrethrins and pyrethroids was not made until materials radiolabelled at relatively high specific activity were available. Even with such preparations, the treatment of insects with labelled compounds at toxicologically relevant doses has made the identification of metabolites difficult. In some studies, this difficulty has been circumvented by the initial identification of primary biotransformation products using enzyme preparations *in vitro*.

Another problem encountered in these studies involves the effects of the absolute and stereochemical configurations of pyrethroids on their metabolic fate. Although the natural pyrethrins are single isomers, most synthetic esters are commonly used as mixtures of enantiomers and diastereomers, the biological activity of which can differ profoundly (Elliott *et al.*, 1974a; Elliott and Janes, 1979). Since most radiosynthetic techniques produce isomer mixtures and most optical resolution procedures are not applicable to the resolution of minute amounts of labelled compounds, the majority of insect metabolism studies have examined the fate of isomer mixtures and thus may not accurately describe the fate of the toxic components.

Pyrethroid structures and nomenclature

All of the most active pyrethroids developed to date are esters of 3-substituted-2,2-dimethylcyclopropanecarboxylic acids (or isosteric acids lacking the cyclopropane ring) and an appropriate alcohol containing one or more unsaturated moieties. The acid and alcohol moieties of the pyrethroids used in insect metabolism studies are shown in Figure 1. Specific esters named in the text are keyed to this figure by the number and letter of the component acid and alcohol moieties. In addition, we have adopted trivial nomenclature systems for certain esters and their metabolites that have proven useful in describing complex metabolic pathways.

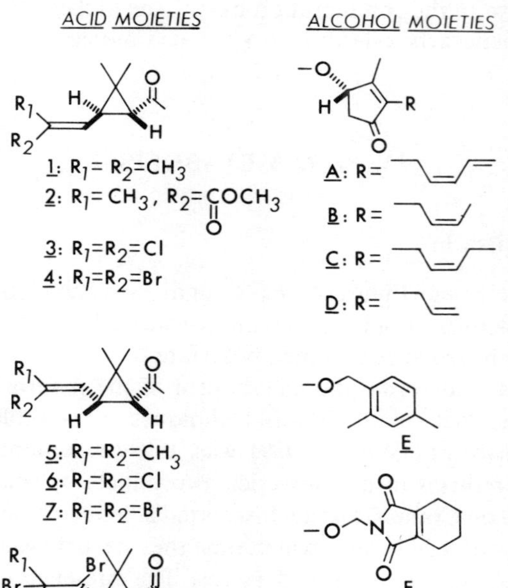

Figure 1 Structures of the acid and alcohol moieties of pyrethrins and pyrethroids used in insect metabolism studies

Most active pyrethroids have two or three chiral centres, so that each ester may have from four to eight optical and geometrical isomers that differ in both biological activity and metabolic fate. A system of nomenclature for pyrethroid isomers is clearly needed for a discussion of the effects of pyrethroid structure on metabolism. In this review, we follow the hybrid nomenclature for cyclopropanecarboxylates originally proposed by Elliott *et al.* (1974b), in which pyrethroid isomers are described by the absolute configuration at cyclopropane C-1 (*R* or *S*), the stereochemistry across the plane of the ring (*trans* or *cis*) resulting from the configuration at C-3, and the absolute configuration of a chiral centre, if present, in the alcohol moiety. For example, the isomer of allethrin (Figure 1, *1D*) corresponding to the naturally-occurring pyrethrins is designated as [1*R*,*trans*,4′*S*]-allethrin. The acid and alcohol moieties shown in Figure 1 are drawn in the configuration having the highest insecticidal activity: 1*R* for cyclopropanecarboxylates *1–9*, *S* for acid moiety *10* and *S* for alcohols *A–D* and *H*.

IN VIVO METABOLISM

Pyrethrins and allethrin

The natural pyrethrins (Figure 1, *1A–C* and *2A–C*) and the closely-related synthetic ester allethrin (Figure 1, *1D*) are considered together because of their similarities in both structure and metabolic fate.

Early attempts to describe the metabolism of the pyrethrins were severely hampered by the lack of analytical techniques of suitable specificity and sensitivity. Nevertheless, Woke (1939) was able to demonstrate the *in vivo* conversion of pyrethrins to non-insecticidal products by southern armyworm (*Spodoptera eridania* Cramer) larvae. Insects fed a defined amount of pyrethrum dust were held for 6 to 24 h, after which time they were dissected for analysis of individual tissues, gut contents and faeces. The presence of pyrethrins was determined by the toxicity of aqueous suspensions of ground tissues, gut contents and faeces to mosquitoes (*Culex fatigans* Wied.) larvae. By this bioassay there were no detectable toxic residues of the pyrethrins in any of the tissue extracts or in extracts of gut contents and excreta.

More precise studies of the degradation of pyrethrins and allethrin in insects were made possible using early radiolabelled preparations. Pellegrini *et al.* (1952) isolated randomly labelled pyrethrins from plants grown in a $^{14}CO_2$ environment. Winteringham *et al.* (1955) synthesized [^{14}C]-[*RS*]-allethrolone from [^{14}C]acetone and esterified it with synthetic [1*RS*,*trans*]-chrysanthemic acid to give fully synthetic [^{14}C]-*trans*-allethrin specifically labelled in the alcohol moiety. Acree *et al.* (1954) synthesized [^{14}C]allethrin (mixture of eight isomers) specifically labelled in the 2-position of the acid moiety. Hayashi *et al.*

(1968) obtained [^{3}H]-allethrin (isomer content not specified) by tritium exchange.

Several groups used these preparations to study the absorption and metabolism of the pyrethrins and allethrin in insects. Zeid *et al.* (1953) demonstrated that 25–34% of a dose of [^{14}C]pyrethrins applied topically to American cockroach (*Periplaneta americana* L.) adults was absorbed and translocated to internal tissues within 4 h after treatment. Most of the internal radiocarbon was associated with the foregut (17–20%) and hindgut (3–7%), but levels were barely detectable (<3%) in the midgut, brain, thoracic ganglia and muscles and the fat body.

Winteringham *et al.* (1955), Bridges (1957), Hopkins and Robbins (1957), Chang and Kearns (1964) and Hayashi *et al.* (1968) used one or more of the labelled pyrethrins and allethrin samples to describe the absorption and degradation of these compounds in adult houseflies (*Musca domestica* L.). All of these studies documented the rapid absorption of rethrolone esters. The rate of absorption was shown to be the same in living and freshly-killed flies (Winteringham *et al.*, 1955) and was reduced when the insecticide was co-applied with a synergist (Winteringham *et al.*, 1955; Chang and Kearns, 1964; Hayashi *et al.*, 1968). The chemical nature of internal residues was examined by paper or column chromatographic fractionation of extracts of flies previously rinsed with solvent to remove external residues. In all studies, labelled pyrethrins and allethrin were extensively metabolized. The synergists piperonyl cyclonene (Figure 2, *A*; Winteringham *et al.*, 1955; Bridges, 1957) and sesamex (Figure 2, *B*; Chang and Kearns, 1964) clearly reduced the degradation of the pyrethrins and allethrin, thus implicating inhibition of metabolism as the mechanism of synergist action.

These authors also attempted to characterize the metabolites of the pyrethrins and allethrin found in extracts of solvent-rinsed insects and dissected insect tissues. Zeid *et al.* (1953) claimed tentative identification of keto-alcohols, chrysanthemic acid and chrysanthemumdicarboxylic acid by paper chromatography in extracts of cockroach tissues, suggesting hydrolysis as a major biodegradation pathway. Studies in houseflies (Winteringham *et al.*, 1955; Bridges, 1957; Hopkins and Robbins, 1957; Chang and Kearns, 1964; Hayashi *et al.*, 1968) were unable to confirm the presence of major acidic metabolites, but instead characterized the non-metabolite radiocarbon as behaving like an alcohol on paper chromatography even when the preparation was specifically labelled in the acid moiety (Hopkins and Robbins, 1957). These results suggested modification of the intact ester, rather than hydrolysis, as the major biodegradative pathway. Further support for this hypothesis was obtained by Chang and Kearns (1964), who separated five metabolite fractions derived from [^{14}C]pyrethrin I (Figure 1, *1A*) and [^{14}C]cinerin I (Figure 1, *1B*; both obtained by fractionation of [^{14}C]pyrethrins) by column chromatography. Of these metabolites, three yielded intact [^{14}C]chrysanthemic acid upon hydrolysis,

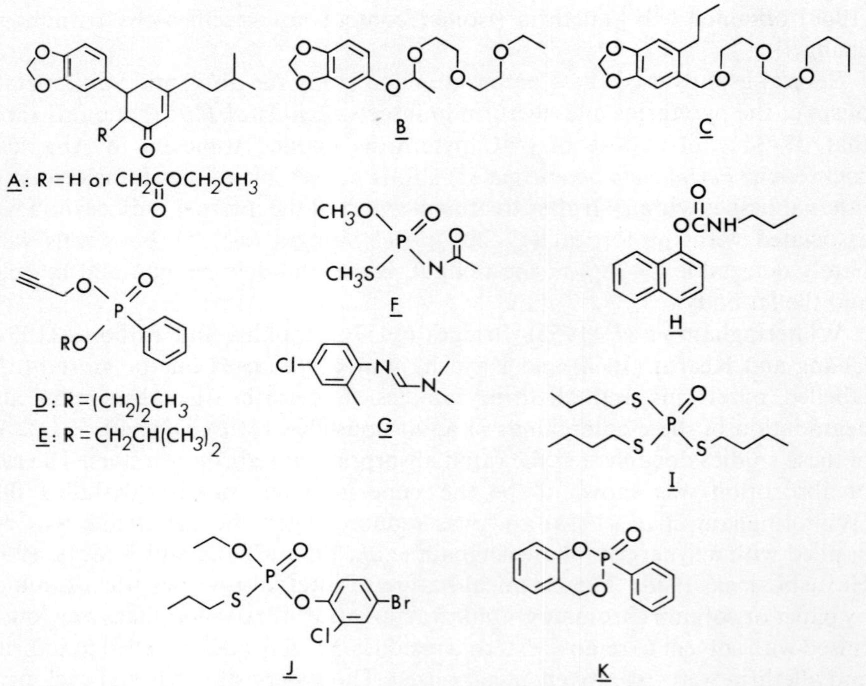

Figure 2 Structures of compounds used as synergists or enzyme inhibitors in metabolism studies: *A*, piperonyl cyclonene; *B*, sesamex; *C*, piperonyl butoxide; *D*, NIA 16388; *E*, MPP; *F*, acephate; *G*, chlordimeform; *H*, NPC; *I*, DEF; *J*, profenofos; *K*, PSCP

suggesting that the metabolites were esters of modified alcohols. Any further characterization of the metabolites generated in these studies was severely hampered by the low specific activities of the labelled esters, the low resolution of chromatographic systems available at the time and the lack of synthetic standard samples of potential metabolites other than the alcohols and acids expected from hydrolysis.

The availability of optically pure [^{14}C]pyrethrin I and [1*R*,*trans*,4′*S*]-allethrin labelled in the acid moiety and of [*1R*,*trans*,4′*RS*]-allethrin labelled in the racemic alcohol moiety (Nishizawa and Casida, 1965; Yamamoto and Casida, 1968), all at much higher specific activities than the earlier preparations, made possible the first definite identification of the metabolites of these compounds in insects. Yamamoto *et al.* (1969) used these preparations to identify several metabolites of allethrin and one metabolite of pyrethrin I in houseflies by thin-layer co-chromatography of the recovered labelled metabolites or their derivatives with synthesized standards.

Both acid- and alcohol-labelled allethrin were converted by houseflies into at least ten labelled metabolites. The identical metabolite patterns obtained

Table 1 Radiocarbon distribution among metabolites and metabolite fractions 1 h after topical treatment of adult houseflies with ^{14}C-labelled pyrethrin I, allethrin and tetramethrin[a]

Compound	Fraction	Recovered radiocarbon (%) Acid-^{14}C	Alcohol-^{14}C
Pyrethrin I	Organosoluble	85.5	
	Pyrethrin I	55.7	
	ω_t-oic acid[b]	2.0	
	Other	27.8	
	Water soluble	14.5	
Allethrin	Organosoluble	81.8	74.2
	Allethrin	54.9	42.0
	ω_t-al[c]	2.1	1.8
	ω_t- and ω_c-oic acids	2.9	3.0
	ω_t-ol	7.8	6.0
	ω_c-ol	4.2	6.8
	Other	9.9	14.6
	Water soluble	16.0	23.7
Tetramethrin	Organosoluble	82.1	74.0
	Tetramethrin	54.0	42.3
	ω_t-oic acid[d]	2.4	2.4
	Other	25.7	29.3
	Water soluble	16.2	19.1

[a] Data from Yamamoto *et al.* (1969).
[b] For metabolite structure see Figure 3[R = pyrethrolone (Figure 1, *A*)].
[c] Structures of allethrin metabolites are shown in Figure 3.
[d] Metabolite structure shown in Figure 4.

from these preparations showed that all of the metabolites retained the ester linkage. When the mixture of metabolites from [alcohol-^{14}C]allethrin was derivatized with 2,4-dinitrophenylhydrazine in methanol, most of the recovered radiocarbon chromatographed with allethrolone methyl ether dinitrophenylhydrazone, whereas cleavage of the metabolites arising from [acid-^{14}C]allethrin gave little if any chrysanthemic acid. These results suggested that allethrin metabolism was limited to modification of the acid moiety of intact esters. Six of these metabolites were tentatively identified on the basis of chromatographic comparison with synthesized standards as products of the progressive oxidation of either the *trans* or *cis* methyl group of the isobutenyl substituent to the corresponding alcohol, aldehyde and carboxylic acid (Table 1; structures and trivial designations given in Figure 3). The major metabolic pathway involved hydroxylation of the *trans* methyl group, followed by either conjugation or further oxidation to aldehyde and carboxylic acid derivatives. A similar sequence resulting from *cis*-methyl hydroxylation was of less importance.

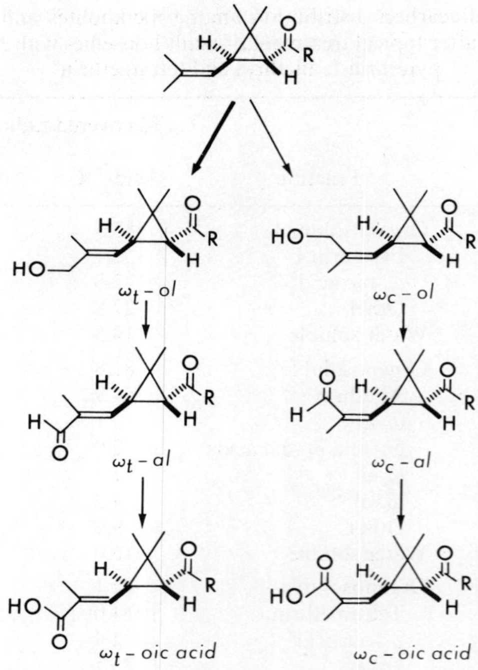

Figure 3 Metabolic pathways for allethrin in houseflies (redrawn from Yamamoto *et al.*, 1969)

Lack of synthetic standards and a radiolabel in the alcohol moiety prevented a similarly detailed study of pyrethrin I metabolism by these authors. Nevertheless, the carboxylic acid derivative resulting from *trans*-methyl hydroxylation was identified as a major metabolite of pyrethrin I (Table 1). No evidence was obtained in this study for ester metabolites of pyrethrin I having an altered alcohol moiety [inferred from the results of Chang and Kearns (1964)], but ester metabolites of allethrin and pyrethrin I oxidized in both the acid and alcohol moieties were subsequently identified from rats (Elliott *et al.*, 1972).

Although Zeid *et al.* (1953) suggested that hydrolysis was involved in pyrethrin metabolism in cockroaches, none of the subsequent work with houseflies or rats substantiated this finding. From these results, the metabolism of the pyrethrin I and allethrin in insects involves oxidative attack, but little or no hydrolytic attack, on the parent ester.

Dimethrin

Yamamoto *et al.* (1969) reported briefly on the fate of acid-labelled [1*R*,*trans*]-dimethrin (Figure 1, *1E*) in houseflies. A major metabolite was shown to be the carboxylic acid formed by oxidation of the *trans* methyl group in the isobutenyl moiety with the ester bond intact, presumably resulting from a reaction sequence similar to that identified for allethrin. Several more polar organosoluble metabolites were resolved on thin-layer chromatograms but not identified.

Tetramethrin

The initial studies of tetramethrin (also called phthalthrin; Figure 1, *1F*) metabolism in insects were performed using a tritium-labelled preparation of unspecified isomer content obtained by catalytic exchange (Hayashi *et al.*, 1968). These authors found that [^{3}H]tetramethrin was rapidly absorbed and metabolized by houseflies. Paper chromatography of internal extracts showed radioactive zones with R_f values corresponding to chrysanthemic acid and tetrahydrophthalimidomethanol, suggesting that ester cleavage had occurred (Figure 4).

Yamamoto *et al.* (1969) examined the fate of both acid- and alcohol-labelled [1*R*,*trans*]-tetramethrin (Yamamoto and Casida, 1968) in houseflies in conjunction with their studies on allethrin, pyrethrin I and dimethrin. Small amounts (2.4% of the applied dose) of the carboxylic acid metabolite arising from oxidation of the *trans* methyl of the isobutenyl moiety in the intact ester were found from both the alcohol- and acid-labelled esters (Table 1; Figure 4), but the majority of non-parent radiocarbon was associated with several more polar metabolites.

Miyamoto and Suzuki (1973) also examined the metabolism of acid- and alcohol-labelled [*1RS*,*trans*]tetramethrin in houseflies in an attempt to

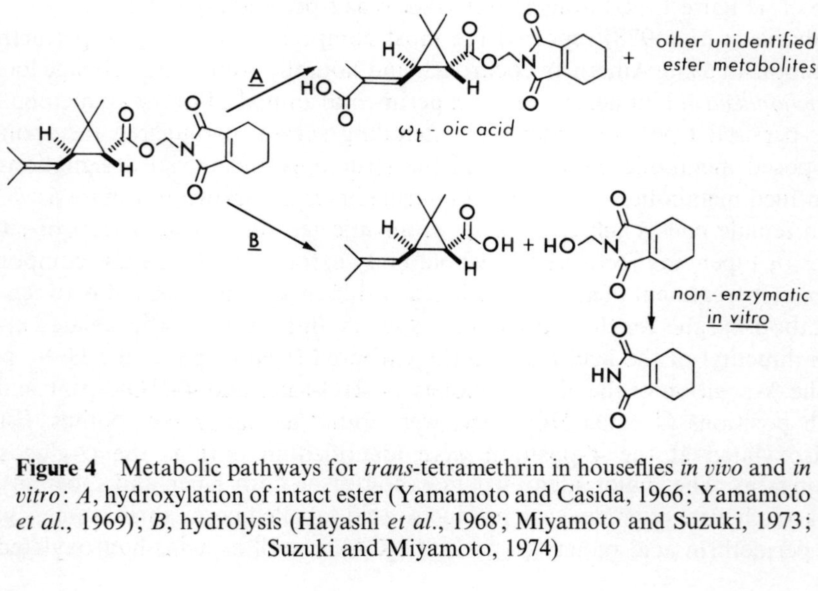

Figure 4 Metabolic pathways for *trans*-tetramethrin in houseflies *in vivo* and *in vitro*: *A*, hydroxylation of intact ester (Yamamoto and Casida, 1966; Yamamoto *et al.*, 1969); *B*, hydrolysis (Hayashi *et al.*, 1968; Miyamoto and Suzuki, 1973; Suzuki and Miyamoto, 1974)

characterize more completely the *in vivo* degradation pathways. In these studies, ester metabolites with oxidized isobutenyl moieties were minor metabolites (<10% of non-parent radiocarbon) and the major metabolites identified by co-chromatography were chrysanthemic acid (20–40% of non-parent radiocarbon) from [acid-^{14}C]tetramethrin and tetrahydrophthalimidomethanol (30–50% of non-parent radiocarbon) from [alcohol-^{14}C]tetramethrin. The oxidase-inhibiting synergist piperonyl butoxide (Figure 2, *C*) had little influence on the rate of tetramethrin metabolism or on the distribution of metabolites, but the *O*-propynyl-*O*-alkyl phenylphosphonate NIA 16388 (Figure 2, *D*) progressively inhibited tetramethrin metabolism at increased intervals between synergist pretreatment and tetramethrin treatment. NIA 16388 also markedly reduced the amount of acid and alcohol metabolites formed relative to the oxidized ester and uncharacterized metabolites. These results provided the first clear indication that ester cleavage was an important pathway for pyrethroid metabolism in insects.

Permethrin

Studies of the metabolism of permethrin provide the most comprehensive picture of pyrethroid metabolism in insects to date. Three laboratories have examined the fate of permethrin in six insect species, and the total number and proportion of metabolites identified from permethrin are far greater than for any other compound. These studies made use of permethrin labelled with carbon-14 in either the acid or alcohol moiety at high specific activities and compared the metabolic fate of the separated 1*RS*,*trans* (Figure 1, *3G*) and 1*RS*,*cis* (Figure 1, *6G*) isomers (referred to as *t*-per and *c*-per below).

Shono *et al.* (1978) reported the most comprehensive study of permethrin metabolism using American cockroach and housefly adults and cabbage looper (*Trichoplusia ni* Hübner) larvae as experimental animals. Forty-two metabolites of *c*-per and *t*-per were identified including several conjugated metabolites. Proposed metabolic pathways and the structures and trivial designations of identified metabolites and inferred intermediates are shown in Figure 5.

In female cockroaches, 19–25% of an injected dose (1 μg/insect) of either *t*-per or *c*-per was recovered in whole body extracts as the parent compound 24 h after treatment (Table 2). Little if any difference was observed between the metabolism rates for the *trans* and *cis* isomers. Intact esters hydroxylated at the *gem*-dimethyl of the acid *trans* to the carbonyl (*t*-HO-*t*-per and *t*-HO-*c*-per), at the 4′-position of the alcohol moiety (4′-HO-*t*-per and 4′-HO-*c*-per), and at both positions (4′-HO,*t*-HO-*c*-per) were found as minor metabolites. Esters hydroxylated at the 4′-position were identified in part as the *O*-glucoside conjugates. The major identified metabolites of both *t*-per and *c*-per in the cockroach were ester cleavage products. Acid-labelled permethrin isomers gave the permethrin acid moiety (*t*- and *c*-Cl_2CA) as well as acids hydroxylated at

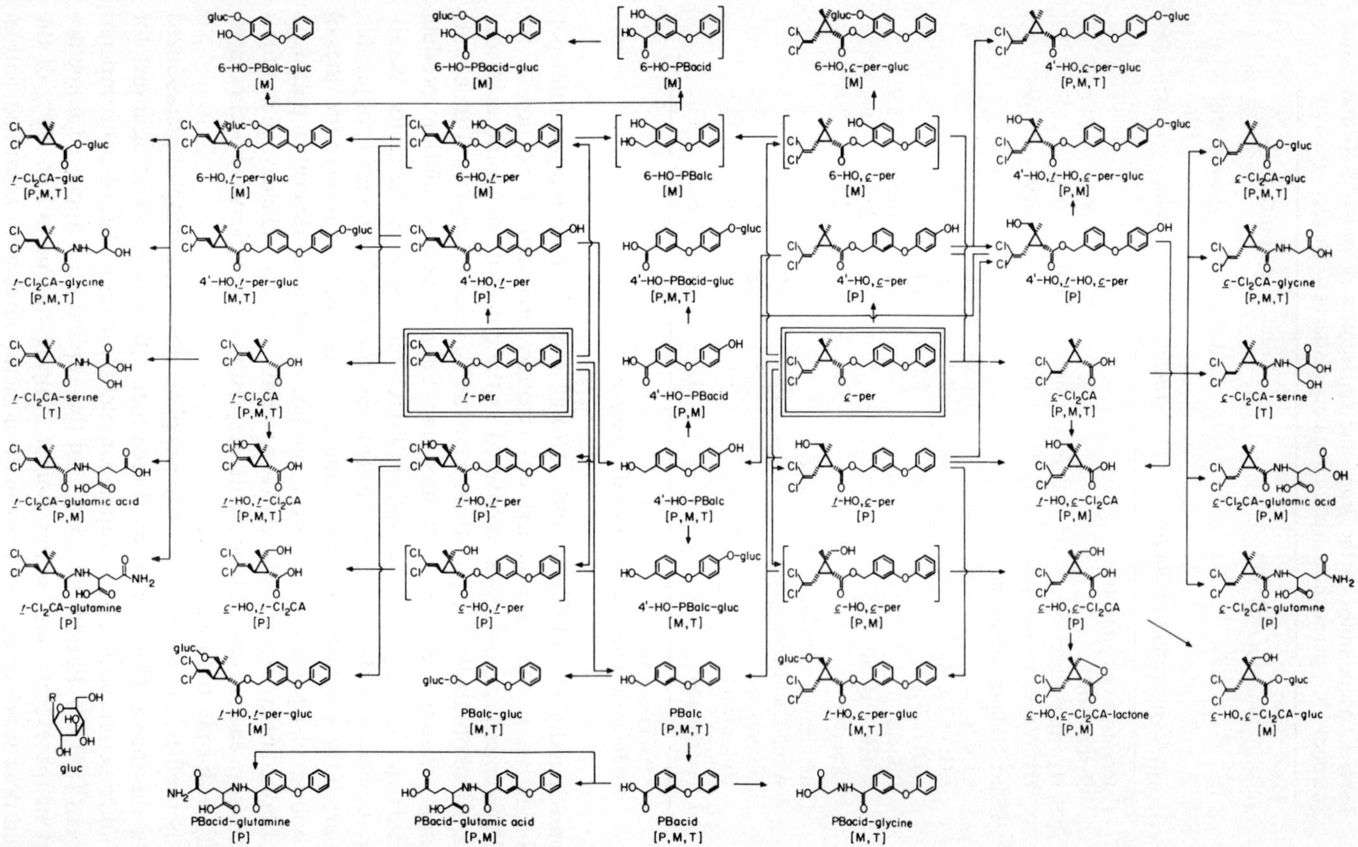

Figure 5 Metabolic pathways for *t*-per and *c*-per detected with adult American cockroaches (P), adult houseflies (M) and cabbage looper larvae (T). Although the 1*R* isomers are shown, metabolites are 1*RS* isomers (from Shono *et al.*, 1978; reproduced by permission of Academic Press)

Table 2 Metabolites of the *trans* and *cis* isomers of [acid-^{14}C]- and [alcohol-^{14}C]-permethrin in adult female cockroaches 24 h after injection[a]

	Injected radiocarbon (%)			
	acid-^{14}C		alcohol-^{14}C	
Compound or fraction	*trans*	*cis*	*trans*	*cis*
Esters, free or conjugated				
Permethrin	22.5	24.7	19.1	24.8
Hydroxyesters	0.3	0.9	0.3	0.7
Hydroxyester conjugates	0.0	0.5	0.0	1.0
Ester cleavage products, free				
Cl_2CA	26.2	5.8		
t-HO-Cl_2CA	12.4	8.9		
c-HO-Cl_2CA	2.0	2.1		
PBacid			19.8	6.9
4′-HO-PBacid			25.7	10.4
Others[b]	3.7	1.3	1.8	2.1
Ester cleavage products, conjugated				
Amino acid conjugates	19.5	9.0	13.3	17.8
Glucosides	2.5	3.8	5.7	8.6
Unknowns	5.1	32.3	3.9	14.8
Unextractable	4.7	8.1	8.2	13.5

[a] Data from Shono *et al.* (1978).
[b] Minor identified metabolites and unknowns.

the *gem*-dimethyl (*t*-HO-Cl_2CA and *c*-HO-Cl_2CA). With both *c*-per and *t*-per, hydroxylation at the *trans*-methyl was strongly preferred. A portion of the Cl_2CA from either isomer was recovered as the glucoside or amino acid conjugate. Alcohol-labelled preparations gave the free alcohol moiety (PBalc) only in small amounts, but the corresponding benzoic acid oxidation product (PBacid) and its 4′-hydroxy derivative (4′-HO-PBacid) were the major metabolites. A portion of the recovered PBacid was found as amino acid conjugates, and part of the 4′-HO-PBacid was found as the glucoside. Although the rates of metabolism of *t*-per and *c*-per were similar, evidence of divergent pathways was obtained by comparing the relative amounts of esters and cleavage products formed from each isomer. These results suggest that ester cleavage occurs more readily with *t*-per than with *c*-per.

In houseflies, the proportion of the applied dose remaining unmetabolized within the insect 48 h after topical treatment (0.01 and 0.0075 μg/insect for *t*-per and *c*-per, respectively) was approximately twice as high for *c*-per compared to *t*-per (Table 3). In addition, 5–8 % of the applied dose appeared unmetabolized in the excreta with either isomer. During this period 66–80 % of the applied dose was excreted and permethrin metabolites were identified only in

Table 3 Metabolites of the *trans* and *cis* isomers of [acid-^{14}C]- and [alcohol-^{14}C]-permethrin in adult female houseflies and cabbage looper larvae and in their excreta 48 h after topical application[a]

	Applied radiocarbon (%)							
	Housefly				Cabbage looper			
	Acid-^{14}C		Alcohol-^{14}C		Acid-^{14}C		Alcohol-^{14}C	
Fraction or compound	*trans*	*cis*	*trans*	*cis*	*trans*	*cis*	*trans*	*cis*
In insects								
Permethrin	1.5	5.2	1.9	8.8	4.4	36.0	5.6	47.1
Unknown or unextractable	10.4	21.1	7.9	20.2	6.9	6.3	11.4	6.2
In excreta								
Esters								
Permethrin	6.4	7.7	5.1	6.6	0.3	1.2	2.1	0.8
Hydroxyester conjugates	4.4	13.1	4.8	13.3	2.6	4.8	1.2	4.0
Ester cleavage products, free								
Cl_2CA	2.6	1.7			9.0	0.9		
t-HO-Cl_2CA	3.6	1.0			1.7	0.0		
PBalc			0.2	0.1			1.3	0.6
PBacid			1.2	0.5			0.0	0.5
4′-HO-PBalc			2.1	0.9			0.0	0.0
4′-HO-PBacid			4.4	2.0			0.0	0.0
Other[b]	2.8	5.1	2.0	1.8	0.0	0.5	1.5	0.0
Ester cleavage products, conjugated								
Amino acid conjugates	16.0	3.8	8.1	3.8	45.8	22.0	3.5	6.6
Glucosides	9.1	8.6	17.7	14.9	1.6	0.9	50.7	16.0
Unknowns	29.4	24.7	34.7	20.8	5.4	6.1	4.5	7.6
Unextractable	—	—	—	—	7.0	12.7	10.3	8.4

[a] Data from Shono *et al.* (1978).
[b] Minor identified metabolites and unknowns.

the excreted fraction. All of the metabolites identified from cockroaches were also found in houseflies. In addition, a new series of metabolites arising from hydroxylation at the 6-position of the alcohol moiety was found only in flies: 6-HO-*t*-per and 6-HO-*c*-per, 6-HO-PBalc and 6-HO-PBacid. These metabolites, as well as all of the hydroxylated ester metabolites, were identified exclusively as glucoside conjugates, whereas the major ester cleavage products, both intact and hydroxylated, were found both in free form and as amino acid or glucoside conjugates. The proportion of unidentified conjugated ester cleavage products derived from both the acid and alcohol moieties of *t*-per and *c*-per was much higher in houseflies than in the other species examined.

Topically-applied *c*-per was much more stable than *t*-per in cabbage looper larvae. Forty-eight hours after treatment (0.01 μg *t*-per, 0.0075 μg *c*-per), 36–47 % of the administered *c*-per remained unmetabolized in the insect, whereas only 4–6 % of the *t*-per radiocarbon could be recovered as the parent compound (Table 3). Ester cleavage was the dominant pathway for *t*-per: the major metabolites were the glucosides of PBalc and 4′-HO-PBalc from [alcohol-^{14}C]-*t*-per and *t*-Cl_2CA (either free or as the glycine conjugate) from [acid-^{14}C]-*t*-per. The only ester metabolite identified from *t*-per was 4′-HO-*t*-per glucoside, and *trans*-methyl hydroxylation in the acid moiety was detected as a small amount of *t*-HO-*t*-Cl_2CA but not the corresponding ester. In contrast, *c*-per gave significant amounts of 4′-HO-*c*-per and *t*-HO-*c*-per as the glucosides and smaller amounts of the ester cleavage products.

The direct identification of several conjugates in this study provides insight into the nature and species specificity of the secondary metabolic reactions of pyrethroids in insects. Cabbage loopers and houseflies efficiently conjugated metabolite hydroxyl groups and Cl_2CA with glucose, but cockroaches left hydroxyesters unconjugated in part and readily oxidized PBalc and 4′-HO-PBalc to the corresponding acids. These differences may have resulted from the different experimental procedures used for cockroaches (treatment by injection, analysis of whole body residues at 24 h) rather than actual species differences in secondary metabolism. All three species extensively conjugated carboxylic acids with one or more of the following amino acids: glycine (Cl_2CA in all three species, PBacid in cabbage loopers and houseflies); glutamic acid (Cl_2CA and PBacid in cockroaches and houseflies); glutamine (Cl_2CA and PBacid in cockroaches); and serine (Cl_2CA in cabbage loopers). Additional amino acid conjugates were inferred among the unidentified conjugated cleavage products since these compounds resisted hydrolysis by β-glucosidase, acid phosphatase and aryl sulphatase.

The metabolism of *c*-per and *t*-per in larvae of the cotton bollworm (*Heliothis zea* Boddie) and the tobacco budworm (*Heliothis virescens* Fab.) was reported by Bigley and Plapp (1978). Twenty-four hours after the larvae were treated topically with sublethal doses of either *c*-per or *t*-per, the external, internal and excreted radioactivity were analysed for parent compounds and metabolites.

Metabolites tentatively identified in bollworm internal extracts and excreta by thin-layer co-chromatography with standards after chemical hydrolysis of conjugates were: *c*- and *t*-Cl_2CA, PBalc, 2′-HO-PBalc, 4′-HO-PBalc, PBacid and 4′-HO-PBacid. Additional metabolites, recognized as ester cleavage products of the acid moiety by their appearance only in insects treated with acid-labelled preparations, were suggested on the basis of comparisons of relative chromatographic behaviour with literature values to be hydroxylated Cl_2CA isomers and their lactones. These results confirm in the bollworm the pathways proposed by Shono *et al.* (1978) and also demonstrate a new pathway, hydroxylation of both isomers at the 2′ position of the alcohol moiety, not previously found in insects but known to occur in the metabolism of *c*-per and *t*-per by rats (Gaughan *et al.*, 1977). Although the metabolites of *c*-per and *t*-per in the budworm were not rigorously examined in this study, the data suggest that the same pathways are found in both *Heliothis* species.

Bigley and Plapp (1978) also examined the relative rates of permethrin isomer metabolism in the tobacco budworm and the cotton bollworm, together with the effects of acephate (Figure 2, *F*) and chlordimeform (Figure 2, *G*) on these rates. In general, the tobacco budworm metabolized the permethrin isomers more rapidly than did the cotton bollworm and excreted *t*-per more rapidly than *c*-per; the bollworm degraded *t*-per more rapidly than *c*-per but excreted radiocarbon derived from both isomers at similar rates. There were no quantitative differences between the insect species with respect to the relative amounts of internal and external radioactivity. When budworms or bollworms were pretreated with either acephate or chlordimeform, the rate of permethrin metabolism was reduced and the excretion of the unchanged pyrethroid was increased. Both compounds had a greater effect on *t*-per metabolism than on that of *c*-per and inhibited permethrin metabolism to a greater degree in the budworm than in the bollworm. The effects of the synergists on metabolite patterns could not be clearly related to the inhibition of specific primary biotransformation pathways.

Holden (1979) also examined the metabolism of *t*-per and *c*-per using adult male American cockroaches and larvae of permethrin-tolerant and susceptible cotton leafworms (*Spodoptera littoralis* Boisd.) as experimental animals. Cockroaches were treated by injection with 1 μg of acid- or alcohol-labelled *t*-per or *c*-per and whole body residues were analysed by methanol extraction and chromatography at various times up to 17 h after treatment. *c*-Per was found to be much more stable than *t*-per so that 17 h after treatment the amount of unmetabolized *c*-per remaining was three times higher than that of *t*-per. The metabolites tentatively identified from either isomer were ester cleavage products: PBalc, PBacid and 4′-HO-PBacid from alcohol-labelled preparations; and Cl_2CA, *t*-HO-Cl_2CA, and *c*-HO-Cl_2CA from acid-labelled preparations. These results confirm the major pathways defined by Shono *et al.* (1978). However, no hydroxyester metabolites were identified with either *c*-per or *t*-per under these analytical conditions. The large difference in permethrin isomer metabolism rates observed in this study contrasts with the similar metabolism rates measured for *t*-per and *c*-per when administered to female American cockroaches by the same route at the same dose (Shono *et al.*, 1978).

Metabolism studies in cotton leafworm larvae were restricted to comparisons of the rates of metabolism of *c*-per and *t*-per in the susceptible and tolerant strains. In contrast to the results of studies with other lepidopterous species (Shono *et al.*, 1978; Bigley and Plapp, 1978) no difference in metabolism rates between *c*-per and *t*-per was observed with either strain. The tolerant strain degraded both isomers somewhat more rapidly than the susceptible strain.

These three studies reveal both quantitative and qualitative species differences in the metabolism of the permethrin isomers. *t*-Per and *c*-per are metabolized at similar rates by cotton leafworm larvae (Holden, 1979), but in all other species *t*-per is degraded more rapidly than *c*-per. The difference in metabolism rates is either small (Shono *et al.*, 1978) or moderate (Holden, 1979) in cockroaches, moderate in houseflies (Shono *et al.*, 1978) and in both cotton bollworm and tobacco budworm larvae (Bigley and Plapp, 1978) and very large in cabbage looper larvae (Shono *et al.*, 1978). In all cases where isomer degradation rates are different the more rapid metabolism of *t*-per is correlated with more extensive ester cleavage. The *cis* isomer undergoes significant hydroxylation as the intact ester in some species, and it is likely that some of the hydroxylated cleavage products of *c*-per are formed by cleavage of hydroxyesters rather than by hydroxylation of *c*-Cl_2CA and PBalc or PBacid. Of the species in which metabolites were identified in some detail, houseflies produced the most complex pattern of metabolites, involving hydroxylation of three (*t*-per) or four (*c*-per) different sites and the production of a variety of unidentified metabolites. Cabbage loopers gave the least complex pattern, with only two sites of hydroxylation observed among the identified metabolites.

Cypermethrin

Holden (1979) examined the fate of the 1*RS*,*trans*,α*RS* and 1*RS*,*cis*,α*RS* isomers of cypermethrin (Figure 1, *3H* and *6H*, respectively; designated below as *t*-cyper and *c*-cyper) in American cockroach adults and cotton leafworm larvae in comparison with the fate of the corresponding permethrin isomers (see above). Compared with permethrin, cypermethrin was more slowly absorbed after topical application and more slowly metabolized after injection in adult male cockroaches. Twenty-four hours after topical application 60% of a dose of *c*-cyper could be recovered in an external solvent rinse, whereas only 20% of an equivalent dose of *c*-per was recovered at the same time. Seventeen hours after injection, 75% of the radiocarbon in methanol extracts of insects was unmetabolized *t*-cyper, whereas only 28% of the parent compound was recovered from insects treated with an equivalent dose of *t*-per. The identification of cypermethrin metabolites was not pursued, but preliminary analyses of the metabolites of acid-labelled *t*-cyper showed a compound co-chromatographing with *t*-Cl_2CA, suggesting that ester cleavage is an important pathway for cyano-substituted esters as well. In cotton leafworm larvae, the cypermethrin isomers could not be distinguished from the permethrin isomers with regard to penetration, but both *t*-cyper and *c*-cyper were metabolized somewhat more slowly than *t*-per and *c*-per, particularly in the permethrin-tolerant strain.

Tralomethrin and tralocythrin

Tralomethrin (Figure 1, *9H*) and tralocythrin (Figure 1, *8H*) are bromine addition products prepared from deltamethrin (Figure 1, *7H*) and [1*R*,*cis*,α*S*]-cypermethrin (Figure 1, *6H*). When applied topically to houseflies or ingested by cabbage looper larvae, both compounds undergo extensive debromination to the corresponding dihalovinyl-substituted compounds (Figure 6) both on

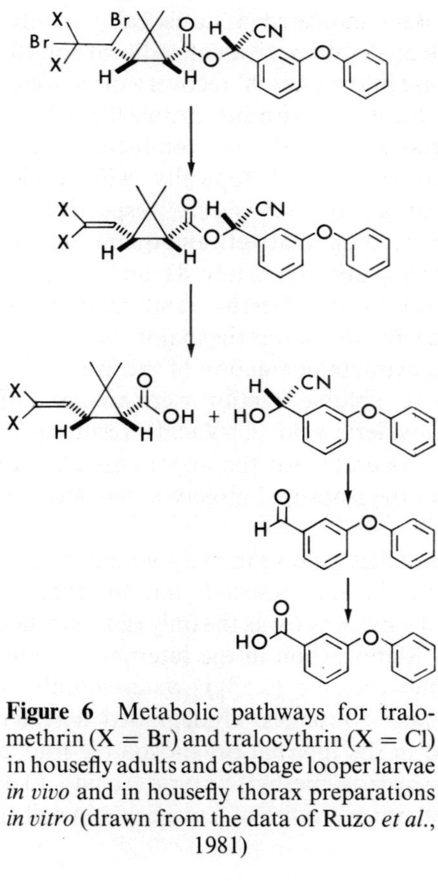

Figure 6 Metabolic pathways for tralomethrin (X = Br) and tralocythrin (X = Cl) in housefly adults and cabbage looper larvae *in vivo* and in housefly thorax preparations *in vitro* (drawn from the data of Ruzo *et al.*, 1981)

the insect surface (houseflies) and in internal extracts (Ruzo *et al.*, 1981). Analysis of housefly extracts and cabbage looper excreta extracts 24 h after treatment revealed 30–41 % debromination of the initial dose, as well as further metabolism to uncharacterized compounds, so that little if any of the parent compounds or their insecticidal debromination products were present after 24 h. These studies, together with *in vitro* studies described below, suggest that tralomethrin and tralocythrin may act in part as 'propyrethroids', undergoing debromination to give non-parent toxicants.

Fenvalerate

The penetration and metabolism of [acid-^{14}C]fenvalerate in insects was examined using both susceptible and permethrin-resistant housefly strains (Lee *et al.*, 1982a). Bioassay data indicated an approximately 1000-fold difference in susceptibility to fenvalerate between the susceptible and resistant strains. Penetration data indicated a statistically significant increase in the external recovery of fenvalerate equivalent radiocarbon (88 versus 74 %) along with a fourfold decrease in the internal recovery of radioactivity (6 versus 25 %) in resistant flies when compared with susceptible flies 4 h after topical treatment.

A comparative metabolism study was conducted using the susceptible and resistant housefly strains treated topically with [acid-^{14}C]fenvalerate at equitoxic doses of 0.02 μg and 0.67 μg/fly, respectively. The metabolism of fenvalerate was determined qualitatively and quantitatively 3 h after treatment. In susceptible houseflies, approximately 81 and 17 % of the applied radioactivity was recovered in the external and internal fraction, respectively (Table 4). Unchanged fenvalerate was the major constituent in both the external wash and the organic-extractable fraction of the internal recovery. Conjugated metabolites in the water soluble fraction were analysed after acid hydrolysis. 2-(4-Chlorophenyl)-isovaleric acid (Cl-Vacid), resulting from the cleavage of the ester linkage of fenvalerate, was the major conjugated material. Oxidation of the methyl group of the isopropyl moiety of fenvalerate was not detected in this strain.

In the resistant houseflies, approximately 90 and 6 % of the applied radioactivity was recovered in the external and internal fraction, respectively (Table 4). Undegraded fenvalerate was the only radiolabelled material recovered in the external wash. Radiocarbon in the internal fraction was either organosoluble (2.5 %) or unextractable (3.5 %); water-soluble conjugates were not detected in the resistant houseflies. Undegraded fenvalerate was the major component recovered in the internal organosoluble fraction. The β-isomer of 2-(4-chlorophenyl)-3-(hydroxymethyl)-butyric acid (3-HO-Cl-Bacid) was also detected. In metabolism studies of fenvalerate in soil (Lee *et al.*, 1981) and

Table 4 Metabolism of [acid-^{14}C]fenvalerate in pyrethroid-susceptible and pyrethroid-resistant adult houseflies 3 h after topical application[a]

Fraction or compound	Applied radiocarbon (%) Susceptible	Resistant
External wash		
Fenvalerate	81.0	90.0
Other[b]	0.3	0.2
Internal extract, organosoluble		
Fenvalerate	3.0	1.5
3-HO-Cl-Bacid	0.0	0.5
Unknowns	0.3	0.5
Internal extract, water soluble		
Cl-Vacid-conj[c]	9.5	0.0
Unknown conjugates	1.9	0.0
Unextractable residues	2.1	3.5
Loss	1.9	3.8

[a] Data from Lee *et al.* (1982a).
[b] Uncharacterized non-parent radiocarbon.
[c] Conjugating moiety not characterized.

rats (Lee *et al.*, 1982b), the α-isomer of 3-OH-Cl-Bacid was transformed into the corresponding δ-lactone, neither of which were observed in the insect metabolism study. The pathways of fenvalerate metabolism in resistant and susceptible houseflies based on identified metabolites are summarized in Figure 7. Since these studies involved the use of acid-labelled material only, no information was obtained concerning the fate of the fenvalerate alcohol moiety.

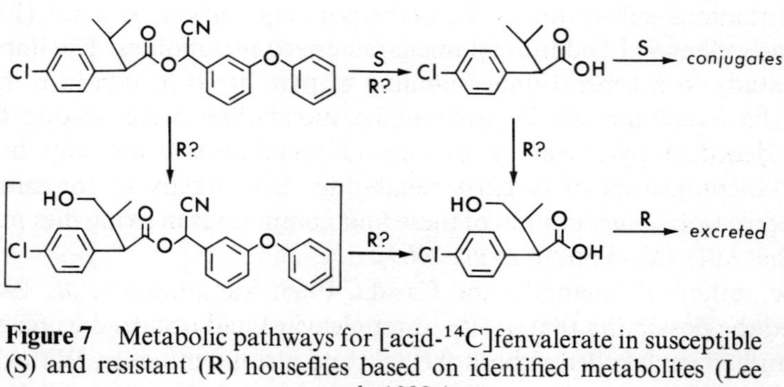

Figure 7 Metabolic pathways for [acid-^{14}C]fenvalerate in susceptible (S) and resistant (R) houseflies based on identified metabolites (Lee *et al.*, 1982a)

IN VITRO METABOLISM

Enzyme systems used in *in vitro* studies

The earliest attempts to identify enzyme preparations capable of metabolizing the pyrethrins and pyrethroids focused on the supposed lability of the ester bond to hydrolytic attack (Chamberlain, 1950; Bridges, 1957). Following the discovery that insects possessed NADPH-dependent oxidative enzymes, analogous to mammalian liver microsomal monooxygenases, that metabolized numerous insecticides (Yamamoto and Casida, 1966; Tsukamoto and Casida, 1967; Wilkinson and Brattsten, 1972), attention shifted to insect monooxygenase preparations for *in vitro* studies. The increased awareness of the importance of both oxidative and hydrolytic reactions in pyrethroid metabolism, particularly with the more recent synthetic esters, has led to the broad use of two types of insect tissue preparations for *in vitro* studies: tissue homogenates or homogenate subfractions fortified with NADPH and having both oxidative and hydrolytic activity; and such preparations lacking NADPH or otherwise modified for the study of hydrolytic activity only. A third type of preparation, fortified with NADPH and pretreated with an esterase inhibitor to allow examination of oxidative reactions only, has been used infrequently in insect metabolism studies.

Metabolism by preparations having both monooxygenase and esterase activity

Yamamoto and Casida (1966) reported the first insect enzyme system capable of metabolizing pyrethrins and pyrethroids *in vitro*. These authors used a housefly abdomen homogenate system fortified with NADPH (Tsukamoto and Casida, 1967) to examine the oxidative metabolism of pyrethrin I, allethrin, dimethrin and tetramethrin. This system formed an acidic metabolite resulting from oxidation of the *trans*-methyl group of the acid isobutenyl substituent of each of the four substrates to the corresponding carboxylic acids (Figures 3 and 4) as well as 9–12 additional uncharacterized metabolites. The importance of this study to a general understanding of pyrethroid metabolism in insects cannot be overemphasized; these acidic metabolites were among the first firmly identified pyrethrin or pyrethroid metabolites from any biological system. Identification of *in vitro* metabolites led directly to the subsequent characterization of metabolites of these four compounds in living flies published somewhat later (Yamamoto *et al.*, 1969).

These authors (Yamamoto and Casida, 1966; Yamamoto *et al.*, 1969) also explored the possibility that *in vitro* ester cleavage had occurred using allethrin and tetramethrin labelled in both the acid and alcohol moieties. With allethrin, the chromatographic patterns of the separated metabolites from both labelled preparations were identical, indicating that no ester cleavage had occurred.

With tetramethrin, the two labelled preparations gave ten ester metabolites, but two additional metabolites were found from acid-labelled tetramethrin only, suggesting that some ester cleavage had occurred.

The finding (Miyamoto and Suzuki, 1973) that tetramethrin was extensively cleaved *in vivo* in some housefly strains prompted a re-examination of the metabolism of this compound in the housefly abdomen homogenate *in vitro* system (Suzuki and Miyamoto, 1974). Unfortified abdomen homogenates extensively metabolized both acid- and alcohol-labelled [1*RS*, *trans*]-tetramethrin, and the addition of NADH, NADPH or other cofactors stimulated metabolism rates only slightly. The synergist NIA 16388, as well as several other organophosphorus and methylcarbamate insecticides, strongly inhibited tetramethrin metabolism. In contrast, piperonyl butoxide showed only slight inhibition, and the addition of oxidase cofactors did not overcome organophosphate inhibition of tetramethrin metabolism. Most of the tetramethrin metabolizing activity in fly abdomen homogenates was found in the microsomal fraction, but activities in other fractions were considerably higher than the relative monooxygenase activities of those fractions as determined by aldrin epoxidation rates. The major metabolites of tetramethrin in this system were chrysanthemic acid and 3,4,5,6-tetrahydrophthalimide (Figure 4) a non-enzymatic product of the tetramethrin alcohol moiety in this system. Oxidized tetramethrin and chrysanthemic acid metabolites were found in significant quantities only in one fly strain having exceptionally high monooxygenase (aldrin epoxidase) activity. These results provided evidence that the ester cleavage reactions observed for tetramethrin *in vitro* by these authors involved hydrolytic pathways catalysed by enzymes distinct from the monooxygenase activity involved in the metabolism of pyrethrin I and allethrin.

The *in vitro* metabolism of the *trans* and *cis* isomers of permethrin was examined in 10 000 *g* supernatants of cabbage looper larval gut homogenates and in microsomes prepared from housefly thorax-plus-abdomen homogenates (Shono and Casida, 1978; Shono *et al.*, 1979). These tissue preparations contained both monooxygenase and pyrethroid-hydrolysing esterase activities. The rates of *t*-per and *c*-per metabolism by oxidases alone, esterases alone, or the combined activities were measured, and metabolites were identified from preparations having either the oxidase system only or both enzyme systems active. The esterase activities of both insect preparations hydrolysed *t*-per more rapidly than *c*-per. Housefly preparations oxidized both isomers at similar rates, but cabbage looper preparations oxidized *c*-per much more rapidly than *t*-per. The combined esterase and oxidase activities of both preparations metabolized *t*-per much faster than *c*-per.

The metabolites formed *in vitro* in these preparations agree with previous *in vivo* studies (Shono *et al.*, 1978; see above) with regard to the greater extent of *t*-per hydrolysis and the major sites of oxidative attack on both isomers in both species. The cabbage looper preparations were virtually devoid of oxidative activity toward *t*-per, but possessed the necessary cofactors and conjugating moieties to oxidize PBalc to PBacid and to conjugate PBacid and both Cl_2CA isomers with glycine.

Metabolism by insect esterase preparations

Studies with esterase systems have focused on the properties and toxicological significance of the enzyme activities that hydrolyse pyrethroids. Jao and Casida (1974a) reported the first successful studies with specific pyrethroid esterase preparations from insects. These authors initially detected pyrethroid esterase activity of whole insect homogenates or their 20 000 *g* supernatant fractions, but they were unable to demonstrate linearity of hydrolysis with respect to protein concentration. In addition, the 20 000 *g* supernatants of some insect homogenates were shown to contain factors capable of inhibiting pyrethroid hydrolysis by highly active mouse liver microsomal esterases. These difficulties were overcome by using acetone powder preparations of insect homogenate supernatants and mouse liver microsomal suspensions, a technique previously used (Chamberlain, 1950; Bridges, 1957) in attempts to demonstrate the *in vitro* hydrolysis of the pyrethrins and allethrin.

Acetone powders prepared from homogenates of adult male milkweed bugs (*Oncopeltus fasciatus* Dallas), male German cockroaches (*Blattella germanica* L.) and female houseflies, as well as those from cabbage looper and yellow mealworm (*Tenebrio molitor* L.) larvae, hydrolysed the 1*R*,*trans* and 1*R*,*cis* isomers of both resmethrin (Figure 1, *1I* and *5I*) and tetramethrin. All preparations except that derived from houseflies hydrolysed [1*R*,*trans*]-resmethrin faster than [1*R*,*cis*]-resmethrin but the magnitude of the *trans*/*cis* rate difference in the insect preparations was much smaller than that found for the resmethrin isomers in acetone powders of mouse liver microsomes. The insect preparations hydrolysed the tetramethrin isomers at rates similar to or slightly greater than those found for the corresponding resmethrin isomers and showed a different pattern of specificity: rates of [1*R*,*trans*]-tetramethrin hydrolysis were higher than those for the 1*R*,*cis* isomer in milkweed bug, housefly and cabbage looper preparations but were slightly slower in cockroach and mealworm preparations. None of the preparations hydrolysed [1*R*,*trans*,4′*S*]-allethrin. Estimates of Michaelis–Menten kinetic parameters for [1*R*,*trans*]-resmethrin hydrolysis revealed that the insect preparations had higher affinities but much lower maximum velocities than mouse liver microsomal esterases (Table 5). Insect preparations were more sensitive to inhibition (Table 5) by NPC (Figure 2, *H*) and DEF (Figure 2, *I*) than mouse liver esterases, but MPP (Figure 2, *E*) was a relatively poor inhibitor of all of the esterase preparations.

This study demonstrated the existence of pyrethroid-hydrolysing activity in several insect species but was unable to provide information on the tissue distribution of esterase activity or the possible multiplicity of pyrethroid

Table 5 The hydrolysis of (+)-*trans*-resmethrin and its inhibition in acetone powder esterase preparations from insects and mammals[a]

	(+)-*trans*-resmethrin hydrolysis			I_{50} values for inhibition of (+)-*trans*-resmethrin hydrolysis[b]		
Enzyme source	relative rate[c]	K_m[b]	V_{max}[d]	NPC	DEF	MPP
Insect						
Milkweed bug	100	25	66	0.23	4.3	13
Cockroach	63	83	58	0.08	0.10	250
Housefly	30	63	30	—	—	—
Cabbage looper	38	111	60	0.28	0.51	280
Mealworm	13	—	—	—	—	—
Mammal						
Mouse liver	1667	125	2083	13	1.0	500

[a] Data from Jao and Casida (1974a).
[b] $M \times 10^{-7}$.
[c] The rate for milkweed bug esterases under standard assay conditions (12.0 pmole/min/mg protein) is set equal to 100.
[d] pmole/min/mg protein.

esterases in insect tissues. Soderlund (1979) briefly reported esterase activity in homogenates of adult male American cockroach fat body and central nerve cord capable of hydrolysing *t*-per but not *c*-per. No esterase activity was detected in the haemolymph, but other tissues were not examined and the activity in fat body and nerve cord preparations was not further characterized. Abdel-Aal and Soderlund (1980) described in greater detail the tissue distribution and properties of the permethrin-hydrolysing activities of southern armyworm larval tissues (Table 6). Whole homogenates of gut, fat body, Malpighian tubules, silk glands, integument and larval head capsules hydrolysed both *t*-per and *c*-per at detectable rates, but no hydrolytic activity was found in armyworm haemolymph. The active tissues hydrolysed *t*-per 2 to 69 times more rapidly than *c*-per. The highest esterase activities per larval equivalent of tissue were found in fat body, gut and integument for *t*-per and in gut and integument for *c*-per. Ishaaya and Casida (1980) also described esterase activity in cabbage looper larval gut and integument preparations that hydrolysed *t*-per and *t*-cyper more rapidly than the corresponding *cis* isomers. The gut esterases hydrolysing the permethrin isomers were more active on a larval equivalent basis than those in the integument preparations, but both tissues had similar activities toward the cypermethrin isomers.

These studies of southern armyworm and cabbage looper larval esterase activities also demonstrated the existence of multiple pyrethroid-hydrolysing esterases with differing substrate specificities and inhibitor sensitivities in insect

Table 6 Hydrolysis of *trans*- and *cis*-permethrin by tissues of southern armyworm larvae[a]

Tissue or region	*trans*-permethrin Relative rate[b]	K_m[c]	V_{max}[d]	*cis*-permethrin Relative rate	K_m	V_{max}
Gut	100	64	2628	9	47	140
Fat body	129	122	2875	2	4	14
Integument	82	40	3622	10	87	1328
Head capsule	9	—	—	3	—	—
Malpighian tubules	15	—	—	2	—	—
Silk gland	9	—	—	4	—	—
Haemolymph	ND[e]	—	—	ND	—	—

[a] Data from Abdel-Aal and Soderlund (1980).
[b] The rate for *trans*-permethrin hydrolysis in gut preparations (2.16 nmoles/min insect equivalent of tissue) is set equal to 100.
[c] $M \times 10^{-6}$.
[d] pmoles/min/mg protein.
[e] No detectable hydrolysis.

tissues. Ishaaya and Casida (1980) used the organophosphate insecticide profenofos (Figure 2, *J*) to inhibit selectively the gut and integument esterases hydrolysing the *trans* and *cis* isomers of permethrin and cypermethrin. Differences in sensitivity to profenofos inhibition with these different substrates demonstrated the presence of multiple pyrethroid esterases in each tissue (Table 7). In gut preparations, separate esterase activities clearly differing in profenofos sensitivity hydrolysed *t*-per and *t*-cyper, whereas one or more

Table 7 Differential inhibition of pyrethroid esterases in cabbage looper larval gut and integument preparations by profenofos[a]

Tissue	Profenofos (M)	Inhibition of control activity (%) Permethrin *trans*	Permethrin *cis*	Cypermethrin *trans*	Cypermethrin *cis*
Gut	10^{-9}	1	6	0	0
	10^{-8}	28	66	0	67
	10^{-7}	55	87	82	91
	10^{-5}	82	91	96	96
Integument	10^{-9}	14	0	27	14
	10^{-7}	59	37	98	72
	10^{-5}	85	63	98	68

[a] Data from Ishaaya and Casida (1980).

activities having greater profenofos sensitivity hydrolysed *c*-per and *c*-cyper. In integument preparations, profenofos inhibition revealed distinct activities hydrolysing each of the substrates. The integument activities hydrolysing *c*-per, *t*-cyper and *c*-cyper also showed clearly different profenofos sensitivities than the corresponding gut activities, suggesting that integument esterases are different from those found in gut preparations.

Abdel-Aal and Soderlund (1980) also employed a selective inhibitor in describing the esterase activities of southern armyworm integument, gut and fat body homogenates. Each tissue was shown to have two *t*-per-hydrolysing activities, only one of which was sensitive to NPC inhibition (Figure 8). The NPC-sensitive activity was predominant in gut and fat body homogenates but was less abundant than the insensitive activity in the integument. The I_{50} values for NPC inhibition of the sensitive esterases of all three tissues were similar, implying either that the same esterases are found in each tissue or that NPC is not a selective inhibitor of the sensitive esterases. The NPC-sensitive activities apparently do not hydrolyse *c*-per, since NPC concentrations giving maximum inhibition of *t*-per hydrolysis had little influence on *c*-per hydrolysis (Figure 8). These results do not show, however, whether *c*-per is hydrolysed by a third esterase activity or by the NPC-insensitive *t*-per hydrolysing activity. Differences between southern armyworm tissues in *c*-per hydrolysing activity are also evident in the apparent K_m and V_{max} values measured for the hydrolysis of permethrin isomers in each tissue (Abdel-Aal and Soderlund, 1980; Table 6).

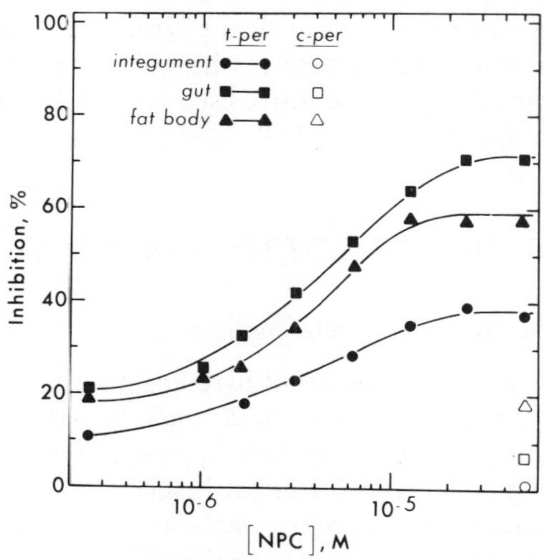

Figure 8 Inhibition of *t*-per and *c*-per hydrolysis by NPC in homogenates of southern armyworm larval integument, gut and fat body (redrawn from Abdel-Aal and Soderlund, 1980)

The differences in apparent K_m and V_{max} values for *c*-per hydrolysis by these preparations are sharply contrasted with the very consistent kinetic behaviour of *t*-per in the same preparations.

Ishaaya and Casida (1981) found extremely high levels of pyrethroid esterase activity in larvae of the green lacewing (*Chrysopa carnea* Stephens). This activity showed unusual substrate specificity, hydrolysing the *cis* isomers of permethrin and cypermethrin more rapidly than the corresponding *trans* isomers and hydrolysing deltamethrin more rapidly than *c*-cyper. The lacewing larval esterases easily represent the most active pyrethroid esterase preparation obtained from insects to date, hydrolysing pyrethroids at rates similar to cabbage looper gut and integument preparations (Ishaaya and Casida, 1980) having sevenfold higher protein levels. This exceptional level of esterase activity appears to be a key feature in the insensitivity of lacewing larvae to pyrethroids (see below).

Additional evidence for pyrethroid esterase activity in houseflies has been obtained by Ruzo *et al.* (1981; Figure 6). Tralomethrin and tralocythrin were converted to deltamethrin and [1*R*,*cis*,α*S*]-cypermethrin upon reaction with glutathione or other tissue thiols in 15,000 *g* supernatants of housefly thoraces. The resulting dihalovinyl-substituted esters were hydrolysed to the corresponding acid and the cyanohydrin, and further metabolism of the cyanohydrin to 3-phenoxybenzaldehyde and PBacid was also noted. The formation of hydrolysis products was blocked by both *N*-ethylmaleimide (NEM) and tetraethylpyrophosphate (TEPP) when tralomethrin was the substrate but only by TEPP when deltamethrin was the substrate. Since NEM was shown to block the debromination of tralomethrin and tralocythrin, these results demonstrate the presence of an organophosphate-sensitive esterase activity in this preparation that hydrolyses deltamethrin and [1*R*,*cis*,α*S*]-cypermethrin but not their brominated analogues.

EFFECTS OF METABOLISM ON BIOLOGICAL ACTIVITY

Metabolism and structure-activity relationships

Much of the effort invested in studies of pyrethroid metabolism in insects has been directed at understanding the ways in which metabolism limits or determines observed insecticidal activity. This approach has also considered the effect of synergists on toxicity, exploring the biochemical mechanisms involved in the action of synergists and providing the basis for the use of synergists both as diagnostic tools in bioassays and as potential additives to enhance pyrethroid activity in the field.

Chang and Kearns (1964) noted that the lower toxicity of cinerin I against houseflies compared with pyrethrin I was correlated with the greater metabolic lability of the former compound. Several early metabolism studies (Winteringham *et al.*, 1955; Hopkins and Robbins, 1957; Chang and Kearns, 1964; Hayashi *et al.*, 1968) demonstrated the *in vivo* inhibition of pyrethroid metabolism by methylenedioxyphenyl synergists (Figure 2*A–C*), and these results were later confirmed in *in vitro* systems (Yamamoto *et al.*, 1969). The action of these synergists as inhibitors of oxidative metabolism has become so widely accepted that in a large body of bioassay literature the existence of toxicity-limiting oxidative metabolism of insecticides is inferred solely on the basis of synergism by methylenedioxyphenyl compounds.

The recognition of the importance of hydrolysis in the metabolism of many pyrethroids led directly to the assessment of esterase inhibitors as modifiers of both metabolism and toxicity. The synergism of tetramethrin in houseflies by NIA 16388 (Miyamoto and Suzuki, 1973) was shown to result from the inhibition of tetramethrin hydrolysis by this compound (Suzuki and Miyamoto, 1974). Evaluation of suspected esterase inhibitors having low insecticidal activity led to the identification of NPC as a synergist for the 1*R*,*trans* isomers of resmethrin and tetramethrin in several species (Jao and Casida, 1974b). The efficacy of NPC as a synergist was related to its potency as an inhibitor of pyrethroid hydrolysis by insect acetone powder preparations (Jao and Casida, 1974a). The linkage between esterase inhibition and synergism is clearly demonstrated in the effect of profenofos on the toxicity and metabolism of permethrin and cypermethrin isomers in cabbage looper larvae (Gaughan *et al.*, 1980; Ishaaya and Casida, 1980). Profenofos is a more effective synergist for *c*-cyper than for *t*-per and is also a more effective inhibitor of *c*-cyper hydrolysis than of *t*-per hydrolysis by cabbage looper gut preparations. Similarly, the high esterase activity in green lacewing larvae (Ishaaya and Casida, 1981) is correlated strongly with the insensitivity of these larvae to pyrethroids, particularly *cis* isomers. Pretreatment of lacewing larvae with PSCP (Figure 2, *K*), a potent *in vitro* inhibitor of lacewing esterases, synergized the toxicity of *t*-per 68-fold.

On the basis of such findings, many workers have adopted the use of esterase inhibitors as synergists in bioassay studies to assess the impact of hydrolysis on toxicity. Unfortunately, the existence in insects of multiple pyrethroid esterases with different inhibitor and substrate specificities (Ishaaya and Casida, 1980; Abdel-Aal and Soderlund, 1980) makes interpretation of these synergism data difficult. Although synergism by such compounds provides preliminary evidence for hydrolytic metabolism, the absence of synergism cannot be interpreted as evidence for the absence of significant hydrolytic metabolism, and estimates of the relative importance of hydrolytic detoxification for a series of compounds cannot be reliably obtained. Despite the limitations of known pyrethroid esterase inhibitors as general pyrethroid synergists, the possibility remains that other esterase inhibitors can be identified or developed specifically as synergists for pyrethroids in practical use.

Pharmacokinetic studies in insects offer a different approach to identifying the quantitative influence of metabolism on pyrethroid structure-activity relationships. Soderlund (1979) described the internal distribution of NRDC 157 (Figure 1, 7*G*) and its insecticidally inactive 1*S* enantiomer in adult male American cockroaches after topical application of doses equal to twice the LD_{50} of the toxic enantiomer. The enantiomers had nearly identical time courses of distribution in the haemolymph, ventral nerve cord and fat body, and the apparent steady-state levels attained in the haemolymph and nerve cord were also the same. These results demonstrated the feasibility of pharmacokinetic studies in insects and provided direct evidence that the large toxicity difference observed between 1*R* and 1*S* enantiomers is due to stereospecificity at the site of action rather than pharmacokinetic specificity. These studies have recently been expanded (Soderlund, 1981) to include NRDC 163 (Figure 1, *4G*), the *trans*-substituted analogue of NRDC 157, which is less insecticidal and is expected to undergo more extensive hydrolysis on the basis of studies of permethrin metabolism in this species (Shono *et al.*, 1978; Holden, 1979). When NRDC 163 was applied at twice its LD_{50}, a dose more than threefold higher than the equitoxic dose of NRDC 157, steady-state levels were similar to those for NRDC 157 in the haemolymph but considerably lower in both fat body and nerve cord samples (Figure 9). These results show the influence of the more rapid degradation of NRDC 163, presumably by hydrolysis, on internal distribution patterns compared to the more stable *cis* isomer. If steady-state

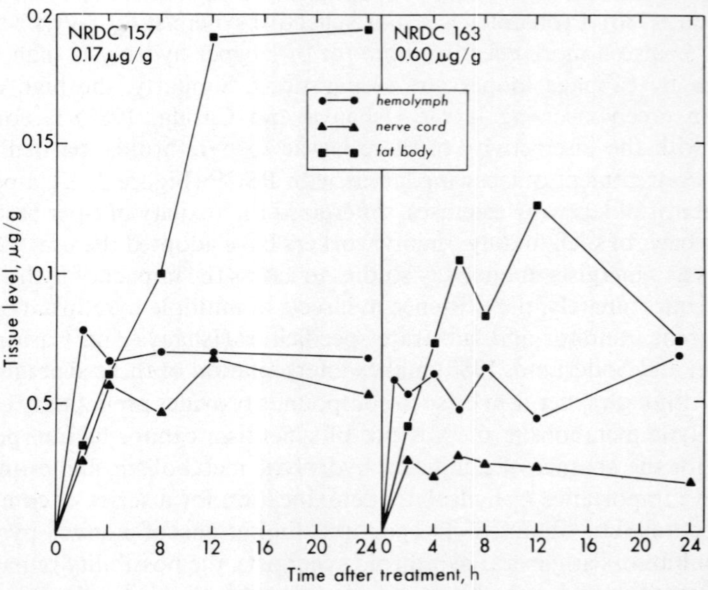

Figure 9 Time courses of accumulation of NRDC 157 and NRDC 163 in the haemolymph, nerve cord and fat body of adult male American cockroaches after topical application at equitoxic (2 × LD_{50}) doses (Soderlund, 1979; 1981)

levels in the nerve cord obtained after treatment with equitoxic doses are an index of relative intrinsic potency, then these data suggest that NRDC 163 is more than twice as potent as NRDC 157 at the level of the nerve. The reversal of this relationship at the level of insecticidal activity clearly illustrates the influence that metabolism can exert on structure–activity relationships.

Ford *et al.* (1981a) have proposed a pharmacokinetic model to describe the movement of insecticides from the surface of insects to internal sites of action and have applied their model to the pharmacokinetics of a series of [1*RS*,*cis*,*trans*]-chrysanthemates based on the penetration curves obtained after topical application to adult mustard beetles (*Phaedon cochleariae* Fab.; Elliott *et al.*, 1970; Ford *et al.*, 1981b). Although the model itself is rigorously derived, it is limited in its power to describe or predict accurately the effects of metabolism and other pharmacokinetic processes on pyrethroid structure–activity relationships. The model describes only two compartments, one internal and one external, but the studies in cockroaches suggest that quantitative relationships between levels of a pyrethroid in key tissues comprising the 'internal' compartment vary from compound to compound. Also, the use of data for isomer mixtures makes it impossible to consider the large differences observed in the fate of pyrethroid isomers in metabolism studies. Much more quantitative information on the disposition of xenobiotics in insects is required before an effective model for the delivery of pyrethroids and other lipophilic insecticides to sites of action in the nervous system can be formulated.

Metabolism and resistance

The development of resistance in pest populations is a serious concern of those involved in the development and agricultural use of pyrethroids. The ability of metabolism to influence the observed toxicity of pyrethroids suggests that metabolic mechanisms might play an important role in pyrethroid resistance, but evidence for this is extremely limited. In pyrethrin-selected or resmethrin-selected houseflies, one of four genetically isolated pyrethroid resistance mechanisms confers resistance that can be abolished by treatment with sesamex, thus implying the existence of an oxidative detoxification mechanism (Farnham, 1971; 1973). However, a non-metabolic resistance mechanism of greater concern involves a reduction in the sensitivity of the nervous system to both pyrethroids and DDT (Farnham, 1973; 1977). This mechanism, called *kdr* in houseflies, has also been proposed as a mechanism resistance in [1*R*,*trans*]-permethrin-selected mosquitoes (*Culex pipiens quinquefasciatus* Say; Priester and Georghiou, 1978), [1*R*,*trans*]-permethrin-selected houseflies (DeVries and Georghiou, 1981a), field-collected permethrin-resistant cotton leafworm larvae (Gammon, 1980) and DDT-selected German cockroaches (Scott and Matsumura, 1981). In each of these cases, the investigators used either bioassay data with synergists (Priester and Georghiou, 1978) or direct comparisons of the *in vitro* or *in vivo* degradation of pyrethroids in resistant and susceptible strains (DeVries and Georghiou, 1981b; Holden, 1979; Scott and Matsumura, 1981) to demonstrate that increased metabolism is not the major resistance mechanism.

Of particular concern are the mechanisms of resistance that may be elicited in lepidopterous species through the intensive and increasing agricultural use of pyrethroids. Relatively low levels of pyrethroid tolerance or resistance have already been reported in tobacco budworm populations in the southwestern United States (Crowder *et al.*, 1979; Twine and Reynolds, 1980; Plapp, 1981), but nothing is known of the underlying mechanism(s) involved. Intensive use of pyrethroids, particularly fenvalerate, to control diamondback moth larvae (*Plutella xylostella* L.) on cruciferous vegetables in Taiwan has produced some of the highest levels of resistance resulting from the agricultural use of pyrethroids reported to date (Liu *et al.*, 1981). The most resistant field-collected strain shows high resistance (67 to 208-fold) to deltamethrin, cypermethrin and fenvalerate, but considerably lower resistance (14-fold) to permethrin. Piperonyl butoxide significantly reduced the degree of resistance for deltamethrin, cypermethrin and fenvalerate, but had little influence on permethrin resistance, suggesting that rapid oxidative metabolism is involved in at least a portion of the observed resistance. DEF had little or no effect on resistance levels, but the absence of DEF synergism alone does not demonstrate the absence of resistance due to hydrolytic metabolism (see above).

It is clear from the foregoing discussion that the significance of increased metabolism as a general mechanism of pyrethroid resistance remains to be established. However, the recent studies on the influence of hydrolytic metabolism on insecticidal activity suggest that further examination of esterase activity in resistant insects is worthwhile, particularly since increased esterase activity is clearly implicated in several cases of organophosphate insecticide resistance (Plapp, 1976). Continued monitoring for possible hydrolytic pyrethroid resistance mechanisms may be particularly important in lepidopterous larvae in view of the dominance of hydrolytic metabolism observed for several pyrethroids in the lepidopterous species examined to date.

CONCLUSIONS

Studies of the metabolism of pyrethrins and pyrethroids in insects *in vivo* and *in vitro* reveal a complex picture. A large array of metabolites is generated by both oxidative attack at one or more sites and hydrolysis of the central ester bond. The relative significance of these processes varies greatly between isomers of the same pyrethroid, between pyrethroids of different structure, and between insect species. Further complexity is introduced in the species diversity of secondary oxidation and conjugation reactions.

A considerable body of information about the metabolic fate of pyrethroids in insects and the nature of the enzymes involved in metabolism has been generated over the past fifteen years, but the literature remains incomplete in many areas. Much of what we know about metabolic pathways for pyrethroids is based on the study of permethrin alone. Although the number and diversity of metabolites identified from the permethrin isomers is impressive, a significant proportion of the total metabolites generated *in vivo*, particularly from the *cis* isomer, remain to be identified. Other important pyrethroids are conspicuous in their absence from this literature. Aside from their use as substrates in one *in vitro* esterase study, there is no information on the metabolic fate of the isomers of resmethrin in insects. Also, the metabolism of pyrethroids such as cypermethrin, deltamethrin and fenvalerate that contain the α-cyano-3-phenoxybenzyl moiety remains to be characterized. Although the metabolic fates of the acid moieties of these compounds are suggested in some cases by studies with the analogous 3-phenoxybenzyl esters, the influence of cyano substitution on preferred sites of oxidative attack in the alcohol moiety in insects remains to be determined. Altered patterns of alcohol moiety hydroxylation in α-cyano-substituted esters compared to their unsubstituted analogues are already known in mammalian metabolism studies (Casida *et al.*, 1979). In addition, the continued development of pyrethroids has already resulted in several compounds nearing commercial availability for which no information on insect metabolism is available.

In spite of the limited number of compounds examined in detail, the major biotransformation events for pyrethroids have been established. However, the quantitative impact of these events on observed insecticidal activity remains unclear and no single satisfactory method has emerged to obtain this information. Studies with synergists are not fully satisfactory for the reasons described above. Pharmacokinetic studies are able to provide more detailed information, but the tedious experimentation involved and the necessary dependency on radiolabelled compounds of high specific activity severely limit both the scope of this approach and the rate at which information can be compiled. More information about quantitative structure–biodegradability relationships for pyrethroids in insects is clearly needed in order to address fundamental questions concerning the intrinsic structure–activity relationships for pyrethroid neurotoxicity and to understand the practical implications of the role of metabolism in resistance and the use of synergists to enhance the efficacy of pyrethroids.

ACKNOWLEDGEMENTS

The authors thank Professor J. E. Casida for providing the original drawing of Figure 5 for reproduction and Ms R. M. McMillen for preparing the remaining figures. Studies reviewed here originating from the senior author's laboratory were supported in part by grants from the National Institute of Environmental Health Sciences (number R01 ES02160) and the Mobil Foundation, Inc., Edison, NJ.

REFERENCES

Abdel-Aal, Y. A. I., and Soderlund, D. M. (1980). 'Pyrethroid-hydrolyzing esterases in southern armyworm larvae: tissue distribution, kinetic properties, and selective inhibition', *Pestic. Biochem. Physiol.*, **14**, 282–289.

Acree, F., Jr., Roan, C. C., and Babers, F. H. (1954). 'The synthesis and chromatographic purification of radioactive allethrin', *J. Econ. Entomol.*, **47**, 1066–1070.

Bigley, W. S., and Plapp, F. W., Jr. (1978). 'Metabolism of *cis*- and *trans*-[^{14}C]permethrin by the tobacco budworm and the bollworm', *J. Agric. Food Chem.*, **26**, 1128–1134.

Bridges, P. M. (1957). 'Absorption and metabolism of [^{14}C]allethrin by the adult housefly, *Musca domestica* L.' *Biochem. J.*, **66**, 316–320.

Casida, J. E., and Ruzo, L. O. (1980). 'Metabolic chemistry of pyrethroid insecticides', *Pestic. Sci.*, **11**, 257–269.

Casida, J. E., Ueda, K., Gaughan, L. C., Jao, L. T., and Soderlund, D. M. (1975/76). 'Structure-biodegradability relationships in pyrethroid insecticides', *Arch. Environ. Contam. Toxicol.*, **3**, 491–500.

Casida, J. E., Gaughan, L. C., and Ruzo, L. O. (1979). 'Comparative metabolism of pyrethroids derived from 3-phenoxybenzyl and α-cyano-3-phenoxybenzyl alcohols', in *Advances in Pesticide Science* (ed. H. Geissbuehler), pp. 182–189, Pergamon, Oxford.

Chamberlain, R. W. (1950). 'An investigation of the action of piperonyl butoxide with pyrethrum', *Am. J. Hyg.*, **52**, 153–183.

Chambers, J. (1980). 'An introduction to the metabolism of pyrethroids', *Residue Rev.*, **73**, 101–124.

Chang, S. C., and Kearns, C. W. (1964). 'Metabolism *in vivo* of C^{14}-labelled pyrethrin I and cinerin I by house flies with special reference to the synergistic mechanism', *J. Econ. Entomol.*, **57**, 397–404.

Crowder, L. A., Tollefson, M. S., and Watson, T. F. (1979). 'Dosage-mortality studies of synthetic pyrethroids and methyl parathion on the tobacco budworm in central Arizona', *J. Econ. Entomol.*, **72**, 1–3.

DeVries, D. H., and Georghiou, G. P. (1981a). 'Decreased nerve sensitivity and decreased cuticular penetration as mechanisms of resistance to pyrethroids in a (1*R*)-*trans*-permethrin-selected strain of the house fly', *Pestic. Biochem. Physiol.*, **15**, 234–241.

DeVries, D. H., and Georghiou, G. P. (1981b). 'Absence of enhanced detoxication of permethrin in pyrethroid-resistant houseflies', *Pestic. Biochem. Physiol.*, **15**, 242–252.

Elliott, M. (1976). 'Properties and applications of pyrethroids', *Environ. Health Perspec.*, **14**, 3–13.

Elliott, M. (1977). 'Synthetic pyrethroids', in *Synthetic Pyrethroids* (ed. M. Elliott), pp. 1–28, American Chemical Society, Washington, DC.

Elliott, M. (1980). 'Established pyrethroid insecticides', *Pestic. Sci.*, **11**, 119–128.

Elliott, M., and Janes, N. F. (1978). 'Synthetic pyrethroids—a new class of insecticide', *Chem. Soc. Rev.*, **7**, 473–505.

Elliott, M., and Janes, N. F. (1979). 'Recent structure–activity correlations in synthetic pyrethroids', in *Advances in Pesticide Science* (ed. H. Geissbuehler), pp. 166–173, Pergamon, Oxford.

Elliott, M., Ford, M. G., and Janes, N. F. (1970). 'Insecticidal activity of the pyrethrins and related compounds. III. Penetration of pyrethroid insecticides into mustard beetles (*Phaedon cochleariae*)', *Pestic. Sci.*, **1**, 220–223.

Elliott, M., Janes, N. F., Kimmel, E. C., and Casida, J. E. (1972). 'Metabolic fate of pyrethrin I, pyrethrin II, and allethrin administered orally to rats', *J. Agric. Food Chem.*, **20**, 300–313.

Elliott, M., Farnham, A. W., Janes, N. F., Needham, P. H., Pulman, D. A., and Stevenson, J. H. (1973). 'A photostable pyrethroid', *Nature*, **246**, 169–170.

Elliott, M., Farnham, A. W., Janes, N. F., Needham, P. H., and Pulman, D. A. (1974a). 'Insecticidally active conformations of pyrethroids', in *Mechanism of Pesticide Action* (ed. G. K. Kohn), pp. 80–91, American Chemical Society, Washington, DC.

Elliott, M., Janes, N. F., and Pulman, D. A. (1974b). 'The pyrethrins and related compounds. Part XVIII. Insecticidal 2,2-dimethylcyclopropanecarboxylates with new unsaturated 3-substituents', *J. Chem. Soc. Perkin Trans. I*, 2470–2474.

Elliott, M., Janes, N. F., and Potter, C. (1978). 'The future of pyrethroids in insect control', *Ann. Rev. Entomol.*, **23**, 443–469.

Farnham, A. W. (1971). 'Changes in cross-resistance patterns of houseflies selected with natural pyrethrins or resmethrin (5-benzyl-3-furylmethyl(±)-*cis,trans-chrysanthemate*)', *Pestic. Sci.*, **2**, 138–143.

Farnham, A. W. (1973). 'Genetics of resistance in pyrethroid-selected houseflies, *Musca domestica* L.', *Pestic. Sci.*, **4**, 513–520.

Farnham, A. W. (1977). 'Genetics of resistance of houseflies (*Musca domestica* L.) to pyrethroids', *Pestic. Sci.*, **8**, 631–636.

Ford, M. G., Greenwood, R., and Thomas, P. J. (1981a). 'The kinetics of insecticide action, Part I: The properties of a mathematical model describing insect pharmacokinetics', *Pestic. Sci.*, **12**, 175–198.

Ford, M. G., Greenwood, R., and Thomas, P. J. (1981b). 'The kinetics of insecticide action, Part II: The relationship between the pharmacokinetics of substituted benzyl (1*RS*)-*cis*, *trans*-chrysanthemates and their relative toxicities to mustard beetles (*Phaedon cochleariae* Fab.)', *Pestic. Sci.*, **12**, 265–284.

Gammon, D. W. (1980). 'Pyrethroid resistance in a strain of *Spodoptera littoralis* is correlated with decreased sensitivity of the CNS *in vitro*', *Pestic. Biochem. Physiol.*, **13**, 53–62.

Gaughan, L. C., Unai, T., and Casida, J. E. (1977). 'Permethrin metabolism in rats', *J. Agric. Food Chem.*, **25**, 9–17.

Gaughan, L. C., Engel, J. L., and Casida, J. E. (1980). 'Pesticide interactions: effects of organophosphorus pesticides on the metabolism, toxicity, and persistence of selected pyrethroid insecticides', *Pestic. Biochem. Physiol.*, **14**, 81–85.

Hayashi, A., Saito, T., and Iyatomi, K. (1968). 'Studies on the increment of the efficacy of insecticides (VIII). Metabolism of ^{3}H-pyrethroids in the adult house fly, *Musca domestica vicinia* Macq.', *Botyu-Kagaku*, **33**, 90–95.

Holden, J. S. (1979). 'Absorption and metabolism of permethrin and cypermethrin in the cockroach and the cotton-leafworm larvae', *Pestic. Sci.*, **10**, 295–307.

Hopkins, T. L., and Robbins, W. E. (1957). 'The absorption, metabolism, and excretion of C^{14}-labeled allethrin by house flies', *J. Econ. Entomol.*, **50**, 684–687.

Hutson, D. H. (1979). 'The metabolic fate of synthetic pyrethroid insecticides in mammals', in *Progress in Drug Metabolism* (eds J. W. Bridges and L. F. Chasseaud), Vol. 3, pp. 215–252, Wiley, Chichester.

Ishaaya, I., and Casida, J. E. (1980). 'Properties and toxicological significance of esterases hydrolyzing permethrin and cypermethrin in *Trichoplusia ni* larval gut and integument', *Pestic. Biochem. Physiol.*, **14**, 178–184.

Ishaaya, I., and Casida, J. E. (1981). 'Pyrethroid esterase(s) may contribute to natural pyrethroid tolerance of larvae of the common green lacewing', *Environ. Entomol.*, **10**, 681–684.

Jao, L. T., and Casida, J. E. (1974a). 'Insect pyrethroid-hydrolyzing esterases', *Pestic. Biochem. Physiol.*, **4**, 465–472.

Jao, L. T., and Casida, J. E. (1974b). 'Esterase inhibitors as synergists for (+)-*trans*-chrysanthemate insecticide chemicals', *Pestic. Biochem. Physiol.*, **4**, 456–464.

Lee, P. W., Vanderlinden, H. F., Silveira, E. J., and Burton, W. B. (1981). 'Aerobic and anaerobic soil degradation of Pydrin[R] Insecticide', Paper No. 22, Pesticide Chemistry Division, American Chemical Society, 182nd National Meeting, New York, NY.

Lee, P. W., Stearns, S. M., and Sanborn, J. R. (1982a). 'Penetration and metabolism of Pydrin^R Insecticide in susceptible and resistant houseflies', Paper No. 69, Pesticide Chemistry Division, American Chemical Society, 183rd National Meeting, Las Vegas, Nev.

Lee, P. W., Stearns, S. M., Powell, W. R., Silveira, E. J., and Burton, W. B. (1982b). 'Rat metabolism of Pydrin^R Insecticide', Paper No. 68, Pesticide Chemistry Division, American Chemical Society, 183rd National Meeting, Las Vegas, Nev.

Liu, M.-Y., Tzeng, Y.-J., and Sun, C.-N. (1981). 'Diamondback moth resistance to several synthetic pyrethroids', *J. Econ. Entomol.*, **74**, 393–396.

Miyamoto, J. (1976). 'Degradation, metabolism and toxicity of synthetic pyrethroids', *Environ. Health Perspec.*, **14**, 15–26.

Miyamoto, J., and Suzuki, T. (1973). 'Metabolism of tetramethrin in houseflies *in vivo*', *Pestic. Biochem. Physiol.*, **3**, 30–41.

Nishizawa, Y., and Casida, J. G. (1965). 'Synthesis of *d-trans*-chrysanthemumic acid-1-C^{14} and its antipode on a semimicro scale', *J. Agric. Food Chem.*, **13**, 525–527.

Pellegrini, J. P., Jr., Miller, A. C., and Sharpless, R. V. (1952). 'Biosynthesis of radioactive pyrethrins using $C^{14}O_2$', *J. Econ. Entomol.*, **45**, 532–536.

Plapp, F. W. (1976). 'Biochemical genetics of insecticide resistance', *Ann. Rev. Entomol.*, **21**, 179–197.

Plapp, F. W., Jr. (1981). 'Toxicity of synthetic pyrethroids to laboratory and field populations of the tobacco budworm in central Texas', *J. Econ. Entomol.*, **74**, 207–209.

Priester, T. M., and Georghiou, G. P. (1978). 'Induction of high resistance to permethrin in *Culex pipiens quinquefasciatus*', *J. Econ. Entomol.*, **71**, 197–200.

Ruzo, L. O., and Casida, J. E. (1977). 'Metabolism and toxicology of pyrethroids with dihalovinyl substituents', *Environ. Health Perspec.*, **21**, 285–292.

Ruzo, L. O., Gaughan, L. C., and Casida, J. E. (1981). 'Metabolism and degradation of the pyrethroids tralomethrin and tralocythrin in insects', *Pestic. Biochem. Physiol.*, **15**, 137–142.

Scott, J. G., and Matsumura, F. (1981). 'Characteristics of a DDT-induced case of cross-resistance to permethrin in *Blatella germanica*', *Pestic. Biochem. Physiol.*, **16**, 21–27.

Shono, T., and Casida, J. E. (1978). 'Species-specificity in enzymatic oxidation of pyrethroid insecticides: 3-phenoxybenzyl and α-cyano-3-phenoxybenzyl 3,-(2,2-dihalovinyl)-2,2-dimethylcyclopropanecarboxylates', *J. Pestic. Sci.*, **3**, 165–168.

Shono, T., Unai, T., and Casida, J. E. (1978). 'Metabolism of permethrin isomers in American cockroach adults, house fly adults, and cabbage looper larvae', *Pestic. Biochem. Physiol.*, **9**, 96–106.

Shono, T., Ohsawa, K., and Casida, J. E. (1979). 'Metabolism of *trans*- and *cis*-permethrin, *trans*- and *cis*-cypermethrin, and decamethrin by microsomal enzymes', *J. Agric. Food Chem.*, **27**, 316–325.

Soderlund, D. M. (1979). 'Pharmacokinetic behavior of enantiomeric pyrethroid esters in the cockroach, *Periplaneta americana* L.', *Pestic. Biochem. Physiol.*, **12**, 38–48.

Soderlund, D. M. (1981). 'Effects of non-neural mechanisms on pyrethroid structure–activity relationships', Paper No. 13, Pesticide Chemistry Division, American Chemical Society, 182nd National Meeting, New York, NY.

Suzuki, T., and Miyamoto, J. (1974). 'Metabolism of tetramethrin in houseflies and rats *in vitro*', *Pestic. Biochem. Physiol.*, **4**, 86–97.

Tsukamoto, M., and Casida, J. E. (1967). 'Metabolism of methylcarbamate insecticides by the $NADPH_2$-requiring enzyme system from houseflies', *Nature*, **213**, 49–51.

Twine, P. H., and Reynolds, H. T. (1980). 'Relative susceptibility and resistance of the tobacco budworm to methyl parathion and synthetic pyrethroids in southern California', *J. Econ. Entomol.*, **73**, 239–242.

Wilkinson, C. F., and Brattsten, L. B. (1972). 'Microsomal drug metabolizing enzymes in insects', *Drug Metabol. Rev.*, **1**, 153–223.

Winteringham, F. P. W., Harrison, A., and Bridges, P. M. (1955). 'Adsorption and metabolism of [^{14}C]pyrethroids by the adult housefly, *Musca domestica* L., *in vivo*', *Biochem. J.*, **61**, 359–367.

Woke, P. A. (1939). 'Inactivation of pyrethrum after ingestion by the southern armyworm and during incubation with its tissues', *J. Agric. Res.*, **58**, 289–295.

Yamamoto, I., and Casida, J. E. (1966). '*O*-Demethyl pyrethrin II analogs from oxidation of pyrethrin I, allethrin, dimethrin, and phthalthrin by a house fly enzyme system', *J. Econ. Entomol.*, **59**, 1542–1543.

Yamamoto, I., and Casida, J. E. (1968). 'Syntheses of ^{14}C-labeled pyrethrin I, allethrin, phthalthrin, and dimethrin on a submillimole scale', *Agric. Biol. Chem.*, **32**, 1382–1391.

Yamamoto, I., Kimmel, E. C., and Casida, J. E. (1969). 'Oxidative metabolism of pyrethroids in houseflies', *J. Agric. Food Chem.*, **17**, 1227–1236.

Zeid, M. M. I., Dahm, P. A., Hein, R. E., and McFarland, R. H. (1953). 'Tissue distribution, excretion of $C^{14}O_2$ and degradation of radioactive pyrethrins in the American cockroach', *J. Econ. Entomol.*, **46**, 324–336.

Index

Contents—Volume 1

Contents—Volume 2

Contents—Volume 3